Advances in 21st Century Human Settlements

Indexed by SCOPUS

This Series focuses on the entire spectrum of human settlements – from rural to urban, in different regions of the world, with questions such as: What factors cause and guide the process of change in human settlements from rural to urban in character, from hamlets and villages to towns, cities and megacities? Is this process different across time and space, how and why? Is there a future for rural life? Is it possible or not to have industrial development in rural settlements, and how? Why does 'urban shrinkage' occur? Are the rural areas urbanizing or is that urban areas are undergoing 'ruralisation' (in form of underserviced slums)? What are the challenges faced by 'mega urban regions', and how they can be/are being addressed? What drives economic dynamism in human settlements? Is the urban-based economic growth paradigm the only answer to the quest for sustainable development, or is there an urgent need to balance between economic growth on one hand and ecosystem restoration and conservation on the other – for the future sustainability of human habitats? How and what new technology is helping to achieve sustainable development in human settlements? What sort of changes in the current planning, management and governance of human settlements are needed to face the changing environment including the climate and increasing disaster risks? What is the uniqueness of the new 'socio-cultural spaces' that emerge in human settlements, and how they change over time? As rural settlements become urban, are the new 'urban spaces' resulting in the loss of rural life and 'socio-cultural spaces'? What is leading the preservation of rural 'socio-cultural spaces' within the urbanizing world, and how? What is the emerging nature of the rural-urban interface, and what factors influence it? What are the emerging perspectives that help understand the human-environment-culture complex through the study of human settlements and the related ecosystems, and how do they transform our understanding of cultural landscapes and 'waterscapes' in the 21st Century? What else is and/or likely to be new vis-à-vis human settlements – now and in the future? The Series, therefore, welcomes contributions with fresh cognitive perspectives to understand the new and emerging realities of the 21st Century human settlements. Such perspectives will include a multidisciplinary analysis, constituting of the demographic, spatio-economic, environmental, technological, and planning, management and governance lenses.

If you are interested in submitting a proposal for this series, please contact the Series Editor, or the Publishing Editor:

Bharat Dahiya (bharatdahiya@gmail.com) or
Loyola D'Silva (loyola.dsilva@springer.com)

More information about this series at http://www.springer.com/series/13196

Le Thi Thu Huong · George M. Pomeroy
Editors

AUC 2019

Proceedings of the 15th International Asian Urbanization Conference, Vietnam

AURA

15th INTERNATIONAL ASIAN URBANIZATION CONFERENCE:
"URBAN FUTURES: CRITICAL TRANSFORMATIONS IN ASIAN CITIES"
27-30 NOVEMBER 2019
VIETNAMESE-GERMAN UNIVERSITY, HO CHI MINH CITY, VIETNAM

VGU

ASHUI·asia

Springer

Editors
Le Thi Thu Huong
Faculty of Engineering
Vietnamese-German University (VGU)
Binh Duong, Vietnam

George M. Pomeroy
Department of Geography-Earth Science
Shippensburg University
Shippensburg, PA, USA

ISSN 2198-2546 ISSN 2198-2554 (electronic)
Advances in 21st Century Human Settlements
ISBN 978-981-15-5607-4 ISBN 978-981-15-5608-1 (eBook)
https://doi.org/10.1007/978-981-15-5608-1

This Springer imprint is published by the registered company Springer Nature Singapore Pte Ltd.
The registered company address is: 152 Beach Road, #21-01/04 Gateway East, Singapore 189721, Singapore

Organization

Organizing Committee

1. Chairman of Local Organizing Committee: Dr. Ha Thuc Vien, Vietnamese-German University, Vietnam.
2. Vice-Chairman of Local Organizing Committee: Assoc. Prof. Dr.-Ing. Pham Van Song, Vietnamese-German University, Vietnam.
3. Member of Local Organizing Committee:
    - Dr. Nguyen Ngoc Hieu, Vietnamese-German University, Vietnam
    - Dr. La Vinh Trung, Vietnamese-German University, Vietnam
    - Dr. Le Thi Thu Huong, Vietnamese-German University, Vietnam
    - Dr. Pham Thai Son, Vietnamese-German University, Vietnam
    - Dr. Vu Anh Tuan, Vietnamese-German University, Vietnam
    - Dr. Nguyen Viet Long, BECAMEX IDC Corp., Vietnam.
4. Conference Secretariat:
    - Dr. La Vinh Trung, Vietnamese-German University, Vietnam
    - Ms. Tran Thuan Anh, Vietnamese-German University, Vietnam
    - M.Sc. Nguyen Thi Hoai, Vietnamese-German University, Vietnam
    - M.Sc. Tran Hoang Nam, Vietnamese-German University, Vietnam
    - M.Sc. Kieu Thi Le, Vietnamese-German University, Vietnam.
5. Scientific Board:
    - Prof. Dr. George Pomeroy, Professor, Shippensburg University, United State
    - Prof. Dr. Debnath Mookherjee, Professor Emeritus, Western Washington University, United States
    - Prof. Dr. Braj Raj Kumar Sinha, Banaras Hindu University, India
    - Assoc. Prof. Dr. Le Anh Duc, Ho Chi Minh City College of Construction, Vietnam
    - Assoc. Prof. Dr. Rini Rachmawati, Gadjah Madah University, Indonesia

- Assoc. Prof. Dr.-Ing. Pham Van Song, Vietnamese-German University, Vietnam
- Dr. Diganta Das, National Institute of Education of Singapore, Singapore
- Dr. Le Thi Thu Huong, Vietnamese-German University, Vietnam
- Dr. Pham Thai Son, Vietnamese-German University, Vietnam
- Dr. Nguyen Ngoc Hieu, Vietnamese-German University, Vietnam
- Dr. Vu Anh Tuan, Vietnamese-German University, Vietnam
- Dr. Le Thi Hong Na, Ho Chi Minh City University of Technology, Vietnam
- Dr. Ngo Le Minh, Ton Duc Thang University, Vietnam
- Dr. Nguyen Luu Bao Doan, Ho Chi Minh City University of Economics, Vietnam.

Sponsors

1. Becamex IDC Corp.
2. Technical University of Darmstadt
3. Shippensburg University
4. SEANNET (Southeast Asia Neighborhoods Network)
5. Groups of the American Association of Geographers.
6. ASHUI.COM (Architecture/Sustainability/Housing/Urban Initiatives/Interior Vietnam)
7. Vietcombank
8. IAB (Weimar institute of Applied Construction Research)
9. VGU association of friends and supporters
10. Vietnam airlines.

VGU
ASSOCIATION OF FRIENDS
AND SUPPORTERS

TECHNISCHE UNIVERSITÄT DARMSTADT

SHIPPENSBURG UNIVERSITY

AAG
AMERICAN ASSOCIATION of GEOGRAPHERS

SEANNET

UKNA Urban Knowledge Network Asia

ASHUI.COM

Vietcombank

IAB

IAB – Weimar Institute of Applied Construction Research
Über der Nonnenwiese 1
99428 Weimar, GERMANY
phone +49 3643 8684-0, fax: +49 3643 8684-113
kontakt@iab-weimar.de, www.iab-weimar.de

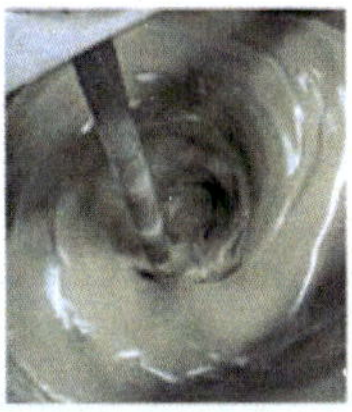

Preface

Although cities, metropolitan areas, megapolitan regions, and urban areas constantly undergo transformations, one may argue, particularly with respect to issues of climate change, that we are truly at a critical juncture, and thus the conference theme "Urban Futures: Critical Transformations in Asian Cities" is especially appropriate. One year ago, the Intergovernmental Panel on Climate Change (IPCC) issued a 700-page report warning that we have just 12 years to make a massive and unprecedented change in the ways we live in order to limit global warming to more moderate levels and to prevent the most devasting impacts. Given this, we are truly at a critical pivot point in terms of how, as urbanists, city planners, urban managers, and scholars, we societally consume and manage resources. There are dire implications for how we mitigate and adapt to climate change. Other transformations are be reaching critical junctures, too. For example, revolutions in computing technologies and programming are enabling "smart cities" as we have only imagined. Also, income polarization brought on through globalization has implications for housing affordability. Across these and other issues, planners must make pivotal choices with long-range implications. Indeed, demographically, it has been roughly three years since Asia crossed the psychological threshold of having more people living in urban areas than not. The conference theme then is as fitting for this time as for any other.

Another critical transformation is reflected by changes in this conference series itself. The initial conference was organized in 1985, by four Asianists, each with a focus on cities, urbanization, and planning, at the University of Akron, in the United States. On the heels of that successful conference, a conference series slowly evolved under the leadership of Dr. Ashok K. Dutt, with increasing regularity and in a variety of venues. That first conference, along with one in London (1997) and Athens, Georgia, USA (2003), was organized in venues that were obviously not Asian, were less accessible, whether in terms of cost or distance, to Asian based scholars and practitioners. Indeed, early on, participation for many modestly resourced individuals—notably those most directly engaged in studying, planning, and living in Asian cities—was prohibitively costly, thus effectively shutting out those scholars and practitioners who perhaps had both the most to contribute and the most to gain from the participation in the conference. In a sense, the conference series itself has undergone a "critical transformation," that leaves it incredibly accessible to scholars from a wide variety of institutions and geographic settings, and in effect, participation has been democratized. The efforts of incredibly capable local organizers in easily reachable venues and a modest, multiple tier registration systems have been instrumental in this transformation to a conference series that features and mixes both "bottom up" grass-roots scholarly participation from a wide range of institutions with respected expert participants from across the globe.

The mission of this conference and earlier iterations of the Asian Urbanization Conference series is to promote the "exchange of views and experiences on cities, metropolitan areas, urbanization, and closely related social, economic, political, and environmental issues in the Asian context" and to facilitate a "dialog that shares research, planning, and problem-solving."

This 15th Asian Urbanization Conference 2019, attracted more than 240 abstracts, 104 full papers, among which 45 and 36 full papers have been selected to publish in the SCOPUS-indexed conference proceedings and edited book volume accordingly after rigorous peer review process.

Shippensburg, USA Prof. Dr. George M. Pomeroy

Contents

Climate Change and Urban Resilience

Climate Change and Cities in the Mekong Delta, Vietnam

Thi Hong Hanh Vu and Thanh Hai Truong

Abstract Climate change is not simply an environmental problem but also an impacting development across the globe. It strongly impacts the Mekong Delta region in Vietnam. Like many other deltas in the world, the Mekong Delta has strong water-based characteristics, where the living environment is strongly influenced by water, and thus consequently suffers the worst impacts of climate change and sea-level rise according to different projections. Having important roles in the economic development, food security and natural ecology of Vietnam, this delta has been received plenty of international, national and local attentions to seek ways to insulate it from the impacts of climate change, to strengthen the roles, and to maximize its potentials. The research considers integrated relationships of three components (1) city scale, (2) city morphology and (3) impacts of climate change. This is to understand how different urban morphological classifications and scales/grades rationally affected by climate change, focusing on the flood and salinity impacts resulted by average projected future energy-related CO_2 emissions scenario. From those above analytical bases, the research proposes groups of climate change adaptabilities for the Mekong region and classified cities within, from spatial planning strategies, water management and more detailed structural and non-structural solutions.

Keywords Mekong delta · Vietnam · Climate change · Sea-level rise · Salinity · Flood · Adaptation

T. H. H. Vu (✉) · T. H. Truong
University of Architecture Hochiminh City (UAH), Ho Chi Minh City, Vietnam
e-mail: hanh.vuthihong@uah.edu.vn

T. H. Truong
e-mail: hai.truongthanh@uah.edu.vn

L. T. T. Huong and G. M. Pomeroy (eds.), *AUC 2019*, Advances in 21st Century Human Settlements, https://doi.org/10.1007/978-981-15-5608-1_1

1 Introduction

The Mekong Delta of Vietnam is located in the lower Mekong Basin, covers a flat land of around 9.3 million hectares and is 1m above sea level on average. This water-dominated land is highly affected by tides and annual seasonal salinity with an area of 1.7 million hectares of saline soils, which affect both aquaculture and agriculture. The water-based social, cultural and economic characteristics of the majority of local communities, therefore, are highly affected by climate change, projecting the most vulnerable area in Vietnam and worldwide (Fig. 1).

To propose appropriate adaptations to climate change, at the city scale in the Mekong Delta, two components are investigated including (1) the cities and (2) impacts of climate change. Regarding the cities, in relationship with the adaptability, the analysis of aspects regarding scale and morphological characters are highly recommended because of the following reasons:

- First, there has been much research on the adaptations a place to climate change, among which, adaptive land uses are especially crucial (Vu 2018; Perthuis 2019). Unfortunately, the focus has been more on agriculture and aquaculture lands rather than urban lands, where the loss of life is expectedly higher in any case of environmental emergencies (Vu and Le 2016). Urban form is the visual expression of physical structure, land uses, housing, transport infrastructure, greenery, agricultural lands and the ecological network, etc. (Shen and Kawa 2013), therefore, directly influences its adaptability to climatic stress (Lilai 2019).
  In fact, the urban form helps defining the spatial morphology of a place (Vu et al. 2019). Ian Bentley suggested different morphological layers of an urban space, according to different lifecycles or periodic transformations of each element that may vary from centuries to decades. However, the understanding of these layers needs an interdisciplinary approach and involvements (Bentley 2013) (Fig. 2).

At different extents of analysis, selective elements are considered, mainly due to their measurability (Vu et al. 2019; Bentley 2013; Hoang 2017):

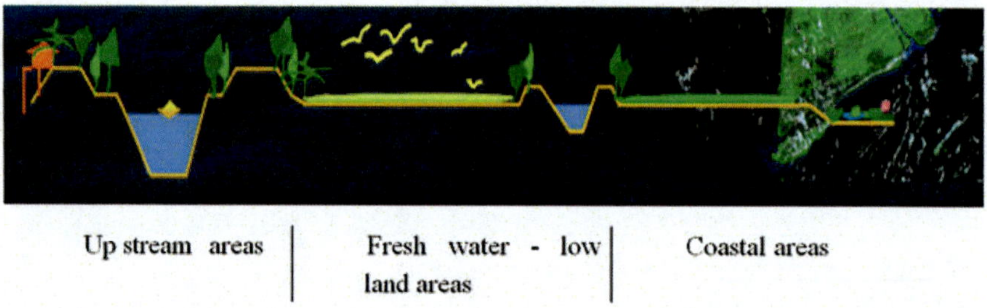

Fig. 1 Diagram shows 3 terrain areas of the Mekong delta, Vietnam (*Source* SIWRP, 2016)

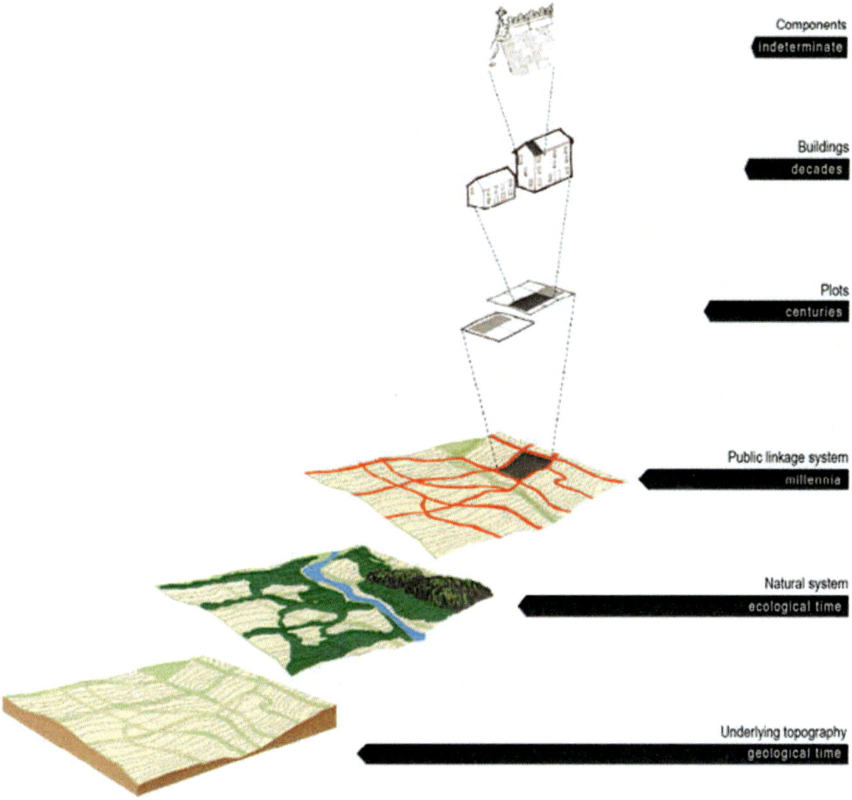

Fig. 2 Morphological layers of a settlement (Bentley 2013)

- **Regional scale**: Natural system (blue and green fields) in relationship with the built fabric (brown field), including the roads that are instrumental to the expansion of the built fabric over time;
- **City scale**: Natural system at greater scale, public linkage system including the circulations (road/street) and open spaces. These networks define the urban structure, from which different land uses are planned: agricultural, industrial, protective and preserved green areas, the CBD, the outskirts, etc.
- **Fragment scale**: The urban form of an area including the street network, block plots and building masses. At this scale, the natural system and public linkages are seen in more detail and in terms of relationships with social-economic activities;
- **Sample scale**: Groups of buildings or smaller areas of a fragment-scale size. This is equivalent to a neighbourhood or a residential unit where not only architecture forms and characteristics but also social-economic activities of a community even of the individual that contributes to the image of the place are analysed (Fig. 3).

Second, climate change impacts are scale- and context-specific and cities are likely to bear some of the greatest costs (Anton et al. 2012); and **city's scale** clearly effects

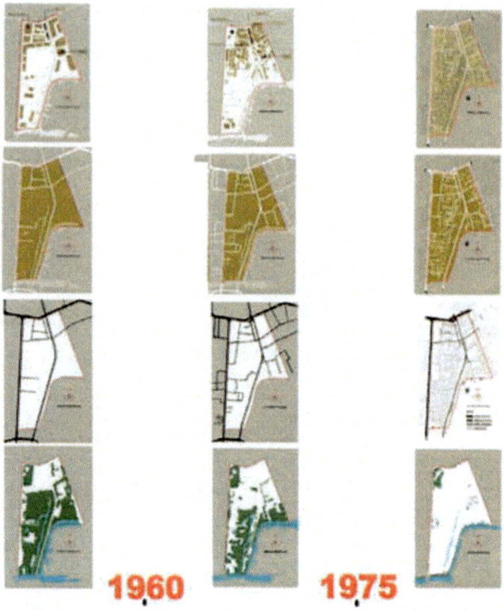

Plots	
Blocks, built up area	
Street network	Crucial roles in protecting it people and urban functions; strategic linking roads must be secured
The blue and green system	Highly relevant to the water/flood: positions, form, flow, capacity in relationship with the built-up areas

Fig. 3 An example of a fragment scale–sample scale morphology (Vu and Thai 2019) (a study at Xuyen Tam canal, Binh Thanh District, HCMC)

of adaptive capacity, however, has been neglected in the climate change literature, and the bulk of the academic literature has been on agriculture and rural impacts (Paterson 2017). **Scale** is recently becoming recognized as a principle characteristic that shapes resilience disaster losses and the governance of disaster risk (Paterson 2017). City scale, in this paper, concerns of the built area(s) and its relationships with two morphological layers mentioned above: blue–green system and road–street network.

The research approach is as the following Fig. 4.

From the above approach, the research is to analyse (Section 1) the cities (scale and morphologies) and (Section 2) the impacts of climate change, focusing on flooding and salinization, and to pro-pose adaptive solutions (Section 3).

2 The Mekong Delta and Its Cities: The Scales and Morphologies

2.1 The Cities and Scales

The Mekong Delta has over 150 cities (2012), classified in different scales/grades according to Urban Planning Law, from grade 1 (Cần Thơ), grade 2 (one to two in each

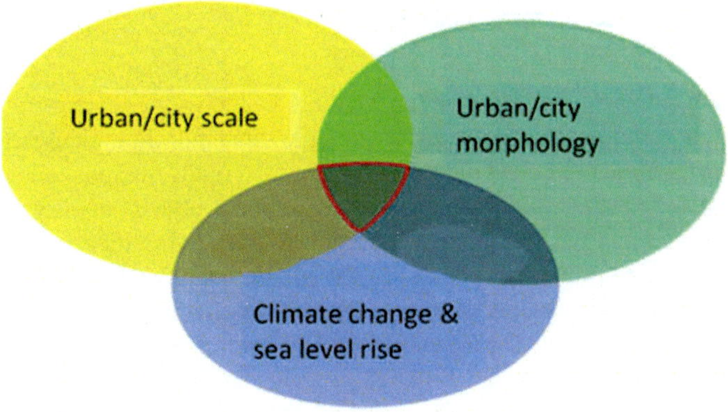

Fig. 4 Research approach (Vu 2018; Perthuis 2019)

province, except Can Tho), grades 3, 4, 5 and unclassified. The classification is based on the scale of population, roles of the city within its province and region/national wide, inner and outer sources for development. There is a tendency of urban upscaling based on its projections of population growth, potential and strategic roles; however, the projections are not always right and not reflect its dynamics including climate change (Shen and Kawa 2013) (Table 1).

Table 1 Classifications of cities/towns in the Mekong Delta (*Source* Statistic Division, 2012) (Vu et al. 2019)

Categories / Grades		**Name**	**Number**
Large cities	Central government City (Grade I)	Cần Thơ	**01**
	Grade II City	Mỹ Tho, Long Xuyên, Rạch Giá	**03**
Medium cities/towns	Grade III City	Cà Mau, Sóc Trăng, Cao Lãnh, Vĩnh Long, Bạc Liêu, Tân An, Bến Tre, Trà Vinh, Sa Đéc, Châu Đốc, Vị Thanh	**11**
	Grade IV Town	Town: Hồng Ngự, Kiến Tường, Vĩnh Châu, Bình Minh Provincial Town (Thị xã): Tân Châu, Ngã Bảy, Hà Tiên, Gò Công	**08**
Small towns	Grade V		**129**
	Un- classified		
Total			**152**

2.2 *The Cities—The Morphologies*

The research analysed 152 cities in the Mekong Delta, trying to classify them into groups of morphological similarities according to three morphological elements and their spatial relationships: the blue–green system, the roads (national and provincial trunk roads) and (3) the built-up areas. The research was able to identify 11 different types as showed in Fig. 5. Among which, the majority are cities with water take up nearly 90%. And, the most popular is the M7 with 61 cities including Can Tho (Fig. 7) shows the ratio of different morphology—classified cities in the Mekong Delta, Vietnam. The actual list of cities along with scale and morphological characters is produced (however, not included in this paper) (Fig. 6).

3 The Mekong Delta and Its Cities: The Impacts of Climate Change

According to the projection by IPCC, in the Mekong Delta, every 100 cm in sea-level rise results in more than 1 million people losing their houses and land. Areas within a hundred kilometres from the coast will be flooded and become saline, with urban infrastructure being heavily affected (20% national and 27% provincial roads).

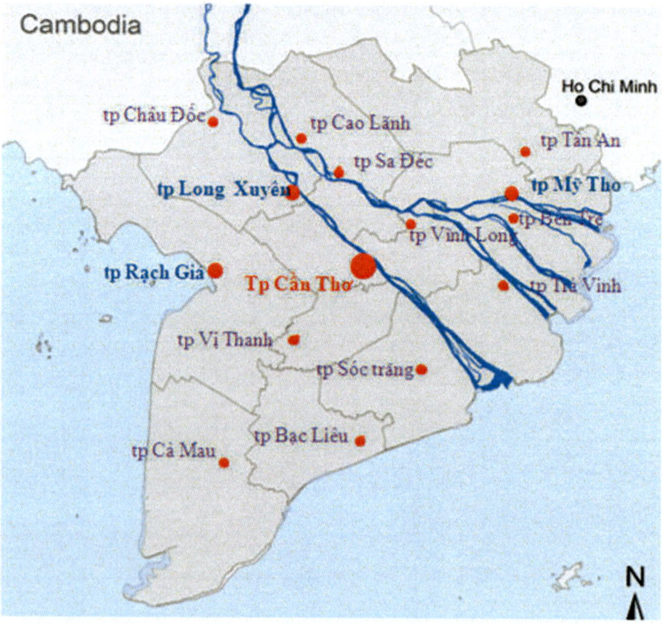

Fig. 5 Distributions of provincial cities in the Mekong Delta (Vu et al. 2019)

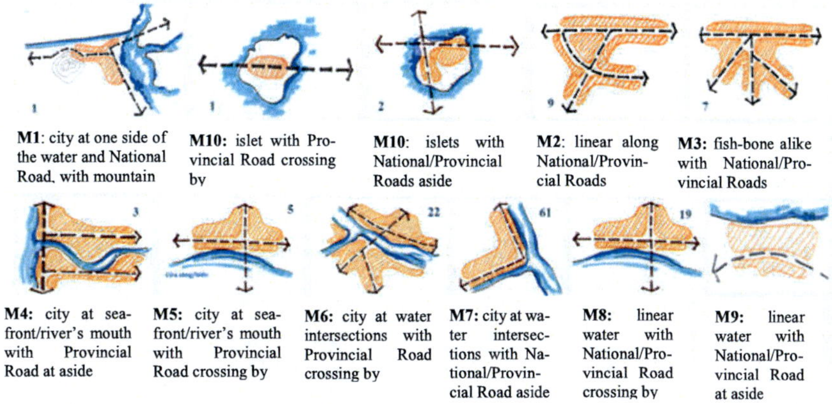

Fig. 6 Eleven different morphologies of cities in the Mekong Delta, Vietnam (Vu et al. 2019). *Note* The number of cities in each morphological category is on the top right corner

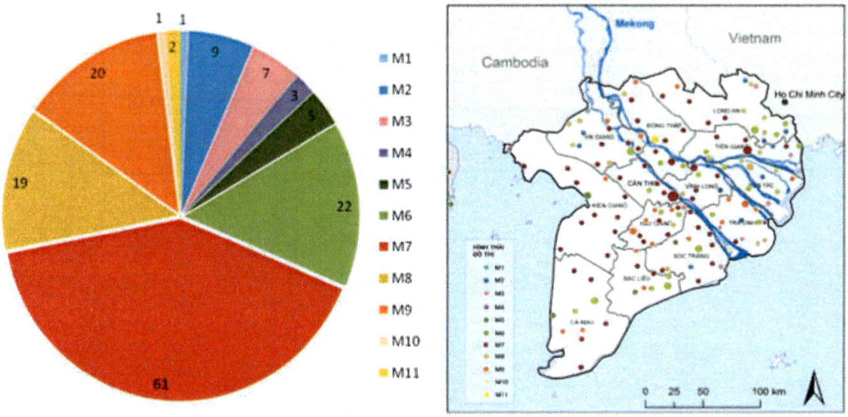

Fig. 7 The ratio (left) and spatial distribution (right) of cities by morphologies in the Mekong Delta, Vietnam (Vu et al. 2019)

Among impacts, it is clear that *Flooding* and *Salinization* are strongly affecting aqua- and agricultural activities in the region, and it is measurable as well. Therefore, the research based on the analytical data projected by the Southern Institute of Water Resources Research on flooding and salinization for the Mekong Delta using the average climate change scenario developed criteria for the impact evaluations (Tables 2, 3 and Figs. 8, 9).

According to the criteria set out above, the research overlaid two impacts and was able to come up with five zones as following:

Table 2 Classifications of cities/towns in the Mekong Delta (*Source* Vu et al, 2019)

	Less	Average	Highly
Flood (mm)	<1.5	1.5–2.5	>2.5
Salinity (g/l)	0–4	4–16	>16

Table 3 Projection of flood and salinity for cities in the Mekong Delta (*Source* Vu H and Le T, 2016)

	Province	City	Flood (m)	Salinity (g/l)
1	An Giang	Chau Doc	> 3.0	0-1.0
		Long Xuyen	2.5 – 3.0	0-1.0
2	Dong Thap	Cao Lanh	2.5 – 3.0	0-1.0
		Sa Dec	2.0 – 2.5	0-1.0
3	Vinh Long	Vinh Long	2.0 – 2.5	0-1.0
4	Can Tho	Can Tho	2.0 – 2.5	0-1.0
5	Hâu Giang	Vi Thanh	1.0-1.5	0-1.0
6	Kiên Giang	Rach Gia	1.0-1.5	16.0-24.0
7	Ca Mau	Ca Mau	1.0-1.5	16.0-24.0
8	Bac Lieu	Bac Lieu	1.0-1.5	16.0-24.0
9	Soc Trang	Soc Trang	1.5-2.0	6.0-16.0
10	Tra Vinh	Tra Vinh	2.0 – 2.5	6.0-16.0
11	Ben Tre	Ben Tre	1.5-2.0	6.0-16.0
12	Tiến Giang	My Tho	1.5-2.0	4.0-6.0
13	Long An	Tan An	2.0 – 2.5	4.0-6.0

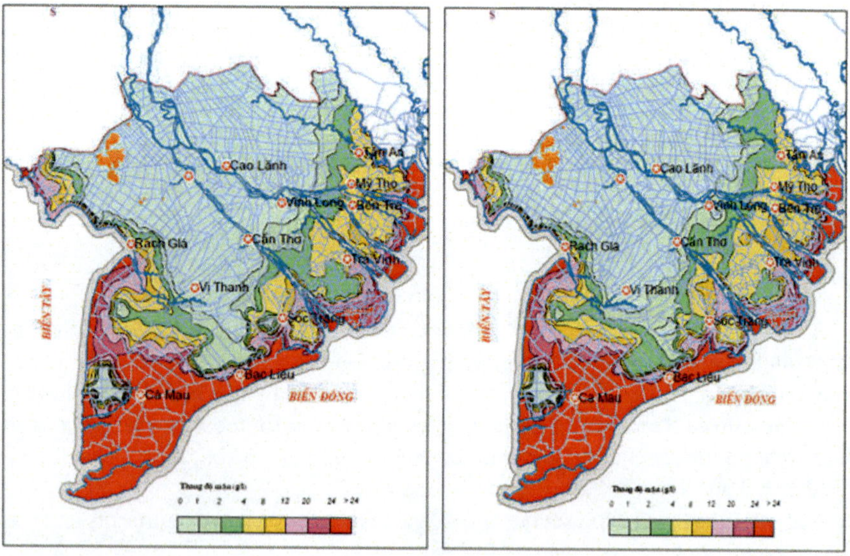

Fig. 8 Salinity in the Mekong Delta, the existing (2000) and provision 2050 (*Source* SIWRP, 2012)

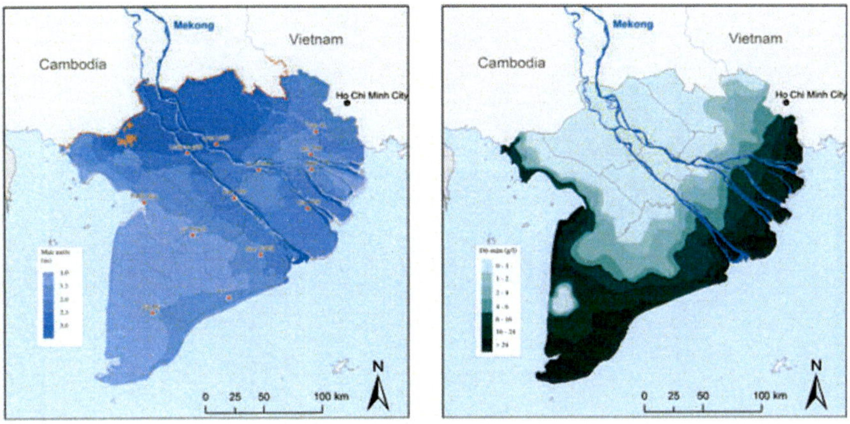

Fig. 9 Projected Flood (left) and Salinity (right) in the Mekong Delta by climate change, Provision 2050 (*Source* Vu H and Le T, 2016)

- Zone I: West coast area, rather highly affected by salinity (only less than the Eastern side) and slightly affected by flooding. Solutions focus on adaptation to salinity;
- Zone II: Eastside coastal area, highly affected by salinity, or rather highly affected by flooding (only less affected than Zone 3 by flood). This zone helps reduce and protect Zone 4 from salinity. Solutions should focus on adaptations to both Salinity and Flood;
- Zone III: Freshwater, highly affected by Flooding, without Salinization;
- Zone IV: Freshwater area, average affected by Flooding, slightly affected by Salinization
- Zone V: Ideal zone, barely affected by Flooding, without Salinization, highly protected area, suitable for urbanization and development of compact cities.

It shows clearly from the research that:

- Similar impacts might occur in different places, regardless of administrative borders and distances. In zone V, in the West coast (and a smaller bit of the far east coast of Ben Tre province), there is an area within Zone 5, whereas zone V scattered the central core of the whole lower basin of the delta from West coast to Ho Chi Minh City;
- Areas were heavily flooded due to seasonal flooding (Zone III); there is not much enlargement of this area by sea-level rise, however, deeper flood, and local people have been well adapted to it (Fig. 10).

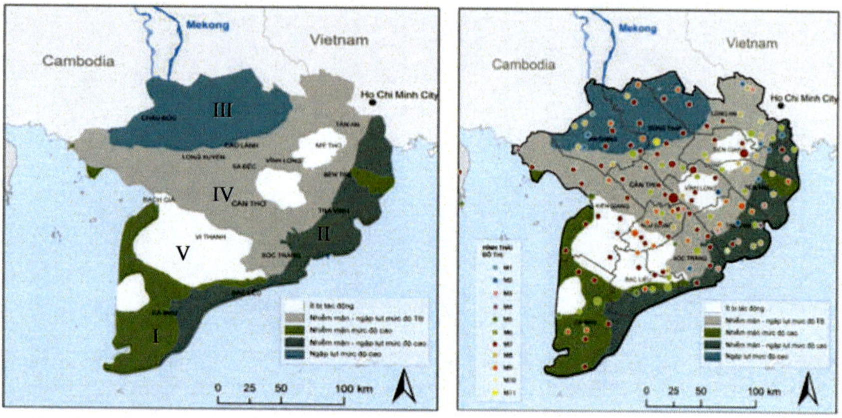

Fig. 10 Climate change-impacted zones in the Mekong Delta (left) overlaid with cities' morphologies and scale (right) (*Source* Vu H and Le T, 2016)

4 Adaptive Solutions for Cities in the Mekong Delta

4.1 General Principles

- Respect nature, especially blue and green systems, increase the capacity of water reservoirs with larger buffer zones to control flooding and protect them from salinity;
- Integrate morphologies, scale and climate change affective classifications of cities/groups of cities;
- Reinsure economic development, social security and eco-environment;
- Enhance local distinctiveness (Fig. 11).

4.2 Planning Principles

- Increase water-absorbable areas, integrated hierarchical water release system;
- Connected infrastructure;
- Scattered and nailed with trunk roads, bases of dikes including hard and climate dikes, provide reliable regional linkages
- North–South road systems are flood-free for the region, sub-regions and cities in the Mekong Delta.

General models of development for each zone:

- *Zone I: Saline eco-cities with fishing farming and exploitation;*
- *Zone II: Flood-adaptive cities (increase ground floor cotes, dikes);*

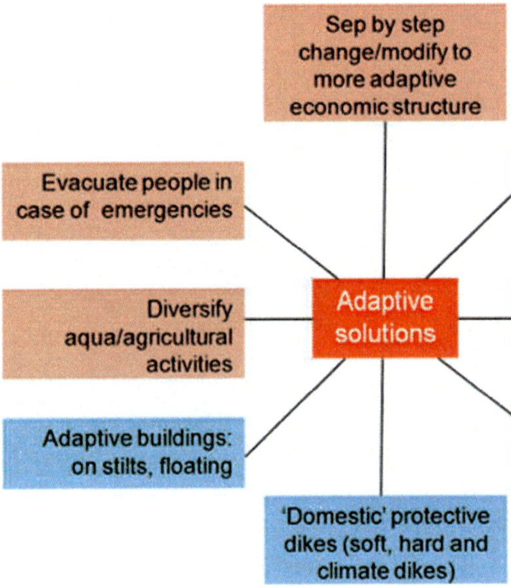

Fig. 11 Structural (blue) and non-structural (brown) adaptive solutions (*Source* Vu H et al, 2019)

- *Zone III: Inherited and further water-based distinctive cities development*
- *Zone IV: Freshwater shortage risk: encourage effective water use and collection. Green cities fostering urban living and garden-based economic development.*
- *Zone V: Suitable for urbanization and dense development of compact cities with moderately built fabric, provide spaces for blue and green systems* (Fig. 12).

4.3 Adaptive Solutions to 11 Cities' Morphologies

Principles:

- Based on the existing
- Multi-functional dikes (dikes = central urban lines)
- Compact, give priority to social infrastructure and disaster—protective elements
- Increase absorbable areas, give rooms to the rivers, give green to cities
- Connect the urban system, especially blue and green systems
- Multifunctional national and provincial roads:
  - Hard dikes: in cases of urban rear
  - Hard dikes integrated with centre urban lines with concentration of social infrastructure (medical, educational structure and social houses, etc.) in cases this roads system cut through cities/towns
  - Circle dikes: to protect the whole city/town: in the case of small-scale city

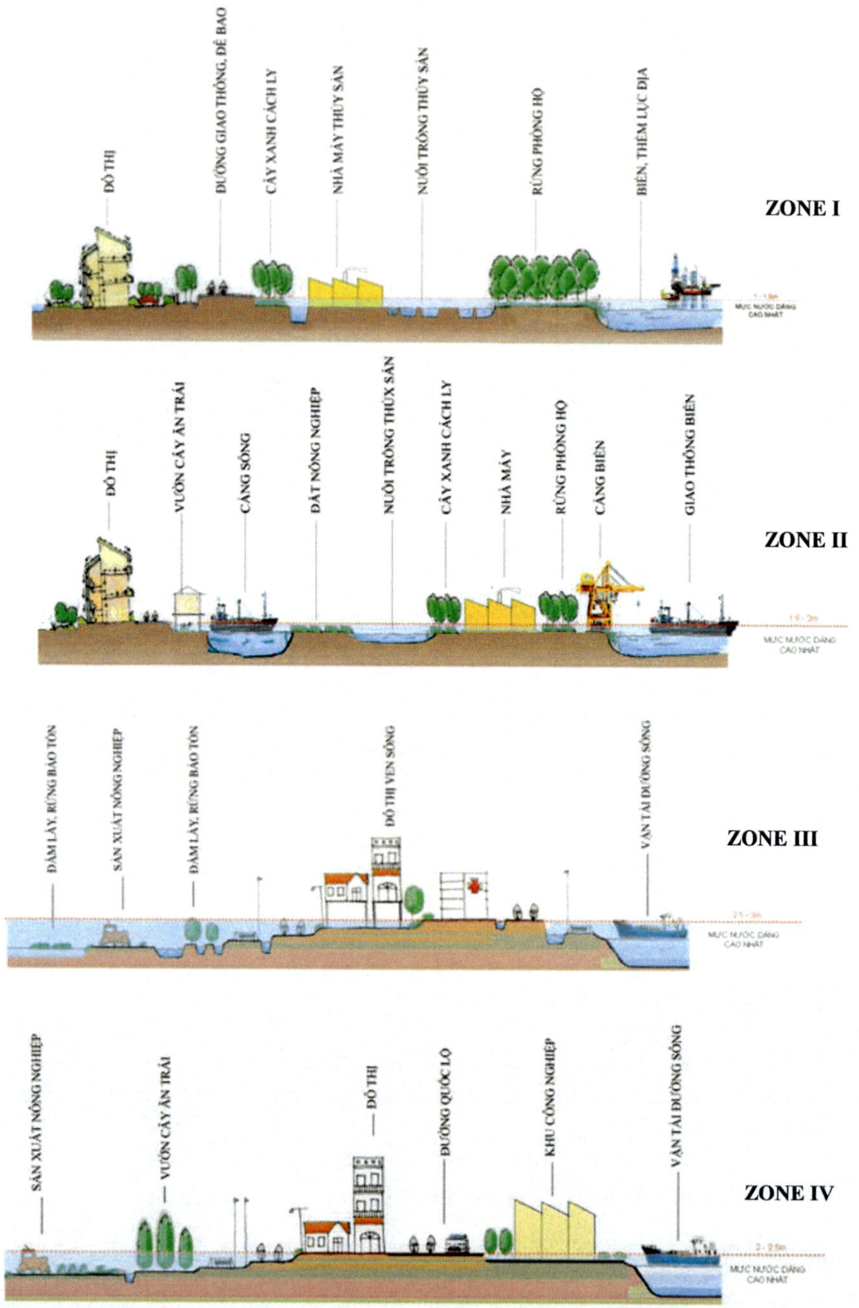

Fig. 12 Adaptive spatial organization for different zones (*Source* Vu H and Le T, 2016)

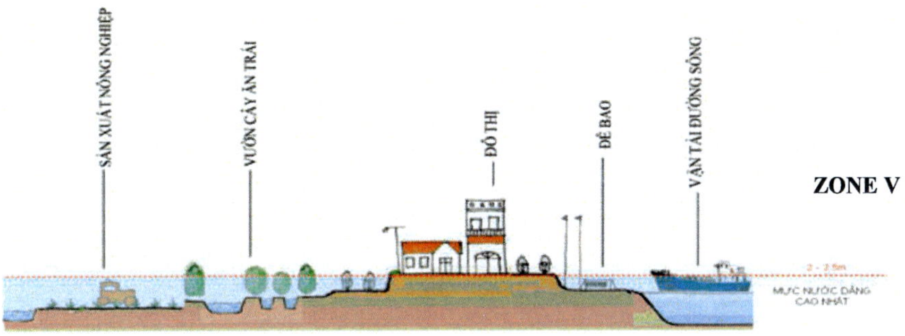

Fig. 12 (continued)

- Lines of hard dike or climate dikes: applied to cities that have scattered blue and green systems (the broken dish pattern of built fabric), such as M6, M7.

Among 11 morphologies, M10 includes only one small town/settlement in the islet surrounded by water. This islet is at high flood risk and in case of emergency, isolated from the in-land supports. The research discourages further development of built-up areas, rather for eco-friendly tourism development, however, with a clear mitigation plan (Fig. 13).

5 Conclusion

The Mekong Delta is one among many places likely to receive severe impacts of climate change and sea-level rise. The region comprises of 13 provinces, over 150 cities/towns ranging from unclassified to Grade 1, home of 17 million populations. The living culture, especially the aqua–agricultural activities are diversified by different communities of different natural conditions: The west and east coasts, upstream border of Cambodia, the freshwater flat plain. In each province, cities/towns were founded more at water intersections than along main roads. Those cities/towns are in the constant state of expansion and upscaling regardless of the vulnerabilities to climate changes, sea-level rise and freshwater shortages.

The research, approaching from scale–morphology–impact of climate change has analysed impacts, identified different morphologies and scales of cities. From that developed adaptive solutions, if only the analysis and evaluation are appropriate, are the solutions reliable. The analysis is only valid with reliable bases, data collections, and if any modeling/experiments which are limited in this research. The research, however propose a method of analysis and integrated approach that might be applied in other relevant research. The identification of 11 morphologies and to certain extent the popularities of each could also provide good sources of references for planning and developing of different climate change adaptations.

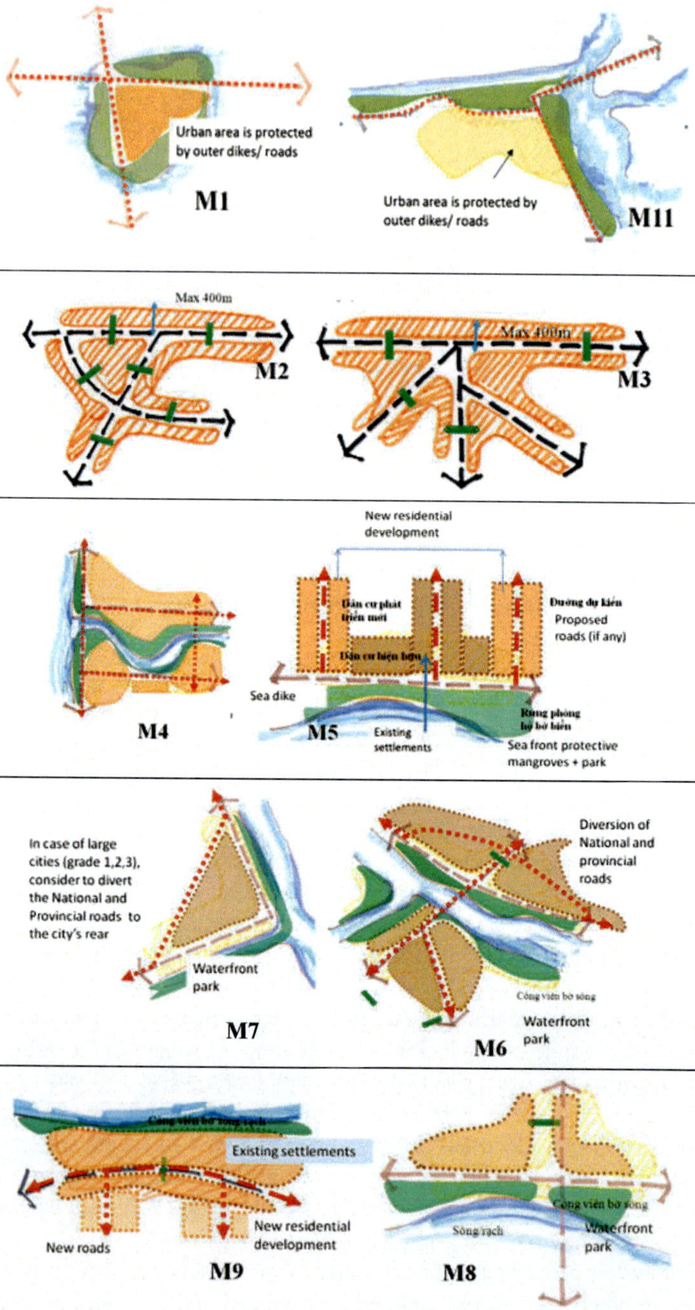

Fig. 13 Adaptive solutions according to the morphologies of cities (*Source* Vu H and Le T, 2016)

References

Anton C, Susan P, Gregg O, Sarah W (2012) Climate change at the city scale: impacts, mitigation and adaptation in Cape Town, 1st edn. Routledge (ISBN-10: 9780415527583)

Bentley I (2013) Responsive urban design, short course program at UAH, Hochiminh City

Hoang L (2017) Urban morphologies of the South East Hochiminh City. PhD thesis, UAH

Lilai X et al (2019) Assessing the adaptive capacity of urban form to climate stress: the case study on urban heat island. IOP Publishing Ltd

Paterson SK et al (2017) Size does matter: city scale and the asymmetries of climate change adaptation in three coastal towns. Geoforum 81:109–119

Perthuis C (2019) Global warming and economic development. In: International conference climate change challenges for the Lower Mekong Basin region: findings, crucial problems and lessons learned from action plans to date, Hochiminh City University of Social Science and Humanities, Hochiminh City

Shen K (2013) Spatial planning and sustainable development, approaches for achieving sustainable urban form in Asian cities. Springer, Netherlands

Vu H (2018) Water-based housing in Southern Vietnam. In: Floating settlement in Mekong Delta. BTU Cottbus, Senftenberg, Germany

Vu H, Le T (2016) Development of climate change adaptations for cities in Mekong Delta, Vietnam, Research Project, Ministry of Construction

Vu H, Thai T (2019) Urban morphology project, Master Class K24, UAH

Vu H, Le T, Truong H (2019) City morphologies in Mekong Delta and climate change adaptations, 1st edn, UAH

Climate Change and Sea-Level Rise Response Solutions for Can Gio District, Ho Chi Minh City: Potential to Adapt Ideas from Selected Developed Countries

Le Minh Ngo and Hai Yen Hoang

Abstract As the impacts of climate change become more and more apparent, it is clear that Vietnam is among the countries that are most heavily affected. Ho Chi Minh City is on the list of ten cities in the world, most threatened by the risks associated with high sea-level rise. According to the estimates of the United Nations, by 2100, the sea level will rise by more than 1 m and nearly 20% of Ho Chi Minh City's area will be flooded. Therefore, finding solutions to respond to climate change in Ho Chi Minh City is very urgent in the current conditions. This research will evaluate climate change phenomenon in Ho Chi Minh City, particularly in Can Gio coastal district. It will then offer solutions to cope with climate change in this area. The proposed solutions include solutions for planning residential clusters and providing architectural design models for housing projects such stilt houses, floating houses, and lightweight concrete houses. Identifying the correct response to climate change-related flooding in Ho Chi Minh City is the key for sustainable development in the future.

Keywords Climate change · Sea-level rise · Site planning solutions · Can Gio District · Ho Chi Minh City

1 Introduction

Ho Chi Minh City (HCMC) is Vietnam's largest city in terms of population and economy, the second in area, and one of the most important economic, political, cultural, and educational centers of Vietnam. Currently, HCMC is a unique urban type in Vietnam. With the rapid population growth rate, infrastructure has not been able to meet planned upgrades overall, urban planning is still inadequate, and part

L. M. Ngo (✉)
Faculty of Civil Engineering, Ton Duc Thang University, Ho Chi Minh City 700000, Vietnam
e-mail: ngoleminh@tdtu.edu.vn

H. Y. Hoang
Faculty of Architecture - Art, Ho Chi Minh City University of Technology (HUTECH), 475A Dien Bien Phu Street, 25 Ward, Binh Thanh District, Ho Chi Minh City, Vietnam
e-mail: hh.yen84@hutech.edu.vn

L. T. T. Huong and G. M. Pomeroy (eds.), *AUC 2019*, Advances in 21st Century Human Settlements, https://doi.org/10.1007/978-981-15-5608-1_2

of the population's awareness of environmental protection is still weak. HCMC is now facing frequent flooding problems. The main reason is due to global climate change, causing saline water to intrude deeply into the land and local climate change causes high tide levels (Huong and Dao, 2018). With the evolution of climate change, HCMC is on the list of ten cities in the world threatened by rising sea-level risks, especially in coastal districts such as Can Gio. According to the estimates of the United Nations, by 2100, the sea level will rise by more than 1 m and nearly 20% of HCMC's area will be flooded (Southern Institute For Spatial Planning, 2015).

Understanding and assessing the impact of flooding due to climate change on houses in Can Gio District (HCMC), from which to develop planning and architectural solutions to cope with flooding in Can Gio District and HCMC in general is very necessary. These solutions contribute to protect residents' lives against flooding caused by climate change, improve the quality of life for the people here, promote the available potentials value in the locality, and toward sustainable development in the future.

1.1 Concept of Climate Change

Climate change is a change of climatic state in the long-term due to natural processes or caused by human effects. Climate change makes the ocean temperature rise, the ice melt at the two poles, leads to increasing droughts, storms, and floods, the sea level gradually increases, and the phenomenon of extreme weather is getting more and more frequent. The main cause of climate change is attributed to human activities through deforestation and too much CO_2 emissions into the atmosphere, causing the Greenhouse Effect. The most serious consequence is global warming, rapid melting of ice, causing sea levels to rise high, resulting in an ecological imbalance and directly threatening the lives of hundreds of millions of people worldwide (Birkmann, 2011). The impacts of climate change along with other human impacts on the natural environment make the consequences more serious. In short, climate change is a change of climate that takes place over in a long period of time caused by natural or human activities.

1.2 Concept of Sea-Level Rise

Sea-level rise is the rise of the global ocean level, which does not include tides, storm surges, etc. Sea-level rise at a certain location may be higher or lower than the global average because of the difference in ocean temperature and other factors.

1.3 *Flooding Caused by Climate Change in HCMC and Can Gio District*

In Can Gio District, the coastal district is in the southeast of HCMC, where the terrain is low with an average elevation from 0 to 1.5 m above sea level, and the location is directly adjacent to the East Sea. This is an area that is expected to become increasingly flooded when the sea level rises higher. The district has strength in fisheries development and the marine economy. In addition, there is a potential for ecotourism. However, natural resources are gradually being exhausted as a direct effect of climate change and sea-level rises in recent years, so the residential areas of Can Gio District are facing the loss of cultivated land, even part of residential land in An Thoi Dong, Ly Nhon, and Thanh An communes. Another difficulty is Can Gio District's need for building the Binh Khanh bridge connecting the center of HCMC with the district. This bridge connects Can Gio District with Nha Be District in the North and will help people travel more conveniently, contributing to promoting tourism development in particular and the general socioeconomic development of Can Gio.

According to statistics in recent years, the main consequences of climate change and sea-level rise for Can Gio District are flooding, saline intrusion, temperature changes, weather changes, and landslides. Two elements stand out: flooding and saline intrusion.

(a) Coastal flooding: Flooding is mainly caused by tides, sea-level rise, and sudden increase in rainfall. Low-lying areas interspersed with high sand dunes along the coast, such as large estuaries, e.g. the Long Tau, the Cai Mep, the Go Gia, the Thi Vai, the Soai Rap, and the Dong Tranh rivers, of Can Gio suburban district are, particularly at risk.

(b) Saline intrusion: Saline intrusion is a consequence of the increasing temperature phenomenon causing sea-level rise high to deeper into the mainland, long time of saltwater intrusion, causing saline intrusion. Temperature and weather changes due to high levels of CO_2 emissions in large industrially constructed cities, like HCMC, have led to changes in rainfall, humidity, and serious impact on the living environment.

In addition, negative impacts such as landslides along riverbanks and coastal erosion also have a direct impact on people's lives in this area. Coastal erosion occurs in most coastal areas, with an intensity of several tens of meters per year. Rising sea levels and river flows are the causes of coastal erosion in Can Gio District.

Flooding occurs both in the center and suburbs of the city, even during dry season. The flooded area is about 140 km^2 with 85% of the flooded area in the center of the city. Damage caused by flood is estimated at about 8 billion VND per year. The reason is that the sewer system was built 50 years ago and has deteriorated. In addition, the construction of industrial and urban areas in the southern region (the drainage area of this city) has made the situation more serious. The drainage in HCMC is based on a system of natural rivers and canals, but about 30% of the canals have been filled by the city authorities. According to a research by the Southern Institute of Water Resources, in the 12 years from 1996 to 2008, in HCMC, there were more than 100 canals and channels with a total area of about 4,000 ha that were occupied and encroached on by people. In addition, the government's mistakes in planning have also made HCMC's flooding situation more serious (Le-Minh 2019).

In summary, flooding is considered a natural disaster, disrupting the activities of the community and society, causing loss of life, property, economy, and environment that the community and society cannot afford to hold it up. People can reduce flood damage by relocating far from rivers and seas. However, economic activities and people's lives are often associated with rivers and the sea. Therefore, whether they like it or not, people still choose to settle near water sources—a source of life, even though there are potential risks of natural disasters such as flooding. Often the value obtained by living near rivers and seas is higher than the cost of forecasting and coping with floods.

The most direct and visible negative impacts of flooding exacerbated by climate change are in the Can Gio area, HCMC. The flooding phenomenon is increasingly damaging civil works, especially the residents' houses, reducing the source of residential land, directly affecting people's lives, and also impacting the economic zones. The inevitable consequence is an increase in poverty, fewer opportunities for agriculture, and forced migration to the central areas, putting even greater pressure on the downtown area. Under the impact of climate change in HCMC, houses are facing the risk of encroachment, even having to relocate, thus greatly affecting the existence and development of residential areas that have existed for a long time (Lusterio, 2007) (Fig. 1).

Fig. 1 Landslide in Can Gio District, HCMC (Asian Development Bank 2010)

2 Research Methodology

This paper is a qualitative study using a mixed-method approach including the aggregating and comparing of data and expert methods. Aggregate and comparison data are applied to give a table of current situation analysis, to build a database of climate change phenomena in Can Gio District, HCMC. The collection of documents and data is conducted directly at the city's agencies, the Research Institute, HCMC Planning and Architecture Department, the Southern Construction Planning Institute—the Construction Ministry, Southern Institute of Irrigation Science, and Urban Management Department of Can Gio District. Based on the summary of climate change impacts on people's houses in the research area, a graphic software is used to show data, the current status of population distribution on maps, affected areas of flooding due to climate change, and provided a forecasting model until 2100. The expert method is applied to consult experts in the fields of architecture, planning, environment, hydrology, and house architecture, thereby establishing a scientific and legal basis based on practical experience on flooding due to climate change in Can Gio area. This research method can be applied and implemented for research on housing adapting to climate change, especially those researching types of houses for resettlement, houses for the poor people, or houses for migrants in HCMC.

3 Experience from Selected Responses to Climate Change and Sea-Level Rise

3.1 Experience of Japan

The Mizuya high-base house that is used to reduce the impact of flooding is the way that Japanese people in the Kanto region adapt to floods. The house has two rooms, two warehouses, and one toilet. Some families also have boats to evacuate in an emergency. Usually, a part of the main house is raised and can turn into a shelter. An advantage is the house is built on high ground, used in case of flooding, to store and protect property. In the first stage, the foundation is 2 m high; after high flooding, the house is raised more than 1.3 m on the old 2-m foundation, the house has been renovated to stay for a long time when natural disasters occur. The downside is that construction is quite expensive (Chung and Dan 2018)

4 European Experiences

(a) The Netherlands

Many house models are designed as a type of boat, which can float on the water in all weather conditions. The designs of Dura Vermeer (Netherlands) offer a "floating village" project for about 12,000 people near Schiphol Airport, not far from Amsterdam.

This project consists of floating houses forming a cluster of houses with the following features: The structure is made by hollow and waterproof concrete and can float with the house. A fixed immersion concrete tunnel is under the water level, where the booster piston is placed. By the time water rises, the house can float on high water. Some other types of houses, such as high-rise houses, high-level houses, were built by soil, and water channels and canals are around, that constitute a network of transportation systems in the whole village. Transportation is by boat. A typical example is the ecological ancient village of Giethoorn in Overjssel Province, Netherlands, 148 km from Amsterdam. This village was formed in 1230 and exists till date. The houses here look very ecological (Chung and Dan 2018).

(a) The United Kingdom

Floating houses were built by the Aquabase Construction Company in Leeds, England. The house is built on a floating platform that allows the house to float freely, even being capable of navigating like a boathouse. The house is designed with the expectation of being able to last over 100 years. The most basic advantage of a floating house is flexibility, which can cope with any rise in water levels and allows the residents easily move their houses away from dangerous areas to safe areas during the flood season. However, it also has disadvantages such as using the floats means the house will not be connected to the grid electricity, and the ability to manage the community is poor. However, this can now be overcome by solar or wind power supply systems with mobile communication technologies, advanced water tanks, and advanced wastewater treatment systems (Chung and Dan 2018) (Figs. 2 and 3).

Fig. 2 Giethoorn village (Overijssel Province, Netherlands) has no roads (https://vnexpress.net/du-lich/ngoi-lang-khong-duong-di-dep-nhu-co-tich-o-ha-lan-3924000.html)

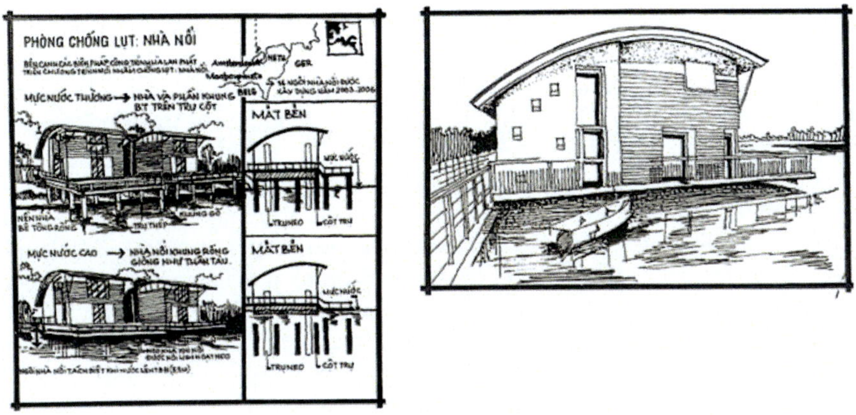

Fig. 3 Housing model adapted to flooding in the Netherlands (Chung and Dan 2018)

5 Response Solutions to Climate Change and Sea-Level Rise for Can Gio District, HCMC

5.1 *Principle and General Requirements*

(a) Principle

The research was conducted to contribute to the protection of people's lives in Can Gio District affected by flooding phenomenon due to climate change. The planning and architectural solutions potentially create better living conditions for people and contribute to maintain and protect traditional cultural and historical values in the locality.

(b) General requirements

Proposing overall solutions on planning to cope with flooding in Can Gio area.

Proposing housing architecture design solutions in Can Gio District to adapt to flooding due to climate change in this area, contributing to improve the quality of life for residents.

5.2 *Residential Planning Solutions*

(a) Ensuring drainage capacity in the urban development process

On average, the ability to penetrate into the soil of rain and flood water for urban land is only 1/5 of natural green space. Therefore, many countries regulate constructing buildings, from people's houses to sidewalks, must reserve a part of natural land, green trees, or use porous materials to increase permeability in place. If a new construction project is large-scale or an urban project reduces the surface drainage capacity of the area, the project must share the burden of infrastructure with the city by constructing a regulating lake or underground storage tanks in the construction area.

In planning and licensing documents, all urban construction and development projects must show the location of the project on the area map that has the risk of flooding. Buildings, lands located in flood-prone areas should be restricted to development and forced to take measures to prevent flooding when renovating buildings. In special cases when the project has economic and landscape value to the locality, the project can still be approved with conditions such as not reducing the water surface area and minimizing the impact on the streamflow (Fig. 4).

Urban areas, where there is high ground, must have a regulating lake with a capacity of about 180–200 m^3 per hectare of construction. For areas with the low ground, a cluster development model and park area with a significant area for flood control buffer zone are proposed. This innovative planning idea is not new elsewhere but it is difficult to apply in Vietnam due to land costs, infrastructure, and residence habits of the people (https://dothivietnam.org).

We can learn from the successful lessons of Curitiba City (Brazil). The city published information about flooding including a map of areas likely to be inundated, causing land prices in those places to drop. At that time, the government easily bought the land to turn them into public buildings and parks. In the dry season, the park is a resting place for the city's people. In the rainy season, these parks, with natural and low ground, will store and absorb a significant amount of water. After such parks are completed, the buildings and land in the area will no longer be in danger of flooding and have beautiful landscapes, thereby increasing the value. Through taxes on real estate values, the government regained the initial investment capital for the park and solved the urban flooding problem (https://dothivietnam.org).

Increasing space for stormwater and trees not only reduces the risk of flooding but also creates a landscape for the city. Increasing space for water is also a more sustainable solution when it does not alter the flow suddenly like dams and embankments. In order to bring this idea to life, we must first expand the thinking space, the perceived space, and the space of responsibility of the government and the community.

(b) Model of an embankment, over-flooded residential clusters for housing construction

These are models for dikes, embankments, and flood control systems. It is understandable that the dikes are firmly constructed roads and higher than the designed flooding water level so that big floods cannot overflow. The dike is often used to protect residential areas, concentrated industrial zones, townships, towns, and areas specializing in growing fruit trees. Also, an embankment is the temporary shoreline

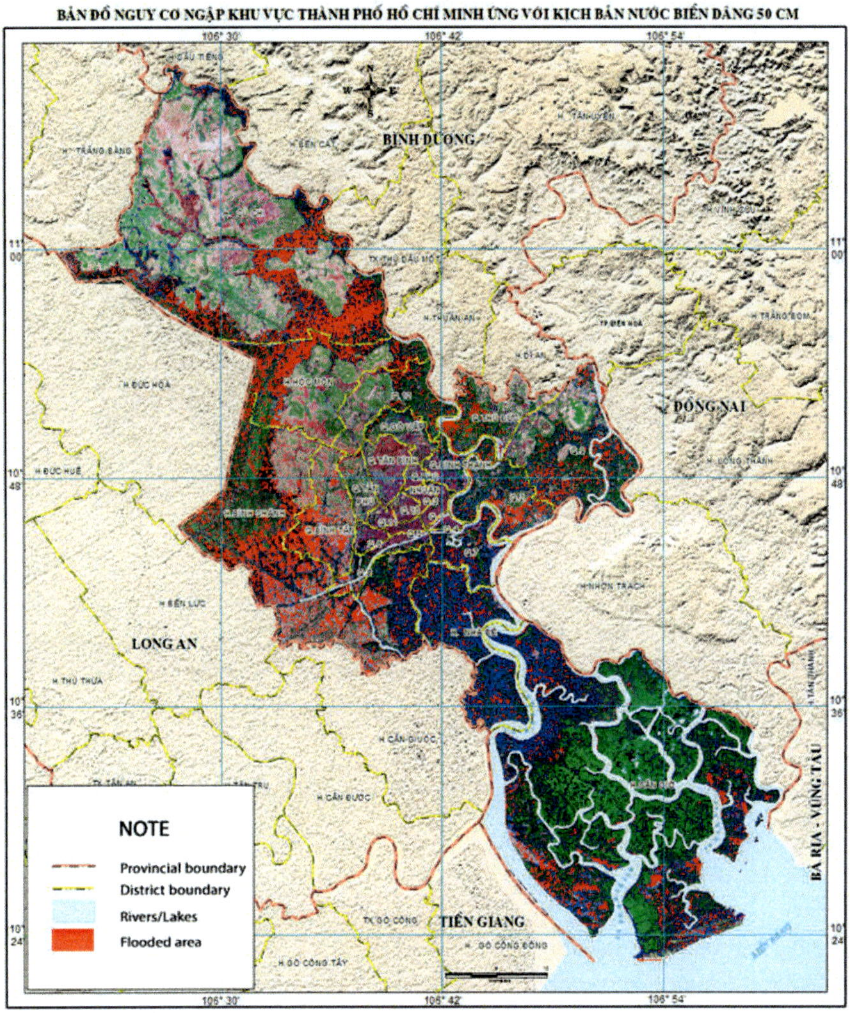

Fig. 4 Map of flooding zoning in HCMC (https://dothivietnam.org)

with altitude not exceeding the floodwater level in August so that when the summer–autumn rice is harvested, flood water will be flooded to get alluvium, adding aquatic resources, acidity to wash alum and wash the field.

The model has advantages: Housing people with their safety from the effects of floods assured. Housing built on clusters, residential lines are built firmly and smoothly to prevent the impact of wind, storms, and cyclones. People's livelihoods are ensured long-term stability, creating conditions for them to feel assured of productive labor, economic development, step by step sustainable development. People have

Fig. 5 Elevated well (Khoa 2015)

conditions to enjoy public welfare buildings, have conditions to improve people's intellectual standards. Many residential clusters suitable to the planning can develop into urban centers in the future (Fig. 5).

The limitation of the model is that the living area is still narrow, so people have difficulties in developing a secondary economy. Large investment capital should be supported by the State and combined with the participation of the people and support of the community.

(c) About the structure of villages—communal space

It is necessary to thoroughly utilize and the existing infrastructure and renovate and raise the flood-proof foundations to develop concentrated population quarters with favorable conditions on common infrastructures; limit the development of small and scattered population quarters; relocate spontaneous, scattered, and small population areas into concentrated areas, creating good conditions for living and production as well as rescue work when floods and storms, flooded water.

Creating 7–10 m strips of green trees along with canals and rivers to protect water drainage ditches. Green fences should be used avoiding the use of hard fences to identify households that have created environmental landscapes and create evacuation routes when necessary. Long-term protection of religious buildings, employing valuable landscape techniques like green trees and water wharves to create landscapes for villages and communes.

5.3 *Solutions for the Design of Houses in Flooded Areas*

(a) **House model on stilts**

- There is a relatively solid structure that saves construction costs compared to houses with foundations.
- Limiting flood water with the contaminants' access to the house.
- Ability to adapt to climate change and withstand storms of level 10–11.

However, this model is not favored by people in the delta because there are some inconveniences during use. (Floating house model to respond to climate change and ensure environmental sanitation 2014).

(b) **Floating house model of Dutch style**

- The house is built by hollow reinforced concrete, steel pillars supporting the house.
- Two poles anchored in the front and rear of the building, anchor the house to a fixed position when it emerges due to water rise.
- This house model can float to the surface to a height of 5.5 m. They are therefore able to adapt to sea-level rise well, allowing households to return to daily life soon after suffering the consequences of flooding.
- Low resistance to storms because there is no bracing against storms (Floating house model to respond to climate change and ensure environmental sanitation 2014).

(c) **Lightweight concrete house model**

This type of floating eco-house is very easy to transport by road and river because it is designed with flexible and convenient structure modules. The resizing of rooms and walls is done in a few simple steps without the presence of experts.

The roof is equipped with solar panels to generate heat and electricity to supply the house using an economical light system: LED and optical fiber ball. The house also has a steel frame structure positioned on the floating foundation system; other parts are used prefabricated components, so limiting waste during construction. The materials used to build the house are non-toxic, locally sourced. (Floating house model to respond to climate change and ensure environmental sanitation 2014).

(d) **Floating house model**

The advantage of this model is that, when the water is submerged, the house floats up to it, so it is not affected by a flood. However, the model has many shortcomings in the process of use. When the water is high, the house is floating. When the water recedes, the house is on the ground. When the house is located on a thick layer of silt because the weight of the house is uneven in every position, the house can be tilted and deviated. In addition, for places with thick silt layers, when it's dry, it will form a hard layer of soil covering and holding tightly to the house, so when there is no flood, the house will be located deep in the land, and when the flood water enters house, it can't stand up on its own. It is not safe to have strong winds, strong water, floating houses can be washed or bumped, so the house is damaged, water spilling into the house will be very dangerous to the life of the whole family. Not suitable for people's living habits. Living on floating houses people are easily isolated from the community. Traveling from one place to another is very difficult for people, especially for children's education and healthcare for old adults. The high cost also makes it not suitable for the economic conditions of the majority of local people in Can Gio District (Figs. 6, 7, 8 and 9).

Fig. 6 Solution of the layout of residential clusters (Le-Minh 2019)

Fig. 7 View of residential clusters (Le-Minh 2019)

Fig. 8 The house's access in flooding (Le-Minh 2019)

Fig. 9 Conventional windows become emer-gency exit in flooding (Le-Minh 2019)

6 Conclusion

Impacts of climate change are increasingly viewed as complex, especially considering the consequences of inappropriate urban expansion and weak land use management, which has led to an increase in the frequency and severity of urban flooding in our country. The Mekong Delta region is also at great risk for saline intrusion, affecting the ability to supply clean water and develop urban infrastructure. In HCMC, rainfall combined with higher and higher tides has lead to the increased expansion of flooding, especially in low-lying areas like in Can Gio District. For some urban areas in the plains and coastal areas, the phenomenon of riverbank erosion and coastlines are added complications. The phenomenon of "heat islands" in big cities affects the living environment. The quality of water resources is reduced, causing difficulties for living and production.

To cope with climate change's more and more complex consequences for HCMC in general and Can Gio District, in particular, we need to have an accurate assessment of climate change scenarios. From that, it offers solutions to urban development planning and architectural design of residential houses in residential clusters in order to adapt to the current climate change situation, creating a living environment with high adaptability and sustainability in the future.

Acknowledgements A portion of the research of this paper was conducted by Dr. Le Minh Ngo, which was made possible by Ho Chi Minh City Department of Science and Technology Grant. Dr. Le Minh Ngo also wishes to thank Ton Duc Thang University for hosting the research activities, and to the students and lecturers who participated in the research.

References

Asian Development Bank (2010) Ho Chi Minh City is adapting to climate change

Birkmann J, Pardoe J (2011) Climate change adaptation and disaster risk reduction: fundamentals, synergies and mismatches. Springer J. www.springer.com/978-94-017-8630-0

Chung DT, Dan LH (2018) New rural coastal housing space adapted to climate change. Vietnamese Architect J. http://www.baoxaydung.com.vn

Dung ND, Solve urban flooding: more space for water. https://dothivietnam.org

Floating house model to respond to climate change and ensure environmental sanitation. Construct Urban J (36) (2014)

https://vnexpress.net/du-lich/ngoi-lang-khong-duong-di-dep-nhu-co-tich-o-ha-lan-3924000.html

Huong LTT, Dao TQ (2018) Housing experience adapting to flood reduction and application opportunities for townhouses in Ho Chi Minh City. Construct J (03):145–148

Khoa LV (2015) Climate change adaptation models in Vietnam. J Environ (3)

Lusterio AC (2007) Living with floods: the settlements of the Vietnam Mê Kông Delta

Le-Minh NGO (2019) Assess the impact of flooding due climate change on houses in Nha Be and Can Gio areas (HCMC), and propose planning and architectural solutions to cope. The scientific research projects. Department of Science and Technology HCMC

Southern Institute For Spatial Planning (SISP) (2015) Develop solutions to adapt to climate change and sea level rise for urban areas in the Mekong Delta. Ministry-level scientific research projects. Ministry of Construction

Understanding the Implications of Urban Heat Island Effects on Household Energy Consumption and Public Health in Southeast Asian Cities: Evidence from Thailand and Indonesia

Sigit D. Arifwidodo, Panitat Ratanawichit, and Orana Chandrasiri

Abstract The study explores the effects of Urban Heat Island (UHI) on urban residents. Using two case studies in Bangkok, Thailand, and Bandung, Indonesia, the study focuses on the effects of UHI on household energy consumption and health and well-being. A survey questionnaire of 400 respondents from each city was employed. The household energy consumption for each respondent was measured using a proxy variable of average monthly electricity consumption. UHI intensity is measured using cooling degree days (CDD) variable constructed from the temperature difference between urban and sub-urban weather stations. The perceived health effect is measured by heat stress, physical health impacts, mental health impacts, and health and well-being outcomes. The data then are analyzed through descriptive and inferential statistics. UHI is found to have a positive correlation with the ownership of air conditioning equipment in Bangkok and Bandung and is found to increase the monthly electricity bill. It is also found that UHI has affected the daily lives of urban residents in terms of increasing household energy consumption for cooling and disruption of activities such as working, sleeping, and general health and well-being.

Keywords Urban heat island · Household energy consumption · Health and well-being · Sustainable urban development · Bangkok · Bandung

1 Introduction

The urban heat island (UHI) phenomenon generally refers to the higher temperature in cities or urban areas than the temperature in surrounding areas (Oke 1995). The UHI conditions increase the risk of climatic and biophysical hazards in urban environments including heat stress and heighten acute and chronic exposure to air

S. D. Arifwidodo (✉) · P. Ratanawichit
Kasetsart University, 50 Ngamwongwan Rd, Bangkok 10900, Thailand
e-mail: Sigit.d@ku.ac.th

O. Chandrasiri
International Health Policy Program, Ministry of Public Health, 88/20 Satharanasuk 6 Alley, Nonthaburi 11000, Thailand

L. T. T. Huong and G. M. Pomeroy (eds.), *AUC 2019*, Advances in 21st Century Human Settlements, https://doi.org/10.1007/978-981-15-5608-1_3

pollutants (EPA 2013). The UHI phenomenon is most observable at night because the roads and other surfaces absorbing solar radiation in daytime release heat in nighttime (Oke 1982). Anthropogenic heat, the heat released by human activities, can be an important factor causing UHI, especially in winter. Urban areas create more heat than rural areas because transportation, population, industrial, and some other activities are higher in urban areas (Kolokotroni et al. 2007).

UHI can have both negative and positive effects on cities. The UHI developments alter the atmospheric characteristics of a region. In tropical cities, UHI will hamper the energy consumption for cooling (Arifwidodo 2015). Ewing and Rong (2008) also found a similar result, that the household energy consumption is associated with the incidence of UHI and the spatial pattern of the metropolitan region. The study finds that the high-density area with less green space experiences higher UHI intensity, which results in higher household energy consumption for cooling in the summertime. Another study by Arifwidodo (2014) in Bandung, Indonesia, also found a significant relationship between household energy consumption for cooling and the increase of local temperature.

UHI can also cause negative health impacts and cause heighten acute and chronic exposure to air pollutants and lower physical health and well-being. Climate change, which is caused by increased anthropogenic emission of carbon dioxide and other greenhouse gases, is a long-term effect with the potential to alter the intensity, temporal pattern, and spatial extent for the UHI in metropolitan regions (Arifwidodo and Chandrasiri 2020). A study conducted by Thawilarp et al. (2015) and Tawatsupa et al. (2012) examines the heat stress and heat-related illnesses in Thailand and finds that there is an increasing prevalence of such illnesses and is associated with the rising temperature.

The study uses two case study areas: Bangkok, Thailand, and Bandung, Indonesia, because these cities represent two different geographical conditions of the rapid-growing city in Southeast Asia. Bangkok is a coastal city with a maximum elevation of 4 m above sea level, while Bandung is a mountainous city with an elevation of 768 m above sea level and surrounded by mountainous volcanic terrain ranging up to 2,400 m. The difference in geographic settings makes the cities experience varied UHI effects. By using two case studies, it is expected that the result can be generalized and the method used in the study can be applied to other cities with similar characteristics.

2 Methodology

In measuring UHI, the study follows Tan et al. (2010) in measuring UHI intensity as the temperature difference (ΔT) between urban area (u) and sub-urban area (s). In both cities, the yearly average temperature from three weather urban stations and one sub-urban station in Bangkok and Bandung was calculated. In linking household energy consumption with UHI, we calculated the Cooling Degree Days (CDD) to investigate the effect of higher temperature on cooling energy consumption in

Bangkok and Bandung. CDD method is a common technique to investigate the effect of higher temperature on cooling energy consumption (Giannakopoulus et al. 2009).

The study also collected the monthly average of electricity consumption as a proxy for household energy consumption. The data then were analyzed using the ordinary least square (OLS) regression model to understand the effect of UHI intensity (measured by CDD) to household energy consumption (measured with average monthly electricity consumption for the year of 2015) as the dependent variable.

In exploring the health effects of UHI, the study obtained survey data on perceived health stress and health outcomes. Previous studies such as Tan et al. (2010) and Thawilarp et al. (2015) have identified that UHI affects health through heat waves and heat stress causing heat-related illnesses. Since Thailand experiences heat waves less frequently, the study uses heat stress as a measure of the health effect of UHI. Heat stress is identified as the uncomfortable feeling when doing daily activities and measured using the following proxy question: "How often did the hot period of this year interfere with the following activities? Sleeping; Housework; daily travel; work; and exercise." The health outcomes were measured by three variables: physical health, mental health, and well-being. The data from the survey questionnaire then were analyzed using descriptive and inferential statistics to understand the patterns and characteristics of the perceived health effects. The study also geocoded the respondents' home addresses to incorporate the temperature data into their responses and create a proxy variable for the UHI intensity. Table 1 summarizes the variables and definitions used in the analysis.

3 Result and Discussion

Bangkok is the capital city of Thailand and located in the central part of the country. It is situated on the low-flat plain of the Chao Phraya River which extends to the Gulf of Thailand. It is the center of industries, manufacture, economy, commerce, and construction. This rapid urbanization has led to several environmental problems such as air pollution, water pollution, and land subsidence as well as problems from the presence of urban heat island, high energy consumption, and biophysical hazards (Arifwidodo and Chandrasiri 2013). In 2015, the maximum temperature difference between urban and rural areas of Bangkok was 7 °C, which is highest than in the last 10 years (Arifwidodo and Chandrasiri 2015).

Bandung is the capital of West Java Province in Indonesia. It is considered as the fourth most populous city in the country. The city is located on a river basin surrounded by volcanic mountains. The higher elevation provides cooler-year-round temperature than most cities in Indonesia, which makes Bandung is famous for tourism. The urbanization in Bandung has caused the change in the microclimate (Arifwidodo 2012). In 1995, the average temperature in the city is 18 °C and rising to 22 °C in 2007 and to 25 °C in 2015, with the maximum temperature difference between urban and rural areas of Bandung was 6 °C, which is highest in the last 10 years (Arifwidodo and Chandrasiri 2019).

Table 1 Variables used in the Analysis

Category	Variable	Definition	Bangkok		Bandung	
			Mean (std dev)	% of 1	Mean (std dev)	% of 1
UHI intensity	CDD24	Cooling degree days using 24 °C as the base temperature	152.68 (32.11)	–	12.1 (6.57)	–
	TEMP	Average monthly temperature difference between urban and sub-urban stations	1.95 (1.97)	–	2.8 (1.8)	
Housing	HOUSETYPE	Type of housing unit, with 1 = detached house; 0 = other	–	46.8		
	AC	Number of Air conditioner unit owned	1.8 (1.2)	–	1.52 (0.56)	–
	TYPEAC	Type of AC unit with 1 = split unit; 0 = other	–	78.3	–	98.4
Energy consumption	USEAC	Frequency of using AC unit in the past year, with 1 = almost every day, 0 = rarely used	–	84.2	–	23.4
	ENRGYSAVE	Energy saving products owned in the house, with 1 = own; 0 = not own	–	76.3	–	65.4
	ENERGY	Total energy consumption, measured with the monthly average of electricity bill for this year in USD	26.69 (13.47)	–	14.45 (12.67)	–
Household	HHMEMBER	No. of household member	4.0 (1.82)	–	5 (2)	–
	INCOME	Average monthly income in the last year (in USD)	627.63 (266.8)	–	197.8 (142.9)	–

(continued)

Table 1 (continued)

Category	Variable	Definition	Bangkok		Bandung	
			Mean (std dev)	% of 1	Mean (std dev)	% of 1
	AVAGE	Average age of respondent	39.6 (14.08)	–	46.15 (22.4)	
	EDUCATION	education of the head of household, with = 1 if respondent has graduated from high school, 0 otherwise	–	70.8	–	54.2
	GENDER	Gender of the respondent, 0 = female, 1 = male	–	38.2	–	65.3
	MARITAL	Marital status of respondent, 0 = single, 1 = married	–	48.3		69.3
	HHSTAT	Status in the household, 0 = head of the household, 1 = other	–	46.7	–	54.8
	TENURE	Housing tenure of the respondents with 1 = own; 0 = rent/other	–	87.4	–	83.5
Health	PHYSICAL	Physical health effect of UHI measured by perceived energy level during the past 4 weeks, with 1 = very low, 5 = very high	–	13.4	–	10.1

(continued)

Table 1 (continued)

Category	Variable	Definition	Bangkok		Bandung	
			Mean (std dev)	% of 1	Mean (std dev)	% of 1
	MENTAL	Emotional problems last 4 weeks experienced by respondents, with 1 = very rarely, 5 = very frequently	–	16.5	–	13.6
	WELLBEING	Life satisfaction for the last four weeks, with 1 = very unsatisfied, 10 = very satisfied	–	11.4	–	9.8
	SMOKE	The respondents smoking addiction, 0 = yes, 1 = no	–	88.7	–	56.4
	ALCOHOL	The respondents alcohol addiction, 0 = yes, 1 = no	–	90.5	–	99.5
	ISCHEMIC	Does the respondent have a ischemic heart disease, 0 = yes, 1 = no	–	99.8	–	98.7
	CEREBROVAS	Does the respondent have a cerebrovascular disease, 0 = yes, 1 = no	–	98.6	–	96.4
	RESPIRATORY	Does the respondent have a respiratory disease, 0 = yes, 1 = no	–	90.9	–	95.9
	CARDIO	Does the respondent have a cardiopulmonary disease, 0 = yes, 1 = no	–	93.7	–	95.9

3.1 Urban Heat Island and Household Energy Consumption

In Bangkok, the average monthly electricity expenditure is 854 THB (27 USD) and 72% of households in the study have air conditioning (AC) equipment in their housing units. There is a positive correlation between income and the number of AC units owned in the house (two-tailed t-statistics, $p < 0.0001$). This is because the higher the income, the households tend to have a bigger floor area in the house. The floor area of the house is also found to have a positive correlation with the frequency of AC use (two-tailed t-statistics, $p < 0.005$).

In Bandung, the average income from our survey is 2,868,267 IDR (198 USD), while the socioeconomic survey is 711,138 IDR (49 USD). On average, households in the study area spend 163,455 IDR (11 USD) per month for electricity expenditure. Most of the respondents live in detached housing with permanent construction (76.3%).

Table 2 summarizes the effect of UHI on household energy consumption in Bangkok and Bandung. After controlling for other variables, the result shows that there is a statistically significant relationship between UHI intensity (measured by CDD) and household energy consumption both in Bangkok and Bandung. The result confirms that UHI magnitude has a positive correlation with household energy consumption in two ways. First, energy consumption for cooling is higher when the UHI magnitude is higher, corresponds to seasonal variations. Second, energy consumption for cooling is higher in the area that suffers a high UHI magnitude. It means in urban areas, where the magnitude of UHI is higher, household energy consumption is also higher compared to the sub-urban area. However, in Bandung, although the average of monthly electricity usage for cooling was 262 kilowatt hours, or 85.34% of the average of total monthly electricity usage of household, there were only 12.42% of respondents who had AC installed in their house. The fact that the variable AC usage is not significant reflects the low percentage of AC equipment and usage compared to the total monthly electricity usage of households which was only 0.22% and 1.34%, respectively.

Table 2 Result of OLS regression model on the effect of UHI to household energy consumption

Variable	Bangkok	Bandung
	Coeff (std. error)	Coeff (std. error)
CDD	0.0049 (0.0012)***	0.0007 (0.0003)**
AC	0.0738 (0.0311)***	0.0738 (0.0256)**
USEAC	0.1316 (0.0409)**	0.1264 (0.0373)
ENRGYSAVE	−0.1014 (0.0988)	−0.0400 (0.0788)
HOUSETYPE	−0.0260 (0.0610)**	−0.0271 (0.0274)**

Note ***$p < 0.001$, **$p < 0.05$. Adjusted R2 for Bangkok = 0.23 ($p < 0.001$), adjusted R2 for Bandung = 0.21 ($p < 0.001$). The model is adjusted with control variables HHMEMBER, INCOME, AVAGE, EDUCATION, GENDER, MARITAL, HHSTAT, TENURE. Dependent variable is ENERGY

Table 3 Correlation between UHI intensity and heat stress

Variable	CDD	
	BKK	BDG
Sleeping 1	0.077**	0.031*
Sleeping 2	0.099**	0.045*
Housework 1	0.135***	0.063***
Housework 2	0.047**	0.064**
Daily travel	0.001**	0.001**
Work	0.006**	0.007**
Exercise	0.119**	0.125**

Note ***$p<0.001$, **$p<0.05$, *$p<0.01$. Adjusted R2 for Bangkok = 0.11 ($p<0.001$), adjusted R2 for Bandung = 0.08 ($p<0.001$). BKK = Bangkok, BDG = Bandung, n.s = not significant. The dependent variables are Sleeping 1, Sleeping 2, Housework 1, Housework 2, Daily travel, work, and exercise

3.2 *Urban Heat Island and Health Outcomes*

To understand the associations between health outcomes, heat stress, and UHI, two models of ordinary least squares OLS regression model are established. The argument is that UHI will disrupt daily activities through heat stress. The heat stress then eventually affects health outcomes. The first model is used to understand the association between UHI intensity (measured with CDD) and heat stress. The second model is to understand the association between heat stress and health and well-being outcomes.

Table 3 summarizes the correlation between heat stress and UHI intensity, measured with CDD. The data suggest that UHI intensity is significantly associated with all heat stress variables. The positive correlation means that the higher the UHI intensity, the higher the disruption of daily activities and causing heat stress. It can be inferred that UHI directly disrupts daily activities of the respondents and causes t heat stress. Only sleeping 1 and sleeping 2 variables shows less significance ($p<0.01$) in Bandung, this is probably because of the weather condition in Bandung during the night which is much colder than in Bangkok.

The correlation between heat stress and health and well-being outcomes is summarized in Table 4. The data suggest that heat stress is associated with health and well-being outcomes, especially life satisfaction. Energy-level variables have a negative correlation with housework activities, daily travel, work, and exercise. Emotional problem variable has a positive correlation with sleeping, housework, daily travel, work, and exercise. Heat interferences with housework 1 are found to have the highest correlation value with all health and well-being outcomes in Bangkok and Bandung. The data suggest that respondents with heat stress problems will have lower life satisfaction lower energy level and experience more frequent emotional problems. These findings share similar result with the literature which found that heat stress significantly reduce health outcomes and well-being. For example, Lan et al. (2010)

Table 4 Correlation between heat stress and well-being outcomes

	Life satisfaction		Energy level		Emotional problem	
	BKK	BDG	BKK	BDG	BKK	BDG
Sleeping 1	−0.03***	−0.05**	−0.078	−0.045	0.196***	0.164**
Sleeping 2	−0.49*	−0.27*	−0.49	−0.23	0.93	0.73
Housework 1	−0.198***	−0.112***	−0.193***	−0.096***	0.215***	0.176***
Housework 2	−0.157***	−0.092***	−0.125*	−0.096*	0.046	0.022
Daily travel	−0.83**	−0.74**	−0.118*	−0.102*	0.136**	0.094**
Work	−0.153**	−0.113**	−0.190***	−0.128***	0.174***	0.153***
Exercise	−0.191***	−0.082***	−0.162***	−0.118***	0.183***	0.098***

Note ***$p < 0.001$, **$p < 0.05$, *$p < 0.01$. Adjusted R2 for Bangkok = 0.121 ($p < 0.001$), adjusted R2 for Bandung = 0.115 ($p < 0.001$). BKK = Bangkok, BDG = Bandung. The model is adjusted with control variables HHMEMBER, INCOME, AVAGE, EDUCATION, GENDER, MARITAL STATUS, HHSTAT, TENURE, SMOKE, ALCOHOL, ISCHEMIC, CEREBROVAS, RESPIRATORY, CARDIO. The dependent variable is life satisfaction, energy level, and emotional problem

found that people working in hot environments had lower motivation to work and experience a negative mood during working. Guo et al. (2012) found that there is an effect of increasing temperature on mortality in Bangkok in Thailand. This study found out that heat stress is not only affecting the working life, but also interferes with other aspects of daily life such as sleeping, daily travel, and exercise.

4 Conclusion

The study explores the effects of UHI on the household energy consumption and perceived health effects in Bangkok, Thailand, and Bandung, Indonesia. The effects on household energy consumption are examined using monthly electricity. The result shows that UHI is associated with the increasing number of AC equipment ownership and the increase of household energy for cooling in both cities. The perceived health effect of UHI is examined using heat stress variables and health outcomes. The analysis of the data suggests that UHI increases heat stress and lowers health outcomes. The study indicates that UHI affects the daily lives of urban residents and the local government needs to pay more attention to it. Integrating UHI adaptation into the decision-making process seems to be the logical implication of the finding. In the cities of developing countries where resources are scarce, a better understanding of the effect of UHI to the urban area can help local government to make a better judgment on allocating resources and amenities, and knowing better when to promote growth or to contain it, while minimizing externalities that may occur.

References

Arifwidodo SD (2012) Exploring the effect of compact development policy to urban quality of life in Bandung, Indonesia. City, Cult Soc 3(4):303–31

Arifwidodo SD (2014) Urban form and household energy use in Bandung, Indonesia. In: Sridhar KS, Wan G (eds) Urbanization in Asia: governance, infrastructure and the environment. Springer, India

Arifwidodo SD (2015) Factor contributing to urban heat island in Bangkok. ARPN J Eng Appl Sci 10(15):6435–6439

Arifwidodo SD, Chandrasiri O (2013) The relationship between housing tenure, sense of place and environmental management practices: a case study of two private land rental communities in Bangkok, Thailand. Sustain Cities Soc 3(2013):16–23

Arifwidodo SD, Chandrasiri O (2015) Urban heat island and household energy consumption in Bangkok, Thailand. Energy Procedia 79(2015):189–1194

Arifwidodo S, Chandrasiri O (2019) Urban heat island and health effects in Bangkok, Thailand. Int J Adv Sci Technol 28(1): 374–382

Arifwidodo SD, Chandrasiri O (2020) Urban heat stress and human health in Bangkok, Thailand. Environ Res 185:109398. https://doi.org/10.1016/j.envres.2020.109398

EPA (2013) Heat island effect (http://www.epa.gov/heatisld/index.htm)

Ewing R, Rong F (2008) The impact of urban form on US residential energy use. Hous Policy Debate 19:1–30

Giannakopoulus C, Le Sager P, Bindi M, Moriondo M, Kostopoulou E, Goodess CM (2009) Climatic changes and associated impacts in the Mediterranean resulting from a 20 C global warming. Global Planet Change 68(3):209–224

Guo Y, Punnasiri K, Tong S (2012) Effects of temperature on mortality in Chiang Mai city, Thailand: a time series study. Environ Health 2012:11–36

Kolokotroni M, Zhang Y, Watkins R (2007) The London heat island and building cooling design. Sol Energy 81:102–110

Lan L, Lian ZW, Pan L (2010) The effects of air temperature on office workers' well-being, workload and productivity evaluated with subjective ratings. Appl Ergonom 24:29–36

Oke TR (1982) The energetic basis of urban heat island. J R Meteorol Soc 108(455):1–24

Oke TR (1995) The heat island of the urban boundary layer: characteristic causes and effects. In: Wind Climate in Cities. Kluwer Academic Publishers, Dordrecht, pp. 81–107

Rosenzweig C, Solecki WD, Parshall L, Chopping M, Pope G, Goldberg R (2005) Characterizing the urban heat island in current and future climate. Global Environ Change Part B: Environ Hazards 6(1):51–62

Tan J, Zheng Y, Tang X, Guo C, Li L, Song G, Zhen X, Yuang D, Kalkstein A, Li F, Chen H (2010) The urban heat island and its impact on heat waves and human health in Shanghai. Internat J Biometeorol 54:75–84

Tawatsupa B, Yiengprugsawan V, Kjellstorm S, Sleih A (2012) Heat stress, health and well-being: findings from a large national cohort of Thai adults. BMJ Open, (2)

Thawilarp S, Thammawijaya P, Praekunnatham H, Siriruttanapruk S (2015) Situation of heat-related illness in Thailand and the proposing of heat warning system. Outbreak Surveil Investi Reports 8(3):15–23

The Need for Understanding Disaster Risk for Resilient City Development

Mumita Tanjeela and Md. Anisur Rahman

Abstract Rapid urbanization has been a common phenomenon in most developing countries. Asia is no exception. Many major Asian cities are fast-growing but without proper physical planning guidelines. For instance, development is taking place in areas that are prone to risk from natural hazards and could be seen as a serious concern. With a view to reduce damage and encourage resilient development, the Sendai Framework on Disaster Risk Reduction (SFDRR) has emphasized on understanding disaster risk. This paper has tried to understand the current situation in Asian cities with regard to resilient development. Publicly available data and information published, are used to understand the development patterns of the major cities in Bangladesh, Myanmar, and Thailand. This research finds that Asian cities are at a very early stage in understanding disaster risk and climate change effects ahead of city planning and physical expansion. As a result of unplanned development without considering potential risks, citizens are becoming vulnerable to natural hazards. Therefore, it is important to understand the potential disaster risks and climate change impacts for resilient city development and for effective urban planning. The outcome of the paper will help in formulating policy guidelines for the respective cases.

Keywords Disaster risk · Resilience

1 Introduction

Urbanization in developing countries has been a challenge for several decades, particularly problematic is rapid urbanization and expansion of developed land in Asian cities. About 55% of the world's population currently lives in urban areas and this proportion is expected to increase to 68% by the year 2050. It is estimated that about 2.5 billion people will be added to urban areas globally and 90% of them will be in

M. Tanjeela (✉)
Assistant Professor, Department of Sociology, East West University, Dhaka, Bangladesh
e-mail: mumita_tanjeela@yahoo.com; mtanjeela@ewubd.edu

Md. A. Rahman
Director, Urban Research Initiatives (URI), Dhaka, Bangladesh

L. T. T. Huong and G. M. Pomeroy (eds.), *AUC 2019*, Advances in 21st Century Human Settlements, https://doi.org/10.1007/978-981-15-5608-1_4

the urban areas of Asia and Africa (UNDP 2018; ADB 2008). There is a growing concern with regard to the spatial trend of urbanization on disaster risk and climate change effects. With many of the cities rapidly growing in Asia the situation is such that these cities are highly vulnerable to catastrophic events mainly due to unplanned development on the coasts, deltas, and floodplains thereby becoming highly exposed to risk and vulnerabilities both to the human population as well as to the existing infrastructure. The pace of urbanization and population growth is such that both the proportion and absolute numbers of the world's population residing in urban areas will increase, with the equivalent of the total current global population living in cities by 2050 (UN 2018).

Therefore, it is essential to take preparedness initiatives for sustaining the pace of development for resilient urbanization. This paper presents a review of the cases from three major cities in the south and Southeast Asia based on the challenges faced due to rapid urbanization. The paper also discusses the necessity of disaster risk assessment for resilient city development.

2 What is Disaster Risk and Resilient City?

It is important to understand the situation or critical issues ahead of any urban development, especially in the developing countries where investment cost is high. Identification of hazard, exposure, and vulnerability lies at the core of a holistic approach to disaster risk management. Risk identification, conducted through risk assessment at various levels helps for emergency preparedness, financial protection measures, and sustainable recovery and reconstruction (World Bank 2013). According to UNDRR (2015), disaster risk is the potential losses, in lives, health status, livelihoods, assets, and services, which could occur to a particular community or a society over some specified future time period. Urban or city resilience refers to the capacity that city or urban area has to respond, or "bounce back" from, and adapt to challenges faced as a result of experiences that interrupt or complicate the economic, social, and environmental functions found in cities and their surrounding areas (ADPC 2015). The concept of a vulnerability assessment is critical and significant for the resilient city and for its contribution to the spatial and socioeconomic mapping of future risks and vulnerabilities. The role of the vulnerability analysis is to analyze and identify types, demography, intensity, scope, and spatial distribution of environmental risk, natural disasters, and future uncertainties in cities. In the case of Bangkok, over 40% of the city is projected to be flooded by 2030 due to increased precipitation and a 15-cm sea-level rise (WB 2018).

2.1 *What is Happening in the Cities with Development?*

Spatial development is taking place in most of the cities both in South and Southeast Asian countries. This section discusses spatial development patterns in relation to disaster risk for the cities of Dhaka, Mandalay, and Bangkok. The city of Dhaka is among the largest megacities in the world (WB 2018), where development taking place in low-laying areas, there is a serious concern on the city's safety. After the devastating flood of 1988, the government has taken numerous initiatives to construct embankment around the city to protect it from seasonal floods. A 30-km-long road cum embankment was proposed along the western bank of the Balu River in order to protect the entire area from flood. However, after this plan, about 19 housing projects have been initiated in the eastern fringe over the last one and half decades (Rahman 2007). Since then spatial development took place at a faster pace. Between 2009 and 2014, an average one and a half square km low-laying areas were filled in for housing estate development (Please refer to Fig. 1). However, the areas where these developments are taking place are marshy and remain underwater during the rainy season because of its natural setting. At the same time, there is a concern about earthquake liquefaction. An earthquake risk assessment conducted by the Comprehensive Disaster Management Program in 2009 has identified the areas that are highly susceptible to the effects of soil liquefaction (Fig. 2).

The eastern part of the city where development is taking place for over a decade falls under the high-risk zone of soil liquefaction effect due to recent fillings. Being located on the soft soil, Dhaka's newly infill areas may experience tragedies such as the Kobe Earthquake of 1995 and Mexico City earthquake of 1985, in case of any earthquake occurs in the future.

Recent political and economic reforms in Myanmar have led to faster urbanization in the country. Mandalay, the second-largest city in Myanmar, is expected to experience great economic growth in the near future as development opportunities are already taking place throughout the city and its surrounding areas. Realizing the fact, in 2011, the Department of Human Settlement and Housing Development under the Ministry of Construction took the initiative for preparing a perspective plan for the city (DHSHD 2011). The 30-year perspective plan indicates its future city growth strategy and installations. However, the plan did not consider any prevailing or prospective hazard or risk for the city (Fig. 3).

Thailand along with Myanmar Earthquake Society conducted an Earthquake Risk Assessment for the city of Mandalay. The assessment result reveals that the area proposed for the city center of Mandalay is highly vulnerable to earthquake hazards.

Don Mueang Airport, the second-largest airport of Thailand was underwater for more than six weeks, the Aythuuya Industrial Estate, the home of hundreds of industries remained inundated for about 2 months and the Navanakorn Industrial Estate submerged in flood water of about 1.5 m for several weeks. The industrial estate which is the world's second-largest producer of computer hard disk drives accounting for approximately 25% of the world's production remained closed for months resulting in the increase of prices in the world market. One-third of the country including a

Fig. 1 Aftabnagar residential area in the eastern part of Dhaka in 1996 (above) and 2014 (below). Red Circled areas in 1996 image denote the water bodies which are disappeared after the development of Housing estate by 2014 (referred areas circled in blue)

Aftabnagar Residential Area in 1996

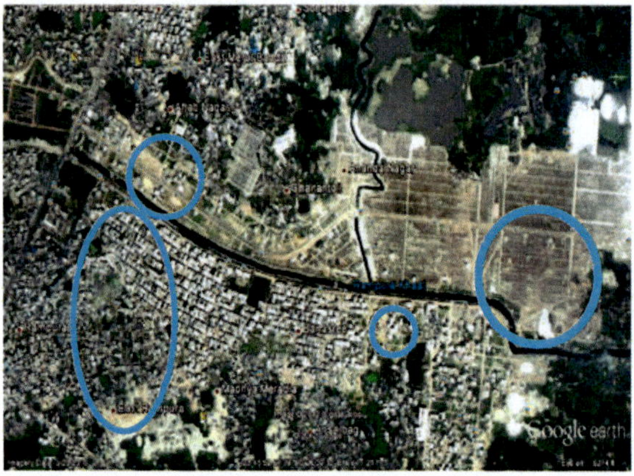

Aftabnagar Residential Area in 2014

major part of the Capital Bangkok were affected by severe flood in 2011 which is the most severe flood to have occurred in more than six decades (Bangkok Post 2011).

At the height of the flooding in Bangkok, 470 locations in 32 northern, western, and eastern sections of the city were impacted as water levels rose to a minimum depth of 80.0 cm (2.6 feet). The failure of several barriers led to the inundation in the affected districts. In total, more than two million residents (20% of the city's population) were given mandatory evacuation notices as floodwaters forced the closure of Don Mueang Airport and multiple main highways in and out of Bangkok (AON 2012).

According to Ouyyanont (1997) from 1900, following five decades, Bangkok was larger than the second-largest city in Thailand, thus making it a classic example of a primate city. In 1980, Bangkok was larger than Chiang Mai by 51 times. Currently, Bangkok is larger than its second-largest city by 60 times in terms of its official

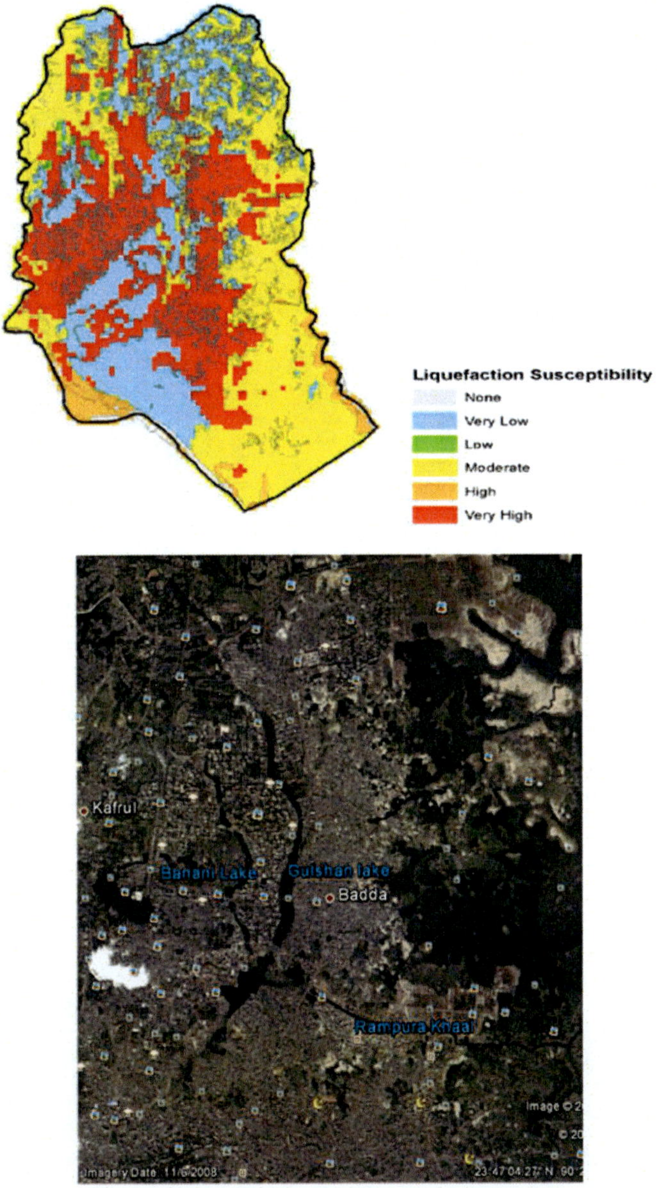

Fig. 2 **a** Liquefaction Susceptibility Map of Dhaka City and Surrounding Areas (above) shows Liquefaction susceptibility ranges in different areas. Below is the aerial view of Dhaka City represents the areas with possible Liquefaction Susceptibility effect to earthquake where development is taking place on marshy lands (*Sources* Seismic Hazard Map of Dhaka, CDMP, 2009 and Google Earth Map)

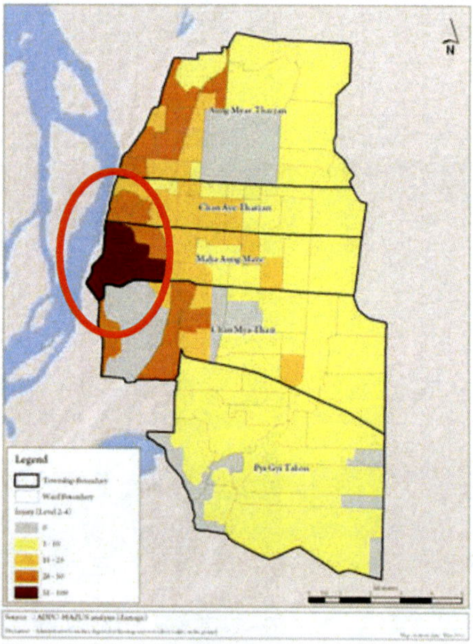

Fig. 3 Red circle in Mandalay Perspective Plan denotes the proposed city center (Above). While the same area of the city in Earthquake Risk map (Below) indicates the high-risk area to Earthquake Hazards

population. Although during the 1960s, there were several recommendations by the urban and regional planners to shift the focus from Bangkok to other regions of the country, but it failed to bring attention to policy and decision-makers. Since much attention has been paid to one city of the country, development has taken place in all directions emanating from Bangkok onto agricultural lands, waterways, marshy lands, and flood plains. The evidence of it could be seen from the inundation of

Fig. 4 The photo Illustrates flood preparedness initiatives by three residents at Ratchathewi (Near Victory Monument) area during the Thailand 2011 flood. The house owner of the left-hand side photo estimated the flood to be 5 feet height and it can be protected by putting plastic cover on his boundary wall, next house owner (middle photo) expected strong flood at 2 feet height and he constructed a permanent concrete wall for that, next house owner (rightmost) expected medium velocity flood which could be prevented by sandbag barrier

the industrial estates in Pathumthani and Ayuthuuya Provinces that are located very much in the low-laying areas and within the vicinity of Bangkok.

During 2011, the Thailand flood, the uncertainty of being inundated by floodwater was much among the communities across Bangkok since there was no proper flood risk map available for the city. As a result, communities and individuals took their own initiatives for flood preparation which was different from each other although being at the same location (Refer to Fig. 4). Similarly, a number of industries, residential areas, academic institutions, the airport went under flood water while there was no adequate understanding of the situation.

3 Discussion and Conclusion

In the case of Dhaka City, development initiatives or urban planning process did not consider the prevailing/future disaster risk of the city. As a result, spatial development taking place on the areas that are vulnerable to earthquake hazards. The proposed perspective plan for the city of Mandalay did not consider the disaster vulnerability of the city to natural hazard, and the proposed future city center is on the location which is susceptible to earthquake liquefaction effect. There was no flood risk assessment map/information for the city of Bangkok which could not foresee the level of effect that the city installations and dwellers had to face. Therefore, it is clear that vulnerability/risk assessment is important and required for resilient development. It requires the integration of Disaster Risk Assessment in the development planning process that will help the cities to understand the prevailing and future risk of hazards and preparedness initiatives accordingly.

Cities in Asia will continue to grow but most cases development may take place without proper guidance leading to disaster. The process of urban development should be in accordance with the proper understanding of possible disaster risk associated with the respective arena. For effective disaster risk management and preparedness, it requires proper risk assessment which is key for avoiding risk-prone areas from sustainable development. It is therefore important to understand the potential risk ahead of urban development initiatives. As the development process and trend of urbanization will continue, initiatives for investment should be in a logical way for ensuring resilient city development. Disaster risk assessment should be part of the physical planning process for effective investment as a major step towards disaster risk reduction.

References

ADB (2008) Managing Asian cities: sustainable and inclusive urban solutions. Philippines, Manila

ADPC (2015) Resilient cities framework. Asian Disaster Preparedness Center, Bangkok, Thailand

AON (2012) 2011 Thailand floods event recap report. AON Corporation, Bangkok, Thailand

Bangkok Post (2011) Floods: hard drives jump in price. https://www.bangkokpost.com/learning/learning-news/263456/floods-hard-drives-jump-in-price. Accessed on 24 July 2020

DHSHD (2011) Structure plan for the city of mandalay. Department of Human Settlement and Housing Development, Mandalay, Myanmar.

Friend RM, Thinphanga P, MacClune K, Henceroth J, Tran PVG, Nghiem TP (2015) Urban transformations and changing patterns of local risk: lessons from the Mekong Region. Int J Dis Resili Built Environ 6(1):30–43

Ouyyanont P (1997) Bangkok's population and the ministry of the capital in early 20th century thai history. Southeast Aisan Studies, vol 35, No 2, 1997, Nonthaburi, Thailand.

Rahman A (2007) Environmental effects of housing estate development in Dhaka City, Bangladesh. unpublished M.Sc. thesis, AIT, Bangkok, Thailand.

The World Bank (2013) Strong, safe and resilient: a strategic policy guide for disaster risk management in East Asia and the Pacific. Washington DC, USA

The World Bank (2018) The turn down heat: climate extremes, regional impacts, and the case for resilience. Washington DC, USA

UN (2018) 68% of the world population projected to live in urban areas by 2050. https://www.un.org/development/desa/en/news/population/2018-revision-of-world-urbanization-prospects.html. Accessed on 24 July 2020

UNDP (2018) Revision of world urbanization prospects. Reviewed on https://www.un.org/development/desa/publications/2018-revision-of-world-urbanization-prospects.html

UNDRR (2015) Sendai framework for disaster risk reduction (SFDRR) 2015–2030. New York, USA

Architectural Solutions with Regards to Climate Change for the Rural Housing of Vinh Long Province, Vietnam

Le Thi Hong Na, Dang Hai Dang, and Nguyen Dang Hoang Nhat Truong

Abstract In the Mekong River Delta, Vinh Long Province is currently suffering the consequences of sea-level rise, saline intrusion, landslides, erratic floods, and thunderstorms, due at least in part to the effects of climate change. On the other hand, rural urbanization has negatively influenced the overall image of the rural social structure, of which, the culture, customs, habits, and living conditions of the people and, especially, the morphology housing architecture of the area are all being affected. In this study, the status of rural housing architecture in Vinh Long is systematically evaluated according to the criteria of sustainable architecture. Survey results show that nearly all of the rural house types in Vinh Long are suffering from the effects of climate change. Among the five leading types of rural housing of Vinh Long, only villas have largely escaped the impacts. Traditional housing is a type not significantly affected by climate change. Street houses, however, are dealing with the effects of river erosion and high tides. In particular, pure-agricultural and simple housing are heavily affected by landslides, tidal surges, saline intrusion, and tornadoes. Based on the relationship with the characteristics of indigenous architecture, economic, social and environmental conditions of Vinh Long Province, architectural solutions with regards to climate change have been proposed to improve the quality of life, to minimize the negative impacts on the environment and to meet the needs of future sustainable development.

Keywords Climate change · Rural housing · Sustainable development

L. T. H. Na (✉) · N. D. H. N. Truong
Ho Chi Minh City University of Technology, VNU-HCM, Ho Chi Minh City, Vietnam
e-mail: na.bmkt@hcmut.edu.vn

D. H. Dang
Investment Promotion and Enterprise Support Center, Vinh Long, Vinh Long Province, Vietnam

L. T. T. Huong and G. M. Pomeroy (eds.), *AUC 2019*, Advances in 21st Century Human Settlements, https://doi.org/10.1007/978-981-15-5608-1_5

1 Current Situation and Future Scenarios of Climate Change

Nowadays, climate change has become the biggest global problem-posing serious environmental and socioeconomic threats to the planet. The anomaly alteration in climate patterns has exerted adverse influence on human and natural systems including increasing frequencies of severe weather events, sea-level rise, and biodiversity loss (IPCC 2018). These phenomena are causing natural hazard risks to individuals and groups who are exposed and vulnerable, especially from the Third World countries. Vietnam is not an exception when it is considered to be among the five most affected countries by climate change due to its long coastline, low-lying, and densely populated delta regions (World Bank 2018). As one of the most downstream parts of the Mekong River, in particular, Vinh Long Province is revealed to be suffered from the negative consequences of climate change such as floods, drought, subsidence, erosion, and rising sea levels. Consequently, such issues have put significant pressure on the technical frame of urban infrastructure, social security, and human health. Moreover, according to the IPCC (Intergovernmental Panel on Climate Change) Fifth Assessment Report on a 1.5 °C warmer world in the upcoming decades, climate change is exacerbating faster than the predicted scenarios (IPCC 2014). Because of this, sustainable development emerges as an important factor in the creation of a planning agenda to mitigate detrimental environmental impacts and improve the quality of life in rural area (Le and Chinvanno 2011).

As a result of economic growth, urbanization and industrialization, the construction industry is inarguably a dominant contributor to global climate change due to its greenhouse gas emission, energy and natural resources consumption. The relationship between this industry and climate change is particularly important issue, as Vietnam is under pressure from rapidly increasing population with the high demand for buildings in the near future. In the context of encouraging the global response to the threat of climate change, solutions in architecture and building practice are most essential, especially in rural areas where are easily vulnerable to natural disasters with low adaptive capacity. Hence, this study aims to evaluate the status of rural house architecture, in case of Vinh Long, to provide sustainable architectural solutions to meet future development needs, in accordance with the economic, cultural, social and climatic characteristics of the region. This paper also provides researchers, authorities, designers and citizens with a more detailed look at the values of rural houses and, in doing so, encouraging sustainable development to not only help people improve their quality of life and minimize their impact on the surrounding environment, but also, to promote a more suitable and more distinguishable architectural style to the Mekong Delta countryside.

2 Architectural Assessment of Rural Housing in Vinh Long Province

In this section, an overview of the current status of rural–urban pattern of Vinh Long is indicated to show the changes related to urbanization and climate change over the past few years. Besides, the main features of different housing types are shown after conducting a categorization of typical rural dwelling types. In addition, the architectural characters in relation to the impacts of climate change, which can be solved by architectural measures, are identified.

In general, rapid rural urbanization has stayed indifferent, leading to a lack of orientation for development, leading to spontaneous construction with a lack of management from the authorities. According to the survey, there is a sparse distribution and uneven development in housing across the regions (Fig. 1). Besides, residence forms along the canal have been gradually changed to urban dwellings, which are street houses and concentrated residential areas. The house fronts which used to face river traffic now face into road traffic because of increases in trade demands. A number of traditional houses have been destroyed and replaced by modern houses of various styles. Based on the statistics of housing quality status in Vinh Long, there is a vast majority in the number of permanent houses with 85.32%, while the figures for semi-permanent houses, unstable house, and temporary house are only 8.21%, 4.66%, and 1.81%, respectively (Fig. 2).

In addition, rural urbanization has drastically changed the urban fabric and the overall image of rural areas, affecting all areas such as economy, culture, community,

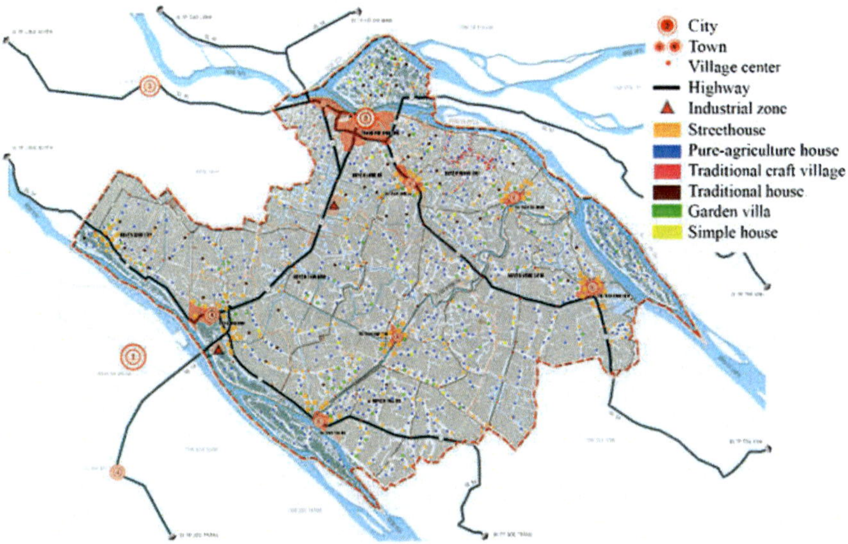

Fig. 1 Housing distribution in Vinh Long Province (*Source* Authors)

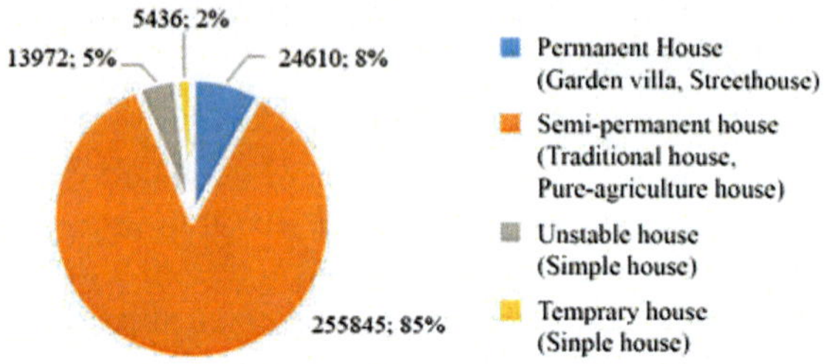

Fig. 2 Proportion of dwelling types according to house type in Vinh Long as of December 12, 2017

people, and architecture. The most noticeable of these are the form of residence of these regions. Urban development has caused changes in the layout of functional space inside the house. The functional spaces in the house are no longer kept horizontally like the traditional folk house with five spans including main part (living room, worship, sleeping area for men) and secondary part (food store, sleeping area for women); instead, it is a vertical layout along the length of the house like street houses in cities. Sorting by a spatial organization and household economic conditions, Vinh Long Province has five types of rural housing including villas, street houses, traditional houses, pure-agricultural houses, and simple houses as shown in Table 1.

Street houses are a dwelling type with combined functions of residence and commerce. Due to the conditions of socioeconomic development and urbanization process, this is a form of spontaneous housing type often clinging to the roads or adjacent to the river, so it is directly affected by climate change (landslides and river surges). The spatial organization can be divided into three different layouts: (i) road–house; (ii) road–house–river; (iii) river–road–house. The house is 1–3 floors high, with a sloped or flat roof. The house's orientation primarily follows the main traffic axis. The house has a structure with a solid reinforced concrete frame system, firmly covered with new materials. Because of arrangement along the length of the house, the functional spaces are inflexible, which does not ensure the natural ventilation. There is no space for greenery due to the high density of construction.

Traditional houses are scattered around the area. It is located along the traffic routes or canals and is highly concerned about the direction of the house. Whereas, traditional craft villages are distributed centrally along the river routes, mainly in Mang Thit district. The house plan still retains the traditional 5-span layout, including the main part and the secondary part. The spatial organization can be divided into three different layouts: (i) road–yard–house–garden–canal–field; (ii) river–road–garden, yard–house–pond–field; (iii) road–canal–yard–house–pond, garden–field. The main part usually faces the main traffic direction (road or canal), oriented toward the south, southwest, or southeast to take advantage of natural ventilation. The roof is 45° pitched, quick-draining, leak-proof with low solar radiation angle. Wide and tall louvered doors are utilized to improve ventilation and lighting. Trees are planted around the house. The wooden frame structure of the house, which has no deep foundation, has columns placed on rocks or cement, columns are buried directly below the ground in some cases. Besides, in Cai Thay Canal area in Mang Thit

Table 1 Comparison of different dwelling types in Vinh Long

	Street house	Traditional house	Pure-agriculture house	Garden villa	Simple house
Overall spatial layout	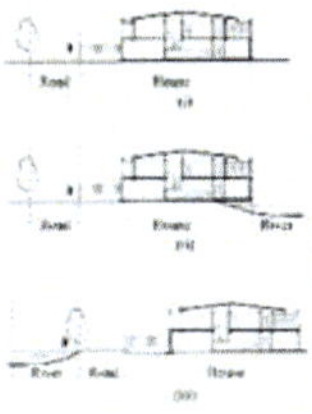	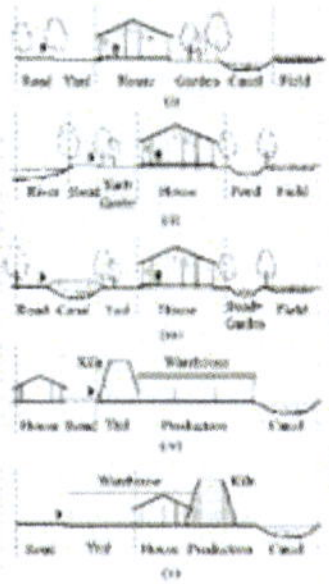	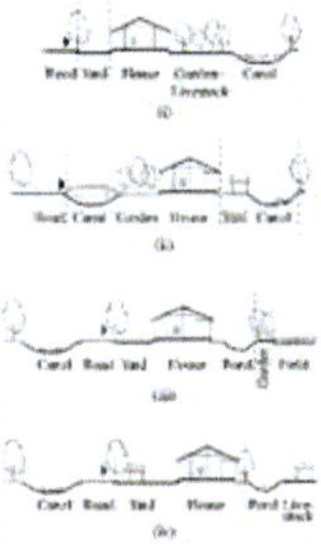	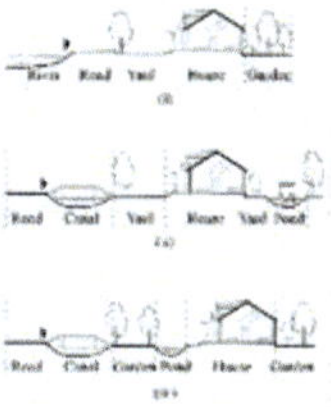	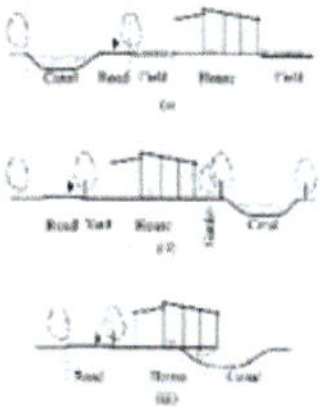
House plan	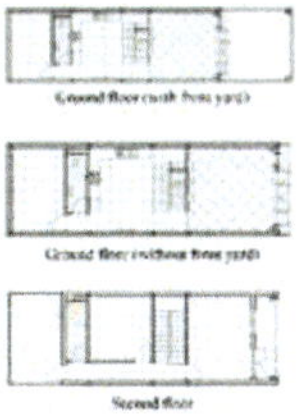	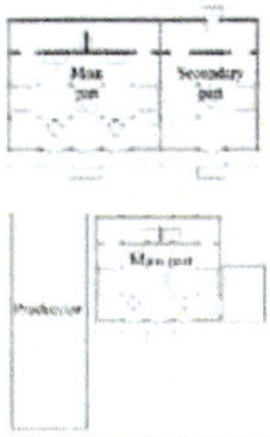	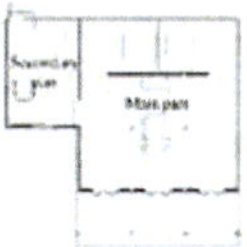	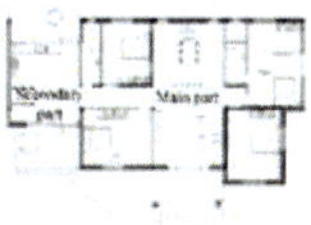	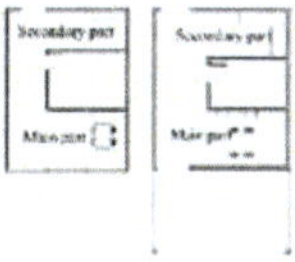
Housing 3D model	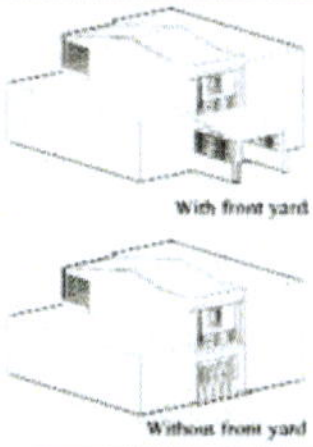	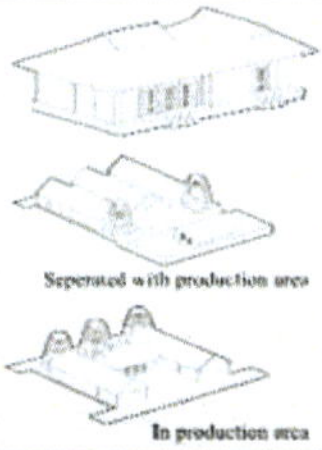	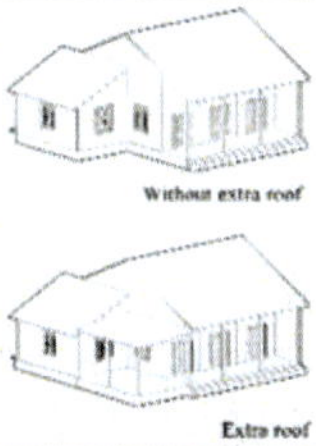		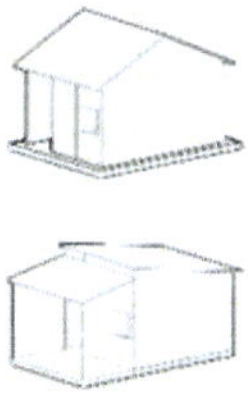
Photos					

Source Authors

District and Long Ho District, houses in traditional craft villages have extra functions of producing ceramic tiles, so there are additional kiln areas for manufacturing. The main house can be located separately or in the production area. Local materials are mainly used such as fired brick and clay. There are two types of overall layout: (i) road–yard–house–production–canal; and (ii) house–road–yard–production–canal. The indoor space is not ventilated as it is adjacent to the manufacturing area. Emission from brick kilns has a direct impact on the surrounding environment.

Pure-agricultural house is for families living by agricultural production. Accounted for the largest number of rural houses in Vinh Long, this housing type is mainly distributed along the river, canal, or traffic routes. Same as the traditional house, the house plan includes main and secondary parts, combined with modern architecture style due to the impact of urbanization. There are four typical types of layout including: (i) road–yard–house–garden, livestock–canal; (ii) road–canal–garden–house–yard–canal; (iii) canal–road–yard–house–pond–garden–field; (iv) canal–road–yard–house–pond–livestock. The house has one floor with a pitched or flat roof covered by corrugated sheets or tiles. A wooden frame structure and brick wall without painting provides poor thermal insulation and may leak in the rainy season. It may be easily flooded during the rainy season as the ground cote is quite low. In recent times, due to climate change, saline intrusion and tidal surges have directly affected the irrigation water and daily life of people. Therefore, it is necessary to provide solutions to help people improve water sources for living and storage in the future.

Garden villas are often built in large areas accessible by water or road traffic. This dwelling type is scattered from the commune center to sparsely populated suburban areas. The house plan is in a centralized manner, with secondary space located around the main house. The main function of this housing type is a place of residence. The house plan is diverse with courtyards, landscaped gardens, orchards, and water surfaces. There are three types of spatial layout: (i) river–road–yard–house–garden; (ii) road–canal–yard–house–yard–pond; (iii) road–canal–yard, garden–pond–house–garden. Living spaces are fully equipped with a larger area compared to urban houses. The house has 1–3 floors with a modern architectural style. The solid reinforced concrete frame structure and strong building envelope help it completely withstand the great influence of weather, flood, tide, and impacts from climate change.

Simple houses are located in residential areas encroaching on rivers and canals or scattered on fields in remote areas. The house plan is relatively small in a square or rectangular shape. The spatial organization can be divided into three layouts: (i) canal–road–field–house–field; (ii) road–yard–house–garden–canal; (iii) road–house–canal. The living space is quite limited, including the living room combined with worship, bedroom, and kitchen, without secondary space. It is first floor high with a sloping roof and simple architecture. The structure of the house is quite temporary with leaf or wood walls, temporary corrugated sheetss and no insulation ability. Recently, irrigation water sources and daily life of people living in this type of housing have been directly affected by the phenomenon of saline water intrusion and tidal surges due to climate change effect.

3 Architectural Solution Strategies to Climate Change

Studies published in the IPCC's report (AR5) have presented that human-caused warming reached approximately 1oC (likely between 0.8 and 1.2 °C) above pre-industrial levels in 2017 and the warming scenarios seem far worse than previously estimated by the end of this century. It is also predicted by IPCC that the sea level will increase by the range between 0.18 and 0.59 m at the end of twenty-first century. Thus, it is extremely urgent to come up with solutions to mitigate and adapt to the impacts of climate change. Mitigation and adaptation strategies, specifically with respect to the built environment, are required to deal with this issue.

Mitigation strategies are actions related to the reduction of long-term risks of climate change to human life in terms of reducing the CO_2 footprint. In the construction industry, we can achieve the reductions by various solutions including advanced technology; material and resources; designing and building methods. Energy renovation is the main contributor by integrating new effective technology in building energy production, reducing the transport of building materials by using local resources, taking advantage of renewable energy, and less embodies energy materials. In addition, integrated building design with active and passive techniques can optimize the living comfort of users and reduce the energy requirement for cooling, heating, and lighting. This requires considerations on local characteristics including climate conditions, typographical factors, and regional ecology.

While mitigation tackles the causes of climate change, adaptation tackles the effects of this phenomenon (Ho 2018). **Adaptation strategies** refer to actions taken to alleviate possible damage, deal with consequences, and exploit any potential benefits. As shown in the previous section, street house, pure-agriculture house, and simple house are considered to be among the housing types affected the most by climate change, namely landslides, tidal surges, saline intrusion, and floods. Affordable solutions that are suitable to local culture as well as regional topography and climate are required to make rural areas resilient to the changing conditions. Firstly, to ensure the safety of shelters when heavy storms occur, spontaneous and scattered residential areas near river and sea need to be shifted to inland areas with high ground level. Concentrated residential areas with synchronous infrastructure systems and solid public buildings as a shelter for evacuation will improve the conditions for living and manufacturing as well as rescue work. Secondly, communes and towns should strengthen the protection of forest planting, "greening" riverbank by creating 7–10 m strips of trees along canals and rivers to protect water drainage ditches and prevent landslides. Besides, roads and alleys are designed with an efficient sewer system to ensure the fastest drainage in the rainy season. Thirdly, rainwater collection and exploitation systems are also considered to help people take initiative in water sources when a saline intrusion occurs. Lastly, housing architecture is considered as the most influential solution in adapting the fluctuation of climate. The building envelope is encouraged to use local ecofriendly materials such as unburnt brick, heat-resistant corrugated sheet with the ability to allow natural ventilation and lighting. Housing structure needs to be more solidly designed for newly built houses and supported with

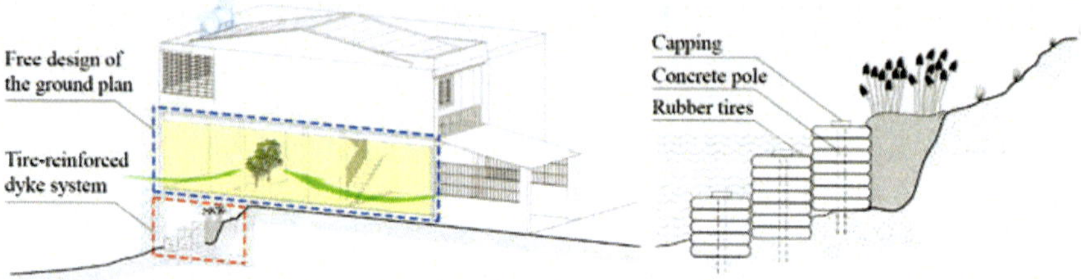

Fig. 3 Vacant-ground-floor housing model and tire-reinforced dike system (*Source* Authors Dang 2019)

an additional bracing system for the existing house to withstand in case of storm and flood. Furthermore, flexible organizations of functional spaces are adapted to adjust the living environment properly when disaster happens; for example, house with loft, floating house, or vacant-ground-floor model. In particular, the table below illustrates the solution model for pure-agriculture house, simple house, and street house, the dwelling types affected the most by climate change.

The proposed solutions for street houses need to adapt to the changes in the weather and reduce the negative impacts from high tide surges, riverbanks, and high solar radiation… For houses in lowland areas in Long Ho, Binh Tan, Tam Binh, and Mang Thit districts (Fig. 1), the ground floor spaces should be vacant, based on the lesson of Le Corbusier (Fig. 3). There are various advantages to be concerned such as (i) upper floor would be sheltered in case of a flood; (ii) natural ventilation is utilized into the house, reducing the obstruction of the airflow to other houses; (iii) the density of greenery is increased for the area, connecting ecological system between the inside and outside environment; (iv) small commercial or production are taken advantage on the ground floor. For the residential area with the backside adjacent to the river affected by the landslide, it is necessary to reinforce dike systems by simply using tires. Although tire-reinforced dike can be considered a temporary solution, it makes use of recycled materials and available local materials, which helps to reduce costs when compared to the construction of concrete dikes.

As being affected by flooding and saline intrusion, a pure-agriculture house can apply the solution of raising the elevation of the house combined with the rainwater collection and treatment system (Fig. 4). The base of the house would be raised to 0.8 m and pushed inside 1 m to prevent flooding when a storm occurs. It not only helps to keep insects, mice, snakes… away but also achieves ventilation for the low part of the house as well as creating soft boundaries for internal and external space. In addition, when applying the rainwater collection and treatment system, occupants can make the most of natural water sources, contributing to improving water for living and irrigation water sources and in the dry season and saline intrusion.

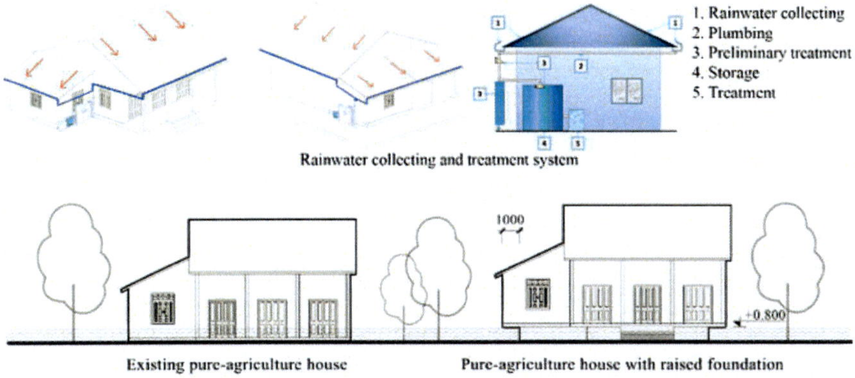

Fig. 4 House foundation raising model and rainwater collection and treatment system (*Source* Authors Dang 2019)

4 Conclusion

The study results showed that rural housing architecture in Vinh Long is facing numerous issues in terms of living comfort and spatial organization, especially the impacts of climate change. Without orientation or management on living quality conditions, residents around the area are suffering from various natural hazards as a consequence of climate change. For these reasons, architectural mitigation and adaptation strategies have been proposed in an effort to minimize the adverse impact of such challenges; however, there are difficulties and limitations to be mentioned. Therefore, more detailed investigations and experiments are needed for future adoption of the findings and proposed solutions presented in this paper. This study would serve as the initial point in an attempt to improve step-by-step the human living environment and protect ourselves as well as the next generation from the alteration of the climate.

References

Dang HD (2019) Sustainable architectural solutions for rural housing in Vinh Long. Master thesis. Ho Chi Minh City University of Architecture

Ho C (2018) The climate change in Vietnam and its impact on agricultural sector in Vietnam. Conference paper in SESAM, UPLB, At UPLB, Philippines

Intergovernmental Panel on Climate Change (IPCC) (2018) Special report on global warming of 1.5 °C

Intergovernmental Panel on Climate Change (IPCC) (2014) Fifth assessment report (AR5) on climate change 2014; mitigation of climate change; impacts, adaptation and vulnerability

Le AT, Chinvanno S (2011) Climate change in the Mekong River delta and key concerns on future climate threats. Advances in global change research, pp 207–217
Sergio A (2009) Climate change and architecture: mitigation and adaptation strategies for a sustainable development. J Sustain Dev 1(1)
World Bank Group (2018) The Vietnam climate risk profile

Application of Artificial Intelligence to Assist in Mapping for Flood-Prone Areas in the Bantul Regency, Yogyakarta

Aditya Wisnugraha Sugiyarto and Achmad Ramadhanna'il Rasjava

Abstract Flooding is one of the natural disasters that poses a serious threat to people who live near watersheds and coastal areas. Floods may affect various sectors of socioeconomic life of a society, such as sectors of the economy, agriculture, and education. This led to the need for an accurate method for predicting flood-prone areas so the public and the government can prevent and minimize the negative impacts and able to assist the recovery process in the sectors affected by the floods more optimal. In the Industrial Revolution 4.0 era, science and technology develop very quickly, one example is the development of Artificial Intelligence (AI) where the system adopts the way of human thinking, such as planning, learning, reasoning, and self-correction which is then manifested in a mathematical form so it can be applied for solving real problems. Therefore, this study used the AI system in the process of prediction and mapping of disaster-prone areas. The data used in this study are rainfall, land altitude, watersheds, river depth, and distance of the settlement to the seashore or river. These data were then processed using the method of deep learning for prediction function and Fuzzy C-Means (FCM) for mapping function which is a technique in AI systems. The final results of this study were obtained three criteria for disaster vulnerability, namely low, medium, and high, which can be used to predict the mapping of flood-prone areas until the coming years in the Bantul Regency area.

Keywords Floods predictions · Floods mapping · Disaster mitigation · Artificial intelligence · Deep learning

---

A. W. Sugiyarto (✉)
Department of Mathematics, Universitas Negeri Yogyakarta, Yogyakarta, Indonesia
e-mail: aditya.wisnugraha2016@student.uny.ac.id; adityawisnugraha@gmail.com

A. R. Rasjava
Department of Chemistry, Universitas Negeri Yogyakarta, Yogyakarta, Indonesia

L. T. T. Huong and G. M. Pomeroy (eds.), *AUC 2019*, Advances in 21st Century Human Settlements, https://doi.org/10.1007/978-981-15-5608-1_6

1 Introduction

Flooding is one of the most frequent and dangerous disasters in the world (Tingsachali 2012). Flooding poses a serious threat, especially for people who live near the watersheds and coastal areas (Aris and Lorenz 2008). Floods commonly take place around but are not limited to rivers, coastal areas, or even plains (Aris and Lorenz 2008; Jiang et al. 2009). Floods sometimes are the product of the interaction between social and environmental processes (Parker 2000). Natural factors causing floods include heavy rainfall, high tides, and high flood, while human factors include improper land use, heavy deforestation, and blocked or damaged drainage channels (Tingsachali 2012). Floods can cause massive losses, such as property damage and disruption of economic activity, thus worsening society. Floods have negative impacts on health because floods sometimes pose a high risk of pathogens and bacteria. Furthermore, floods can cause death or injury to the victims (Jiang et al. 2009; Van Voorst 2015). One of the examples is the 2019 Bantul flood. Bantul is a regency in Yogyakarta, Indonesia. In March 2019, heavy floods occurred in the area. Thirteen of 17 districts in Bantul Regency were flooded, with more than 4400 people evacuated and 2 deaths.

Risk management has been discussed for minimizing the loss caused by natural or human-made disasters around the world (Plate 2002). One of the disaster risk management is flood management. Many studies shown that losses related to floods disaster are strongly linked to populations and assets value, especially in a densely populated area (Aris et al. 2015). Flood disaster risk management can be done for minimizing the losses caused by floods, to simplify the support and rebuilding the society (Price et al. 2008). In floods disaster management, relocating the settlement is the last and the worst option (Chan 2006). One of the steps in disaster management is prevention by studying and analyzing the factors causing the floods disaster.

One way of studying and analyzing the floods disaster is by making a flood forecasting method. Deep learning can be used for making predictions (Sugyarto et al. 2019). In the previous study, (Campolo et al. 1999) predicting the rain intensity using ANN then using the data for predicting floods disaster. A study has been conducted by (Jiang et al. 2009) for assessing and mapping the floods disaster risk using FCA, FSE, and FSM. While mapping the floods disaster risk can be done using flood coverage maps, forecasting floods disaster can't only be done by predicting the rain intensity. In this study, we propose a deep learning method for forecasting and mapping the floods disaster risk using weather forecasting data, rainfall, land altitude, watersheds, river depth, and distance of the settlement to the seashore or river in each district in Bantul Regency. Using Fuzzy C-Means (FCM), we are predicting the floods disaster in Bantul Regency using variables above than by the results, we are mapping the floods disaster risk in each district in Bantul Regency as three categories: low, medium, and high risks.

2 Theoretical Review

2.1 *Deep Learning*

In this research, we used deep learning to predict the rainfall firstly, then we use the prediction to mapping. There are many types of deep learning (Alom et al. 2019), such as deep supervised learning (DNN, CNN, RNN, LSTM, and GRU), deep semi-supervised learning (such as generative adversarial networks (GAN)), deep unsupervised learning (Auto-Encoders, Restricted Boltzmann Machines (RBM), and the development of GAN), and deep reinforcement learning.

LSTM one type of deep learning method that has been popular in these years. LSTM is a special kind of recurrent neural network, which might be able to connect previous information to the present task (Hong et al. 2019). A recent study by Sugiyarto and Abadi (2019) shows better results compared to other methods for time series data prediction, so in this study, we use LSTM-RNN deep learning method. The architecture of LSTM-RNN model is shown (see Fig. 1).

2.2 *Fuzzy C-means*

Fuzzy clustering is a powerful unsupervised method for the analysis of data and the construction of models (Suganya and Shanthi 2012). In many situations, fuzzy clustering is more natural than hard clustering. Objects on the boundaries between several classes are not forced to fully belong to one of the classes, but rather are assigned membership degrees between 0 and 1 indicating their partial membership. Fuzzy c-means algorithm is the most widely used. A recent study by Singh and Mahajan (2014) giving good results was obtained by using FCM for clustering compared to by using another clustering algorithm; so in this study, we use FCM method for clustering and mapping processes.

According to Kusumadewi and Purnomo (2004), FCM algorithm process is as follows:

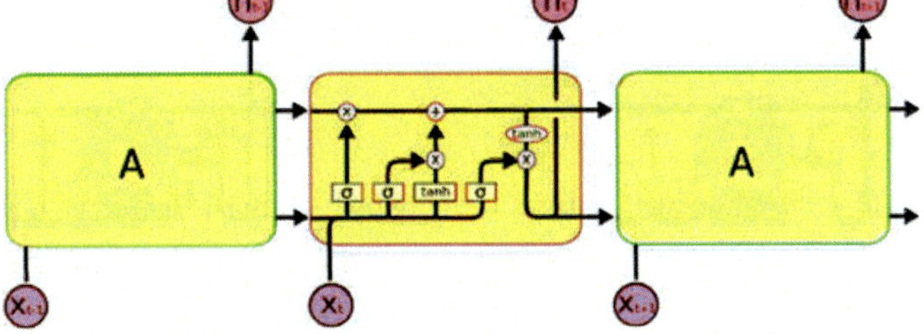

Fig. 1 Architecture deep learning LSTM-RNN model

First, input X as the data that will be clustered, as a matrix of $n \times m$ (n = total data sample, m = each data attribute). X_{ij} = sample data i ($i = 1, 2, \ldots, n$), attribute j ($j = 1, 2, \ldots, m$).

Then determine the number of clusters (c), power (w), maximum smallest error iteration expected, initial objective function ($P_0 = 0$), and initial iteration ($t = 1$).

In the third step, generate random number μ_{ik}, $i = 1, 2, \ldots, n$; $k = 1, 2, \ldots, c$; as initial partition matrices U elements then calculate the value of every column (attribute) using (1)

$$Q_j = \sum_{k=1}^{c} \mu_{ik}, \text{ with } j = 1, 2, \ldots, m, \text{ for } \mu_{ik} = \frac{\mu_{ik}}{Q_j} \tag{1}$$

In the fourth step, calculate the center of the k-cluster: V_{kj}, with $k = 1, 2, \ldots, c$; and $j = 1, 2, \ldots, m$ using (2)

$$V_{kj} = \frac{\sum_{i=1}^{n} \left((\mu_{ik})^w * X_{ij} \right)}{\sum_{i=1}^{n} (\mu_{ik})^w} \tag{2}$$

In the fifth step, calculate the objective function on the t-iteration using (3)

$$P_t = \sum_{i=1}^{n} \sum_{k=1}^{c} \left(\left[\sum_{j=1}^{m} \left(X_{ij} - V_{kj} \right)^2 \right] (\mu_{ik})^w \right) \tag{3}$$

In the sixth step, calculate the changes to the partition matrix using (4)

$$\mu_{ik} = \frac{\left[\sum_{j=1}^{m} \left(X_{ij} - V_{kj} \right)^2 \right]^{\frac{-1}{w-1}}}{\sum_{k=1}^{c} \left[\sum_{j=1}^{m} \left(X_{ij} - V_{kj} \right)^2 \right]^{\frac{-1}{w-1}}}, dengan i = 1, 2, \ldots, n; dan k = 1, 2, \ldots, c \tag{4}$$

After that, check the stopping condition: if ($|P_t - P_{t-1}|$ < smallest expected error) then stops, if not: $t = t + 1$, repeat step 4.

3 Method of Research

In general, the method used in this study is shown in the research flowchart (see Fig. 2).

Fig. 2 Research flowchart

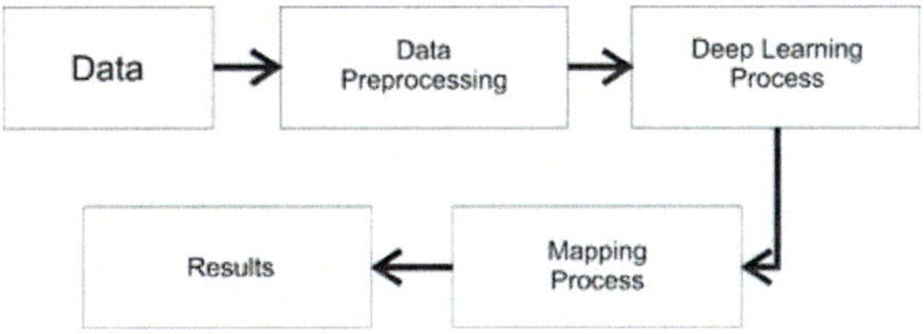

3.1 Data

The data used in this research were areas height (m), shortest settlement distance to the river (m), average river depth (m), average river discharge (m^3/s), area by height (km^2), and average monthly rainfall (mm) from January 2012 to December 2017 (BPS—Statistic of Bantul Regency 2013, 2014, 2015, 2016, 2017, 2018). All the data were collected from the Statistic Central Agency of Bantul Regency.

3.2 Data Preprocessing

After the data were collected, data preprocessing will be conducted, which is the data normalization. Not all of the data were normalized, only a handful of the data will be used for the prediction, which is the average monthly rainfall data, where the data is then used for the prediction. The normalization process was conducted using (5).

$$x' = \frac{x - x_{\min}}{x_{\max} - x_{\min}} \tag{5}$$

3.3 Deep Learning Process (Prediction)

After the data were normalized, the next process is conducting the prediction for the next data. In this process, one of the deep learning methods, the LSTM-RNN, was used. The prediction was conducted by searching the best LSTM-RNN model from each data type because there are 12 weather monitoring posts in Bantul Regency, the prediction was done based on one by one post. From these, average monthly rainfall data will be obtained one best model for every weather monitoring post, so there are 12 models for predicting the average monthly rainfall. To get the best model, a trial and error experiment was conducted by paying attention to the Mean Squared Error (MSE) value from each obtained model. The best model was obtained based on the smallest MSE. The calculation of MSE was done using (6).

$$\mathrm{MSE} = \frac{\sum_{i=1}^{n}(f_i - y_i)^2}{n} \tag{6}$$

3.4 Mapping Process

After obtaining monthly average rainfall prediction, the data will be used together with another data for mapping the flood-prone areas. FCM method was used in this study. There are three clusters of flood-prone areas obtained as the result of the mapping process, the low-risk, medium-risk, and high-risk flood-prone areas that will be visualized in the Bantul Regency map.

4 Results and Discussion

4.1 Prediction Using Deep Learning

The first step of this study is conducting the average monthly rainfall data. These data then used for predicting the average rainfall in the next years. The prediction process was conducted by using the average monthly rainfall data and resulting in the prediction of the next average monthly rainfall data. After the prediction data were obtained, then average rainfall data for the next year was calculated. The data will be used as the mapping data for the next year.

The first step is the data normalization, data from every weather monitoring post was normalized using Eq. (5). On the training process for searching the best model of deep learning, the data are divided into two groups, data training and data testing with 75% dan 25%, respectively. The architecture of the deep learning model used in this study was shown (see Fig. 3).

In the process for searching the best model, two input variables were used, there are Y_{t-2} (average rainfall data in the same month as 2 years before) and Y_t (average

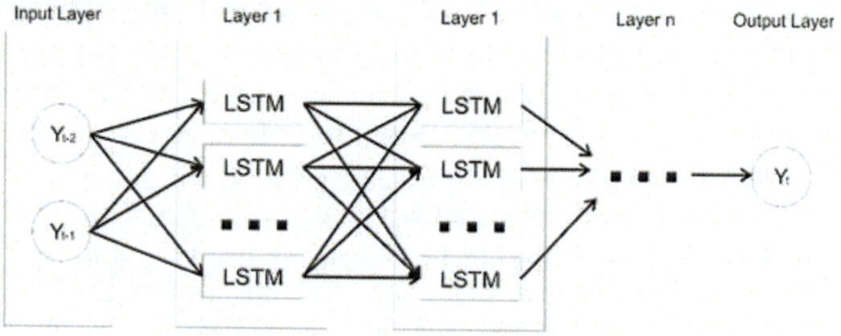

Fig. 3 Architecture of LSTM-RNN model

Table 1 Prediction data with deep learning method

Number	Weather monitoring post	MSE training	MSE testing
1	Ringinharjo	0.00087735	0.0096
2	Nyemengan	0.00017783	0.0693
3	Gandok	0.00089664	0.04
4	Kotagede	0.000066032	0.045
5	Pundong	0.0015	0.0466
6	Barongan	0.000087729	0.0328
7	Ngetak	0.00082795	0.0422
8	Gedongan	0.00034698	0.117
9	Piyungan	0.00018407	0.0216
10	Sedayu	0.00022793	0.1028
11	Ngestiharjo	0.00016664	0.0402
12	Dlingo	0.00085828	0.0438
Average		0.000518119	0.050908333

rainfall data in the same month as 1 year before). In this process, trial and error experiments also conducted for determining the number of LSTM neurons for every layer, or how many LSTM layers used. An indicator was used for determining whether it is good or not by using MSE was done using Eq. (6). After repeating the process, the smallest MSE model was obtained as shown in Table 1.

4.2 Mapping Process

Six variables determined the mapping of the flood-prone areas, that are area, distance to the river, river depth, river discharge, area by height, and average annual rainfall. From these six variables, we get profit and cost variables. Profit variable is a variable when the value is bigger, it has positive results, while cost variable is a variable when the value is bigger, it has negative results. In this study, there are five profit variables and one cost variable, the average rainfall variable per year, so it is necessary to transform the data.

In the data transformation process, fuzzification of data is first performed or makes the data in the range 0–1 with (7).

$$\mu = \frac{x - x_{\min}}{x_{\max} - x_{\min}} \tag{7}$$

After obtaining new data ranging from 0 to 1, a transformation is performed with $\mu' = 1 - \mu$ for the cost variable. As an example for 2012, data prior to the data fuzzification and transformation as in Table 2.

Table 2 Mapping data of 2012

Number	District	Area by height (m above sea level)	Closest settlement distance to the nearest river (m)	River depth (m)	River discharge (m^3)	Area (km^2)	Rainfall (mm)
1	Srandakan	12	5.5	5	58.5	1834	110.17
2	Sanden	55	4.3	10	100	2327	110.17
3	Kretek	56	6	5	11.11	2449	110.22
4	Pundong	20	5.5	8	12.35	2177	148.92
5	Bambanglipuro	18	4.7	2	9.87	2282	199.65
6	Pandak	34	8.5	3.65	33.585	2429	214.56
7	Bantul	42	7	2	9.87	2199	226.84
8	Jetis	37	7	4.75	11.275	2549	169.03
9	Imogiri	58	6	7.5	10.83	3509	135.2
10	Dlingo	200	6	7	9.31	815	164.17
11	Pleret	52	5.8	8	12.35	1783	158.45
12	Piyungan	86	5	8	12.35	1965	164.17
13	Banguntapan	78	6	10	100	2154	141.81
14	Sewon	69	7	2.9	9.58	2676	190.35
15	Kasihan	81	9	2.15	9.27	2608	172.44
16	Pajangan	38	10	3.65	33.585	2867	193.45
17	Sedayu	58	13	5	58.5	3262	161.79

These data then fuzzified according to the Eq. (7) and the cost variable was transformed by $\mu' = 1 - \mu$. The results are shown in Table 3.

The data then used as FCM input for mapping flood-prone areas in Bantul Regency and to get the cluster centers and partition matrices. The results of the cluster center with three clusters from the data in Table 3 are as follows:

$$\begin{bmatrix} 0.2223 & 0.4401 & 0.0158 & 0.1362 & 0.6458 & 0.3257 \\ 0.2922 & 0.1131 & 0.9479 & 0.9638 & 0.5259 & 0.8554 \\ 0.2331 & 0.1884 & 0.0462 & 0.0738 & 0.5044 & 0.6450 \end{bmatrix}$$

From the cluster center, the average value of each cluster is sought and obtained three average values in each cluster: 0.2977, 0.6164, and 0.2818. This value interprets that the smallest average cluster value (third cluster) is the group of the most disaster-prone areas, then the first cluster is the group of areas with a moderate level of vulnerability, and for the second cluster with the largest average cluster value is the safest area from floods in Bantul Regency. Map visualization of disaster-prone areas on Bantul Regency is presented (see Fig. 4).

Table 3 Fuzzified mapping data of 2012

Number	District	Area by height (m above sea level)	Closest settlement distance to the nearest river (m)	River depth (m)	River discharge (m^3)	Area (km^2)	Rainfall (mm)
1	Srandakan	0	0.1379	0.375	0.5426	0.3782	1
2	Sanden	0.2287	0	1	1	0.5612	1
3	Kretek	0.234	0.1954	0.375	0.0203	0.6065	0.9995
4	Pundong	0.0426	0.1379	0.75	0.0339	0.5056	0.6679
5	Bambanglipuro	0.0319	0.046	0	0.0066	0.5445	0.233
6	Pandak	0.117	0.4828	0.2063	0.268	0.5991	0.1053
7	Bantul	0.1596	0.3103	0	0.0066	0.5137	0
8	Jetis	0.133	0.3103	0.3438	0.0221	0.6437	0.4955
9	Imogiri	0.2447	0.1954	0.6875	0.0172	1	0.7854
10	Dlingo	1	0.1954	0.625	0.0004	0	0.5372
11	Pleret	0.2128	0.1724	0.75	0.0339	0.3593	0.5861
12	Piyungan	0.3936	0.0805	0.75	0.0339	0.4269	0.5372
13	Banguntapan	0.3511	0.1954	1	1	0.497	0.7288
14	Sewon	0.3032	0.3103	0.1125	0.0034	0.6908	0.3128
15	Kasihan	0.367	0.5402	0.0188	0	0.6656	0.4663
16	Pajangan	0.1383	0.6552	0.2063	0.268	0.7617	0.2862
17	Sedayu	0.2447	1	0.375	0.5426	0.9083	0.5575

Fig. 4 Flood-prone areas in Bantul Regency on 2012

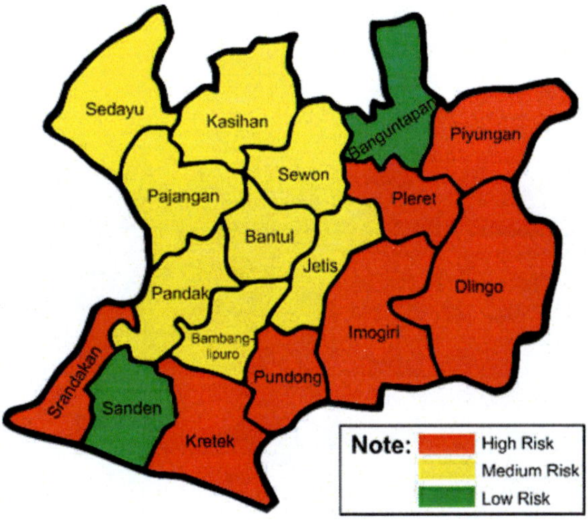

The same method was also conducted in the data of 2013, 2014, 2015, 2016, and 2017. Map visualization of flood-prone areas in Bantul Regency is presented (see Fig. 5).

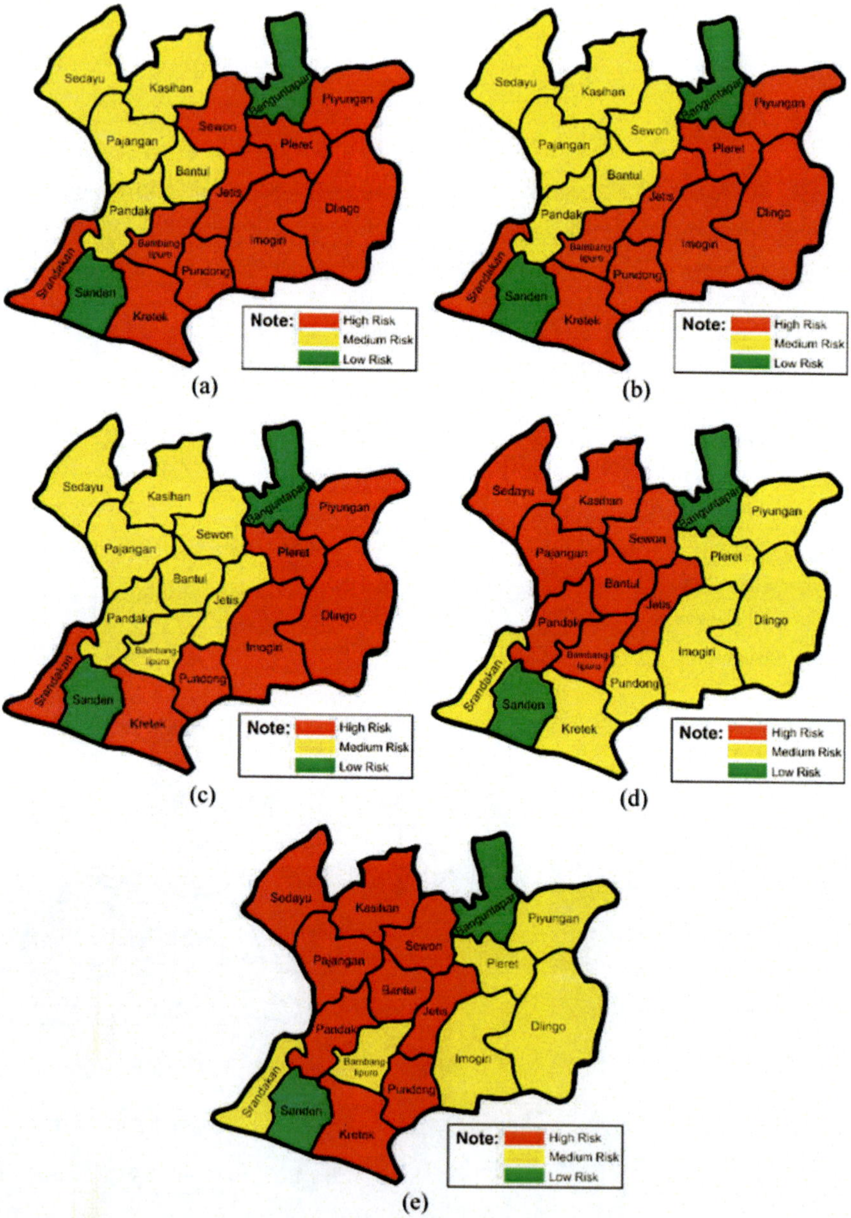

Fig. 5 Flood-prone areas in Bantul district **a** 2013 **b** 2014 **c** 2015 **d** 2016 **e** 2017

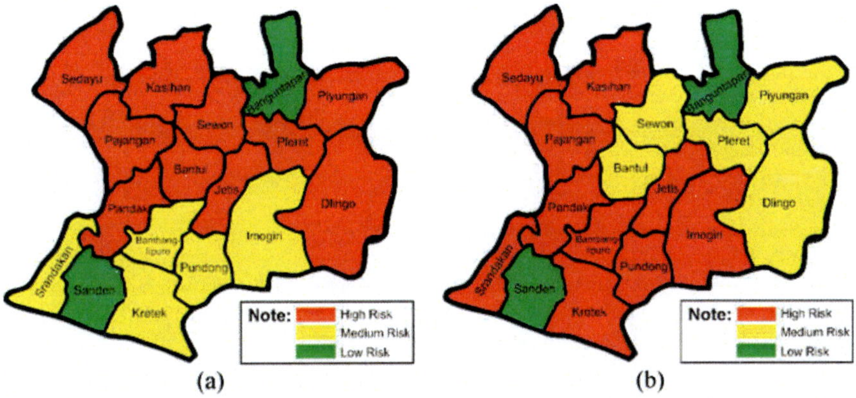

Fig. 6 Prediction of flood-prone areas in Bantul Regency **a** 2018 **b** 2019

From these figures, it was found that there are differences in flood-prone areas in Bantul Regency, one of the changes is due to differences in average annual rainfall in each district in Bantul Regency. For 2018, flood-prone area mapping prediction is shown (see Fig. 6).

5 Conclusion and Recommendation

From the result of this study, it can be concluded that the deep learning method is great for predicting average monthly rainfall which will be the basis for predicting annual mapping of flood-prone areas in the Bantul Regency. The results of the mapping conducted with the FCM method are three disaster-prone levels, they are low, medium, and high vulnerability of flooding. From these prediction results, it is expected that the community will be able to prepare themselves for the possibility of the coming flood disasters.

It is undeniable that this study still has many shortcomings, so further research is expected by adding variables or using other methods for improving the accuracy of the prediction of flood-prone areas in Bantul Regency.

References

Alom MZ, Taha TM, Yakopcic C, Westberg S, Sidike P, Nasrin MS, Hasan M, Essen BCV, Awwal AAS, Asari VK (2019) A state-of-the-art-survey on deep learning theory and architectures. MDPI Electron 8(3):292

Aris MM, Lorenz K (2008) Coastal flood management in Semarang, Indonesia. J Environ Geol 55:1507–1528

Aris MM, Andung BS, Ward P (2015) Community responses and adaptation strategies toward flood hazard in Jakarta, Indonesia. J Nat Hazards 75:1127–1144
BPS—Statistic of Bantul Regency (2013) Bantul in figures 2013. BPS—Statistics of Bantul Regency, Bantul
BPS—Statistic of Bantul Regency (2014) Bantul in figures 2014. BPS—Statistics of Bantul Regency, Bantul
BPS—Statistic of Bantul Regency (2015) Bantul in figures 2015. BPS—Statistics of Bantul Regency, Bantul
BPS—Statistic of Bantul Regency (2016) Bantul in figures 2016. BPS—Statistics of Bantul Regency, Bantul
BPS—Statistic of Bantul Regency (2017) Bantul in figures 2017. BPS—Statistics of Bantul Regency, Bantul
BPS—Statistic of Bantul Regency (2018) Bantul in figures 2018. BPS—Statistics of Bantul Regency, Bantul
Campolo M, Andreussi P, Soldati A (1999) River flood forecasting with a neural network model. J Water Resour Res 35(4):1191–1197
Chan NW (2006) Flood disaster management in Malaysia: an evaluation of the effectiveness of government resettlement schemes. Disaster Prev Manag Int J 44(4):22–29. http://dx.doi.org/10.1108/09653569510093405
Hong X, Lin R, Yang C, Zeng N, Cai C, Gou J, Yang J (2019) Predicting Alzheimer's disease using LSTM. IEEE Access 7:80893–80901
Jiang W, Deng L, Chen L, Wu J, Li J (2009) Risk assessment and validation of flood disaster based on fuzzy mathematics. Prog Nat Sci [Internet] 19(10):1419–25. http://dx.doi.org/10.1016/j.pnsc.2008.12.010
Kusumadewi S, Purnomo H (2004) Fuzzy logic application for decision support, 1st edn. Graha Ilmu, Yogyakarta
Parker DJ (2000) Introduction to floods and flood management. Floods, vol 1. Routledge, London
Plate EJ (2002) Flood risk and flood management. J Hydrol 267:2–11
Price RK, Vojinovic Z, Urban ZV (2008) Urban flood disaster management. Urban Water J 5(3):259–276. https://doi.org/10.1080/15730620802099721
Singh T, Mahajan MM (2014) Performance comparison of fuzzy C means with respect to other clustering algorithm. Int J Adv Res Comput Sci Softw Eng 4(5):89–93
Suganya R, Shanthi R (2012) Fuzzy C-means algorithm—a review. Int J Sci Res Publ 2(11):1–3
Sugiyarto AW, Abadi AM (2019) Prediction of Indonesian palm oil production using long short-term memory recurrent neural network (LSTM-RNN). In: Proceedings 1st international conference on artificial intelligence and data sciences (AiDAS). IEEE, Malaysia, pp 53–57
Sugyarto AW, Hendiyani N, Wutswa DU, Rasjava AR (2019) Optimization of genetic algorithms on backpropagation neural network to predict national rice production levels. In: 2nd international conference on applied information technology and innovation. IEEE, Bali, pp 104–108
Tingsachali T (2012) Urban flood disaster management. Procedia Eng [Internet] 32:25–37. http://dx.doi.org/10.1016/j.proeng.2012.01.1233
Van Voorst R (2015) Formal and informal flood governance in Jakarta, Indonesia. Habitat Int [Internet] 6–11. http://dx.doi.org/10.1016/j.habitatint.2015.08.023

Urban Vegetable Gardening Brings Greening to Slum Environment and Helps Mitigate Climate Change Effects

Siddharth Agarwal, Shabnam Verma, Neeraj Verma, Kailash Vishvakarma, and Kanupriya Kothiwal

Abstract Urban Health Resource Centre's (UHRC) social facilitators encourage families to grow vegetables in small spaces in slum houses. Seeds were provided. Facilitators motivate families to tend plants as they grow. During 2018–2019, 495 families grew beans, bottle gourds, round gourds, sponge gourds, tomatoes, brinjals, small green peppers, and pumpkins in small spaces, broken buckets. A total of 495 families shared with 1485 neighbor families, thereby benefitting 9,900 population. Costs saved over the 3 years total to INR 8,251,932. Slum families can be motivated and overcome space constraints of small slum houses to nurture household vegetable gardens. Despite each home garden being small, these have the potential to mitigate carbon emissions a key climate change challenge affecting the world. Vegetable plants release oxygen, sequester carbon in the soil, and reduce atmospheric carbon. Vegetable gardens contribute to the cooling effect in urban spaces. Growing vegetables despite small spaces and sharing with neighbors promote a sense of psychological well-being, accomplishment, and enhances social cooperation all crucial to the well-being of vulnerable city populations. Home-based vegetable gardens save costs for the family. Methods used in this program research to grow vegetables in small slum houses and other lessons learned of perseveringly motivating slum families have the potential of their replication or adaptation in cities of India and other LMICs.

Keywords Household · Vegetable-gardening · Climate change mitigation · Slums · India

S. Agarwal (✉)
Urban Health Resource Centre, 136, Humayunpur, Safdarjung Enclave, New Delhi 110029, India
e-mail: a.siddharth.in@gmail.com

S. Verma · N. Verma · K. Vishvakarma · K. Kothiwal
Urban Health Resource Centre, J-133/A, First Floor, Flat No. 4, Kanchan Apartments, MIG Colony, Indore 452001, India

L. T. T. Huong and G. M. Pomeroy (eds.), *AUC 2019*, Advances in 21st Century Human Settlements, https://doi.org/10.1007/978-981-15-5608-1_7

1 Introduction and Background

1.1 Rapid Urbanization in India

India's urban population growth of 31.8% during 2001–2011 stands in stark contrast to the national population growth of 17.6% (Census of India 2011) during the same decade. India's urban population nearly doubled between 1991 (217 million) and 2011, increasing rapidly from 217 million in 1991 to 285 million in 2001 (Agarwal et al. 2007) to 377 million (31%) in 2011 (Census of India 2011). Rapid urbanization and widespread urban sprawl have depleted green cover and increased urban vulnerability to climate change and accompanying warming (Iman and Banerjee 2016).

The rapid increase in urban population, growth of cities in terms of infrastructure, housing, market spaces, and increase in economic activities and urban sprawl (or extension of city limits over time) are concomitant phenomena taking place in most million-plus cities in South Asia, India, and more broadly in LMICs. The extent of urbanization and activities that take place in urbanizing spaces contribute to a change in land use patterns. More and more land earlier used in cities for agriculture, growing vegetables, trees, green stretches is getting reduced and being converted into built-up areas (Sudhira et al. 2004).

1.2 Increase in Built-up Area and Change in Land Use and Character in Cities

The built-up areas of most million-plus cities are increasing fast. The expansion of the urban area of Gwalior has been quantified by deriving data for four decades (1972–2013) from the Landsat images. The results show that the urban built-up area increased by 08.48 km^2 during the first 18 years (1972–1990) which has increased to 16.28 km^2 during the next 16 years (1990–2006). The built-up area of Gwalior city has further gone up to 23.19 km^2 in the next 7 years (2006–2013). Overall, during the last 40 years, the growth of the urban built-up is nearly three times the built-up areas in 1972. The average decadal growth rate of the population is 27.28%, while that of built-up land is 36.29% (Mohapatra et al. 2014).

With urbanization, there occur significant changes in landscape patterns and land cover of the area concerned. This phenomenon leads to several undesirable (or negative) effects in terms of the physical environment such as loss of agricultural land, of orchards and vegetable gardens, surface and groundwater depletion, changes in geomorphic features, and flooding (Mohapatra et al 2014).

1.3 Shrinking of Green Spaces in Slums and Cities

In addition to tenure insecurity, the constant risk of eviction, a sub-optimal formal supply of basic infrastructure and services, slum dwellers lack green areas (UN Habitat 2012).

The area urban greens should be preserved for the maintenance of the environmental quality of the urban area. The consequences of gradual disappearance or shrinkages of effective urban greens (EUG) are manifold. This phenomenon is causing the deterioration in the quality of life of urban dwellers, particularly slum populations, which indirectly is the bane of this malice-ridden urbanization (Sarkar 2015). In India, the level of urbanization increased from 25.7% in 1991 to 27.8% in 2001 (Agarwal et al. 2007) and further to 31.2% in 2011 (Census of India 2011). This rapid urbanization has also led to the conversion of green spaces to built urban spaces and to more dense housing particularly in slums and informal settlements. It is crucial to green slums that constitute 33% of Asia's population and are growing rapidly.

1.4 Why This Research Program Was Undertaken: Household Vegetable Gardening Can Help Slum Families and the Environment

Literature tells us that household-level vegetable gardening by slum families helps through plants releasing oxygen and sequestering carbon in the soil, enhancing soil fertility and reducing atmospheric carbon. It has been estimated that vegetable gardening covering an area of 0.4 acres (or 4000 m^2 or 17,422 ft^2) has the potential to sequester 19 tons of carbon over 10 years (Okvat and Zautra 2011). Consuming of homegrown vegetables reduces carbon footprint of purchased vegetables transported to urban markets and also saves cost. Vegetable gardens contribute to the cooling effect in urban spaces (Dubbeling et al. 2009). Growing vegetables despite small spaces and sharing with neighbors promote a sense of psychological well-being, accomplishment, and enhances social cooperation, all crucial to vulnerable city populations (Wakefield et al. 2007).

2 This Research Program

This research program shows (a) how slum families can overcome the adversity of living in small houses and utilize the small spaces creatively with zest and grow vegetables (b) for their own consumption. This paper also demonstrates the (c) extent of green cover such a slum-level vegetable gardening initiative contributes to the greening of the slum neighborhoods and (d) has the potential to mitigate climate change effects by sequestrating carbon.

Seeds are provided by the Urban Health Resource Centre during the months of June and July every year to families who are willing to grow vegetables. Information on vegetables grown, approximate yield in weight (kilograms) is asked from vegetable-growing families during group discussions and triangulated to arrive at an estimated weight of different vegetables that they grow. This is followed up from the month of July to the month of April of the next year when most vegetable plants begin to dry up. This project is nested in UHRC's ongoing multidimensional, self-reliance program for 250,000-slum population.

3 Study Context

This research was taken up in slums in the Northwest part of Indore, which is a rapidly growing city in Central India. Indore is the 14th largest city of India by the population. It is the most vibrant hub of commercial and industrial activities and also the largest city of Madhya Pradesh. Indore's metropolitan area has a population of 2.17 million (Office of the Registrar General, Census of India, 2011), a decadal increase of 44% from 1,506,062 in 2001 (Office of the Registrar General and Census Commissioner, India 2001) While Bhopal is the administrative capital of Madhya Pradesh, Indore, and its surrounding areas constitute the economic epicenter of the state. The commercial growth of Indore has attracted migrant workers from other districts of Madhya Pradesh as well as states like Bihar and Uttar Pradesh (Agarwal 2016) adding to the population density. The population density of Indore city was 11,569 persons per square kilometer as per the District Census Handbook Indore Census 2011 (Directorate of Census Operations Madhya Pradesh, Census of India 2011).

Indore has a semi-arid climate. In the summer of 2019, the temperature in Indore touched a high of 43 °C (accuweather.com 2019). Projections suggest that temperatures by 2030 may reach close to 50 °C on some summer days (TARU 2012). With fast-paced infrastructure development the green cover is getting depleted, temperatures are rising and the water table is receding.

4 Methods and Program Approach

We assessed the vegetable gardening in slum houses through transect walks and household observations in all 18 slums where vegetables were growing during 2018–19. During earlier years, fewer slums were visited. During transect walks, we sought the permission of the adult family member and observed the vegetable plants at different stages of maturation. Group discussions were conducted during 2018–19 with representatives of 10 families of each of the 18 slums where vegetables were grown. Respondents from the 18 slums included 120 women and 40 youth. After seeking their permission, their vegetable plants were observed to understand the spaces and pots they creatively use to grow vegetables.

Approximate green cover produced by household vegetable gardens was assessed through observation of green cover in 8 households in each of the 18 slums and a consensus average of green cover per household growing vegetables as estimated.

Slum families in Indore were motivated by UHRC's social facilitators to grow vegetables in small spaces in front and behind houses or in paint cans, partially broken buckets, or similar containers that the families could arrange. Seeds were provided. Facilitators motivated families to tend plants as they grow.

5 Results

5.1 Perseverant Efforts Built the Confidence of Slums Families to Grow Vegetables

With gentle motivation and providing seeds over 4 years from 2012 to 2016, few slum families started growing vegetables demonstrating the socio-behavioral trait of being early adopters. Gradually, the number of families who developed the confidence to grow vegetables in their small houses year after year increased from 350 during 2016–17 to 495 in 2018–19.

5.2 How They Overcame the Challenge of Small Spaces

They overcome the challenge of small spaces and find ways to grow vegetables: (a) use small not constructed patch within the house, or just behind the house, (b) use broken buckets, paint buckets, other suitable containers that can be turned into pots of growing vegetables, and c) they utilize the rains that water the plants as they sprout during the months of June, July, August, September, and some years a few showers in early October.

5.3 What Was the Experience of Slum Families Growing Vegetables

Respondents reported that nurturing the growth of vegetable plants and creepers and seeing vegetables growing on the plans gave (a) sense of positivity, (b) fostered greater hope, (c) motivation to grow every year, (d) a feeling of joy, (e) enhanced self-worth, and (f) more positive feeling about their abode where they spend most of the hours each day.

5.4 Vegetables Grown and Quantity in Kilogram

With stimulation, families developed incremental confidence to grow beans, bottle gourd, round gourd, ribbed gourd, sponge gourd, bitter gourd, tomatoes, and brinjal (eggplant) on small spaces, broken buckets and other containers they could manage. Tiny open spaces were utilized to grow creepers. Creepers were made to climb on the terrace of the houses so the weight of the vegetables can be held by the terrace as they grow. Table 1 shows the estimated yield of different vegetables grown during the 3 years for which information is compiled.

5.5 Number of Families and Population Benefitted

Based on the listing of families maintained by UHRC team members in Indore, from September 2016 to April 2017, 350 slum families grew vegetables. This number increased to 428 in September 2017 to April 2018 and further to 495 in September 2018 to April 2019. This demonstrates the gradual increase in confidence about the ability to grow vegetables despite small spaces in small slum houses. Respondents reported that they used broken buckets, large polythene bags, paint buckets, pots, and small unbuilt spaces in or behind their house to grow vegetables. They sowed seeds provided by Urban Health Resource Centre during June and July and watered the saplings and plants during days when it did not rain and after the rainy season which usually lasts till September end.

5.6 Slum Families Saved Costs on Vegetables Over 6 Months Each Year

Based on the reported market cost of different vegetables during the high-yield and low-yield seasons, we estimated the average cost saved per family per month. This value was multiplied by 6 months (during which vegetables were available for consumption, then by the number of families growing vegetables. The average cost saved over 6 months was multiplied by the number of families with whom vegetables were shared and divided by two based on the qualitative assessment that for these neighbor families about half the requirement of vegetables was fulfilled. These values were added to arrive at the aggregate cost saved each year. Table 2 provides details of the cost saved during each of the 3 years that have been studied in this program research. In aggregate terms, the vegetable gardening initiative enabled slum families to save INR 1,375,322 (equivalent to USD 19,647) over the 3 years.

Consumption of homegrown vegetables by vegetable growing families and neighbors also contributed to reduced dependence on buying vegetables from the market and reducing carbon footprint of the slum families. Transportation of vegetables

Table 1 Weight of different vegetables grown in 16 slums from September 2016 to April 2019

Year	Tomatoes (kg)	Beans (kg)	Bottle gourd (kg)	Round gourd (kg)	Brinjal (kg)	Ribbed gourd (kg)	Sponge gourd (kg)	Bitter gourd (kg)	Green pepper (kg)	Pumpkin (kg)	Total (kg)
September 2016–April 2017	549	623	671	705	741	657	621	558	408	664	6197
September 2017–April 2018	770	789	840	812	739	670	795	712	553	802	7482
September 2018–April 2019	864	828	1059	1057	774	775	780	685	566	1106	8494
Total (kg)	2183	2240	2570	2574	2254	2102	2196	1955	1527	2572	22,173

Table 2 Families growing vegetables, total yield of vegetables, neighbor families with whom vegetables were shared, population benefitted, cost saved and green cover in 16 slums from September 2016 to April 2019

Year	Families who grew vegetables	Total yield of vegetables (kg)	Neighbour families vegetables shared with	Total population benefitted	Total cost saved INR	Equivalent cost saved USD	Total green cover (ft^2)
September 2016–April 2017	350	6197	1050	7000	INR 2,110,500	USD 30,150	31,500
September 2017–April 2018	428	7482	1284	8560	INR 2,831,220	USD 40,446	38,520
September 2018–April 2019	495	8494	1485	9900	INR 3,310,212	USD 47,289	44,550
Aggregate values over 3 years	–	18,022	3819	25,460	INR 8,251,932	USD 117,885	–

from the location of production to city's markets entails energy consumption and CO_2 emissions from the transport vehicles, which bring vegetables to the central distribution market and peripheral locality-level selling points in the city.

5.7 *Green Cover Contributed by Home Gardens*

We assessed the approximate green cover produced by household vegetable gardens by observing the green cover produced in 8 households in each of the 18 slums and arrived at a consensus average green cover per household through group interaction and triangulation with 5–6 representatives of 5–6 families in each slum. The average green cover per family was assessed as 90 square feet. During the year 2018–19 with 495 families, the estimated total green cover of home gardens came to 44,550 feet^2, while it was 38,520 feet^2 during 2017–18. We used an estimate of carbon sequestration that has been provided for a 0.4 acres, organic, communal garden that had increased its soil organic matter from 1 to 7.7% in 10 years (Meadows 2000), and a follow-on conservative estimate based on the observation that soil organic matter was confined to the top 8 inches, a garden with a green cover of 0.4 acres had sequestered 19 tons of carbon from the atmosphere over 10 years (in Okvat and Zautra 2011). The green cover in the present study in Indore, India for 2018–19 translates to 1.02 acres which would sequester about 48.45 tons of carbon from the atmosphere over 10 years.

Based on the per capita carbon emission of an Indian of 1.18 tons each year into the atmosphere (Boutabba 2014), the experience with household vegetable gardens in Indore shows that these small gardens have the potential to offset the annual carbon emissions for nearly 50 Indians over 10 years. The above calculation must be interpreted with caution given the simplification of the arithmetic and the complex nature of factors associated with carbon sequestration from the soil by plants. It definitely provides a positive example of measures that can contribute to climate change mitigation and also have several other benefits for resource-poor slum populations.

6 Discussion

6.1 Methods to Overcome Space Constraint in Slum Houses Have Potential for Adaptation

The methods used by slum families to overcome the adversity of living in small houses and not having space described in this study are adaptable for other cities of India and LMICs. Among space optimizing methods, creepers are made to climb to the roofs and fences, thus occupying less space (Niñez 1985). Innovative methods of gardening in small spaces as learned from the present study have also been reported from Bangalore. These including plants in plastic bags, paint cans, old partly broken buckets, indicating the importance given to planting in environments with limited finances (Gopal and Nagendra 2014).

6.2 Early Adopters Played the Role of Motivators and Spread Confidence

This study demonstrates that early adopters serve as motivators for subsequent adopters (Newman 2008) who begin growing vegetables as they observe neighbors and friends growing vegetables and sharing with them. The sight of tomatoes or ribbed gourds, bottle gourds, beans, pumpkins beginning to grow on the plants and creepers lends greater energy to the families growing them. Children are naturally active and effective motivators. Their energy lends more zest and tenacity to the family to grow vegetables.

6.3 *Small Household Vegetable Gardens Provide Supplemental Food for Consumption*

The present research reinforces the concept of the household vegetable garden as a small-scale production system supplying plant and animal consumption and utilitarian items either not obtainable, affordable, save income from wage-earning. The home gardens serve an important role in supporting nutrition and in promoting health through homegrown vegetables (Gopal and Nagendra 2014). The mini household gardens encouraged in this program research are located within the dwelling and thus can receive the required special care. They occupy small land or space (Browning 1985) and labor marginal to major household economic activities. Production is supplemental rather than a main source of family consumption or income; are a production system that the poor can easily enter at some level (Marsh 1998). Household gardens are marked by low capital input and simple technology (adapted from Niñez 1985; Michelle and Hanstad 2004).

This study demonstrates the benefits of home gardens documented in the literature. For poor and marginalized families unable to afford expensive vegetables to fulfill their nutritional needs, home gardens are a source of nutritive vegetables (Ali et al. 2008).

6.4 *Psycho-emotional and Tangible Benefits Experienced*

The positivity described by respondents in this study is similar to the different feelings of psycho-emotional well-being reported in earlier studies. Studies have described that home-based vegetable growers experienced a feeling of relaxation, sense of accomplishment to see vegetable plants and vegetables grow, being able to grow vegetables in their house, and share with neighbors, people rated vegetable gardening highly as a valuable way to utilize time, experienced relaxation, and felt a sense of accomplishment (Kaplan 1973; Stuart 2005; Kingsley et al. 2009). The most important and tangible benefits are the supply of fresh vegetables, savings on food expenses (Oluoch et al. 2009; Kaplan 1973). Nearly 10,000 population benefited from 495 families growing vegetables showing the multiplier effect of the approach. These results hold promise to adapt the approaches of encouraging slum families in cities of India and other LMICs to grow vegetables.

6.5 *Climate Change Benefits of Household Vegetable Gardens*

Gardening assists in the effort to mitigate the negative effects of climate change via both direct pathways (greenhouse gas, GHG mitigation) and indirect pathways (urban lifestyle change and education). One direct pathway is carbon sequestration. This entails the removal of existing carbon from the atmosphere and has been called a "reverse greenhouse effect" (Meadows 2000), an important adjunct to efforts to decrease new carbon emissions. Plants take in CO_2 and separate the carbon and oxygen. The oxygen is released, while carbon is captured (sequestered) in the soil, enhancing soil fertility and reducing atmospheric carbon. The Intergovernmental Panel on Climate Change (1990) identified the sequestration of carbon in soils as one of the main options for GHG mitigation (cited in Okvat and Zautra 2011). The experience from the present study reinforces the significance of this methodology of encouraging household vegetable gardens in slums in LMIC cities. Another GHG mitigation pathway is the reduction of carbon emissions associated with the transportation of vegetables to the city and subsequently within the city. Vegetation reduces ambient air temperature leading to a cooling effect. Indirect pathways include people particularly children and youth learning about climate change processes, understanding the value of plants and trees in protecting the environment, improving groundwater conservation, sustainability.

7 Conclusion

Slum families can be motivated and helped to develop the confidence of overcoming space constraints of small slum houses to develop household vegetable gardens. Despite each home garden being very small, these have the potential to mitigate carbon emissions which are a key climate change challenge affecting the world. A tangible value contributed by the home-based vegetable gardens is that they save valuable family income through the provision of vegetables for the family's consumption. Growing vegetables despite small spaces and sharing with neighbors promote a sense of psychological well-being, accomplishment, and enhances social cooperation all crucial to the well-being of vulnerable city populations. The methods used in this program research to grow vegetables creatively in small spaces of slum houses and other lessons learned of perseveringly motivating slum families have the potential of their replication or adaptation in cities of India and other LMICs.

Acknowledgements The authors are grateful to slum families who overcame adversities of their small and sub-optimal living spaces and grew vegetables and enabled the UHRC team to learn from their experience of growing vegetables in slums. We thankfully acknowledge the slum residents' openly sharing their experiences with us and providing insights into how they grew vegetables in their small slum houses and how the children helped in tending to the plants. We are thankful for the support provided by Neha Mandloi and Ankush Rathore of UHRC's Indore team.

References

Accuweather temperature data for Indore for May and June, 2019. https://www.accuweather.com/en/in/indore/204411/june-weather/204411?year=2019. Accessed 30 Sept 2019

Agarwal S (2016) Urban migration and social exclusion: study from Indore slums and informal settlements. https://pubs.iied.org/10777IIED/

Agarwal S, Satyavada A, Kaushik S, Kumar R (2007) Urbanization, urban poverty and health of the urban poor: status, challenges and the way forward. Demogr India 36(1)

Ali MY, Ahmed MM, Islam MB (2008) Homestead vegetable gardening: meeting the need of year round vegetable requirement of farm family. In: National workshop on multiple cropping held at Bangladesh Agricultural Research Council, Farmgate, Dhaka-1215, Bangladesh, pp 23–24

Brownrigg L (1985) Home gardening in international development: what the literature shows. The League for International Food Education, Washington DC, USA

Boutabba MA (2014) The impact of financial development, income, energy and trade on carbon emissions: evidence from the Indian economy. Econ Model 40:33–41. Last accessed on July 20, 2020 from https://www.univ-evry.fr/fileadmin/mediatheque/ueveinstitutionnel/03_Recherche/laboratoires/Epee/wp/13-05.pdf

Directorate of Census Operations Madhya Pradesh, Census of India 2011. District Census handbook Indore. http://censusindia.gov.in/2011census/dchb/DCHB_A/23/2322_PART_A_DCHB_INDORE.pdf. Accessed 30 Sept 2019

Dubbeling M, Caton Campbell M, Hoekstra F, Van Veenhuizen R (2009) Editorial: building resilient cities. Urban Agric Mag 22:3–11

Gopal D, Nagendra H (2014) Vegetation in Bangalore's slums: boosting livelihoods, well-being and social capital. Sustainability 6(5):2459–2473

Government of India (2015) Census of India 2011. http://censusindia.gov.in/2011-prov-results/paper2/data_files/Gujrat/6-pop10-28.pdf. Accessed 15 Sept 2019

Government of India. Ministry of Home Affairs. Office of the Registrar General & Census Commissioner (2011) Census of India 2011 primary census abstract: figures at a glance India. Office of the Registrar General & Census Commissioner, New Delhi. http://www.censusindia.gov.in/2011census/PCA/PCA_Highlights/pca_highlights_file/India/5Figures_at_glance.pdf. Accessed 29 Sept 2019

Imam AU, Banerjee UK (2016) Urbanisation and greening of Indian cities: problems, practices, and policies. Ambio 45(4):442–457

Kaplan R (1973) Some psychological benefits of gardening. Environ Behav 5(2):145–162

Kingsley JY, Townsend M, Henderson-Wilson C (2009) Cultivating health and wellbeing: members' perceptions of the health benefits of a Port Melbourne community garden. Leisure Stud 28(2):207–219

Marsh R (1998) Building on traditional gardening to improve household food security. Food Nutr Agr 22:4–14

Meadows D (2000) How does your garden grow?: two brothers talk carbon sequestration. Grist [Seattle], p 1. https://grist.org/article/how/. Accessed 29 Sept 2019

Mitchell R, Hanstad T (2004) Small homegarden plots and sustainable livelihoods for the poor. LSP working paper 11, Rome, Italy

Mohapatra SN, Pani P, Sharma M (2014) Rapid urban expansion and its implications on geomorphology: a remote sensing and GIS based study. Geogr J

Newman L (2008) Extreme local food: two case studies in assisted urban small plot intensive agriculture. Environ J Interdiscip Stud 36(1)

Niñez V (1985) Working at half-potential: constructive analysis of home garden programmes in the Lima slums with suggestions for an alternative approach. Food Nutr Bull 7(3):1–9

Okvat HA, Zautra AJ (2011) Community gardening: a parsimonious path to individual, community, and environmental resilience. Am J Community Psychol 47(3-4):374–387

Oluoch MO, Pichop GN, Silué D, Abukutsa-Onyango MO, Diouf M, Shackleton CM (2009) Production and harvesting systems for African indigenous vegetables. In: Pasquini CM, Drescher MW,

Shackleton AW (eds) African indigenous vegetables in urban agriculture. Earthscan, London, pp 145–170

Sarkar GD (2015) An assessment of declining urban greens under Patna municipal corporation based on normalized difference vegetation index. Univ J Environ Res Technol 5(5)

Stuart SM (2005) Lifting spirits: creating gardens in California domestic violence shelters. In: Barlett PF (ed) Urban place: reconnecting with the natural world. Cambridge, MA, MIT Press, pp 61–88

Sudhira HS, Ramachandra TV, Jagadish KS (2004) Urban sprawl: metrics, dynamics and modelling using GIS. Int J Appl Earth Obs Geoinf 5(1):29–39

UN Habitat (2012) Streets as tools for urban transformation in Slums. A street-led approach to citywide slum upgrading. UN Habitat, Nairobi

Wakefield S, Yeudall F, Taron C, Reynolds J, Skinner A (2007) Growing urban health: community gardening in South-East Toronto. Health Promot Int 22(2):92–101

Inclusive, Implementable Urban Governance, and Sustainable Urban Finance

Collaborative Approach for Resource Mobilization Transformation in Vietnam

Hieu Nguyen Ngoc and Dao Tran Quang

Abstract This article discusses the barriers in transforming the resource mobilization for urban infrastructure development in Vietnam. Depleting Official Development Aids (ODA) and shrinking public budgets since 2016 urged authorities to look for alternative financial resources. Troubles in the Build and Transfer (BT) model and difficulties of land use rights sales helped decision makers to supplement the traditional approach; however, there are essential barriers that should be lifted to support the better collaboration. The author used institutional and background data analyses to unveil how key barriers hampered motivation of local authorities to seek for co-financing mechanism. The findings advocated further renovation on urban autonomy besides improving capacity and appropriate tools to foster better collaboration among key stakeholders.

Keywords Financing infrastructure development · Urban budgeting and finance transformation · Collaborative development approach · Urban autonomy

1 Introduction

Vietnam is facing increasing demand for urban infrastructure development. Over past three decades, high urbanization rate transformed the urban sector significantly, from 20% in 1988 to over 36.6% in 2018 (Chr 2019; MOC 2019; World Bank 2011). The demand for infrastructure development is surmounting in both new and old areas. For example, to accommodate one million new urban citizens may cost $3.1 bn. per year from 2016 to 2020 period (VR and MNTT 2016); meanwhile, redeveloping an aging or underinvested infrastructure hosting 34 million urban residents may cost even higher amount, especially where two largest mega clusters need huge investment to

H. Nguyen Ngoc (✉)
Vietnamese German University, Binh Duong, Vietnam
e-mail: Hieu.nn@vgu.edu.vn

D. Tran Quang
Ho Chi Minh City Transport University, Ho Chi Minh City, Vietnam
e-mail: Dao.tran@gmail.com

L. T. T. Huong and G. M. Pomeroy (eds.), *AUC 2019*, Advances in 21st Century Human Settlements, https://doi.org/10.1007/978-981-15-5608-1_8

restructure these regions with express ways, metro lines, and flood protection system. Further, the compelling demand from a younger and richer urban generation and a burgeoning automobile culture could virtually inflate actual urban infrastructure upgrading demand in the long term.

It is critical for Vietnam to seek for diversified financial resources for infrastructure spending. Estimates from the World Bank showed that during the period 2011–2015, the government's recurrent expenditures were growing fast (10.5% per annual) at both central and provincial level, exceeding the investment growth rate (6.1%). The country needs approximately US$25–30 billion per year for infrastructure development during the period 2011–2020, while public sources were able to contribute only US$16 billion per year (World Bank 2013; Campanaro and Duc Dang 2018). The shrinking access to concessional loans from Official Development Aids (ODA) was also compelling. Becoming a middle-income country from 2015 implied that less expensive loans for infrastructure will reduce rapidly.[1] It is noted that since 1993 Vietnam has disbursed about $64 bn out of about $93 bn committed ODA funding, mostly for developing physical infrastructure.

Over past few decades, different policies to mobilize private funding for infrastructure have been introduced and developed. The rule of "*Urban growth should fuel itself*" has been announced since early 1990s. A policy introduced in 2006 on new urban area development (Decree 02/2006) bounds developers with the cost of most of the physical infrastructure, in exchange for access to cheap farming/vacant land by using a compulsory purchase mechanism.[2] Foreign investment policy encouraged foreign investment on Built–Operate–Transfer (BOT) major infrastructure projects. From 1994 to 2009, 32 projects finance $6.7 bn out of $9.5 bn committed. Since 2010, the Public–Private Partnership mechanism (PPP) (stipulated in Decree 71/2010) extended this policy to domestic investors. Since then, this policy attracted over a hundred of projects invested in electric power plants, toll roads, bridges and tunnels, airport terminals, and major seaports with national partners (state-owned companies at the national level or national institutions). At the local level, 41 provinces have initiated urban infrastructure projects such as arterial roads, water supply plants, water drainage system, wastewater treatment, and garbage treatment plants. However, all of them chose the Built and Transfer (BT) model, which is land in exchange for building.

There were concerns about BT model at local level. Implementation of BT model triggered a number of corruption cases as developers targeted only undervalued land in exchange for infrastructure whereby project was waived off from bidding. Recent National Assembly's dialogues condemned this model as cheap land sales,

[1]International Development Association (IDA—World Bank) terminated favorable terms for Vietnam from July 2017 and Asian Development Bank (ADB) terminated concessional ordinary capital resources by January 2019. Note that 30% of total favorable Vietnam's ODA comes from the World Bank.

[2]Developers can request authority to compulsorily purchase (mostly out of urban area) land for development.

not mobilizing private resources, and create rich targets for corruption.[3] There were a number of projects under criminal investigations.[4] Besides BT, waves of brownfield development during the urban consolidation period (after the boom period 2000–2010) reduce both contribution to infrastructure development and revenues on land use fees/land conversion charge. It seems that existing model weakens the recycle of value creation by infrastructure development within existing framework.

In the context of fast growth in Vietnam, it is critical to enable a better value-sharing mechanism to boost investment contributions to urban infrastructure. If development management framework is better aligned with urban finance, and if local government could foster innovation in synergizing talents and resources of the whole society to respond to higher needs, there should be a lot of efforts to fuel the institutional change to cope with different kinds of barriers.

2 Literature Review

Over past decades, the value-sharing approach became common approach to cope with increasing needs for infrastructure investment demand. The basic concept of community's approach is to value sharing of investment cost. From the Middle Age, betterment levy was introduced as early attempts to common goods[5] to bind beneficiaries from returning deserving values created from investors/creators. Many instruments were developed during the twentieth century in the UK,[6] US,[7] and other countries with increasing innovation for capturing increasing value of property or benefits of property owners and residents. Nowadays, Community Infrastructure Levy, Development Impact Fee, Capital Gain Tax, Transferable Development Rights (TDR), or Land Readjustment (LR) became a family of instruments, where local government recapture or recycle the value increase from planning activity, utilities investment, or environment improvement (Joe 2009; Smolka 2013; Smolka and Furtado 2002; Ingram and Hong 2012; Hong 2007). Land Value Capture (LVC) or Value Capture Finance represents for a group of strategic and sustainable tools of local government to cope with emerging needs for (urban) infrastructure development and upgrading.

[3]https://vietnamnet.vn/vn/ban-tron-truc-tuyen/doi-dat-lay-ha-tang-loi-kep-mat-ngan-ty-413444.html.

[4]https://nhadautu.vn/gs-dang-hung-vo-thanh-tra-du-an-bt-nao-rung-du-an-do-d11694.html.

[5]Henry VI imposed a betterment levy in 1497 to capture the increase in private land values arising from flood defense works.

[6]UK adopted Community Infrastructure Levy (CIL), City Development Corporation as well as betterment levy for capturing the value increases from residential development to reinvest in local infrastructure through land ownership.

[7]US has a number of tools such as special assessments, developer exactions, betterment contributions, linkage fees, improvement districts, community benefit agreements, the transfer of development rights, and land assembly or land readjustment. Besides, the affordable housing tool could be understood as land value capture as developer should return a part of its benefit to the local community.

Mobilizing resources using LVC instruments can be divided into upstream and downstream financing that link to the country's context. The upstream financing refers to instruments requesting the sharing value from developers (at early stage) and downstream one refers to cost distribution to users or property owners via taxing instruments at later stage. In practice, each country (state or city) adopted and created new instruments depending on their development context.[8] Developed countries could apply both upstream and downstream financing; however, even states in US and developed countries are facing limited capacities to implement complicated policies that relates to revaluing property (Walters 2012).

Upstream financing instruments are growing and more popular in both fast growing countries and developed ones. Pressure from fast growth explained increasing number of countries seeking for reducing up-front cost. Successful economies such as Japan, Taiwan, and South Korea became phenomena along with urban peripheries developed by Land Readjustment (LR) tool (Ingram and Hong 2012). LR then was "exported" to many developing countries such as Thailand, the Philippines, India, Bhutan, and Brazil (Souza et al. 2018). Especially, the rise of smart growth paradigms boosted up diverse upstream financing instruments such as Transferable Development Rights (TDR), sales of air rights, Joint Development programs (JD), developer contribution from the US or advanced economies (US, UK, Japan, Singapore, South Korea, Hong Kong, Taiwan) to fast urbanizing countries (India, Brazil, Poland, Thailand, Philippines) to support new waves of brownfield development, compact city development, and to support Transit-Oriented Development (Cervero 2003; Suzuki et al. 2013; McIntosh 2015). Rich countries such as US also observed shifting (downstream) tax vehicle from federal or state fuel tax to (upstream) localized instruments such as special assessment districts, property impact fees, development right sales, and land readjustment (Suzuki and Murakami 2015). Britain moved from individual betterment levy to community betterment levy and planning gain, which impose contribution to developers (Booth 2012).

Each country has its own development context that drives them toward particular approach. Review from the literature showed that each country may have their own distinctive legal and institutional setting that drives the way to mobilize resources. For example, France interferes in future development area with infrastructure tax and regulatory tools due to its traditional imperium state's approach. Meanwhile, Britain uses contractual approach to request for contribution between community/local government and developers. Federal status in the US allowed this country to have diverse approaches with different instruments in different states and urban regions. Strong tax powers and law enforcement in the US also enabled them to exploit downstream financing using taxes, charges to local businesses owners, and wider users besides property owners or developers (McIntosh 2015). The chosen approaches reflect UK as both established democratic system and discretion planning system. Meanwhile, China makes use of its United State model (authoritarian style) and public ownership of land to acquire cheap farmland and sell serviced land to private developers under leasing contract. Hong Kong authority authorized Metro

[8] Author's expression about the concept.

Transit Authority (MTRC) to apply Rail Plus Property (R + P) program to sell development rights to property owners/developers in the TOD zone or develop. This model relies upon strong integrity and strong government established on democracy and strong business sector. Brazil developed City Statute (2001) to introduce the concept "social function of property" that enabled local governments to ensure "*just distribution of the costs and benefits of urbanization*" by selling CEPAC (Certificate of Potential Additional Construction) and transfer of this rights, pre-emption rights, compulsory parceling, and building and utilization (Sandroni 2011; Cities Alliance, N.S.f.U. Programmes, and B.M.o. Cities 2010). India uses Town Planning Scheme (TPS) to make benefit from farmer's contribution of cheap farmland in return of smaller serviced land (Sanyal and Deuskar 2012) (Table 1).

Understanding momentum of change in resource mobilization requires proper analytical framework. The approaches highlighted by many scholars over development planning framework (Philip Booth), land use and property rights (Susan Feinstein), institutional arrangement and business model for partnership (Murakami and Suzuki), tax, charges, and fees (Lawrence C. Walters, Yu-Hung Hong) are valuable lessons to any interested country. However, as most discussions assumed, local/municipal authorities were granted autonomous status. In practice, autonomous

Table 1 Showcase models and instruments for capturing value increase

Case study	Name of implemented project	Organization/Unit in charge	Value capture mechanism and financing instrument
Changsha (capital of Hunan province), China	An outer-ring circumferential highway	The municipality transferred to a public–private agency—the Ring Road Investment Corporation	• Land leasing sales • Financing from China Development Bank and commercial banks (borrow against the future anticipated land value)
Hong Kong SAR, China	Hong Kong Metro	Hong Kong Mass Transit Railway Corporation (MTRC)	• Direct development of government property • Joint development • Sale of surplus property/development rights/air right
Mumbai, India	Mumbai Metro	Mumbai Metropolitan Regional Development Authority (MMRDA)	• Issue of municipal bonds • Advertisements and auction for the naming of station • Non-fare box instruments: premium FAR, TDR, and land sales

Source Adapted from Jillella and Newman (2016), UNDP (2007), McIntosh (2015), Peterson (2006)

status is not a universal concept, as the influence from upper authorities over land, planning, budgeting, and taxing may differ from place to another. These could be knowledge gaps to consider in designing proper analytical framework in resource mobilization.

In the context of Vietnam, there were insufficient analyses to unveil the development framework that conditions for changes. Over past decade, the introduction of land use fee policy (to capture value increase by changing purpose from 2009), the implementation of Build and Transfer (BT model focusing on land in exchange for building from 2010), and the special laws for certain cities (Ho Chi Minh and Hanoi) should have become good sources for discussions. At the same time, there was lack of debate on why recognized community-based initiatives (land readjustment and joint facility development) have not been legalized (ACVN 2014; Vinh 2014). There is a gap to cover both development management and urban finance perspectives with the link to the governing structure.

This article develops an appropriate analytical framework to clarify momentum/barriers for institutional change toward collaborative financing urban infrastructure. Discussions will approach institutions from three perspectives, including (1) development (planning, land and property, and development management), (2) public financing (taxing, budgeting, and public investment), and (3) referencing to governing model (institutional and organizational framework). The data and illustrations are collected from national statistics and Ho Chi Minh City, the largest city of the nation playing the key role of the most dynamic urban region in Vietnam.

3 Urban Planning, Property Rights, and Development Management Models

3.1 Urban Planning System

Vietnam urban planning has transformed from a tool of controlling resources in centrally planned economy toward a supporting growth in market-oriented one. "*Doi Moi*" or "economic renovation" launched in Vietnam from early 1990 triggered a transformation in small-scale housing improvement and then larger scale urban development. Particularly, from early 2000s, the state introduced a number of policies on land use (Land Law 2003), Building and Planning (Construction Law 2004), and Local Government Law 2004 (Law on People's Committees and People's Councils) to support decentralization where local authority recognized real estate markets and increasing property values will drive the city growth. During the process, unfair compensation and double pricing compensation have been revised to abate farmers opposition together with more engagement of community over participatory planning (Coulthart et al. 2006). Decree 02/2006 played a central role to mobilize resources from private sector (developers) to build neighborhood level infrastructure. However, this mechanism relied upon State's compulsory purchase of farming land at the crop's yield price.[9]

The planning system responded to new "*land lead*" with a number of legal and institutional changes. After the passage of Construction Law of 2004 and then the Urban Planning Law of 2009, there were a number of regulations to push urban growth toward land-based development (Table 2). The land-related revenues became important sources for local budget that provincial authorities were open for further horizontal growth. During 2000 to 2012, the spatial growth inflated three times higher than population growth (Dong 2013) showed that spatial planning shifted from controlling function to market-enabling property development, where general plan was prepared to satisfy possible opportunities with a long-term vision. For example, land allocated for urban development at the year of 2015 was 19 person per hectare, five times lower than the target of 100 person/hectare specified in the Decision 445 (MOC 2019).

There are some concerns with this new planning model. First, the swing from plan-led system to market-led one showed a weak link between plans (land for development) and infrastructure development plan (Chr 2019). Although Decree 11/2013 (Government 2013) urged localities to restrain from sprawl with more programed expansion, little change was observed (MOC 2019). Secondly, the planning system pays little attention to building rights as the Planning code from the Building code (2008) does not cover the floor area ratio issues (MOC 2008). In practice, provincial plans may regulate how to build in particular zoning and detailed plans; however, the influence of land-based control gives little rooms for maneuvering building density.

3.2 Land Use and Property Rights

Property owners inherited growing of rights over past decades. The bundle of rights from Land Law 1993 to 2003 and then 2013 has extended and clarified these rights. From recognition of long-term land use rights through land use title certificate (1993), land users can use, build houses, inherit, mortgage, and trade (1993). From 2003, regulations over compulsory purchase were modified toward fairer compensation where there were some restrictions from abusing compulsory purchase. From 2013, people are more accessible to land use plan and they can contribute land as share to development projects with longer term of use for agricultural land extended to flat rate of 50 years (instead of 20 in 2003). The new user's rights to contribute land as share could trigger community initiatives development; however, we have not observed follow-up framework to enable collaboration at both small group and project level.

[9] Land Law 1993 and 2003 equals 30 times of annual yield level of the rice crops, which is very cheap compared with "current" market price. Land Law 2013 and other regulations adjusted farmland toward market price, but still lower than actual trading price.

Table 2 Major planning regulations affecting collaborative resource mobilization from 2000

Year	Law/regulations	Nature of document	Impacts on resource mobilization
2004	Construction Law (Construction Law 2004)	Decentralized planning control system	Local authority can utilize land as resource for urban development
2006	New Urban Development Regulation (Decree 02) (Government 2006)	Incentive policy for serviced urban land development	Local authority can support private developers selling serviced land using State's power
2009	Urban Planning Law (National Assembly 2009)	Better use of planning as tool for controlling development, zoning, and permits	More control over environmental impact, community participatory, and control of development via zoning and detailed plans
2009	National Orientation of Urban Growth (Decision 445, Prime Minister) (Prime Minister 2009a)	National guide the urban growth	Resource allocation to region and mobilization from all sources, including private sector and foreign loan
2012	Urban Upgrading (Decision 758) (Prime Minister 2009b)	Enabling ODA loan for poor city	Allocate resources for poor cities or disadvantaged districts in strategic regions
2012	National Urban Development Program (Prime Minister 2012)	Allocation of resource to develop city by ranks following National Orientation of Urban Growth	Mobilize resource from budget and foreign loan (ODA) as well as private by city and priorities
2014	Revised Building Law	Separation of private and public development projects	Further control of public projects, deregulate private development projects to support resource mobilization
2017	Planning Law	Control spatial plans to align to resource plan	Improve planning integration, control abuse of planning over land release as major resource for development

Source Authors

Land Laws used limited approaches to mobilize resources. The existing approaches to capture land value increase are taxes on transaction and changing "purposes." Taxes on property transaction target capital gain tax as income of (25% value difference) (National Assembly 2007); however, it seems that due to a poor state valuation and verification system, tax collectors accepted to charge 2% of the

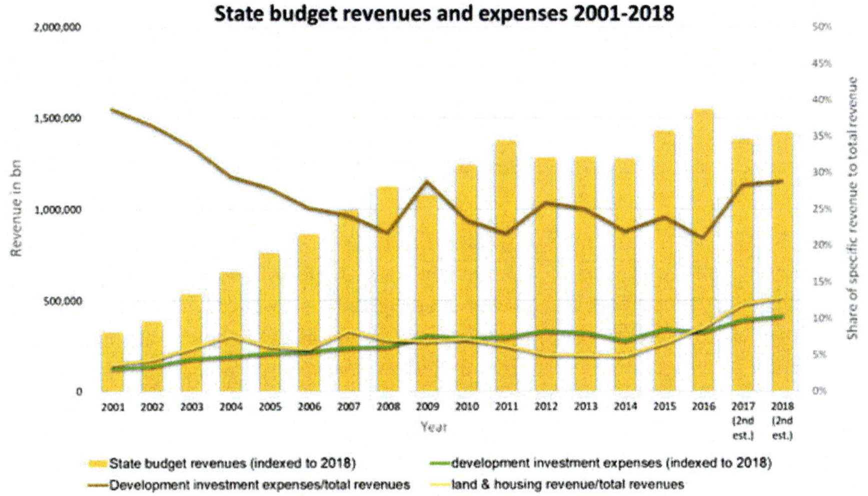

Fig. 1 State revenues, development investment, and its dependency on land revenues. *Source* Authors, compiled from Statistical Year Book 2000–2018

state price. As land use purposes play central role of land policy, the State monitors and charges whoever changing purposes at comparatively high cost (40% of difference between two land by purpose—before 2013 and 50% to 100% of difference—after 2013). From 2013, land auction regulations, introduction of land use fee, and revision of land value framework (after 5 years remain unchanged) helped local authority to lift land-related revenues, even after the burst of the property market from 2010 to 2013 (*See* Fig. 1). A new issue over property tax has been discussed for a decade; however, the current Draft (2017) has not been adopted due to many reasons, including the difference from the ambition of Ministry of Finance over redistribution among wealthy group and expectation of having revenue to improve the welfare of local community (Hieu 2017).

Besides land-based charges, changing of business ownership also mobilizes resources for urban development. Selling the state share from State-Owned Enterprise (SOE) and state's institution initiated over 20 years with the target to transform the economic ownership structure. As large share of SOE's assets are land value in either land lease rights or buildings, an objective to sell $10 bn in the market of the SOE equitization plan from 2016 to 2020 (out of the mass $55 bn) implies that State could mobilize more underutilized resources. However, this mission met only 27% of the target due to many challenges (Do 2016). From the market perspective, whoever owns undervalued property will keep its profit of renting or buying back land with State's low price to resell to the market. Besides SOE, the sales of land and property from the arm forces or public institutions also exposed to abuse of the existing oversight control of planning, poor land valuation, and weak law enforcement.[10] Nevertheless, the restructuring of land and property ownership could contribute substance for urban development.

3.3 Development Management Models

Urban development projects are managed in two tiers with three groups to return investment. Provincial authority manages large-scale development that requires preparing project document and investment proposal. Large-scale projects converting large-scale development should consult or even adopt central authority. Sub-provincial authority (District or City under province) manages smaller (including individual buildings) projects. In principle, public infrastructure projects are divided into three groups: (1) non-return investment cost, (2) partly return, and (3) full-return. PPP projects are designed to link the group (3) via bidding process with some exception of urgency. PPP is the growing segment, while public only projects are on decrease.

There are several models of managing growth to capture value in Vietnam so far. Development model relying upon private initiatives is the normal one; however, there are three well-known urban regions in Vietnam having specific development models, namely: Da Nang city, Binh Duong province, and Ho Chi Minh City (with development ribbon of Southern—Phu My Hung new city). Da Nang and Binh Duong became rising star for livable cities, smart cities where land and human resources were mobilized effectively to sustain higher growth over past two decades; while Southern Ho Chi Minh proved its quality of growth and investment efficiency. Three cases showed how land value was captured for financing infrastructure; however, each region has its own feature with key players from the State power, Business backup by State, and public–private partnership using BT. Upstream financing instruments have been adopted and implemented in Vietnam with innovation and dedication of local authorities (Table 3).

PPP model using land in exchange for building (BT) grew fast, but currently stalled for reinforcement. PPP for domestic investors observed successes in several sectors (electricity, water supply, transport) with over $70 bn committed up to January 2019 in over 300 projects (188 BT and 140 BOT[11]). Investigations over most of BT projects uncovered that they were embedded with problems from land valuation and project monopoly (Khiet 2019). New framework boosting competitive bidding and valuing land at proper phasing are expected to attract appropriate investors. It may take time for such patchy solutions to take effect in the existing working conditions.

[10]Truong Tan San, http://www.tapchicongsan.org.vn/Home/Binh-luan/2018/53380/Day-manh-dau-tranh-ngan-chan-day-lui-tinh-trang-tham-nhung.aspx.

[11]Report from Ministry of Planning and Investment for National Assembly, 2019.

Table 3 Successful resource mobilization cases in three urban regions using different land value capture instruments

#	Location	Model	Explanation
1	Da Nang	Land exchange for infrastructure at community level Strong leadership to implement plan using private developer's capacity Regulatory approach combined with community discussion	Capture value increase by State's power to resell in the market Saving land revenues for infrastructure development and community support Property-led development focusing resort and service sector Cheap disconnected land increases value through bridges and tourism industry
2	Binh Duong	Land developed by State-Owned professional developer (Becamex Corp) Integrated urban and industrial parks development Strong leadership, strategic planning, and implementation from the capable authority	Capture raw land conversion by State's investment and diversified business company Combined and mutual reinforcing property value and job creation from connecting industrial park Cheap forest land, low density in strategic location near Ho Chi Minh city
3	Ho Chi Minh city South Saigon—Phu My Hung	PPP—Joint Venture (IPC and TC&D (Taiwan)) to build infrastructure exchange for raw land Long-term commitment on managing growth from developers/managers model	Raw land conversion capture by developers Building completed public infrastructure (Nguyen Van Linh) and long-term commitment, strong planning, and control Combined and mutual reinforcing property value and job creation from connecting industrial park Large-scale investment and internalization of value increase by developer's ownership

Source Authors

4 Infrastructure Financing

4.1 Budgeting and Public Investment

The design of Vietnam Budget Law of 2015 (which also provides some reforms to the tax system) is centered on central–local relations to distribute wealth among localities. The tax lines are centrally specified and controlled; while fees are decentralized. Expenses by purposes are regulated and assigned to implement by tiers (National Assembly 2015). Central and local collect and use separately and share some group of taxes (Table 4).

While inter-governmental fiscal relations have improved, however, there continues to be a higher demand for change. The central policy of redistribution relies upon investment policy/subsidy and inter-governmental fiscal relation, of which the actual shared ratio of shared revenues (for each location) is decided from central authority. In 2018, only 16 provinces/cities have surplus budgets and. 47 others received subsidies. The redistribution of wealth is conducted via several tools, including national target programs (improving living conditions for poorer provinces). However, this politically driven mechanism was blamed for eroding the motivation of local authority in saving and generating higher revenues for both richer and poorer localities (Nga 2017). Even decentralization improves autonomy over spending; there is little or no change in revenue generation, which lays crucial foundation to any resource mobilization solutions (World Bank 2015).

Local authorities could not avoid generating more revenues not only because of debt or national control. Ho Chi Minh which is the richest city is facing with severe shortage of infrastructure finance where revenues from land could not finance flood control alone. The revenues from land is shrinking with the land stock being exhausted (Fig. 2). Although auction and BT could help lifting the revenues; however, it could not play stronger role like previous decades. If local authorities keep exploiting BT

Table 4 Basic division between central and local level Vietnam Budgeting Law

Revenues by tier	Types of revenues	Types of expenses
Central only	VAT tariff, special consumption tariff, environmental protection from tariff, business income tax on import/export area, ODA, revenues from central-owned businesses, natural resources (oil and gas)	National impact expenses
Local only	Land-related revenues, natural resources (except oil and gas), stamp duty, lottery, revenues from central-owned businesses	Local impact expenses
Shared in between	VAT taxes, business income (except oil and gas), domestic special consumption taxes, individual income tax, domestic environmental protection taxes	Subsidies and redistribution

Source Authors

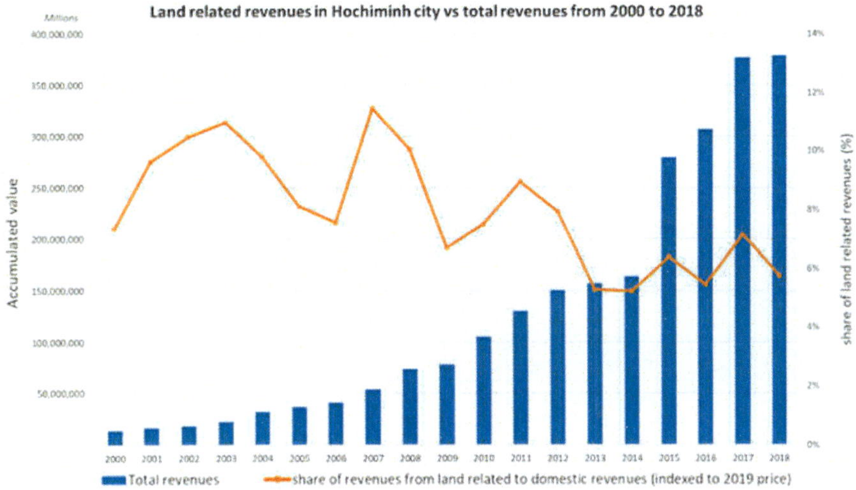

Fig. 2 Shrinking land-based revenues in Ho Chi Minh city. *Source* Authors computed from Ho Chi Minh City Statistical Year Book

for land lease and one-off sales, they will cause further sprawling that will weaken their financial situation in the long run.

4.2 *Public Investment and Financing Management*

The shrinking of ODA and debt-ceiling alarm from 2014 urged central authority to introduce new regulations that squeezed public investment. However, these may bring about higher stimulation to collaborate with other sector. The recent Public Investment Law compels cities to control their investment expenses by mid-term (5 year and 3 year) priority framework (National Assembly 2014, 2017). Together with the Debt Management Law, this may result in trimming-off of unnecessary projects, which were blamed for spreading in subsidized provinces. The more programs linked to longer term of public investment with debt level, the higher responsibility to spend in proper projects using local-generated revenues (Fig. 3). Therefore, localities are being motivated to seek for local solutions such as partnership with private sector or community. However, changing the inertia of existing system may require time, efforts, and leadership champions.

Within local government, municipalities have a limited share of revenue and investment power over infrastructure. Land-related revenues are divided between provincial and lower tiers. Provincial authorities, however, are responsible to pay for physical infrastructure, while districts only pay for local basic social infrastructure. The share between land-related revenue from using land is normally dedicated for district and lower tiers; while revenues (land use fee and taxes of developers and

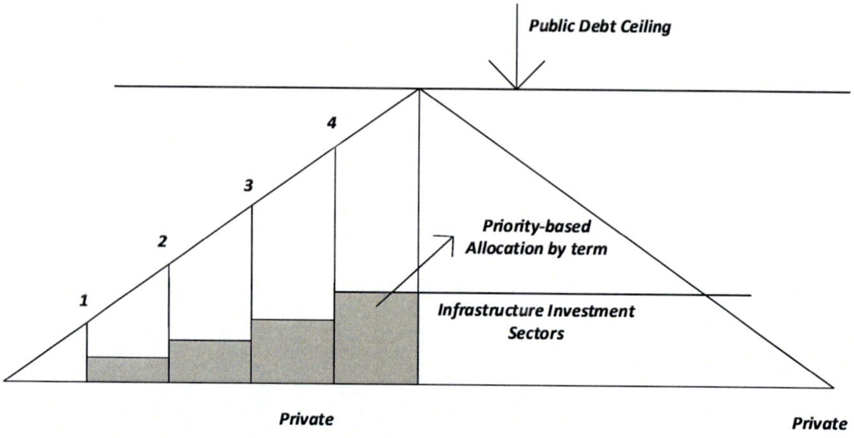

Fig. 3 New mid-term investment and debt management mechanism. *Source* Authors

property transaction) are often controlled by provincial one. As most of the demand for early growth concentrates on physical infrastructure, sub-provincial authorities play minor role to mobilize and utilize resources from tax and fees.

5 Urban Growth Model and Governing Structure

Current urban growth relies upon governing structures that are influenced by provincial authority. As authorities' over planning, land charges, and development management concentrate to provincial authorities, it is understandable that municipalities at the sub-provincial level cannot play a role in financing urban growth. Land-based development model, taxes from developers, and VAT sales of property governing structure go to provincial authority. Public reform over past decade only empowers provincial authorities and not yet sub-provincial ones.

It seems that existing governing model does little to support localized value-sharing model. We may conceptualize the existing land-based development model with tiers of government and stakeholders in a triangle model (Fig. 4). This model relies upon a central relationship between developer and provincial authority in financing urban infrastructure. As neighborhood infrastructure is developed by developers and city infrastructure development via provincial's authority, there is no role for the municipalities to play. Besides, communities can pay their fees to support developers (could also be managers in many new urban areas), but they could hardly work with city authority, as they have little expenses and very little revenue autonomy. This design does not (or very little supportive) make use of upstream financing of sharing value at local level.

If we look closer to existing organizational structure, we may find other problems. Firstly, the urban administrative units have little difference from rural ones;

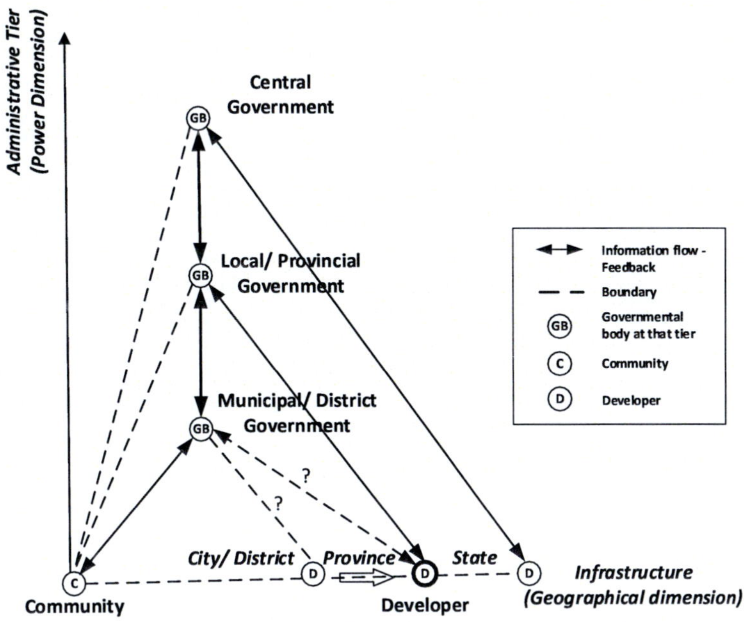

Fig. 4 Governing structure and urban infrastructure financing model. *Source* Authors

while tasks and roles differ due to the higher demand for public services and coordinating activities for growth (Canh 2013). Besides, the design of tiers formed a three layers of government authority working over the same site, with lack of function and role of each level. Newly piloted municipal proposal (in Hanoi) requested to abandoned People's Council at Wards level (Tri 2019). However, compared with a more radical proposal (Fig. 5), where departments are grouped by function, grass-root level becomes administrative units, new towns are self-organized with autonomous status, and the gaps between actual situation and anti-silo management model are discernible.

6 Discussions

This paper explored how existing legal and institutional framework is influencing the current situation as well as momentum of change toward collaborative resource mobilization. The design of analytical framework provided better connection among different factors to govern urban infrastructure development management and financing aspects in a changing context of Vietnam.

Evidence from Vietnam and Ho Chi Minh City reinforces the need to adopt "*beneficiary pays rule.*" The threat of sprawl with decreasing public resources urges more comprehensive transformation of development toward supporting compact,

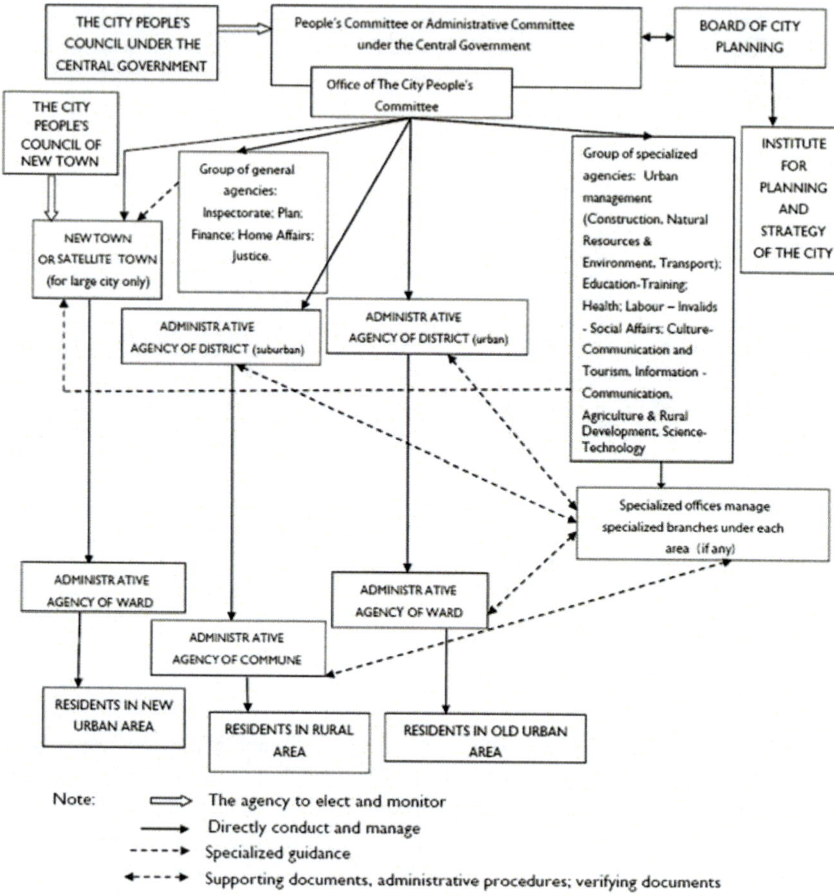

Fig. 5 Proposed design of municipality for Ho Chi Minh City. *Source* (Canh 2013)

TOD, and smarter growth. Upstream financing is an appropriate strategy to cope with the shortage of resources; however, we need to strengthen both development management and financing for development.

The momentum of change exists with positive signals from resource-linked planning from the new Planning Law, PPP enabling policies, public investment, and governing structure. However, central and local authorities should cope with concrete barriers for legal and institutional changes over each issue such as integrated and spatial-based planning (instead of land based), property rights clarity with development rights, and fiscal autonomy with collaborative governing model.

Legal transformation should start-off with development rights and community value-sharing approaches. Land valuation should be established from both land and building perspectives. Land valuation, planning regulation, and development control tools, therefore, should have been reviewed to address the value of contributors and beneficiaries over building rights and property ownerships.

Organizational change or institutional change should focus on enabling local autonomy, including local revenue autonomy. Successful cases in Da Nang or Binh Duong showed cities could capture raw land value increase to finance the infrastructure. However, this land-based model will no longer be applicable for top densely populated cities. The needs for improving living condition at smaller scale could not fit to city level control. It should be launched from community level.

Growing to new level with collaborative approach is challenging for any city authority; however, Vietnamese culture emanates from a strong sense of community, so sharing value concept could be familiar. To adopt and implement new rule as City Statue in Brazil, we will need authority's dedication and community's engagement. In the coming decade, we expect to see new Binh Duong or Da Nang to lead the cities to prosper with new collaborative model.

References

ACVN (2014) Affordable housing. In: Thi VV, Dieu AL (eds). Hanoi, Vietnam

Booth PA (2012) The unearned increment: property and the capture of betterment value in Britain and France. In: Ingram GK, Hong Y-H (eds) Value capture and land policies. The Lincoln Land Institute of Land Policy, Cambridge, MA, pp 74–93

Campanaro A, Duc Dang C (2018) Mobilizing finance for local infrastructure development in vietnam: a city infrastructure financing facility. The World Bank

Canh NT (2013) Municipal government in Vietnam: case studies of Ho Chi Minh City and Da Nang. Hanoi

Cervero R (2003) Growing smart by linking transportation and land use: perspectives from California. Built Environ 29(1):66

Chr E (2019) Phát triển đô thị bền vững và thích ứng. Đánh giá nhanh và phân tích khoảng cách

Cities Alliance, N.S.f.U. Programmes, and B.M.o. Cities (2010) The city statute of Brazil—a commentary

Construction Law (2004)

Coulthart A, Quang N, Sharp H (2006) Urban development strategy. Hanoi

Do ND (2016) Some challenges on equitisation of state own enterprise in Vietnam. 1

Dong HV (2013) Land use status from 3 national census perspective. Economy Forecast Rev 2013(15)

Government (2006) Decree on new urban development regulation. Hanoi

Government (2013) Decree 11 2013 on urban development management. Government, Vietnam

Hieu NN (2017) Property tax—which options for the urban finance? The Builder 2017(312–313)

Hong YH (2007) Assembling land for urban development. In: Hong YH, Needham B (eds) Analyzing land readjustment. Lincoln Institute of Land Policy, Cambridge Massachusetts, USA, pp 3–36

Ingram GK, Hong Y-H (2012) Value capture and policies. The Lincoln Land Institute, Cambridge, MA

Jillella SSK, Newman P (2016) Innovative value capture based rail transit financing: an opportunity for emerging transit cities of India. J Sustain Urban Plan Prog 1(1):56–65

Joe H (2009) Value capture finance. In: Notay A, Clark G (eds) Making urban development pay its way. Urban Land Institute, London, UK

Khiet NT (2019) Bat cap trong trien khai cac du an PPP o Vietnam va cach khac phuc. Tai chinh 2019
McIntosh JR et al (2015) Framework for land value capture from investments in transit in car-dependent cities. 10(1)
McIntosh JR et al (2015) Framework for land value capture from investments in transit in car-dependent cities. J Transp Land Use 10(1)
MOC (2008) Vietnamese building code. In: 01/2008. MOC, Vietnam
MOC (2019) Draft urbanisation assessment report from 2011 to 2020 and orientation of targets towards 2030 with first 5 year plan 2021–2025
National Assembly (2007) Individual income tax law. In: 04/2007/QH12. Vietnam National Assembly, Vietnam
National Assembly (2009) Luat quy hoach do thi (urban planning law). In: 30/2009/QH12. Vietnam
National Assembly (2014) Public investment law. In: 49/2014/QH 13. National Assembly, Vietnam
National Assembly (2015) National budget law. In: 83/2015/QH 13. National Assembly, Vietnam
National Assembly (2017) Public debt management law. In: 20/2017/QH 14. National Assembly, Vietnam
Nga DT (2017) Quan he giua ngan sach trung uong va dia phuong: thuc trang va mot so de xuat. Tai chinh 2017
Peterson GE (2006) Land leasing and land sale as an infrastructure-financing option. The World Bank
Prime Minister (2009a) Decision 445/TTg/2009 on orientation of planning and development for Vietnam urban system towards 2025, vision 2050
Prime Minister (2009b) Decision 758 on adoption of national urban upgrading program. Government, Vietnam
Prime Minister (2012) Decision 1659 on national urban development program
Sandroni PH (2011) Recent experience with land value capture in Sao Paulo, Brazil. Landlines
Sanyal B, Deuskar C (2012) A better way to grow?: town planning schemes as a hybrid land readjustment process in Ahmadabad, India. In: Sanyal B, Deuskar C, (eds) Value capture and land policies. The Lincoln Land Institute of Land Policy, Cambridge, MA, pp 149–183
Smolka M (2013) Implementing land value capture in Latin America—policies and tools for urban development. In: Smolka M (ed). Lincoln Land Institute, Cambridge, MA, USA
Smolka M, Furtado F (2002) Mobilising land value increments for urban development: learning from the Latin America experience. IDS Bull 33(3):84
Souza FFD, Ochi T, Hosono A (2018) Land readjustment: solving urban problems through innovative approach. JICA Research Institute, Tokyo, Japan,p 255
Suzuki H, Murakami J (2015) Theory of land value capture and its instruments. In: Financing transit-oriented development with land values: adapting land value capture in developing countries, pp 43–67
Suzuki H, Cervero R, Iuchi K (2013) Transforming cities with transit: transit and land-use integration for sustainable urban development. Urban development. World Bank Publications
Tri B (2019) De an quan ly theo mo hinh chinh quyen do thi tai Hanoi, in 46/KL-TW. In: Chinh B (ed). Hanoi
UNDP (2007) Paying for urban infrastructure and services: a comparative study of municipal finance in Ho Chi Minh City, Shanghai and Jakarta
Vinh VT (2014) Land pooling & readjustment applicability in Vietnam. In: Thi VV (ed) Land pooling & readjustment in urban upgrading projects in Vietnam. Hanoi, Vietnam
VR B and MNTT (2016) Bao cao chinh sach nhu cau dau tu ha tang ky thuat giai doan 2016–2035. MOC, Hanoi
Walters LC (2012) Are property-related taxes effective value capture instruments? In: Walters LC (eds) Value capture and land policies. The Lincoln Land Institute of Land Policy, Cambridge, MA, pp 187–215
World Bank (2011) Vietnam urbanization review

World Bank (2013) Assessment of the financing framework for municipal infrastructure in Vietnam: final report

World Bank (2015) Making the whole greater than the sum of the parts: a review of fiscal decentralization in Vietnam. WB, Hanoi

Governance Characteristics of Dhaka City for Ensuring Implementation of Land Use Planning

Musfera Jahan

Abstract Governance has become a fashionable term in recent decades, and the concepts of governance and urban governance are defined in a variety of ways. Governance of basic urban services has become a much-debated issue in the last one decade because of their inadequate status for the Dhaka city. Governance of urban basic utilities in Dhaka city is a collective business—around fifty organizations are involved in the provisioning of numerous services/utilities. Based on aforementioned analysis, it might be mentioned that Bangladesh has been facing a number of challenges in the path of democratic or good governance. These issues comprise a lack of accountability and transparency, a lack of government effectiveness and regularity quality, a lack of rule of law, inefficient leadership, ineffective political institutions, rampant corruption, and widespread poverty. The present practice of fragmented governance has ameliorated the surfacing of some critical service problems, namely non-coordination in service management, wastage of resources, inefficient delivery, and public inconveniences. Land use planning as a decision-making tool creates an enabling environment for sustainable development of land resources which meets people's needs and demands that can be ensured by effective governance system.

Keywords Land use planning · Governance · Urban services · Dhaka city

1 Introduction

Governance has become a trending term in recent decades, and the concepts of governance and urban governance are defined in a variety of ways. However, the concept has remained difficult to define, with attempts leading to as many definitions as there are researchers. However, based on usage, governance can be distinguished into three groups of approaches, viewing governance as a synonym for Government, a normative framework, and an analytical framework for non-hierarchical coordination systems. In Dhaka, governance of basic urban services has become a much-debated

M. Jahan (✉)
NKY-PROTEK-SHELTECH JV, Dhaka, Bangladesh
e-mail: musferajahan@gmail.com

L. T. T. Huong and G. M. Pomeroy (eds.), *AUC 2019*, Advances in 21st Century Human Settlements, https://doi.org/10.1007/978-981-15-5608-1_9

issue in the last decade because of their inadequate status. Governance of urban basic utilities in Dhaka city is a collective business—around fifty organizations are involved in the provisioning of numerous services/utilities.

In present situation, barriers of effective land use planning are the coordination of service management, wastage of resources, insufficient manpower, public inconveniences, for the city. But this scenario can be improved for implanting of land use planning for Dhaka city.

2 Urban Governance

Urban governance refers to partnership in urban development between urban local governments and other stakeholders, such as business leaders and landowners. Hence, urban governance is broader than the urban government. The notion of the partnership itself is vague but apparently is not to be construed as an equal relationship or a state interventionist mode coupled with a minimalist role for other actors. Rather, the partnership is more akin to active participation by extra-state actors and active negotiation among the actors, including the state. The defining feature of urban governance is that the management of cities is not the sole preserve of government or the private sector, but is the preserve of a wide variety of actors that interact with one another to govern cities. The broad meaning of urban governance has led to various interpretations. On the one hand, urban governance is seen as a progressive, all-inclusive process of taking care of cities. On the other hand, it is regarded as a neoliberal, minimalist state approach of running cities as businesses. In practice, the concept is considered to be more nuanced, transcending the state–market, socialist–capitalist frameworks and encompassing the regulation or moderation of the interplay between markets, civil society, and the state (Table 1).

Table 1 Some of the definition of governance

Definition of governance
The exercise of political, economic, and administrative authority to manage a nation's affairs. It is the complex mechanisms, processes, relationships, and institutions through which citizens and groups articulate their interests, exercise their rights and obligations, and mediate their differences
The action, manner, or system of governing in which the boundary between organizations and public and private sectors has become permeable… The essence of governance is the interactive relationship between and within government and non-government forces
Governance usually refers to the coordination, regulation, or steering of affairs between actors, in specific sectors (e.g., environment) or in a city, such as in urban governance

3 Measuring Urban Governance

The Urban Governance Index (UGI) is being developed by UN-HABITAT in line with the campaign's advocacy and capacity building strategies with a two-fold purpose. At the global level, the index will be used to demonstrate the importance of good urban governance in achieving broad development objectives, such as the Millennium Development Goals and those in the Habitat Agenda. At the local level, the index is expected to catalyze local action to improve the quality of urban governance. A UN Inter-Agency meeting in June 2001 reviewed seven principles of sustainability, subsidiarity, equity, efficiency, transparency and accountability, civic engagement, and security and finally recommended the adoption of five UN principles of Good Urban Governance (Table 2).

Table 2 Principles and indicators of urban governance analysis

Principle	Indicators
Effectiveness	1. Major sources of income 2. Predictability of transfers in the local government budget 3. Published performance delivery standards 4. Consumer Satisfaction 5. Existence of a vision statement
Equity	6. Citizens' Charter: the right of access to basic services 7. Percentage of women councilors in local authorities 8. Pro-poor pricing policies for basic services 9. Street vending permitted in central retail areas
Participation	10. Elected Council 11. Election of Mayor 12. Voter Participation by Sex 13. Referenda 14. People's Councils 15. Civic Associations per 10,000 population
Accountability	16. Formal Publication (of contracts/tenders, budgets, and accounts) 17. Control by higher levels of Government 18. Codes of conduct 19. Ombudsman's Office 20. Hotline 21. Anti-corruption Commission 22. Disclosure of income/assets 23. Independent audit
Security	24. Crime Prevention 25. Violence against Women Policies 26. HIV/AIDS Policy

4 Analysis of the Urban Governance in Dhaka

In Dhaka, governance of basic urban services has become a much-debated issue in the last decade because of their inadequate status. The provision of basic services, such as water supply, sewage network, electricity supply, telecommunication, urban roads, and highways, is short in supply and irregular (Rahman 2015). In specific terms, frequent malfunctioning, limited capacity to expand, bureaucratic complexity in availing services, lack of public accountability, pilferage of scarce resources, etc., are some of the common characteristics of basic urban services in Dhaka. In fact, the basic urban services in Dhaka city, at the moment, are constantly failing to assure better living, competitive business environment, and sustainable economic growth.

Governance of urban basic utilities in Dhaka city is a collective business—around forty organizations are involved in the provisioning of numerous services/utilities. Of those forty, this research has kept its scope limited to four services, namely water supply and sewerage, power supply, telecommunication (fixed-line govt. service), and different categories of road that are under the jurisdiction of Dhaka City Corporation. The choice has been made on two considerations. First, these are core services—according to the government's relevant policy documents, i.e., Annual Development Plan. Secondly, the present crisis of city governance (inconvenient living, loss of business potentials and investments) in Dhaka arose due to ailing governance of basic urban services.

Urban basic service operators in Dhaka exude mixed characteristics. There are municipal organizations—Dhaka city corporations headed by elected representatives. Additionally, there are government-controlled service providers, namely special development agencies—RAJUK for city planning and development control, special purpose authority—DWASA for providing water supply, drainage and sewerage service, and special government bodies—BTCL for fixed-line telephone, DESCO/DPDC for power supply, etc. Alongside, private or business organizations have long been involved in transportation service, real estate, mobile telecommunication and various information technology-dependent services, health care, recreation, etc., in Dhaka. NGOs are playing an active role in solid waste collection and disposal, traffic regulation, community policing, and other activities.

5 Deficiencies in Governance in the Land Use Planning Process

Like many other cities in South Asia, Dhaka is characterized by fragmented institutional arrangements. Recent moves toward decentralization have strengthened and expanded the role of local governments in Bangladesh. However, Dhaka's City Corporations (CCs) continue to face limited and overlapping functional assignments, and their capacity for revenue generation is weak. Meanwhile, a large number of central government agencies are mandated to work on almost every key functional area within the jurisdiction of Dhaka CCs. No effective coordination mechanisms are in place across functional areas. Nor do they work well within each functional area.

Consequently, translating transformational ideas into reality in Dhaka is not easy. Numerous development plans and strategies, generated by different government ministries and departments, have appeared over time. However, these plans and strategies have often been at cross-purposes and only partially implemented, at best. Few have led to substantial improvements in the city. And yet there are reasons for optimism. Where mandates have been clear and strong coordination mechanisms have been established, exceptional implementation capacities have emerged. These successful urban development examples, while few and far between, suggest an alternative way of thinking about solutions for Dhaka. In practice, fragmented mode of governance has established itself as a troubling arrangement because of the following reasons.

5.1 *Multiparty Involvement and Lack of Coordination*

About 50 organizations are involved with the delivery and management of basic urban services in Dhaka. These service organizations have been made to operate under different ministries as per the respective ordinances. However, delegated directives in these ordinances have prevented the service providers to develop a much sought integrated and collective "modus operandi" to serve the city dwellers. As a result, a serious quandary of coordination and mutual interaction between the service providers has become evident. This has triggered the emergence of two major consequences in the service governance paradigm. First, of course, is the inefficient service decision making and second has been the opening up of the decision-making process to abuse (Moinuddin 2006). The obvious outcome has been poor service governance—emerging in terms of frequent malfunctioning, limited capacity to expand, and bureaucratic complexity in service provisioning, lack of public accountability, wastage of valuable resources, etc.

According to the opinions of three ward commissioners (ward no. 90, 66, and 13) from Dhaka (with whom the author had face-to-face informal discussions in February 2010), the absence of strong organizational consensus has caused serious stress and strains amongst different service providers resulting in the wastage of valuable resources.

5.2 *Ineffective Coordination Mechanisms*

A number of central agencies are mandated to assume key urban management functions within the jurisdictions of the Dhaka CCs. Some of these agencies are departments of the central government that have a metropolitan scope, such as the Dhaka

Water and Sewerage Authority (DWASA) and the Dhaka Transport Coordination Authority (DTCA). Others are institutions with a national scope but with special relevance to Dhaka, such as the Bangladesh Water Development Board (BWDB) and the Roads and Highways Department (RHD). In taking up their mandates, these agencies rarely involve the Dhaka CCs or coordinate with each other.

The fragmentation of jurisdictions across sectors and functions is not unique to Dhaka. But the extent of confusion and overlapping, the weakness of coordination mechanisms, and the gaps in accountability are rather extraordinary. Coordination problems are not only pervasive across functional areas but also, more unusually, within each of them. Addressing key urban development issues requires the involvement of multiple institutions that have overlapping mandates but no effective mechanism to work together.

Studies have described and analyzed this institutional maze overtime. One of them estimated that as of 1998 over 50 agencies were involved in urban development and the provision of services in the jurisdiction of the Dhaka CCs. Another described the situation as a governance crisis. This institutional fragmentation makes it difficult to address the three key challenges faced by Dhaka: flooding, congestion, and messiness.

6 Participation for Ensuring Governance in Land Use Planning Process

6.1 Participation in Urban Development Planning

It has been seen that urbanization, urban growth, and urban expansion are taking place in Bangladesh at rapid rates. For sustainability, these require proper planning, which implies preparing technically sound urban plans (Master Plans, Detail Area Plans, Action Area Plans, etc.), with adequate participation of citizens and stakeholders. Such plans are being prepared in Bangladesh for cities and towns of all sizes from the Capital or Megacity Dhaka to a small Pourashava. In four large cities (Dhaka, Chittagong, Khulna, and Rajshahi), urban plans are prepared by City Development Authorities (like RAJUK, CDA, KDA, and RDA) with some degree of participation.

In the case of Dhaka, civil society groups and professional bodies have compelled RAJUK to allow greater participation in plan preparation. The government at one stage respected the opinions of such bodies and passed necessary administrative decision such as the Detail Area Plans (in 2009), but subsequently yielded to the pressure exerted by powerful real estate groups. This illustrates the complexity of participatory governance in urban planning. Level of participation in urban planning is considerably lower in other cities, especially in Pourashavas, where plans are prepared under the supervision of central government agencies like the Urban Development Directorate or the Local Government Engineering Department.

Ideally, the responsibility of preparing the plan or any other development plan of a city should rest with local urban authority, such as the City Corporation or Pourashava. These bodies are also legally mandated to perform the planning function. But in the early fifties, Dhaka Improvement Trust (DIT) was created to take exclusive responsibility of developing Dhaka, then Capital of East Pakistan, in a planned manner and with this the planning function of Dhaka Municipality was suspended. A similar development took place in the other three large cities at that time. Dhaka Municipal Corporation, however, got a full-fledged Urban Planning Department in 1990, but without the authority of preparing a Master Plan. Instead, the Department could only undertake small-scale site plans, like shopping centers, or parks. It could, however, maintain liaison with RAJUK in its Master Planning or Detail Area Planning exercises. No other city corporation or Pourashava had its own Urban Planning Department or unit until about the year 2000. Such units have been established gradually in all A Category Pourashavas, basically as obligatory conditions under Urban Governance and Infrastructure Improvement Project (UGIIP) supported by the Asian Development Bank. This and other externally funded projects have indeed brought back some essential functions to the Pourashavas, but not yet to the large city corporations. The projects like UGIIP also inspire greater participation of people in the planning process.

6.2 *Participation in Urban Financial Management*

Urban local authorities, both city corporations, and Pourashavas are themselves responsible for the preparation of their respective annual budgets. By law, they are required to solicit the participation of citizens in the budget preparation process. Normally, people in authority's finance or accounts department prepare the draft budget, which is then presented to the Mayor and the Councilor. Generally, the budget is prepared along with a set structure with some nominal adjustments such as a certain percentage change in each approved head. In recent time, the City Corporations and Pourashavas have invited greater participation of the people. Members of the TLCC also take part in discussions on the budget, but basically, there remains little scope for major change. Urban local bodies are seriously weak in financial strength. Most of the authorities cannot raise enough property or holding taxes. Few of them ever try to reassess or enhance property rates for fear of losing popular support. They are also incapable of innovating new sources of revenue earning. However, many urban local bodies receive one or more development funds from the central government or development partners in which citizens have scope for participation.

6.3 Participation in Urban Infrastructure Development

Major infrastructural projects are prepared and implemented in urban areas by central government agencies or even by city authorities without much of people's participation. Such projects normally have financial and technical support of foreign development agencies. These often lack transparency in the process of planning and implementation. Only when a powerful government agency or civil society group raises an objection, compromises are made in plans.

Opposition by civil society groups may cause a delay in implementation of projects even when these projects are legitimate (a good case in point is the plan for the first metro line in Dhaka). Citizens, in general, show little interest, for or against, even for such mega projects. Political parties, interestingly, remain conspicuously silent on such issues. Major and expensive environmental improvement projects may be taken up, initially on pressure from the civil society, but during implementation hardly any participation of the citizens are solicited, rather more involvement of the army has been ensured. Case in point is the Hatirjheel Lake Development project, Dhanmondi or Gulshan Lake development projects in Dhaka. The Green City movement is not a political force yet. But in future it will be forced by the authorities.

6.4 Transparency and Accountability in Urban Governance

Transparency and accountability in the administration of urban authorities are essential for good governance. The Pourashava Act 2009 makes it obligatory for Pourashavas these days to exhibit Citizen's Charter openly so that people are better aware of developments in their local authorities. The Commission for Right to Information further empowers people to demand information from their local leadership and authorities. Moreover, the election process in local bodies and the formation of the Town Level Coordination Committee with the participation of representatives of civil society is another step forward. The various Standing Committees are expected to process municipal decision making. The existence of a comparatively free and vibrant print and electronic media and watchful eyes of the Transparency International Bangladesh (TIB) further improve the possibility of greater transparency. Despite all these checks and balances, allegations of corruption in urban local bodies are rather common. Mayors of large cities are often charged with serious cases of irregularities. Lack of transparency in development projects and partisanship in appointments to municipal jobs is a common allegation.

6.5 Leadership in Urban Governance

Leadership is one of the most significant factors in good urban governance, as well as any other sectors of development. Local urban governance depends greatly on the quality of local urban leadership, meaning the elected mayor and his/her councilors. Visionary, dynamic, innovative, committed, and honest leadership can make a city more efficient and livable. In general, City Corporations and Pourashavas in Bangladesh lack such leadership. Dhaka was not fortunate in this respect.

There is also greater participation of elected women councilors in Pourashavas, although they themselves are not satisfied with their status and privileges. It may be noted that projects like UGIIP, MDF, or UPPR have played some role in improving the governance style of urban local bodies, particularly through the emphasis on citizen's participation in local decision making. Intelligent local urban leaders have exploited the positive elements of such projects effectively. Strong leadership can run a proper governance system for the cities.

6.6 Coordination in Urban Governance

A key concept in good urban governance for implementing land use planning in the city is that of coordination, particularly among the governmental and para-state institutions that are involved. In the case of Dhaka city, for example, as many as 40 different governmental or autonomous organizations are known to have such involvement, many of which are directly on a day-to-day basis. There is a serious problem of coordination among such agencies resulting in wastage of resources and time, and ultimately causing sufferings to the people.

Occasional attempts were made to solve the problem of coordination in Dhaka city. One attempt was made in the mid-1990s when Mohammad Hanif was the city Mayor. He, in fact, demanded the formation of a City Government with the implication that all relevant governmental service delivery and controlling agencies be brought under the single authority of the City Mayor (or Governor, if such a title might be more acceptable). The Government of that time only allowed a Coordination Committee to be chaired by the Minister for Local Government and co-chaired by the Mayor. The arrangement became dysfunctional in no time. It was formed with over 30 agency chiefs and five civil society representatives. The Committee, like the previous one, could not work properly, and at the request of the Mayor, the committee's responsibility was shifted to the Prime Minister's office with the Principal Secretary to the Prime Minister acting as the Chief Coordinator on behalf of the elected Mayor. It worked for some time, but as it was not constitutionally established, the system discontinued the change of government, and coordination remains a very critical issue in governance in Dhaka city. It is also a serious problem in other major cities. The attempt has, however, been made to improve Coordination of agencies in Pourashavas through the Town Level Coordination Committee (TLCC) formed under UGIIP. Civil society representatives also sit in the TLCC.

6.7 Role of Civil Society Organizations

Civil society organizations, which include NGOs, CBOs, and professional groups, among others, have significant roles to play in urban development and governance. In the nineteenth and early twentieth centuries, the *Panchayets* (neighborhood level social/community institutions) used to contribute significantly to the maintenance of social order in Dhaka. Now there are new and different types of social groups performing such functions. Among these, neighborhood level environmental activist groups in the middle or upper-class areas or the slum dwellers associations as well as city or national level civil society groups are parts of new democratic dynamics of urban development and urban governance. The national level civil society groups such as Bangladesh Paribesh Andolon (BAPA), Bangladesh Environmental Lawyers Association (BELA), and Consumer Association of Bangladesh (CAB), etc., are now important civil society organizations, which raise their voice on urban issues. BAPA, for example, has a Task Force on urbanization and offers both analytical discussions and agenda for action. BELA goes for legal action against public or private projects harming the urban environment. Civil society's urban awareness is much stronger today than ever before, but still not strong enough to stop all kinds of bad projects.

6.8 Role of the Private Sector

Dhaka, the megacity with about 19 million people (in an area of 1528 km^2, including DCC, Narayanganj City Corporation, and other Pourashavas), is becoming distinctive as an unequal city with hardly any evidence of planned effort in reducing the pattern. The private corporate sector plays a strong role in accentuating the inequality. During the past two decades, particularly the private land and housing development companies have primarily invested in luxury residential development rapidly pushing the lower-income groups out of the housing market, and to marginal locations of the degraded physical environment.

Sometimes critical issues in urban governance arise due to motivated actions of the private sector. In the area of industrial development, the private sector sets up factories which produce chemical effluents which harm the environment. The governmental agencies responsible for maintaining environmental standard are either incapable of controlling the unacceptable activities of industries or are in alliance with real estate companies, particularly those involved in land development, who are mostly uncontainable and almost ungovernable. They now have a strong lobby in the Parliament as well as friends in the administration. The victim is the environment, particularly the water bodies and rivers in and around Dhaka and other urban areas.

Weak governance is held responsible for such activities. The Government has enacted a number of relevant acts and rules to safeguard the urban environment but poor implementation results in negative development.

6.9 Role of Development Partners and International Agencies in Urban Governance

Roles of development partners, which include foreign country-specific aid agencies and also international financial and development organizations, have become increasingly significant in urban development and urban governance in Bangladesh since its recent times. UN bodies like UNDP, UN-Habitat, UNCDF, and UNICEF have played an active role in helping build physical infrastructure and provide social support in urban areas, particularly in sanitation, drainage, water supply, waste disposal, and income-generating activities. The role of foreign development agencies like JICA, Danida, DFID, and GTZ has also been significant. Similarly, projects like UGIIP of ADB or Municipal Development Fund of World Bank have also made an impact on both urban structure and urban governance system in Bangladesh. The development partners now have their own lobby and insist on participation in the decision-making process related to urbanization and urban governance in Bangladesh. The Government of Bangladesh has recognized its role and ensured their participation in the newly formed Bangladesh Urban Forum (BUF). They have also participated in the formulation of National Urban Sector Policy, now awaiting government approval. In other words, many of the major urban development decisions will be taken with the participation of foreign and international development partners.

7 Actors and Institutions of Urban Governance for Ensuring Land Use Planning in Dhaka City

The actors and institutions involved in the process of governance include private sector businesses, both corporate and informal; civil society, including community-based organizations, NGOs, political parties, religious groups, trade unions, and trade associations; and the whole range of governmental agencies of national, regional, and local government, including traditional authorities where they exist. These are illustrated in Fig. 1. Urban governance also, crucially, involves individual citizens and households, of all income groups, inasmuch as they have any influence over what happens. Within this, the city (or municipal) government is but one element, albeit often the largest and most obvious.

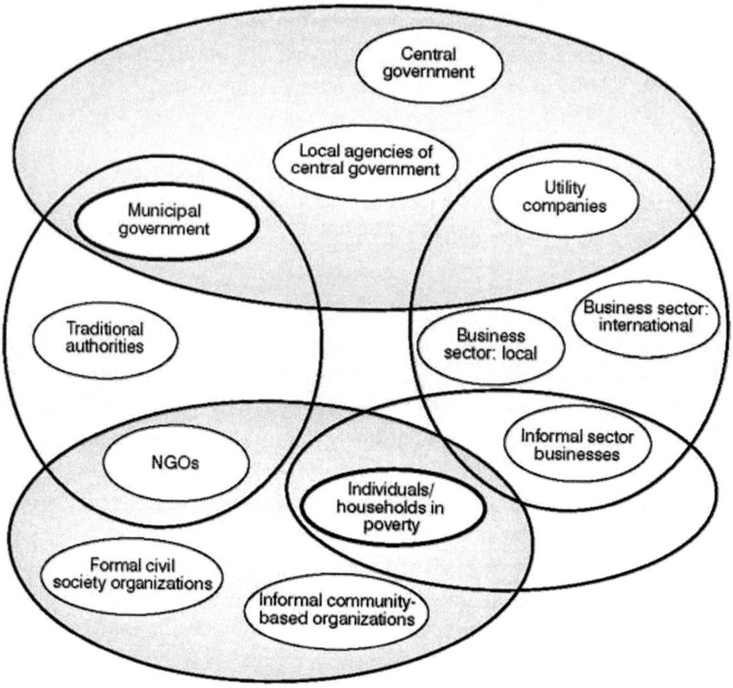

Fig. 1 Actors in urban governance (*Source* Devas 2004: 25)

8 Conclusion

Bangladesh has been facing a number of challenges in the path of democratic or good governance. These issues comprise lack of accountability and transparency, lack of government effectiveness and regularity quality, lack of rule of law, inefficient leadership, ineffective political institutions, rampant corruption, widespread poverty, and so on. However, for ensuring and implementation of land use planning in Dhaka city, urban governance is necessary. Rajuk and other relevant organizations has a vital role in ensuring urban governance as well as other sectors such as housing, transportation, energy, water, agriculture, tourism, and economic development, which also have demands on land use planning.

References

Ahmed ZN. Building regulation changes: effects on residential building design in Dhaka for energy efficiency

Rahman MAU (2015) Coordination of urban planning organizations as a process of achieving effective and socially just planning: a case of Dhaka city, Bangladesh. Int J Sustain Built Environ 4(2):330–340

Urban Planning in Vietnam: Why Gender Matters

Phuoc Dinh Le and Huyen Minh Do

Abstract This paper provides some initial thoughts on urban planning through a gender perspective. Historically and globally, women have been ignored or neglected in the planning domain, which can be seen throughout the development of human civilization where women were literally absent in this field, not until the 1960s when the "second wave" of feminism questioned and challenged the notion of its homogeneity. This issue is clear when it comes to Vietnam's urban planning, which has long been portrayed as a male-orientated profession. Urban planning appears to be separated from gender studies, and indeed there is lack of literature on the connection of the two. It seems that the majority of urban planning educators, experts, and architect planners have taken for granted the notion of gender neutrality in urban planning, and they have been unaware of how gender difference can shape and influence the establishment of urban form and how cities function. This paper, therefore, will take a preliminary effort to offer an indicative account on this matter, showing the context of gender unequal distribution in Vietnam's urban planning, and how deep gender inequality is in this industry. It then provides an analysis on the root causes of gender inequalities. The justification for gender study in urban planning will be discussed as well as possible scopes for further investigations will also be raised.

Keywords Vietnam · Gender difference · Urban planning · Inclusive planning · Feminism

1 Introduction

Urban planning has an essential role to play in creating a platform on which the human prosperity, diversity, and inclusion are centralized and promoted, which subsequently is expected to enhance the sustainable development of towns and cities. Here, diverse and inclusive repercussions of urban planning are compatible with the elicitation of

P. D. Le (✉) · H. M. Do
Hanoi Architectural University, Nguyentrai, Thanhxuan, Hanoi, Vietnam
e-mail: Phuoc.ledinh.hau@gmail.com

L. T. T. Huong and G. M. Pomeroy (eds.), *AUC 2019*, Advances in 21st Century Human Settlements, https://doi.org/10.1007/978-981-15-5608-1_10

equality amongst people regardless their gender, ethnicity, class, and social status. Gender equality in this sense is critically important as society constitutes and is made up by both men and women. It is always unacceptable for the unequal development in which the privilege and prosperity of one gender is built upon the expense of the other. In fact, gender equality is listed as a main goal of among seventeen sustainable development goals introduced by the United Nations.

Since urban planning is powerful and used as a political tool to shape the sustainable prospect and vision of urban spaces, bringing gender equality to the fore is critically necessary. However, it seems that gender has literally been ignored or neglected in Vietnam's urban planning discipline. Both men and women play certain roles and responsibilities in society as well as experience urban spaces differently; however, as a matter of fact, women's voice, life experience, and their potentials do not actually contribute to the establishment and on-going development of urban planning industry, due to gender blindness, reflected on the notion of neutrality and male-orientated ideology. It is in fact a worldwide phenomenon rooted deeply in either developed or developing countries, where issues and dilemma, associated with and resulted from gender neutrality in planning, have been historically studied, mostly by feminists (Greed 1988, 2001; Rahder and O'Neill 1998; Reeves et al. 2012). Furthermore, reports prepared by UN-Habitat have identified contentious issues and trends associated with gender inequality in urban planning in developing countries. In these reports, gender streaming and gender sensitivity approaches have also been introduced as a new way of looking at traditional and current urban planning methodologies. However, it seems like these efforts are not being paid attention, and to some extent are not considered being applicable in Vietnam context and culture.

One explanation for this is that urban planning in Vietnam primarily means construction planning, which focuses more on the creation of physical environment while less focuses on social aspects as Nguyen and Le (2017) point out. In Vietnam, urban planning in fact has long been defined as the creation and organization of architectural spaces, urban landscapes, technical and social infrastructure, and houses in order to establish a comfortable environment for urban dwellers. Research on matters related to urban planning is highly technical. It usually emphasizes that the physical spaces and objects are found to be separated from social world, which shows the lack of interdisciplinary connection. As a consequence, there has been lack of research on gender and its connection and intertwined relationship with urban planning. It seems that the majority of urban planning educators, experts, and urban planners have taken for granted the notion of homogeneity in urban planning, and they have been unaware of how gender difference can shape and influence the establishment of urban form, how cities function, and how urban residents' lives are impacted. Listerborn (2007) states that if the planners are not educated about or bear prejudice toward the people they plan for, the repercussions of planning processes will illustrate just that. In light of that, we argue that if planning experts, educators, and people practicing and studying urban planning have poor knowledge about or are unconscious about the constructed bias they have toward the people and the communities they are planning for, the repercussions are rather dogmatically exclusive and woefully incomplete. Therefore, the awareness of the problematic gender neutrality in urban planning in

Vietnam needs to be raised. This paper first shows the context of gender unequal distributions in Vietnam's urban planning, and how deep gender inequality is in this industry. It then provides an analysis on the root causes of gender inequalities. The justification for gender study in urban planning will be discussed in the last section.

2 How Males and Females Are Distributed in Urban Planning Education and Profession in Vietnam?

In order to answer this question, we looked at some empirical evidence, mainly through open sources, on statistics information about the ratio of men to women of urban planning experts, educators, architectural planners, and planning students in key universities and institutes across the country.

2.1 *Urban Planning Education—Experts, Educators, and Planning Students*

Urban planning courses are integrated into either engineering construction or architecture. Its workforce is mainly educated and graduated from National University of Civil Engineering (NUCE), Hanoi Architectural University (HAU), and the University of Architecture Ho Chi Minh (UAH), with a professional job title as an architect or an engineer. These universities have a long history of development comprising many founding academics, professors, and senior lecturers that play important roles in establishing and shaping Vietnam urban planning industry (see Table 1). The Division of Urban and Regional planning of the NUCE has educated the first architect and engineer generation, laying an initial foundation for urban planning in Vietnam. There are various units integrated in the Division's educational curriculum, including regional

Table 1 Executive dean of NUCE Division of Urban and Regional Planning periodically

Name	Professional title	Term	Gender/sex
Quang Thao Truong	Professor	1961–1965	Male
Phung Vo Nguyen	Architect	1966–1968	Male
Quang Thao Truong	Professor	1969–1989	Male
Dai Ton	Professor	1990–1992	Male
Kim Luyen Nguyen	Professor	1993–1995	Male
Van Khom Tran	Architect	1996–1999	Male
Sy Que Nguyen	Doctor/Architect	1999–2003	Male
Hung Cuong Pham	Doctor/Architect	2004–present	Male

planning, master planning, detailed planning, rural planning, architectural landscaping design, urban sociology, transportation, technical infrastructure and social facility planning, and many different scale planning projects (Lam and Pham 2017).

Initially, educators and senior lecturers working in the Division were totally male-dominated, not until 1990–2000 when female lecturers started to join the professional workforce. Ms. Thu Trang Dam, Ms. Thuy Loan Pham, Ms. Thi Thanh Mai Nguyen, and Ms. Thi My Hanh Phung were the very first female members of the faculty in the Division. The number of female lecturers has been growing slowly since 2000 and reached 15 for the time being, including 40% females and 60% males.

Meanwhile, the Faculty of Urban and Rural Planning at the HAU has been contributing to Vietnam's urban planning workforce through its reputation on providing educational and training services of regional planning, urban planning, rural planning, urban design, architectural landscaping design, and urban eco-environmental management. Urban management used to be a part of the faculty's curriculum, but it has been detached to become a separated faculty. However, for the sake of this paper, some data from the Faculty of Urban Management may be collected. Similar to NUCE Division of Urban and Regional Planning, the number of lecturers working for HAU Faculty of Urban and Rural Planning was predominantly male (about 85%). The proportion of female lecturers is growing to reach 46% at the present. However, those who working in the Executive Board including Vice Dean, Head, and Deputy Head are disproportionately male, accounted over 70% (see Table 2).

At the same time, the UAH Department of Urban Planning is responsible for formulating program, creating curriculum, and delivering educational services on urban planning. The Department comprises of four Divisions including Urban Planning, Urban Management, Urban Landscaping, and Urban Design (Khoa Quy Hoach 2012). Currently, the total number of lecturers at the Faculty is around 36, constituted by 40% female and 60% males, and the Executive Dean of the Department has been predominantly male (see Table 3).

Table 2 Executive dean of HAU Faculty of Urban and Regional Planning each term

Name	Professional title	Term	Gender/sex
Duc Viem Do	Professor	1992–2001	Male
Duc Thang Le	Professor/architect	2001–2011	Male
Xuan Hinh Nguyen	Doctor/architect	2011–2018	Male
Tu Quyen Luong	Professor/architect	2018–present	Female

Table 3 Periodic executive dean of UAH Department of Urban Planning

Name	Profession title	Term	Gender/sex
Kim Mang Huynh	Architect	1976–1980	Male
Khac Van Vo	Architect	1981–1993	Male
Van Muoi Khuong	Architect	1993–2009	Male
Thanh Ha Nguyen	Professor/architect	2009–2011	Male
Phu Hung Do	Doctor/architect	2011–present	Male

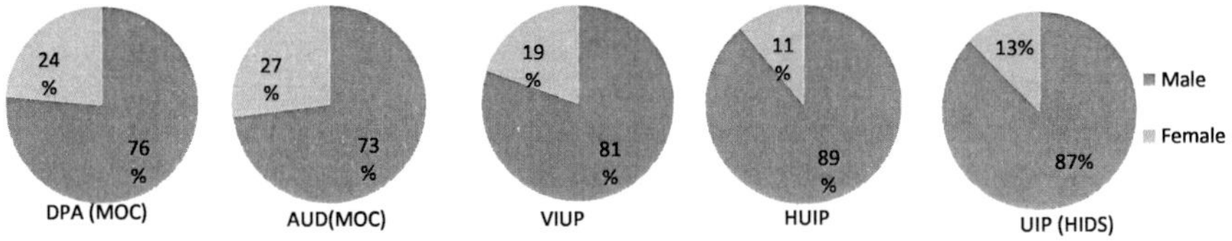

Fig. 1 The percentage of current males and females holding executive positions in five institutes

2.2 *The Planning Profession—Senior Architects Specialized in Urban Planning*

Senior architects specialized in Vietnam's urban planning are officially called architect planners. They work for either governmental institutes or private consultancy companies. However, in this paper, we only collected data from the former due to their overwhelming domination in this industry. Collectively, data gathered from the Department of Planning and Architect (DPA) and the Agency of Urban Development (AUD) of the Ministry of Construction (MOC); Vietnam Institute for Urban and Rural Planning (VIUP); Hanoi Urban Planning Institute (HUPI); and Urban Planning Institution (UPI) of Ho Chi Minh Institute for Development Studies (HIDS) are used for our analysis (refer Fig. 1).

According to the charts, the percentage of males holding executive positions in these five institutes is too far higher than that of the females. The figure for males is nearly four times higher than their opposite sex in DPA (MOC), AUD (MOC), and VIUP, while over four times for HUIP and UIP (HIDS) (89–11% and 88–12%, respectively).

2.3 *Students Undertaking Planning Courses*

Although there are an increasing number of governmental and private universities providing urban and regional planning courses, three universities mentioned previously are still on the lead within the industry. Also, due to time limitation and the shortage of statistics on the number of graduated students, it is almost impossible to acquire data from all universities having planning course throughout the country. Therefore, only data about students who are currently studying and some recently graduated from these three universities is collected in this paper.

Accordingly, the HAU has received around 330 new students every year including about 200 newbies for the Faculty of Urban and Rural Planning and the remainders for the Faculty of Urban Management. Meanwhile, the number of students entering the Divisions of Urban and Regional planning at the NUCE varies from 50 to 80 students each year. At the same time, around 100 new students will enroll in the Department of Urban Planning at the UAH. The male-female ratio of planning students each year is presented as follows (Figs. 2, 3, 4 and 5).

It is clear that male students from the HAU and the NUCE have far outnumbered females regardless of some fluctuation. The percentage of male students is more than three times higher than the figure for female students—often at 70-to-30 ratio. This male-female ratio has reached 80–20, for example, in the year of 2005Q, 2008Q, 2015QL, 62QH, and 63QH. However, the proportion of males at the UAH is slightly higher than that for females, about 55% males and 45% females.

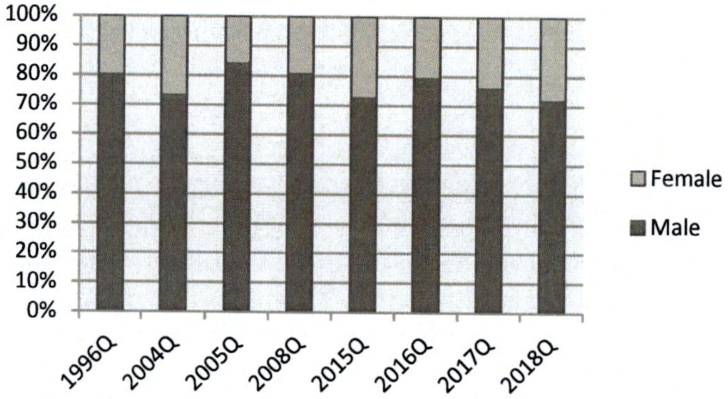

Fig. 2 The male-female ratio of urban and rural planning students at HAU

Fig. 3 The male-female ratio of urban management students at HAU

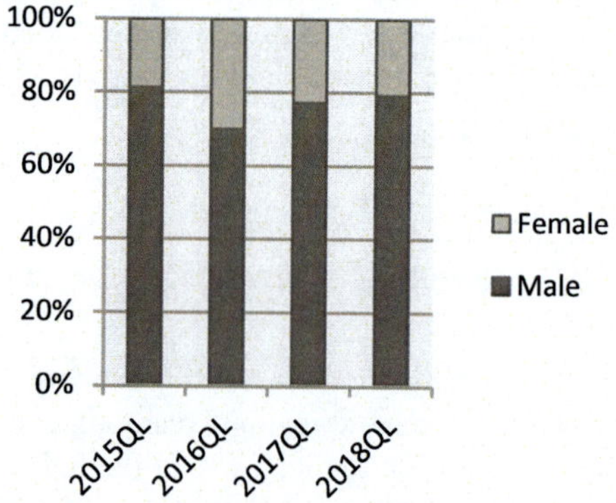

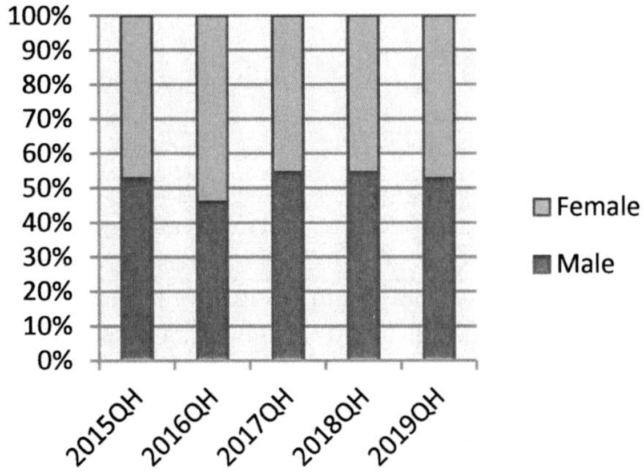

Fig. 4 The male-female ratio of urban planning students at UAH

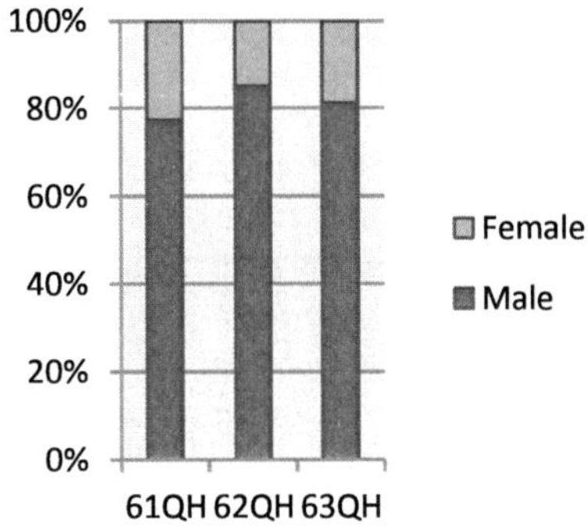

Fig. 5 The male-female ratio of urban planning students at NUCE

3 How Deeply Is Gender Inequality Rooted in Planning Field?

It is undeniable that there are substantial inequalities in the percentage of males and females in the universities and planning institutes in Vietnam. Male experts, educators, lecturers, and architect planners who hold positions of power are almost totally dominant in the early stages of development. They currently have high achievement in both academic and professional life, and have a strong career in planning industry. The majority of them are already professors, holding a Ph.D. degree, senior lecturers, and senior planners. They are also holding much power granted from their positions, including director general, deputy director general, director of institute, director of department, dean of faculty, head of division, so forth and so on. This means that the establishment and introduction of planning theories, the adoption of planning methodologies and mechanisms, the establishment of planning curriculum, decision making, and problem identification and solving are mainly under male responsibilities.

Nonetheless, it does not mean that women have no role to play here. In fact they do. However, it seems like they tap into or try to “fit” in the planning domain quite late in the field—not until the end of the 90s when the number of female architect planners started to grow. While the proportion of lecturers from the universities in question has increased to reach over 40% after the year of 2000, the students taking part in the planning courses remain unchanged with a general male-female ratio of over 70–30%. Nevertheless, women in planning career are holding very little power due to their absence in the executive positions. This results in women’s abilities to get their voice heard causing the limitation in practicing their decision making power. Rahder and O’Neill (1998) also point out the inequalities in distributing male and female educators holding positions of power to educational institutes, which will cause the lack of role models and gender-sensitive support for female students, not to mention their unawareness of current gender inequalities.

It is worth mentioning here that the growing participations of Vietnamese women in planning in the late 1990s are strongly linked to the colonization of Western countries. Women movement in Vietnam has started during the colonization of France military troops from 1945 to 1954. Educational system of Vietnam at that time was strongly influenced by France feminism movement. Prior to this period, Vietnam was perpetually dominated by Confucian tradition with an absolute ideology, rested heavily on patriarchal and patrilineal society that foists different roles and responsibilities upon men and women in every aspect of their lives (Grosse 2015). Accordingly, men are the bottoms of the universe holding enormous power in familial, societal, and political dimensions (Scott and Truong 2007). Although there have been many exhausted efforts undertaken to deal with gender inequality followed with positive results—even ranked at the top countries that have gender equals in many sectors in Asia (Tran 2018), Vietnam has not paid attention to the unequal distribution of males and females in planning field. While the number of women participating in planning education and profession is growing, this figure is inadequate and still fluctuated within one-third as a whole. Also, growing in number does not guarantee to appropriately cope with gender inequality in planning discipline, which has been discussed by many studies worldwide (Greed 1988, 2001; Rahder and Altilia 2004; Rahder and O’Neill 1998) in which the authors worryingly state that, with their dominated and oppressed positions in society, women’s voice and standpoints are vulnerable to abandonment and invisibility, if not constantly fought for.

4 Gender Construction and Urban Planning in Vietnam

The unequal distribution of males and females in planning field in Vietnam is arguably a socially constructed product of gender. Gender is a philosophically contentious term, which is central to the discourse of feminism. In this paper, we hold a feminist standpoint that gender is defined as what societies make of sexual differences including different roles and expectations, bodily representations, and norms and meanings assigned to women and men (Anderson 2017). Feminism, in this manner, is not merely associated with women affairs; rather, it is according to Weedon (1997) a politics, which concerns power relations between men and women in society. The author states, "[t]hese power relations structure all areas of life, the family, education and welfare, the worlds of work and politics, culture and leisure. They determine who does what and for whom, what we are and what we might become" (1997, p. 1).

Of significance to our analysis, feminist theory provides interpretations of the construction of femininity and masculinity, which becomes a steering wheel shaping life experience differences between men and women. These socially constructed genders are expected to conform to different norms of gender roles and expectations (Anderson 2017). Women are the primary subjects of femininity, while masculinity is regarded as a virtue of men. Being a woman equates with nature, often stereotypically portrayed as passive, frail, emotional, and sympathetic; whereas, being a man equates with culture, which is characterized as active, strong, rational, and systematic (Day 2001; Goodey 1997; Jackson 2009).

In addition to these norms, the domination and combination of Confucian and Communist ideas and policies has a permeated impact on gender values in Vietnam (Grosse 2015; Nguyen and Simkin 2017). Vietnamese men are socialized and expected to hold a dominated position and vice versa for women (Higgins 2015). A proverb "đà ông xây nhà, đàn bà xây tổ ấm" (men build house, women build home) is often chanted as an ideal norm foisted upon males and females. This emphasizes the naturalized women's roles in "homemaking, childrearing, and maintaining family stability," while men's roles perceived as rice winner and financial backbone of the family (Nguyen and Simkin 2017, p. 610). These norms and gender values are reiterated, repeated, and educated to become "common sense" and unquestioned teaching rooted deeply in Vietnamese male and female mindset, resulting in the taken-for-granted attitudes toward gender inequality in the society.

Collectively, urban planning field has been entangled in the framework of socially constructed dichotomy of men/women. This field traditionally and globally has been male-dominated with its expertise based on objective, rational, technically derived absolute knowledge and understanding that independent of temporal, spatial, and social identity (Listerborn 2007; MacGregor 1996; Sandercock and Forsyth 1992; Snyder 1995). Since men are socially associated with culture, rationality, objectivity, and systematically less emotion, they are believed to be best suited for planning professions. This notion is even stiffened in Vietnam urban planning industry, which has been classified as construction, architect, or technical disciplines.

Planning educational services have been supplied by architecture, engineering, and construction-based knowledge universities (the HAU, NUCE, UAH to name just a few). Urban planning projects and studies are carried out by well-trained urban planners from governmental institutes administered and managed by the Ministry of Construction (MOC). The approaches for planning research and practice, albeit few modifications, have predominantly been rational, hierarchical, and top-down. Subsequently, these approaches dictate deductive theoretical framework, serving as a platform for the establishment of planning standards, regulations, guide, forecasting methods, and analysis techniques, which in turn are used for both the orientation of planning research and practice, and for planning education in undergraduate and postgraduate courses. This can be seen through the classification of the divisions in faculties, departments, or planning curriculum.

In other words, urban planning in Vietnam is reliant entirely on quantitative methodology that is primarily adopted in natural sciences. The epistemological and intellectual roots of this methodology are featured in the positivist objectivity and value-neutrality of knowledge (Westmarland 2001). It emphasizes the absolute power of experts in testing and extending new scientific knowledge (Snyder 1995; Westmarland 2001) and separates this knowledge from experience. This is because "the 'knowledge' based on experience is not true knowledge, but considered partial, particular, and tainted by subjectivity or emotion" (Code, cited in Westmarland 2001, p. 100). Such experienced, intuitive, empathetic forms of knowledge, according to MacGregor (1996), have been associated with feminine values and identities. Therefore, Vietnamese students and those who practicing planning have been educated and trained with a planning curriculum, conceived to be the most suitable, in order to detach their subjective experiences from their scientific ways of thinking and knowing.

Nonetheless, there has been a significant body of research strongly demonstrating how important it is to take subjective experience and knowledge into account, specifically when it comes to human-related issues. Merely leaning on rationality or objective knowledge will not provide a complete and comprehensive picture on complex or somehow mysterious matters involving human beings (Keller 1983). According to Hall (cited in Truong 2007), studying about and practicing planning are even harder than putting people onto the moon. This is simply because planning is not just about creating the physically suitable environment but also highly concerned about the underlying patterns of human behavior, which are hard to be discovered by only employing scientific methods used in physics or mathematics. In this sense, qualitative methods including storytelling, ideas provocation with the utilization symbolic forms, intuition, or leaning via doing are more suitable (MacGregor 1996). This is crucial to urban planning study because of life differences between men and women in responding to urbanization and interacting with the built environment as many studies have found out, which will be discussed in the next section.

5 Differentiated Gendered Experience, Attitudes, and Perception Toward Urban Environment—The Need for Gender Awareness

An extensive research worldwide has confirmed that men and women perceive the built environment differently, encounter different issues and expect different things that result from the tied knot between their biological differences and their social roles and norms (Adebanjo et al. 2012; Beebeejaun 2017; Fenster 2005). This can be seen through various dimensions resonated with urban planning. Firstly, urban planning either enhances or obstructs employment opportunities for gendered objects.

In their report, Reeves et al. (2012) note the important role of urban planning in creating places and spaces for employment opportunities, which not only includes the attention to land uses but also the integrations of transport and housing that work for everyone. Since men and women in Vietnam have different employment status and types of jobs that are selected to fulfill separated gender roles and expectations, their need and everyday expectations are not the same. For example, women are four times likely to play a role as main caregivers in the family, looking after their husband, children, the elderly, and ill people, and doing the majority of house chores (Higgins 2015; Nguyen and Simkin 2017; Trinh 2019). Everyday issues they come across are in relation to transportation, commuting patterns, and usage of health, educational, and childcare services, whose locations are determined by urban planning. Hence, if urban planning is gender-neutral, assuming and treating everybody the same regarding their needs, it literally reinforces inequalities.

Travel patterns a fundamental aspect of the design of urban transport systems and, at the same time, is highly affected by gender. An extensive body of research elsewhere has men and women have different commuting pattern as well as travel behaviors (Gordon et al. 1989; Hanson 2010; Johnston-Anumonwo 1992; Turner and Niemeier 1997). The findings from these studies mostly agree that women tend to have shorter time commuting to work, but have to engage in many shorter trips or supporting trips resulted from household responsibilities than men. Women's travel behavior is fundamentally much more complex than men regarding timing, duration, location of destinations, frequency and sequencing activities, number of stops, and trip length (Lee et al. 2018). Abidemi (2002) also found that residential density has a significant impact on men and women's travel behaviors. This finding is consistent with Lee, Vojnovic, and Grady's 2018 research in which the authors revealed that urban form has placed different burdens of travel on women, especially in disadvantaged neighborhoods. All of the evidence presented strongly recommends the need for gender study in urban transport planning, patterns, allocations of everyday services, and urban density in Vietnam, where there is an insufficiency of, or lack thereof, such studies.

Another difference in men and women experience affected by urban planning is the concerns over their health. Many studies have pointed out that different gender roles and responsibilities impact both negatively and positively on male and female health status. By reviewing research on urban health and gender, Frye et al. (2008) found that men and women occupy different physical spaces in urban areas reinforcing traditional gendered roles that have different effects on their health. Given roles as mothers, homemakers, and caregivers with different travel behavior, Vietnamese women center around residential areas and its vicinity compared to men. Hence, the influence of occupied places on men and women is not same both physically and mentally. Other researchers found that women spend more time in green space than men due to their responsibilities of taking care of children and the elderly, which resulted in positive impacts on their general health (Caula et al. 2009). Furthermore, men and women's perceptions toward green spaces are different. Women are more active than men in interacting and praising the aesthetic values of urban green spaces, which results from their longer spending time in green spaces resulted from their social roles (Sang et al. 2016). Additionally, while male's main concerns about urban green spaces are merely their quality, female's ones are both the quality of green spaces and their own safety. This is because their perception about safety and risk of victimization is different from men as Richardson and Mitchell (2010) hypothesize.

Perceived lack of safety in experiencing urban spaces, yet, is another crucial concern of urban planning. The built environment undeniably influences the way human perceive security. Physical and environmental factors, decided through urban planning and urban design such as street lighting, plants, the façade of buildings, and general visibility, which form the physical atmosphere of public space, have an association with feeling of safety (Nasar and Jones 1997; Painter 1996). Since women and men are educated to perceive things differently in order to comply with their gender identities, they too conceive safety in urban space differently. Vietnamese men are expected to be brave, fearless, and exertive which give them the power to be dominant in urban spaces, while the majority women are educated to be feminine, gentle, and dependent. Therefore, understanding these differences will help planners to have different approaches to deal with safety concerns of the general public. However, due to the outnumbered figure of men over women in planning and the unequal distribution of executive power between two genders, life experience and opinions of women are underestimated and underrepresented. In other words, women spatial and life experiences are decidedly shaped by men. This is by no mean a standard of a fairness and egalitarian society.

6 Conclusion

The paper has provided some preliminary perspectives on gender in urban planning in Vietnam. It showed the deep inequalities in distributing males and females working in urban planning field including experts, educators, urban planners, and students. Particularly, people holding executive power who have established and shaped urban planning industry is dominantly male. This unequal exertion has its root cause from Vietnamese ideas about gender differences in roles and expectations in every aspect of life. These differences have reflected back on the way both men and women experience urban spaces and places. The paper also raises concerns over the epistemological root of urban planning knowledge and challenges its homogeneity, rationality, and objectivity while giving indicative account on the acceptance of dynamic and nuanced subjectivity. No solution has been suggested due to the fact that we just scratched the surface of this matter. We need to gain more insights into the interrelationship gender inequality and urban planning in Vietnam. Therefore, we urge planning researchers to start to conduct more research on the topic, which will contribute to the creation of better place for inclusion and gender equality promotion.

References

Abidemi AR (2002) Gender differences in intra-urban travel behaviour: a preliminary survey in Ibadan, Nigeria. Paper presented at the Urban mobility for all proceedings of the 10th international Codatu conference Lome, Togo

Adebanjo M, Khosla P, Snyder V (2012) Gender issue guide: urban planning and design. Retrieved from United Nations Human Settlements Programme

Anderson E (2017) Feminist epistemology and philosophy of science. In: Zalta EN (ed) The Stanford encyclopedia of philosophy

Beebeejaun Y (2017) Gender, urban space, and the right to everyday life. J Urban Affairs 39(3):323–334

Caula S, Hvenegaard GT, Marty P (2009) The influence of bird information, attitudes, and demographics on public preferences toward urban green spaces: the case of Montpellier, France. Urban For Urban Green 8(2):117–128

Day K (2001) Constructing masculinity and women's fear in public space in Irvine, California. Gend Place Culture 8(2):109–127

Fenster T (2005) The right to the gendered city: different formations of belonging in everyday life. J Gend Stud 14(3):217–231

Frye V, Putnam S, O'Campo P (2008) Whither gender in urban health? Health Place 14:616–622

Goodey J (1997) Masculinities, fear of crime and fearlessness. Brit J Crim 37(3):401–418

Gordon P, Kumar A, Richardson HW (1989) Gender differences in metropolitan travel behaviour. Reg Stud 23(6):499–510

Greed C (1988) Is more better? with reference to the position of women chartered surveyors in Britain. Women's Stud IntForum 11(3):187–197

Greed C (2001) Women and planning in Britain—25 years on a reflection. Women Environ Int Mag (50/51):7–10

Grosse I (2015) Gender values in Vietnam—between confucianism, communism, and modernization. Asian J Peacebuild 3(2):253–272

Hanson S (2010) Gender and mobility: new approaches for informing sustainability. Gend Place Culture 17(1):5–23

Higgins R (2015) Is it my job to make him care? Middle-class women and gender inequality in Ho Chi Minh city. Anthropol Work Rev 36(2):74–86

Jackson J (2009) A psychological perspective on vulnerability in the fear of crime. Psychol Crime Law 15(4):365–390

Johnston-Anumonwo I (1992) The influence of household type on gender differences in work trip distance. Prof Geogr 44(2):161–169

Keller EF (1983) Interlude: a sketch of the terrain. A feeling for the organism: the life and work of Barbara McClintock. W. H. Freeman, San Francisco, pp 89–106

Khoa Quy Hoach (2012) Planning faculty. http://www.uah.edu.vn/router_detail/khoa-quy-hoach-752.html

Lam QC, Pham HC (2017) Introduction of planning department. http://ktqh.nuce.edu.vn/loi-gioi-thieu-bm-qh.htm

Lee J, Vojnovic I, Grady SC (2018) The 'transportation disadvantaged': urban form, gender and automobile versus non-automobile travel in the Detroit region. Urban Stud 55(11):2470–2498

Listerborn C (2007) Who speaks? And who listens? The relationship between planners and women's participation in local planning in a multi-cultural urban environment. GeoJournal 70(1):61–74

MacGregor S (1996) Rethinking planning–reframing difference. Women Environ 39(40):22–25

Nasar JL, Jones KM (1997) Landscapes of fear and stress. Environ Behav 29(3):291–323

Nguyen TQT, Simkin K (2017) Gender discrimination in Vietnam: the role of personal face. J Gend Stud 26(6):609–617

Nguyen XH, Le XH (2017) Opportunity and challenge in urban and rural planning education in Vietnam. Paper presented at the Integration in the educational system of urban and rural planning in Vietnam, Hanoi

Painter K (1996) The influence of street lighting improvements on crime, fear and pedestrian street use, after dark. Landsc Urban Plan 35(3):193–201
Rahder B, Altilia C (2004) Where is feminism in planning going? appropriation or transformation? Plan Theory 3(2):107–116
Rahder BL, O'Neill K (1998) Women and planning: education for social change. Plan Pract Res 13(3):247–265
Reeves D, Parfitt B, Archer C (2012) Gender and urban planning: issues and trends. Retrieved from United Nations Human Settlements Programme
Richardson EA, Mitchell R (2010) Gender differences in relationships between urban green space and health in the United Kingdom. Soc Sci Med 71:568–575
Sandercock L, Forsyth A (1992) A gender agenda: new directions for planning theory. J Am Plan Assoc 58(1):49–59
Sang ÅO, Knez I, Gunnarsson B, Hedblom M (2016) The effects of naturalness, gender, and age on how urban green space is perceived and used. Urban For Urban Green 18:268–276
Scott S, Truong TKC (2007) Gender research in Vietnam: traditional approaches and emerging trajectories. Women's Stud Int Forum 30(3):243–253
Snyder MG (1995) Feminist theory and planning theory: lessons from feminist epistemologies. Berkeley Plan J 10(1):91–106
Tran TC (2018) Gender equality in Vietnam: theory and practice. http://hpu2.edu.vn/vi/cong-doan-truong/tin-hoat-dong/binh-dang-gioi-o-viet-nam-nhung-van-de-ly-luan-va-thuc-tien-29.html
Trinh TH (2019) Who does house chores? https://nongnghiep.vn/viec-nha-ai-lam-post233709.html
Truong QT (2007) Reflectivity on urban planning definition. From urban and rural planning forum (in Vietnamese)
Turner T, Niemeier D (1997) Travel to work and household responsibility: new evidence. Transportation 24(4):397–419
Weedon C (1997) Feminism and theory. In: Feminist practice & post-structuralist theory, 2 edn. Blackwell Publishers, Oxford, pp 1–11
Westmarland N (2001) The quantitative/qualitative debate and feminist research: a subjective view of objectivity. Forum Qual Soc Res 2(1):1–13

Slum Children-Youth Groups Demonstrate Zest, Collective Confidence and Tenacity to Improve Access to Education and Self-development Opportunities

Siddharth Agarwal, Shabnam Verma, Kanupriya Kothiwal, Neeraj Verma, and Kailash Vishvakarma

Abstract Slum children/youth are excluded from the benefits of India's urbanizing economy. Unequal access to education, self-expression hinder actualizing their potential. This programme research is undertaken in Indore, India to better understand the methods through which slum children (a) overcome gender inequality; (b) overcome hesitation, develop confidence and skills and (c) access government scholarships. Information was collected through Focus Group Discussions with slum children and youth, individual interviews and participant observation during activities. The Urban Health Resource Center (UHRC) organizes slum children's groups in Indore. One boy, one girl leaders per group contribute to a more gender-equitable society. They collectively identify needs for stimulation; ways to enhance confidence, strengthen as groups, contribute to evolution of programme and motivate children in neighbouring slums. Extra-curricular activities by UHRC provide platforms for self-expression (such as street plays), develop teamwork and leadership skills and build self-confidence, zest and tenacity to overcome odds. Children and UHRC's social facilitators urge ward councillors to endorse hand-written applications for income certificates, obtain certificates from District Magistrate's office; submit to schools to access Government scholarship. Child/youth development is fostered through avenues for self-expression, excelling incrementally and cultivating positive self-image. Communication and leadership skills emerge. Children/youth groups' petitions/requests to Municipal Corporation have improved slum lanes, drains. Interruption of children's school education has reduced. Social interactions help children/youth to overcome hesitation, focus on aspirations and acquire ability to

---

S. Agarwal
Urban Health Resource Centre, 136, Humayunpur Safdarjung Enclave, New Delhi 110029, India

S. Verma · K. Kothiwal (✉) · N. Verma · K. Vishvakarma
Urban Health Resource Centre, J-133/A, First Floor, Flat no. 4, Kanchan Apartments, MIG Colony, Indore 452001, India
e-mail: kanupriya381@gmail.com

L. T. T. Huong and G. M. Pomeroy (eds.), *AUC 2019*, Advances in 21st Century Human Settlements, https://doi.org/10.1007/978-981-15-5608-1_11

influence circumstances. Increasing number of youth completes schooling, complements family income; demonstrates gender sensitivity; undertakes socially productive community actions, e.g. improving hygiene and living environment, reduce school-dropout.

Keywords Children · Youth · Education · Slums · India

1 Introduction

1.1 Increasing Population of Urban Slum Youth in India

There is ongoing increase in urban youth population, with lesser opportunities for slum/vulnerable children and youth. In contemporary times, in India, as well as globally there is discourse about the 'power of young persons and youth'. Never before have there been so many young people globally. It is said that there will never again be a time when the world is expected to have such a proportion of young people. The large numbers of young people present a brimming potential for social, education, environmental and economic progress. Today, in absolute numbers, young people add up to 1.8 bin a world population of 7.3 billion. That is up from 721 million people aged 10–24 in 1950, when the world's population totaled 2.5 billion (United Nations Department of Economic and Social Affairs 2014). And in some parts of the world, not only do the numbers of youth grow, but so does their share of the population. With the right policies and investments and the engagement of young people in nurturing their own potential, the largest generation of young people in human history can become the problem-solving producers, creators, entrepreneurs, equity-focused change agents and leaders of the coming decades. There are an estimated 430 million young people in India in the age group of 15–34 years. The youth population (15–32 years) comprises 35 percent of India's urban population (Government of India 2011, NSS 66th round). By this estimate, India has approximately 70 million youth (15–32 years) in urban slums and similarly disadvantaged population out of its urban vulnerable population of about 200 million, which is eligible for food subsidy under Food Security Act (Government of India 2013) that designates 50% urban population as eligible for food subsidy. The demographic trends of urban slum youth population as well as their growing literacy, awareness and education levels warrant attention to the challenges of slum youth in research, program and policy in India. In India, the youth population (15–34 years) increased from 353 million in 2001 to 430 million in 2011, and is expected to continue to increase to 464 million in 2021 and is later projected to decline to 458 million in 2026 (UN HABITAT and IRIS 2012). As awareness, education and skills improve among urban slum youth population, they are anticipated to have a growing potential to represent the slum-dwelling citizenry and contribute to improving the lives of this segment. With 356 million 10–24 year-olds, India has the world's largest population of young persons despite having a smaller population than China (UNFPA 2014).

1.2 Disparity in Education

NFHS-3 analysis by wealth quartiles shows disparities across urban population in India with respect to education. Among the poorest quartile in urban India, 38.7% of respondents in NFHS-3 had received no education as against only 12.3% among the rest of India's urban population. Taking a look at attainment of secondary education, NFHS-3 analysis by wealth quartiles shows that among the poorest urban quartile 29.4% reported having attained secondary education, which nearly half of what the responses of the rest of the urban population show at 49.3% (Agarwal 2017).

1.3 Lack of Facilities

Slum youth and children face a scenario of interruption in schooling and less optimal education facilities, with little access to books beyond school. They also live in an environment which does not offer requisite child and youth development opportunities: no access to sports facilities, insufficient teaching methods, e.g. no or negligible access to pictorial learning aids such as geographical maps, lack of facilities for practical learning by doing available to more fortunate children and youth born in well-to-do families. The life of slum and informal settlement child or youth is typically arduous, the journey to school often involving walking bare feet through a drain. During the afternoons after school, most children/youth spend time working at the neighbourhood incense-sticks making small factory earning INR 10 per work hour or at a shop or an alternate work-place to contribute to the family income. Often there is lack of family support, girls being at a particular disadvantage. Given this scenario, many children and youth resign themselves to a second-best outcome, the situation being more adverse among girls.

1.4 Nurturing the Potential Power of 70 Million Urban Disadvantaged Youth

It is crucial to nurture and stimulate the potential of urban disadvantaged youth for equitable opportunities in urbanizing India. How government, civil society, other stakeholders work to cultivate and nourish the aspirations of this segment of India's youth and address their needs will define our common future. This segment represents a huge source of this 'power' living in underprivileged communities whose contribution is as crucial as that of the more fortunate young populations. There is a need to expedite provision of opportunities towards socio-economic development with this potential 'demographic dividend' where children and youth acquire education, skills, competence, improve society, contribute to economic growth. There is vast potential presented by India's youth force who can gain education, skills, acquire skills to contribute as skilled, semi-skilled, highly skilled vocations in service, IT, manufacturing sectors in increasingly urbanizing economy of India.

The confidence of children and youth to overcome obstacles increases through pursuing activities they enjoy and those they gain knowledge and skills from (Payne 2008). Social interactions help children/youth to free themselves from difficult situations, focus towards their future and acquire the ability to influence it (Vygotsky 1978). Reasoning with one-self develops thinking skills (Hwang and Nilsson 2003). Psychological research has shown that a variety of factors influence the capacity for creativity and innovation, which cannot be reduced to a purely intra-individual concept or to the creation of something 'new and valuable' (Csikszentmihaly 1996: 28). Instead, it should be seen as the interaction between the person and the socio-cultural surrounding. Creativity is defined as any act, idea or product that changes the existing domain or that transforms an existing domain into a new one. It is important to be able to cultivate this capacity in individuals, groups, communities and organizations.

2 Methodology and Programme Approach

2.1 Focus Group Discussions and Individual Interviews

Focus group discussions with slum children and youth were held in 10 slums in Indore involving 150 children and youth respondents which included group member and non-group member children and youth. Individual interviews with children and parents were done. Participant observation during activities was also done. Qualitative open-ended probes were discussed and children and youth were encouraged to share their experiences with respect to (a) extent of opportunities they have in school and home to develop self-confidence; (b) group activities they are able to engage in at school or in the community; (c) avenues for self-expression at school; (d) difficulties faced in pursuing uninterrupted school education and education after class XII; (e) gender inequality at home and in the community with respect to education and opportunities for going out of their community or city.

They were also encouraged to share about the activities they engaged in as part of UHRC's programme and whether they gained in any way from those. They were encouraged to show any certificates they had received on performing well during group activities.

2.2 How Do UHRC's Children-Youth Groups Programme Operate

The approach involves formation and encouraging children's and youth groups in slums mentored by slum-based women's groups and UHRC's social facilitators. The programme's methodology and approach aimed at enabling children and youth develop their confidence, skills, pursue education and all-round development. The focal point of UHRC's efforts is stimulation, motivation, capacity building creating opportunities and platforms for self-expression and confidence building. Children/youth groups mentored by Urban Health Resource Centre's (UHRC) slum women-groups guide project design and are central to implementation. One boy, one girl leader per group facilitates gender equity. They identify opportunity gaps, inadequate stimulation/motivation faced, aspirations, ways to strengthen as groups, contribute to progressive evolution of programme and facilitate youth/children groups in neighbouring slums. What is unique about this programme methodology is that it has been building brick by brick over seven years, slum youth and children are the 'experts' and there is continuous evolution of the approaches over time.

2.3 Fostering Team Spirit, Leadership, Organizing Skills, Gainful Learning

As a proactive measure to encourage gender equity, every group has nearly equal number of boys and girls and two leaders; one boy and one girl. Each group can have 10–20 members. The group meets each month and writes what is undertaken such as book reading, games, discussing possible solution to challenges faced. Children gain knowledge and skills such as learning the geography of India through maps which adds the dimension of using visual learning tools and stimulates right brain faculties. Quiz, drawing contests, skills to write request/petitions to civic authorities enhance knowledge and acquire skills. UHRC has provided a mobile library of storybooks about inspirational personalities and enjoyable reads are available. Children and youth read these books at their meetings. During some meetings, they also write key messages they gather from the books.

2.4 Enabling a Movement Towards Self-reliance

Most children-youth groups save with each member saving small amounts varying from INR 10–100. The savings are primarily used for education-related needs of children and youth. A systematic record of savings is kept and maintained by the group members upon being trained by UHRC's social facilitators. The savings are also used for activities such as buying a football or organizing an event such as sports day. By means of collective savings, the children/youth are also able to do things as a group which perhaps they would not have been able to do individually.

2.5 *Platforms for Self-expression, Building Confidence, Fostering Gender Quality and Pride*

Children and youth shared their experiences of participating in different co-curricular activities organized by UHRC. The methodology proactively creates avenues for self-expression and confidence building through extra-curricular activities they relish, e.g. performing street plays and singing group songs on themes such as 'courage and determination', 'equal social opportunities for girls and boys', serving as compere at annual events of children's groups. Teamwork, gender equality and leadership skills are nurtured through boy and girl children and youth teams managing group pursuits and engaging in activities they enjoy, e.g. sports day, practise sessions for street play. These activities involve planning of activities, identifying location, managing items required such as microphone, play equipment such as cricket bats, balls. Tenacity and zest to overcome odds and not give up develops through continued and perseverant efforts at education, making their living environment cleaner. Activities include poetry recitation, singing of group songs on themes that motivate courage and determination, short essay writing, drawing and painting on subjects such as how can slum environment be improved, origami to promote creative skills and the sense of fulfilment through producing something with one's own hands. Children and youth are encouraged, guided and provided seeds to sow and grow vegetables in slums. This contributes a big sense of hope, of fulfilment and pride when they see the vegetables growing through their efforts. Children and youth also reported that several children were part of UHRC's teams participating in the Mumbai Marathon held each year in January in Mumbai. The Mumbai Marathon is the largest mass participation sporting event in Asia. They shared that participating in such a big sporting event like the Mumbai Marathon in a world-famous city like Mumbai gave a confidence boost to the child who participated, her/his family, friends, neighbours and relatives. They also mentioned that performing street play and singing group song on a social theme in the main market areas of their city and on the beach in Mumbai were confidence boosters for them.

2.6 *Promoting Education*

Children reported that they are often forced to discontinue schooling owing to financial hardships faced by their families. They also shared that many of them were not able to get scholarship owing to difficulty in obtaining documentation that is prerequisite for the scholarship. Children (including non-group members) and UHRC's social facilitators urge the ward councillor (elected Local Government representative from the ward, which is the smallest administrative unit in the city) to endorse hand-written applications for income certificates, obtain certificates from District Magistrate's office, submit them to schools and get Government scholarship. They also helped to obtain caste certificate from the appropriate caste leader to support their scholarship application. During group discussions, children also reported that many mothers of children were members of slum women's groups which also run a collective savings scheme. They were able to borrow from the collective savings of the women's group in their slum to make the expenditure at the time of starting a new school class. This demonstrates practical how-to-do ways slum children-youth are using to overcome social, educational exclusion, social injustice.

2.7 Engaging in Educational Group Activities in Their Neighbourhood

Children reported during group discussions that they were taught geography of India without the use of maps, that they did not have exposure to general knowledge on phenomenon they observe and hear about. They also mentioned that they had very little exposure to storybooks and their reading was confined to school books. The programme team organized different education activities for children during which (a) children who were not members of groups; (b) children from socio-economically weaker families and (c) children who had do drop out of school owing to their family's circumstances were encouraged and drawn to participate. Children and youth participate in quiz, geography sessions using large banner-size map and small individual maps organized by the programme team. After a quiz session and a geography session each question or item is discussed so the children learn more details about it and link the new knowledge to related information for better remembering and better recall. The quiz answer sheets and the small individual maps of India which the children have marked and labelled are left with them so they can discuss with their siblings, parents, neighbours and friends to further spread stimulation and knowledge. A mobile library with over 200 books with stories and bio-sketches is managed to provide opportunity for children to read. Book reading and message writing sessions are also held in different slums so that the children read the non-curricular books and glean key messages from what they read during the session. Two to three children show the book or part of it that they read and speak out what they culled out as key messages and learning from the reading.

2.8 *Contributing to Community Benefit, Engaging in Social Interaction*

The children and youth are motivated to think through stimulating them with key questions: 'How can we collectively move towards addressing a given problem; how can you find out more details about how to go about fixing a problem; which aspects are feasible to try through our own collective efforts and for which parts of the problem do we need to approach some authority'? The programme methodology includes providing avenues for social interactions with peers and older age group persons such as parents, women's group members, volunteers from different walks of life, which helps children/youth to free themselves from difficult situations, focus towards future, acquire ability to influence their circumstances.

3 Results and Outcomes

3.1 *Active Children and Youth in Slums*

Fifteen children/youth groups with 181 youth/children members are active in Indore slums. Child/youth development is fostered through avenues for self-expression, teamwork, excelling in performance, reasoning with self and cultivating positive self-image. Communication and leadership skills emerge. Youth/children's groups have helped other children from their neighbourhoods in writing applications for income certificates and in obtaining caste certificates to support their applications for school education scholarships. Loans from women's savings groups and children's groups helped prevent interruption of schooling of over 1000 non-group member children during 2016–2019.

3.2 *Youth and Children Have Built Confidence Through Presenting Before External Audience*

Utilizing platforms of expressing themselves provided by UHRC's programme methodology youth and children expressed themselves through oral, written communication, through reciting poems, singing songs on themes such as determination and courage further contribute to increased self-confidence. They have gained confidence and zest through performing street plays, singing group songs on socially relevant themes like gender equality in education in their local area, in the city's central locations, in cities beyond their own city such as Mumbai and Delhi.

3.3 Children and Youth Groups Think, Implement Solutions Using Collective Strength

Youth-children groups complement the efforts of women's groups in identifying and implementing solutions to challenges faced. They have written collective petitions to civic authorities for cleaning garbage and drains, written individual applications for 'income certificates' to the local ward-level elected representative and obtained the documents to become eligible for scholarship at school, identified play space and organized a sports day. All these outcomes are the result of thinking practice and through promoting the faculty of reasoning and logical thinking. The ability to reason with one-self as well as with their parents, neighbours and UHRC's social facilitators, which is encouraged during activities that are the core of UHRC's methodology of implementing this programme, develops the thinking process, an approach also recognized in earlier works (Hwang and Nilsson 2003). Social interactions, such as with older community representatives, elected ward-level politician, socially useful activities have helped children and youth to develop the strength to free themselves from the immediate difficult situation that they otherwise face as integral to their slum lives and are instead able to focus towards future and acquire ability to improve their circumstance.

3.4 Children Eagerly Participate in Education Activities

During 2016–2019 over 750 children participated in quiz, geography sessions and book reading and message writing sessions. Respondent children reported that in addition to learning new facts from quiz, developing a geo-spatial and visual understanding of the location of different states of India, major cities that they had heard of and learning about capital cities of States, they found these activities stimulating and interesting.

3.5 Future Potential

With continual mentoring, motivation they will not only improve their own lives but will also contribute to their societies in tangible ways. These efforts will contribute to more equitable improvement of living environment, health and overall well-being in the city. Growing number of youth are expected to gain education, self-confidence and skills. Youth are already beginning to contribute to a healthier and more hygienic living environment and playing a role in other forms of social impact with more children and youth groups being formed, improved gender equity and gender sensitivity in the society, their growing role in addressing challenges. Computer literacy will add value to livelihood skills and in the age of ever expanding reach of the internet for accessing government services (such a Picture ID and Proof of address, birth certificate), will also help underprivileged communities access online forms and services and hence reduce the transactions cost of access. More youth will be educated beyond class 12 and as such will be able to contribute substantially to family income as well as to the collective confidence of the socially disadvantaged community. There is the potential for growth beyond what has been visualized in this paper and with slum youth and children groups transcending conventional boundaries.

4 Discussion

Through fostering the ability to reason with one-self, these efforts facilitate development of the thinking process (Vygtosky 1994) among youth and steadily self-efficacy to overcome obstacles, increases through pursuing efforts that entail engagement with civic authorities, e.g. for income and caste certificates for government scholarship. That the approach of discussing challenges used in this programme research such as how can we (the children and youth) work towards overcoming an obstacle (e.g. difficulty in obtaining government scholarships) helps to stimulate and develop thinking skills among children and youth has been stressed in earlier works (Devereux 2002).

The platforms and avenues created as part of this programme approach for children are analogous to 'place', which 'do not refer simply to a geographic location but also to the opportunities that are available to create meaning within a place' (Wilson 1997). Opportunity-based pedagogies as implemented by this programme are effective in furthering learning and acquiring knowledge and skills 'which have some direct bearing on the wellbeing of the social and ecological places people actually inhabit' (Grunewald 2003). The outcomes of enhanced self-esteem and improved self-image among children and youth reinforce self-confidence in her/his own abilities (Hwang and Nilsson 2003). Pursuing different activities that they enjoy as well as acquire knowledge and skills through those pursuits, have also reinforced their growing self-confidence as noted in earlier works (Payne 2008).

The skills and competencies gained by children are broadly consistent with the outcomes that are most commonly associated with youth intervention activities such as relationships with others, sense of self, values, beliefs and future aspirations (Dickson et al. 2012). Through the different activities and social interactions youth and children engage in as described in the methods section, they steadily build their self-esteem which is a judgment of self-worth (Bandura 2006; Mohajer and Earnest 2009).

5 Conclusion and Implications

This programme research demonstrates how social research gains from practical difficulties of research participants. Group and individual interactions with children and youth helped reinforce the premise that the children and youth are actively engaging in programme activities through their own choice and to overcome the difficulties they face in their lives. The programme is proactively encouraging the participation of non-group member children in various activities to expand inclusion of the weaker social strata and to also foster the spirit of helping all among programme participants. This affirms the trust that this programme follows that, if options and opportunities are gently placed before them, all human beings including children and youth would volitionally engage in activities that they recognize as being of benefit to their self-development and/or enabling them tide over challenges and positively transform their lives. The avenues for self-expression, teamwork, excelling incrementally held the children from slums enhance self-confidence, develop communication and leadership skills and cultivate a positive self-image.

Social interactions help children/youth to overcome hesitation, focus on aspirations and acquire ability to influence circumstances. Increasing number of children and youth complete schooling, pursue education or technical learning after completing class XII, complement family income. Children and youth are also demonstrating gender sensitivity and undertaking socially productive community actions, e.g. reduce school-dropout.

The programme also contributes towards helping them build greater recognition in their schools and in their communities, which acts as an additional motivating factor for them to persevere in programme efforts. They also progressively strengthen their faith in their growing ability to improve their future, that of their neighbourhood as well as that of coming generations of their community. Increasing number of youth completes school education, complements family income; social impact via improved gender sensitivity is visible; improved hygiene and living environment. Growing self-confidence, social interaction with peers, persons external to social milieu, collaborative-skills and leadership add value to their lives and help build livelihood skills. Computer literacy, expanding the reach of internet is expected to help underprivileged communities access online Government schemes, services, thereby reducing the transaction cost of access.

This programme research shows how sustained efforts can be undertaken with socially underprivileged children/youth to enable them to avail government entitlements (such as government scholarships), pursue education and develop their personality. This approach is adaptable in Indian cities home to 200 million socially disadvantaged population.

Acknowledgements The authors acknowledge the active input and energy of the slum youth, children, their parents, women's group members from Indore. The insight gained through interaction with stakeholders from municipal/city health and other departments is thankfully acknowledged.

References

Agarwal S (2017) Slum Children Improve Education by Overcoming Social Disadvantage and Inequality, Self-confidence, Team-skills in Slum Youth-Children Collectives: Paper presented at South Asia Education Summit, New Delhi, 27-07-2017 (Unpublished)

Bandura A (2006) Guide for constructing self-efficacy scales. In: Pajares F, Urdan T (eds) Adolescence and education: self-efficacy beliefs of adolescents, vol V. Information Age Publishing, Greenwhich, pp 307–337

Csikszentmihaly M (1996) Creativity, flow and the psychology of discovery and invention. Harper Collins

Devereux J (2002) Developing thinking skills through scientific and mathematical experiences in the early years. In: Miller L, Drury R, Campbell R (eds) Exploring early years education and care. David Fulton, London

Dickson K, Vigurs CA, Newman M (2012) Youth work: a systematic 'Map' of the research literature. EPPI-Centre, Social Science Research Unit, Institute Education, University of London

Government of India (2011) National sample survey 66th round (2009–10). Ministry of Statistics and Programme Implementation, New Delhi

Government of India (2013) Ministry of Law and Justice. The National Food Security Act 2013. The Gazette of India, New Delhi, 10 Sept 2013. http://indiacode.nic.in/acts-in-pdf/202013.pdf. Accessed 15 July 2019

Grunewald DA (2003) The best of both worlds: a critical pedagogy of place. Educ Res 32(4):3–12

Hwang P, Nilsson B (2003) Utvecklingspsykologi, Natur och kultur, Stockholm. Cited in Sandra Svensson, 'Children and youths in Dharavi´s rise to empowerment—from a NGO perspective'. Thesis, Halmstad University. http://www.diva-portal.org/smash/get/diva2:729798/FULLTEXT01.pdf. Accessed 20 July 2019

Mohajer N, Earnest J (2009) Youth empowerment for the most vulnerable: a model based on the pedagogy of Freire and experiences in the field. Health Educ 109(5):424–438

Payne M (2008) Modern teoribildning i socialt arbete, Natur & Kultur, Stockholm. Cited in Sandra Svensson. Children and youths in Dharavi's rise to empowerment—from a NGO perspective. Thesis, Halmstad University. http://www.diva-portal.org/smash/get/diva2:729798/FULLTEXT01.pdf. Accessed 20 Aug 2019

UN HABITAT and IRIS (2012) State of urban youth 2012. http://www.esocialsciences.org/general/a201341118517_19.pdf. Accessed 06 Aug 2019

United Nations Population Fund (UNFPA) (2014) State of world population 2014: The power of 1.8 billion adolescents, youth and transformation of the future. https://www.unfpa.org/sites/default/files/pub-pdf/EN-SWOP14-Report_FINAL-web.pdf. Accessed 19 July 2020

Vygotsky LS (1978) Mind in society: the development of higher psychological processes. Harvard U.P, Cambridge, Mass

Vygotsky L (1994) The problem of the environment. In: van der Veer R, Valsiner J (eds) The Vygotsky reader. Blackwell, Oxford, pp 338–354

Wilson R (1997) A sense of place. Early Childhood Educ J 24(3):191–194

On the Use of Data Envelopment Analysis to Improve Performance Efficiency of Governmental Management in Big Cities

Hai Dung Dinh and Khoi Minh Le

Abstract In this work, a specific performance measurement method known as Data Envelopment Analysis (DEA) is discussed for application in governmental management and improves the efficiency. DEA is a great model that is applied in many operating environments of the modern economy. According to Business Performance Improvement Resource (BPIR) definition, performance refers to current outputs and outcomes obtained from operating processes that permit evaluation and comparison of information. Performance can be expressed in non-financial and financial terms. Measurement refers to numerical information that quantifies input, output, and performance dimensions of processes, products, services, and the overall organization. In business, performance measurement helps managers improve their decision making and organization performance because it provides essential and quality feedback that the operations may be guided accordingly by allowing managers to achieve the best solution. It is a great way to understand, manage, and improve the overall functioning of the organization, especially to improve business success. If the measurement result is wrong or inaccurate, the users of data will be misled and can make a bad decision. First time introduced in 1970, DEA was quickly recognized as an excellent methodology for performance measurement. Being a "data oriented" approach for evaluating the performance of a set of peer entities called Decision Making Units (DMUs) which convert multiple inputs into multiple outputs, it is ideal for measuring the relative efficiencies of units with similar services or product and gives a big advantage of being able to deal with multidimensional nature of input/output variables.

Keywords Operations research · Data envelopment analysis · Performance measurement · Decision making · Business management

H. D. Dinh (✉) · K. M. Le
Vietnamese-German University, Binh Duong, Vietnam
e-mail: dung.dh@vgu.edu.vn

L. T. T. Huong and G. M. Pomeroy (eds.), *AUC 2019*, Advances in 21st Century Human Settlements, https://doi.org/10.1007/978-981-15-5608-1_12

1 Introduction

Improving public sector efficiency using scientific and analytical methods is one of the keys to deliver solutions for a smarter city. This leads to also an urgent need to apply quantitative approaches for several purposes, where in most cases operations research is inevitable. The argument for improvement of efficiency level in the public sector at a time of total budget reduction and increased pressure on services due to demanding customers is obvious, but the importance of efficiency goes beyond saving money. It is about the satisfaction of the whole ecosystem, and in order to boost economy-wide productivity, the public sector and public services that serve civilians as well as businesses must be at high efficiency. Furthermore, productivity is only one aspect of efficiency when we talk about delivering a smarter city which is capable not only of delivering more for less, but also for better public services. Boosting governmental management productivity for creating a smart city requires both an evidence-based understanding of what public services have been able to deliver so far, the drivers of these outcomes and the different options available.

Subsequent paragraphs, however, are indented. With this in mind, in this work, we introduce Data Envelopment Analysis (DEA), a performance measurement technique, in such a way as to be appropriate to decision-makers with little background in economics and operational research. Our main purpose is to show the adoption of a strong practical approach in order to allow decision-makers to conduct their own efficiency analysis and to easily interpret results as we may see in an explanatory numerical example.

This paper is structured as follows: Sect. 2 provides a brief literature review on data envelopment analysis and the mathematical formulation. Section 3 describes the main idea of implementing DEA for our work, the model, defines the variables used and the application of the proposed model including a numerical example. We also include the discussions of the results achieved from the model and example. Section 4 offers the closing remarks and future possible extensions of the model.

2 Data Envelopment Analysis

2.1 The Basic Idea

Data Envelopment Analysis (DEA) is a non-parametric mathematical programming method that calculates the efficiencies of Decision Making Units (DMUs), considering their resources (inputs) and products (outputs), see Thanassoulis et al. (2008). To keep it simple and to make this paper more understandable, the term "DMU" is used in a generic way. Standard DEA models calculate the multipliers for inputs and outputs of each DMU, so that its efficiency is maximized, following the model´s restrictions. For a correct and useful analysis, it is very important to define the homogeneity of the DMUs. Homogeneous DMUs can be defined as those that perform quite similar jobs under the same general conditions, such that the variables used are the same. The units evaluated need to be sufficiently similar so that the comparison makes sense, but should also be sufficiently different to perform the efficiency evaluation.

$$Efficiency = \frac{output}{input} \quad (1)$$

Thus mathematically:

$$e^j = \frac{\Sigma_{r=1} u_{rj} x_r}{\Sigma_{i=1} v_{ij} y_i} \quad (2)$$

Seiford and Thrall (1990) describe DEA in terms of floating a piece-wise linear surface to rest on top of the observations, i.e., envelop the data. In another work of Charnes et al., the key constructs of a DEA model are described more specifically as consisting of two components: the envelopment surface and the efficient projection path to the envelopment surface (Charnes et al. 1994). The shape of the projection path to the envelope surface shows us whether the model is either output-oriented or input-oriented. If the output is fixed, the model attempts to minimize the input factors and vice versa, with fixed input we would like to maximize the achievable output.

Each DMU's efficiency score is calculated relative to an efficiency frontier. DMUs located on the efficiency frontier have an efficiency score of 1 (or 100%). DMUs operating beneath the frontier have an efficiency lower than 1 and hence have the capacity to improve future performance. No DMU can be located above the efficiency frontier because they cannot have an efficiency score greater than 100%. Those DMUs located on the frontier serve as benchmarks to inefficient firms. The application of DEA models requires executing three steps:

1. Choose the units to be entered into the analysis.
2. Choose the appropriate input and output to be considered.
3. Identify the model orientation as input- or output-oriented and the returns to scale.

The ratio between inputs and outputs is called returns to scale. In the literature commonly mentioned are two returns possibilities in DEA models: Constant Returns to Scale (CRS) and Variable Returns to Scale (VRS). Establishing the relative efficiency of the units chosen is tremendously affected by the variables. Golany and Roll (1989) structured the procedures of choosing the input and output variables in three alternatives: careful judgment, quantitative analysis not based on DEA techniques, and selection based on DEA. For selecting variables from the DEA technique, Norman and Stoker (1991) proposed a systematic procedure for validating pre-selected variables, inspired by the stepwise method. The method suggests establishing an initial input-output pair, calculates the efficiency score of DMUs based on this pair and the correlation coefficients of all the other variables with this score.

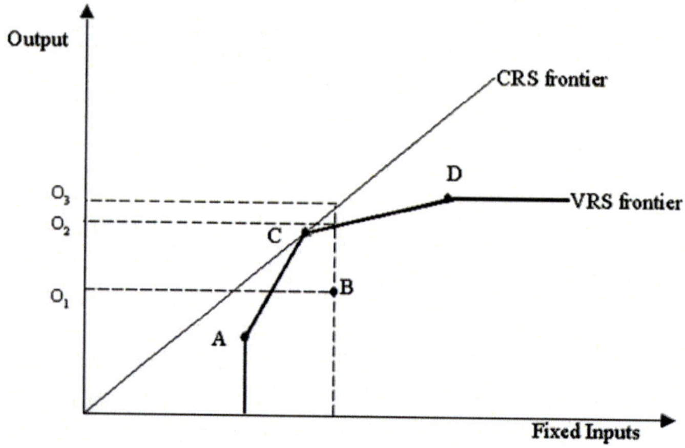

Fig. 1 CRS versus VRS model

To select the next variable to enter the model, the list of variables is covered in a descending order of the correlation coefficient module. This ensures that we can incorporate the variable that allows us to better fit the DMUs to the efficiency frontier. A problem of a DEA model is that it could lead to distorted results, such as efficient DMUs that attribute null multipliers for several inputs and/or outputs, as we may see in the next section of the numerical example (Fig. 1).

2.2 *Application of DEA*

How useful and important the application of DEA is, has been proven in numerous studies, especially for transportation sector. Berechman (1994) argues that transportation infrastructures can impact business productivity, prices of products, and also even affect the location of firms, thus there is the need to analyze efficiency in this sector. Njoh (2009) studied the countries of West and Central Africa and presented similar results, especially for the less developed countries. Another field that receives great attention is the airport management. For example, the analysis of airports and their efficiency has been carried out by a lot of researchers. Yang (2010) estimated the efficiency of 12 airports in the Asia Pacific region from 1998 to 2006. Curi et al. (2011) used the data envelopment analysis to estimate the efficiency of 18 Italian airports from 2000 to 2004. Or in the case of Brazil, the work of Fernandes and Pacheco (2002) examined the efficiency of 35 domestic airports. The DEA technique was straightforward, and identified which airports used their resources efficiently and which demonstrated a weak performance. They found that only two airports are efficient in the country. The other airports have an efficiency score of less than 50%, and so they could affirm that these inefficient airports are significantly underutilized. However, the least efficient airport in their research was even expanded after they obtained the data for this study. Similarly, the concept of DEA can be applied for various sectors in education, hospital management, or public services, what we consider next.

3 The Model and Its Numerical Example

3.1 The Model

In this section, we present the model applied for evaluating efficiency of public servants and an example implemented for explanatory purposes. By calculating an efficiency score, it indicates if a public servant is efficient or has capacity for improvement. By setting target values for input and output, it calculates how much input must be decreased or output increased in order to become efficient. Argumentation regarding public services tends to focus on waste, bureaucracy, and inefficiency. However, in the end, every decision might be seen as a financial decision because every decision has financial consequences. For example, given limited resources with infinite demand requires governmental organizations to continuously reduce or contain costs while maintaining service quality. One way to do this is by improving the performance so that each unit of input delivers more of the intended outputs. Thus, it is one of the ways we can create a smarter city with finite resources and infinite demand.

Without loss of generality, in the coming part, we consider the application of DEA for an organizational unit that consists of some staff as public servants. A public servant, sometimes called a civil servant, is a person employed in the public sector on behalf of a government department or agency. According to this definition, a policeman, a doctor in a public hospital, a teacher, or even a judge can be classified as public servants. This leads to the fact that a lot of models with the same platform and similar formulation of decision variables and objective function as in this paper can be applied for governmental services. A governmental service in most of the cases has the characteristics of a public good that is offered at a fixed price and which may be under-provided by the market. This is also the reason why we formulate our model in this section as an output-oriented DEA model. In most cases, public services are services such as healthcare and education that do not involve manufacturing of goods. To simplify any model, similar units can be understood as those that perform the same functions that operate under the same conditions. The DEA method does not consider the organization size or the structure and features of the workload in order to classify it as efficient, but rather the relationship between the use of resources (inputs) to obtain the product (outputs). Therefore, the efficiency addressed here refers to the relationship between the results obtained and the resources used. The managerial aspects of resource allocation decisions are more important than the features of the investigated units, here the public servants. Table 1 summarizes the descriptive efficiency statistics of 12 public servants. The analysis for some series of

Table 1 An analysis of efficiency of public servants

Public servants	Outputs: Nr. of Customers served	Outputs: Actual cases solved	Outputs: Satisfaction	Inputs: Labor Hrs	Inputs: Background	Weighted Output	Weighted Input	Efficiency		Constraints
1	62	54	80	4.74	6	0.6259	1.0000	0.6259	<=	1
2	55	43	68	6.38	7.42	0.5320	1.2673	0.4198	<=	1
3	77	60	55	5.04	6.35	0.4303	1.0597	0.4060	<=	1
4	98	83	87	3.61	6.34	0.6806	0.9741	0.6987	<=	1
5	86	78	94	3.45	4.43	0.7364	0.7364	1.0000	<=	1
6	79	79	71	5.25	6.31	0.5555	1.0673	0.5204	<=	1
7	93	86	90	5.36	3.23	0.7041	0.7041	1.0000	<=	1
8	95	80	100	7.09	8.69	0.7823	1.4616	0.5353	<=	1
9	81	73	83	6.49	7.28	0.6493	1.2569	0.5166	<=	1
10	74	68	77	7.36	9.07	0.6024	1.5232	0.3955	<=	1
11	88	82	71	5.46	6.69	0.5555	1.1253	0.4936	<=	1
12	90	86	74	6.58	8.75	0.5789	1.4387	0.4024	<=	1
Weights	0.0000	0.0000	0.0078	0.0590	0.1200					

PS		1
Weighted Output	Output	0.6259
Weighted Input	Input	1.0000

periods may show us a trend, i.e., an increase or decrease in the efficiency of each staff and average efficiency of the whole unit.

From the above Table, we can recognize and conclude that, for example, public servant Nr. 5 and public servant Nr. 7 have the same grade of efficiency which is also the maximal achievable efficiency. Our considered public servant Nr. 1 has only an efficiency of 0.6259 compared to the maximal achievable efficiency. In terms of efficiency and considering the variables used, also the other public servants among the group could improve the use of its resources and has the potential to significantly increase the number of passengers processed, while maintaining the input levels observed. Furthermore, if we take a closer look at both optimal cases of public servant Nr. 5 and public servant Nr. 7, we are able to conduct a detailed analysis of the input and output factors of why these lead to the same, highest level of efficiency. This will help to improve the general situation and also give hints for the other employees how to achieve a higher performance, see Table 2.

Table 1 also shows the two least efficient staff in the investigation: the public servant Nr. 10 with 0.3955 and Nr. 12 with only 0.4024. We see that using the data envelopment analysis the potential increase of customer numbers, solved cases, or satisfaction of customers (output) was actually no more so high, but the resources used (inputs) by these two staff are extremely more compared to the whole group.

Similarly, we see that staff Nr. 8, even though having very good output parameters, requires too much resources. According to the Extended Pareto-Koopmans Definition, full (100%) efficiency is attained by any DMU if and only if none of its inputs

Table 2 Input and output values of the most efficient public servants

Public servants	Nr. of Customers served	Actual cases solved	Satisfaction	Labor Hrs	Background
5	86	78	94	3.45	4.43
7	93	86	90	5.36	3.23

or outputs can be improved without worsening some of its other inputs or outputs. Thus, in the case of Nr. 8, the performance is quite half of those most efficient staff.

3.2 *Verification and Scenario Analysis*

We can also create a new table to identify the efficiency of the two main explanatory variables of the model and verify some of the reasons why some staff are considered more efficient than others. For example, if a staff's efficiency depended exclusively on the ratio between the number of customers and the total number of labor hours spent, the public servant Nr. 4 would be the most efficient, since each labor hour served up to 27 customers. Therefore, the least efficient would be the public servant Nr. 2 with only 8 customers/hour.

In the next step, we adjust input and output variables in an empirical case. Suppose that there is now a new public servant joining the group that replaces the older one. For illustration purposes in our example, let us suppose that this is public servant Nr. 12 with the input and output data as in the following (Table 3).

For verifying and explaining reasons, we assume the case that the new staff can ideally do the job, reaching 100% of customers served and solving also 100% the problems of customers leading to a satisfaction level. Solving the mathematical problem gives us a clear result. As for the case of public servant Nr. 12, the data envelopment analysis determined that this is the most efficient staff with regard to using his resources. However, it is also to mention that the relations of weights for input and output are fictional data and illustrate an extreme scenario.

It is clear that the data envelopment analysis considered the best combination of inputs in order to generate better results, respecting the different input scales. Therefore, for each group investigated, an ideal combination was found, which served as a reference for the less efficient staff. An ideal combination means that the resources

Table 3 Adjustment of input and output values for model verification

	-------- Outputs --------			----- Inputs -----		Weighted	Weighted			
Public servants	Nr. of Customers served	Actual cases solved	Satisfaction	Labor Hrs	Background	Output	Input	Efficiency		Constraints
1	62	54	80	4.74	6	0.5626	1.0000	0.5626	<=	1
2	55	43	68	6.38	7.42	0.4782	1.3460	0.3553	<=	1
3	77	60	55	5.04	6.35	0.3868	1.0633	0.3638	<=	1
4	98	83	87	3.61	6.34	0.6118	0.7616	0.8033	<=	1
5	86	78	94	3.45	4.43	0.6610	0.7278	0.9082	<=	1
6	79	79	71	5.25	6.31	0.4993	1.1076	0.4508	<=	1
7	93	86	90	5.36	3.23	0.6329	1.1308	0.5597	<=	1
8	95	80	100	7.09	8.69	0.7032	1.4958	0.4701	<=	1
9	81	73	83	6.49	7.28	0.5837	1.3692	0.4263	<=	1
10	74	68	77	7.36	9.07	0.5415	1.5527	0.3487	<=	1
11	88	82	71	5.46	6.69	0.4993	1.1519	0.4335	<=	1
12	**100**	**100**	**90**	**3**	**3**	0.6329	0.6329	1.0000	<=	1
Weights	0.0000	0.0000	0.0070	0.2110	0.0000					

PS		1
Weighted Output	Output	0.5626
Weighted Input	Input	1.0000

(inputs) were optimized, which means they were better used for the findings, which does not necessarily represent the absolute best use. The design of a model to be an input- or output-oriented model depends upon many factors, but mainly on the working process characterizing the organization (i.e., minimize the use of inputs to produce a given level of output or maximize the level of output given levels of the inputs). For the purpose of calculating efficiency of public servants in governmental agencies, in our example, the output-oriented DEA measures have been considered. This can be easily adjusted into an input model, see Thanassoulis et al. (2004).

4 Conclusion

With the fast-growing economy, big cities have become the leading marketplace for goods and services, including public services. Implementing a reasonable management approach has become a general and important decision for governmental agencies. However, it is always difficult to make a decision on the personnel and organizational matters that balance the performance and related cost. Moreover, the public sectors in certain countries tend to be monopolistic, i.e., there is sometimes no urgent need for improvement. Thus, applying quantitative methods is significantly important and useful because it provides proofs in form of numbers that can be used as argumentation in decision-making situations and change management toward smart cities.

In this paper, we applied Data Envelopment Analysis and showed the great potential of using it for the analysis of various possible governmental services. The DEA is an operational research technique based on mathematical programming, with the objective to comparatively analyze independent units in terms of their relative performance. We employed the concept of DEA to measure the performance of the public services. Results of the numerical example of 12 public servants verify the necessity of considering the personnel decision in improving overall performance. One of the great advantages of the method is the so-called non-parametric feature. It is classified as non-parametric because it does not use a predefined efficiency function, identical for all units in the input-output analysis relationship. Therefore, to use it does not require preparing a fixed weighted formula to measure the efficiency of the analyzed units, because the weights of each variable are determined by the technique.

Future research works can consider a lot of further extensions in this field. Governmental organizations have different characteristics according to every specific sector, and there is the need for detailed analysis of each application. As mentioned by Charnes et al. 1994, the DEA can be regarded as a body of concepts and methodologies embedded within collection models with different possible interpretations, we can combine other techniques of operations research to make useful interpretation of the results. Input and output factors can be varied according to the purpose of management, leading to a quantitative-qualitative potential. This means that the point of view and experience of the managers play a certain role in defining the inputs as well as outputs, leading to the different and interesting interpretation of the final results. We also suggest using the bootstrapping data envelopment analysis in other sectors, as well as applying other methods for optimizing the process of selecting variables.

References

Berechman J (1994) Urban and regional economic impacts of transportation investment: a critical assessment and proposed methodology. Transp Res Part A Policy and Pract 28(4):351–362

Charnes A, Cooper W, Lewin AY, Seiford LM (1994) Data envelopment analysis: theory, methodology, and application. Kluwer Academic Publishers, Boston

Curi C, Gitto S, Mancuso P (2011) New evidence on the efficiency of Italian airports: a bootstrapped DEA analysis. Socio Econ Planning Sci 45(2):84–93

Fernandes E, Pacheco RR (2002) Efficient use of airport capacity. Transp Res Part A Policy and Pract 36(3):225–238

Golany B, Roll Y (1989) An application procedure for DEA. Omega 17(3):237–250

Njoh A (2009) The development theory of transportation infrastructure examined in the context of central and West Africa. Rev of Black Polit Econ 36(3–4):227–243

Norman M, Stoker B (1991) Data envelopment Analysis: the assessment of performance. Wiley, Sussex

Thanassoulis E, Portela MCS, Allen R (2004) Incorporating value judgements in DEA. In: Cooper WW, Seiford LM, Zhu J (eds) Handbook on data envelopment analysis. Kluwer Academic Publishers, Boston, pp 99–138

Thanassoulis E, Portela MCS, Despic O (2008) Data envelopment analysis: the mathematical programming approach to efficiency analysis. In: Fried HO, Lovell CA, Schmidt SS (eds) The measurement of productive efficiency and productivity growth. Oxford University Press, New York, pp 251–420

Seiford Lawrence, Thrall Robert M (1990) Recent developments in DEA: the mathematical programming approach to frontier analysis. J Econ 46(1–2):7–38

Yang HH (2010) Measuring the efficiencies of Asia-Pacific international airports: parametric and non-parametric evidence

Smart and Green Mobility

A Transportation Optimization Model for Solving the Single Delivery Truck Routing Problem with the Alldifferent Constraint in MS Excel

Hai Dung Dinh

Abstract Smart city consists of smart transportation which helps to minimize the use of resources. In this paper, we focus on how to design a mathematical framework with its objective function and constraints in Solver in MS Excel to solve the Single Delivery Truck Routing Problem. The routing optimization problem is particularly useful and finds applicability in various industrial areas, including solid waste management, warehouse order picking, manufacturing, and logistical planning. Let us consider a distribution center, i.e., a factory, at one location that has to send a truck daily with refills to retail centers located in a set of N-1 neighboring locations. The distances or travel times between pairs of neighboring locations are given as an N × N matrix. The distribution center wishes to route the truck from the origin to all other retail centers and back to the origin such that the distance or time traveled is minimized. For example, a so-called greedy algorithm will always find the local move for the next best connection possible route and provides a solution of X_Greedy that is optimal at that move. Can we find a route that is better, i.e., shorter than X_Greedy? Different types of mathematical algorithms, among which heuristics such as genetic algorithm or particle swarm optimization can be used to address the problem. This paper offers another approach in solving the proposed problem with the MS Excel Solver. The numerical example given in this research work is intuitive and can be easily modified to deal with further complex planning problems.

Keywords Operations research · Mathematical optimization · Transportation planning · Network flow optimization · Heuristics

1 Introduction

Transportation optimization problems have been the subject of numerous researches, namely in applying algorithms and techniques for finding the shortest route or the best combination of trips. This paper describes the use of Excel Solver and the

H. D. Dinh (✉)
Vietnamese-German University, Binh Duong, Vietnam
e-mail: dung.dh@vgu.edu.vn

L. T. T. Huong and G. M. Pomeroy (eds.), *AUC 2019*, Advances in 21st Century Human Settlements, https://doi.org/10.1007/978-981-15-5608-1_13

Affdifferent constraint to solve the task. Excel Solver can perform the optimization required, whether nonlinear or linear programming formulations are used. For more complicated problem instances and models, the macro language for Excel, Visual Basic for Application can control the Solver, and a macro can be manually programmed to automate the process and functions. It is argued that making use of the developing power of spreadsheets to implement OR techniques such as optimization can help make OR more accessible to practitioners. At the same time, the capabilities of spreadsheet packages such as Microsoft Excel have been growing rapidly. By making use of features such as the Solver and the Data Analysis tools, many of the statistical analysis, optimization and simulation tasks involved in OR education and practice can now be tackled using a spreadsheet.

Schrijver (1986) described the linear program that can be solved to optimality. In these cases, researchers apply the branch-and-bound methodology. Another research direction focuses on the use of heuristic algorithms. Algorithms can become complicated and require modification for each problem setting, see Nemhauser and Wolsey (1988).

2 The Single Delivery Truck Routing Problem and the Alldifferent Constraint

2.1 The Single Delivery Truck Routing Problem

The Single Delivery Truck Routing Problem is the basic form of the Travelling Salesman Problem (TSP). Because it is NP-hard, for this combinatorial optimization problem, many exact and heuristic algorithms have been developed. Per definition, we need to find the combination of movements that results in a shortest possible route. If the original node must be visited at the end, we have a round trip. Thus, in our later model, we consider a truck that has to travel to all locations as efficiently as possible, each location once. On the basis of the structure of the cost matrix, or distance matrix, a TSP can either be symmetric or asymmetric. In case, we have a j-location symmetric TSP, there are $(j-1)!/2$ possible solutions. Generally, applying a total search is not possible even for an averagely large variable n.

For solving the problem, it is necessary to understand the basic of Constraint Programming (CP). A CP program has the goal to achieve a solution to a given Constraint Satisfaction Program (CSP), mostly by splitting a CSP into sub-programs. This applies also to transportation problem, sometimes by using the Branch-and-Bound principle. The splitting process then results in the construction of a decision tree. Each node of the tree represents a smaller CSP.

For a detailed research on CP, one can see the work of other authors (Apt 2003) and (Dechter 2003). In constraint programming, again, a mathematical model is built and offers solutions that range over their domain, called the search space that consists of many values, and constraints on this space. Each constraint restricts the space in a different way. For comprehensive purpose, let us consider the following example. Given x and y having values in $\{1, 2, 3\}$ and the constraint that the sum of x and y is smaller than or equal to 4 will result into pairs of (x, y) cannot be (3, 2), (2, 3), (3, 3).

2.2 *The Alldifferent Constraint*

In a mathematical optimization problem, there can be several constraint types. One of the famous constraints which have been introduced in a lot of practical problems is the Alldifferent constraint. Simply understood, the variables in this constraint must hold a value that is not taken by another variable.

The Alldifferent constraint has such an important meaning that the various authors have shown the benefits of constraint programming in research works (Gomes and Shmoys 2002). What can be easily observed is that its application has a very wide range, from logistical planning such as air traffic management (Barnier and Brisset 2002) to production planning and scheduling like rostering problems (Tsang et al. 2004). Also, the Alldifferent constraint can be modified to deal with global optimization, for example, the sort constraint (Older et al. 1995), the cycle constraint (Beldiceanu and Contejean 1994), or for example see Beldiceanu et al. (2005) for an overview.

Last but not least, we mention the symmetric Alldifferent constraint that is relevant for this paper. Here there is an assumption that all variables must take different values pairwise, meaning that if the variable representing element i is assigned to the value representing element j, then it is at first a feasible solution for this pair of value. Thus, immediately the variable that represents the element j must then be assigned to the value representing element i. See analysis of different problem settings by Henz et al. (2004) or Lee (2002). If a problem is complicated enough that the decision-maker has to think of ways of relaxing the constraints on the search space, then a method called multiple Alldifferent constraints find its application, see, for example, the work of Appa et al. (2004). We restrict on further insights into the theory of this interesting constraint at this point and focus in the next section on how to apply it as a general approach for our proposed model.

3 The Model and Its Numerical Example

3.1 *The Model and Objective Function*

Consider a manufacturing location in city 1, the capital city of Hanoi, that has to send a truck with deliverables to warehouse location in other cities n = 2, …, 8. The cities in our example are given in Table 1. The distances between pairs of neighboring cities are known and given as a distance matrix in Table 2. In some problem instances,

Table 1 Cities for the round trip

L	M
City no.	Name
1	Hanoi
2	Hai Phong
3	Hai Duong
4	Nam Dinh
5	Ninh Binh
6	Hue
7	VungTau
8	HCMC

Table 2 Distance matrix

	A	B	C	D	E	F	G	H	I	J
1	D_ij Matrix	Trip	1	2	3	4	5	6	7	8
2		1		121	76	84	94	669	1682	1724
3		2	121		68	101	155	730	1735	1785
4		3	76	68		84	108	684	1689	1739
5		4	84	101	84		32	605	1611	1659
6		5	94	155	10S	32		573	1578	1628
7		6	669	730	684	605	573		1001	1049
8		7	1682	1735	1689	1611	1578	1001		96
9		8	1724	1785	1739	1659	1628	1049	96	

some towns may not have a direct road between them, leading to a missing value of distance. In our model, for simplicity, we assume all cities are connected. The road map connecting the eight cities is illustrated as a network flow model.

As the managing director, we need to route the truck from Hanoi to the other seven cities and back to Hanoi. Our requirement is that the truck passes through every single city only once and returns to the departure such that the distance covered is the minimum. The distances in km are real data taken for each pair of cities, as shown in Fig. 1.

We can have very different definitions of optimal results. For some cases, it can be the least travel time, avoiding traffic jams and bad roads. In other cases, it can be the least cost associated with the journey, meaning that fuel cost and road fee is minimized. As a standard objective function, in our example, we define the goal to be the shortest route, i.e., the minimum total km traveled.

Notice that we attempt to choose cities that make up very clear geographical regions. Five of the cities are near to each other in Northern region, and the other three quite far away. But whether or not we realize the groups of cities next to each

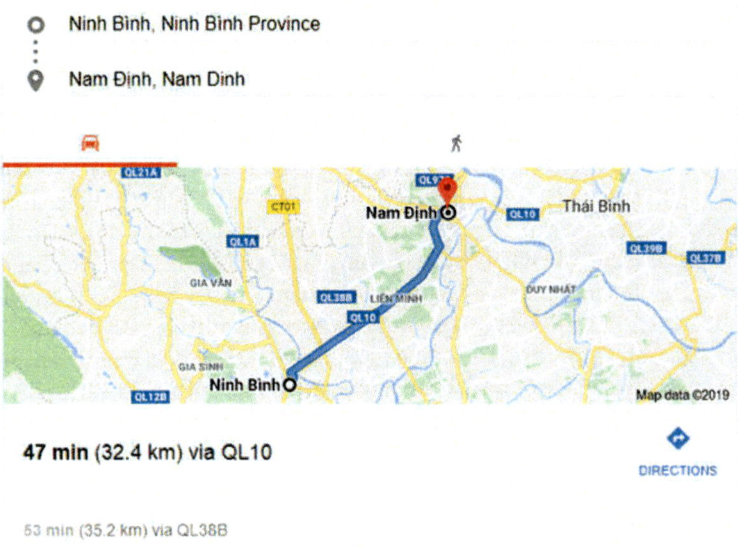

Fig. 1 Distance between cities taken from google maps

other as sub-group, the optimization process can automatically identify the most efficient combination of trips through the cities.

For example, a possible route is given in Table 3 as 1–2–3–4–5–6–7–8. The total distance is then 3699 km. We raise the question if a shorter route is possible? The Solver solution is provided in the next section.

Table 3 A naïve solution

P	Q
City	Distance
1	1724
2	121
3	68
4	84
5	32
6	573
7	1001
8	96
Total	3699

3.2 Constraint and Solution

First, we define a consequence of numbers (1–8) to later interpret the consequence of cities for the travel. The step-by-step setup of the problem in an Excel spreadsheet is given as follows. A name range "distancematrix" is defined for the range from B2 to J9 and will be used in the VLOOKUP function later. In cell Q2, we input the formula = VLOOKUP(P2, distancematrix, P9 + 1, FALSE) because the problem is supposed to be a round trip, i.e., the truck has to travel back to the origin. In the next cell Q3, we have = VLOOKUP(P3, distancematrix, P2 + 1, FALSE). This logic also applies for the further cells. For the total distance will use the SUM function for values in Q2–Q9. The Solver of Excel (Data/Analysis/Solver) is called up, and we set min Q10 as the objective. Changing variables are defined as P2:P9, which is the consequence of trips. The only constraint we need to input in the Solver is now P2:P9 Alldifferent. As for the search engine, we choose evolutionary method (Fig. 2).

The optimal solution of 3667 has been found in less a second of time. However, the evolutionary algorithm of Solver still continues to search for improvement in about 9 s. Table 4 shows the optimal solution. We now have an order of journey as 1–3–2–4–6–8–7–5. As for interpretation, the truck starts from Ha Noi, drives through Hai Duong, Hai Phong, Nam Dinh, then to Hue, arrives later in the South in HCM

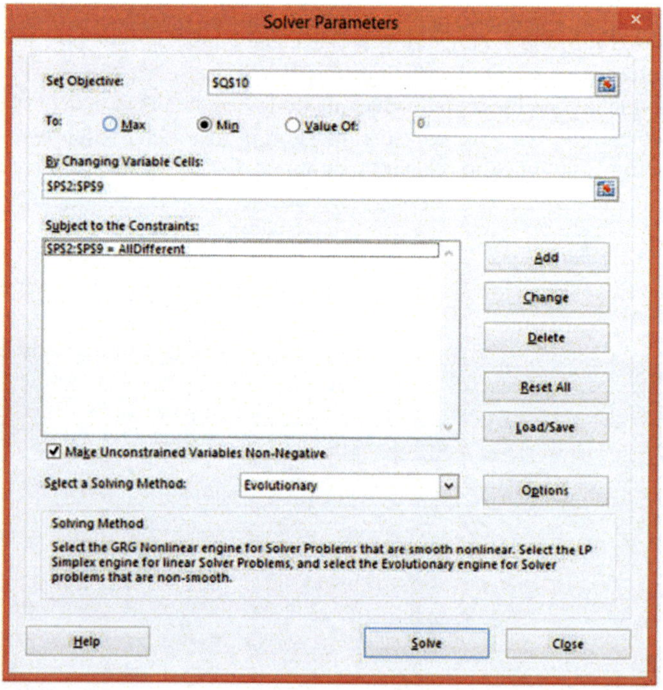

Fig. 2 The solver parameters

Table 4 Optimal solution

P	Q
City	Distance
1	94
3	76
2	68
4	101
6	605
8	1049
7	96
5	1578
Total	3667

City, then Vung Tau, and returns to the North in Ninh Binh and finally to Ha Noi. This confirms again our intuitive solution that the groups of cities next to one another must be visited in a systematic way.

With its search engine and the built-in algorithm, the Solver tool of MS Excel solved the problem in a few seconds. The heuristics applied here is an evolutionary algorithm (Dinh 2020).

4 Conclusion

We showed the advantage of using a simple set of constraints to deal with the transportation problem, the Single Delivery Truck Routing. Applying the technique to tackle with more sophisticated problem is truly recommended. It is not necessarily an advanced way to deal with such kind of optimization tasks, but it offers a convenient approach to build and extend more demanding mathematical models in a comprehensive way. One of the possible further extension of the model is the Multiple Delivery Trucks Routing. In that problem, additional to the distances, the demand of each city is given. Multiple trucks travel from the origin city to other cities in the network, each has a predefined capacity. The objective is then to find the lowest possible number of trucks needed for overall distribution and simultaneously the shortest trip covered by the fleet of trucks. In general, this also leads to the reduction of both cost and the negative impact on the traffic network and environment.

References

Appa G, Magos D, Mourtos I (2004) LP relaxations of multiple all different predicates. In: Regin JC, Rueher M (eds) Proceedings of the 1st international conference on the integration of AI and OR techniques in constraint programming for combinatorial optimization problems (CPAIOR 2004), of LNCS, vol 3011, pp 364–369. Springer

Apt K (2003) Principles of constraint programming. Cambridge University Press

Barnier N, Brisset P (2002) Graph coloring for air traffic flow management. In: Proceedings of the 4th international workshop on integration of AI and or techniques in constraint programming for combinatorial optimization problems (CP-AIOR 2002), pp 133–147

Beldiceanu N, Carlsson M, Rampon J-X (2005) Global constraint catalog. Technical report T2005–8, SICS

Beldiceanu N, Contejean E (1994) Introducing global constraints in CHIP. Math Comput Model 20(12):97–123

Dechter R (2003) Constraint Processing. Morgan Kaufmann

Dinh HD (2020) Chapter 157 modeling and applying heuristics for optimization of solid waste collection under consideration of vehicle capacity. Springer Science and Business Media LLC, 2020

Gomes C, Shmoys D (2002) Completing Quasigroups or Latin squares: a structured graph coloring problem. In: Proceedings of the computational symposium on graph coloring and its generalizations

Henz M, Muller T, Thiel S (2004) Global constraints for round robin tournament scheduling. Eur J Oper Res 153(1):92–101

Lee J (2002) All-different polytopes. J Comb Optim 6(3):335–352

Nemhauser G, Wolsey L (1988) Integer and combinatorial optimization. Wiley

Older W, Swinkels G, Van Emden M (1995) Getting to the real problem: experience with BNR prolog in or. In: Proceedings of the 3rd international conference on the practical applications of prolog (PAP'95). Alinmead Software Ltd

Schrijver A (1986) Theory of linear and integer programming. Wiley

Tsang E, Ford J, Mills P, Bradwell R, Williams R, Scott P (2004) ZDC-rostering: a personnel scheduling system based on constraint programming. Technical report 406, University of Essex, Colchester, UK

An Elicitation Study on Potential Users' Salient Beliefs of Using Future City Light Rail Transit

Simon Dertha Tarigan, Reynaldo Siahaan, and Oloheta Gulo

Abstract The Theory of Planned Behavior (TPB) has been widely used to explore behavioral intention in many disciplines including future travel mode choice. LRT construction in Medan city is due to commence in 2020. We are interested in exploring its potential users' salient beliefs that will drive psychological factors, which eventually generate future intention to use LRT for daily trip. An elicitation study is recommended to first identify the salient beliefs that potential users hold toward the future LRT. Using semi-structured interviews consisting of 9 open-ended questions, we interviewed and obtained data from 150 respondents. In general, the result shows that the potential users dominantly hold positive belief, for example, the ability to improve travel time efficiency and help with congestion. Whereas difficult access to station and possibly stuck in crowding are believed to be negative drive.

Keywords Salient beliefs · Theory of planned behavior · Future LRT system · Commuting trip

1 Introduction

Motorized transportation in Southeast Asian countries, especially by private vehicle, has exploded since the 1990s and continues to expand today. Better income partly contributes to the increase (Senbil et al. 2007). Urban density was also found to affect the phenomena (Acharya and MoRichi 2007), whereas negative experiences and a low level of satisfaction of available public transportation may also play a role (Joewono et al. 2015). The augmentation of private vehicle ownership can be seen in major cities of Indonesia. Located in North Sumatera province, Medan is currently the fourth most populated city of Indonesia (2.5 million). According to report from IPSOS Consulting in 2015, the ratio of motorcycle and car per 1000 persons was 401 and 91, respectively. Given the positive growth trends of annual GDP (4–5%) and population growth of Medan, these numbers are experiencing an

S. D. Tarigan · R. Siahaan (✉) · O. Gulo
Department of Civil Engineering, Saint Thomas Catholic University, Medan 20132, North Sumatera, Indonesia
e-mail: siahaan.reynaldo@gmail.com

L. T. T. Huong and G. M. Pomeroy (eds.), *AUC 2019*, Advances in 21st Century Human Settlements, https://doi.org/10.1007/978-981-15-5608-1_14

increase as well. In the last 8 years, the number of private vehicles (2.7 million in 2011) in Medan has multiplied more than 1.5 times, among which more than 85% is motorcycle (according to central bureau of statistics of Indonesia). The statistic exhibits the reason why more than 92% of urban trip in Medan are made by private vehicle. Consequently, traffic congestion becomes a major problem, which is also worsened by the lack of law enforcement for traffic violation (Tarigan et al. 2017). Local government of Medan forecasted that people would suffer from massive traffic gridlock in the year 2024.

As a developing city in Southeast Asia, Medan entirely lacks adequate public transport facilities and relies mostly on paratransit (Phun and Yai 2016). Paratransit modes in Medan are microbus (*angkot*), motorized rickshaw, and non-motorized rickshaw. Low-to-mid income household members commonly use these types of paratransit mode, whereas higher and much higher income population groups tend to ride motorcycle and car, respectively. In 2014, new Bus Rapid Transit (BRT) system called Trans Mebidang was introduced, which consists of two lines. Each line connects adjacent regency to central part of Medan. Average travel time for each line is ranging from 80 to 90 min with only 23-km (line 1) and 32-km (line 2) travel distances. Considering Medan annual budget size, BRT seemed to fit the mobility solution for Medan in the short term (Zhang 2009). However, lack of hygiene, inappropriate bus stops, and inconvenient fleet conditions led to dissatisfied user. Hence, the BRT's occupancy level is quite low at the present (Panondang 2018).

To fight the congestion, Medan authority planned to construct Light Rail Transit (LRT) system as part of their long-term urban solution along with BRT service improvement and paratransit integration (Medan Bureau of Transportation 2010). The LRT project will operate along 22.7-km elevated rail track, which consists of 17 stations including terminal stations (Bappenas 2016). The early feasibility study is finished, and the government is currently working on the financial structure to support the project. Total travel time for a one-way trip is expected to be approximately 44 min. LRT system possesses the potential to increase public transportation ridership, promote economic growth, and reduce private vehicle ownership (Zhang 2009; Currie and Delbosc 2013; Scherer 2010; Gomez-Ibanez 1985). However, it is also clear that constructing a rail-based public transportation system will require substantial capital cost, and thus merit the arising doubts and debates for city like Medan (Zhang 2009; Pojani and Stead 2015; Gomez-Ibanez 1985). It is important that the government take great care in choosing to implement such rail-based system. In the case of Medan, the question about how people perceive, and more importantly their travel behavior that will respond to this new system should be investigated early in order to prepare the appropriate supporting policies and regulations (Joewono et al. 2007).

In the present study, we used theoretical framework suggested by Theory of Planned Behavior (TPB) to explore underpinning beliefs that affect the travelers' intentions to use the future LRT system (Ajzen 1991). In TPB, elicitation study is known as the first of the two phases and highly recommended to be conducted prior to going into more quantitative analysis and extension of the theory. The main purpose of elicitation study is to investigate through interviews the most common underlying beliefs of the targeted individuals regarding the behavior in question. We are interested in eliciting the salient beliefs of potential users/travelers in Medan in using such LRT system, particularly for daily commuting trip.

2 Theoretical Background

Theory of planned behavior (Ajzen 1991) is an extension of Azjen and Fishbein's 1980 Theory of Reasoned Action (TRA) which is widely known for its application in predicting various behavior based on a set of psychological variables. According to the theory, a particular behavior is generated by the individual's intention, which is driven by psychological constructs such as Attitude (ATT), Subjective Norm (SN), and Perceived Behavioral Control (PBC). Attitude toward behavior refers to individual's evaluation whether certain behavior is good, or bad for them. In the context of this study, attitude is the inclination toward LRT as travel mode. Subjective norm refers to individual's perception regarding social view of their influential people toward certain behavior. Perceived behavioral control refers to individual's perception regarding things that would ease, or hinder the behavior based on related past experiences. The general structure of the theory can be seen in the following picture (Fig. 1).

In addition, TPB stated that a behavior is a function of salient information, or relevant belief to that behavior (Ajzen 1991). The "salient" term represents the notion that an individual will only keep limited belief in their mind related to a particular behavior. TPB suggested that there are three salient beliefs supporting the aforementioned constructs, namely: behavioral beliefs, normative beliefs, and control beliefs. These beliefs are the direct antecedents of the following psychological constructs, respectively: the attitude of an individual toward a particular behavior, the subjective norm, and the perceived behavioral control. Using this framework, it is possible to predict a particular behavior by examining the underlying beliefs of individuals, or intervene such behavior by changing these beliefs.

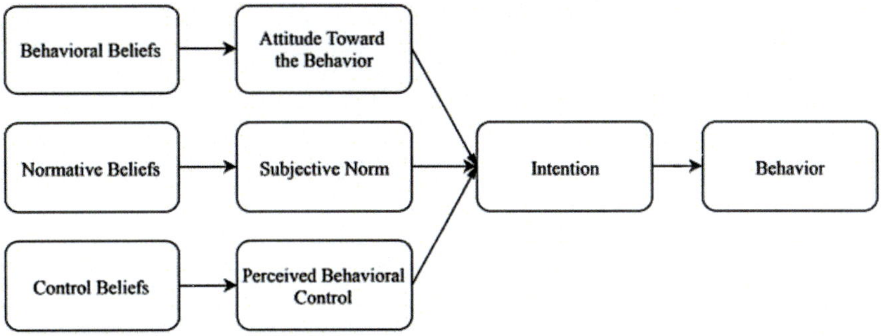

Fig. 1 Theoretical framework of the theory of planned behavior

Since its first emergence, TPB has been widely used in health and medicine field of research (Downs and Hausenblas 2005) to predict and prepare behavioral intervention policy that promotes healthy life. The theory has also become popular in transportation-related research, particularly in travel mode choice (Bamberg et al. 2003; Lanzini and Khan 2017). Several studies have applied TPB to predict psychological determinants that affect mode choice intention especially for existing facilities, but only some have utilized it for a hypothetical facility upgrade, or new facility introduction. In relation to our research, we summarized some of the past studies that specifically applied TPB on the latter case.

There are two types of measurement using TPB, namely indirect measurement (elicitation study) and direct measurement (Francis et al. 2004). Direct measurement method is normally conducted at a later stage, in which fixed-item scaled questions are presented to targeted individuals. Meanwhile, the process of eliciting salient beliefs through indirect method is commonly known as the first phase of TPB analysis. The result from the first phase is essential to design the fixed-item measurement instrument and do quantitative analysis. Even though elicitation study is an important part of TPB (Curtis et al. 2010), majority of past studies primarily used the direct measurement method and skipped the elicitation phase. Table 1 presents some of reviewed past studies, which utilized TPB to predict individuals' intention to perform a future behavior.

Table 1 Summary of past studies using TPB on hypothetical behavior

No.	Author (year)	Targeted behavior	Targeted population	Elicitation study?
1.	Muromachi (2017)	Intention of future car purchase by university students	University students	No
2.	Rowe et al. (2016)	Intention to take risks when becoming motorist	Pre-drivers	Yes
3.	Nordlund and Westin (2013)	Intention to travel by new railway line in the future	Residents nearby the line	No
4.	Sigurdardottir et al. (2013)	Intention to commute by car as adults	Adolescents	No
5.	Choocharukul and Fujii (2007)	Future private car use	Final year university students	No

3 Method

3.1 Defining Behavior of Interest

Following (Ajzen 2006), the behavior of interest in this study had to be made clear to ensure appropriate measurement and reduce ambiguity while interviewing. The most important elements that need to be defined are action, context, and time elements. For this study, we are much more interested in exploring what people think of making LRT their daily mode of choice. Therefore, we determined as follows: "*Using LRT as my primary mode for daily commuting trip*".

3.2 Population and Sample Selection

In general, collecting 80 valid interviews would be appropriate for such studies (Francis et al. 2004). However, considering the length of the LRT trace and the size of land-use that would be affected by it, we opted to follow 25 people interviews recommendations (Godin and Kok 1996). A total of seven future stations location were chosen as interview locations (Fig. 2). At each of these locations, we conducted interview to 25 people using purposive sampling.

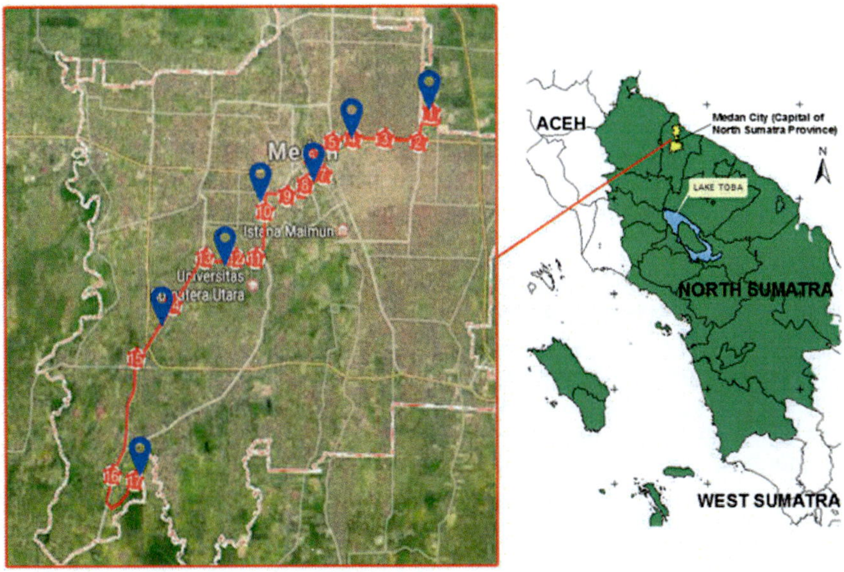

Fig. 2 LRT route map and interview location

In relation to the targeted behavior defined earlier, our population of interest should be those individuals who have daily activity around the determined location. We narrowed down our sample by several required criteria, such as should be older than 16 years old; should live, or work, or go to school in the vicinity; is familiar with the physical and operational characteristic of an LRT system. The former is employed to limit our respondent to those who have independency of making daily trip, whereas the latter ensures that the respondent can relate to our topic of discussion. A total of 170 interviews was made and reported, of which 20 were discarded for validity and consistency consideration.

3.3 Questionnaire Design and Data Collection

Elicitation study in TPB suggested interviews using a semi-structured questionnaire (Ajzen 2006). The questionnaire was written in Bahasa Indonesian and was divided into two sections. The first section was designated for general characteristics of the respondents, such as gender, age, household size, occupation, income, private vehicle ownership, primary mode of travel, and contact.

The second section comprises nine open-ended question related to beliefs. In order to elicit behavioral beliefs, we posed question: (1) *what do you think the advantages of using LRT as your primary mode for daily commuting trip?* (2) *what do you think the disadvantages of using LRT as ...?* (3) *Tell us some things that will attract you to use LRT as ....* (4) *Tell us some things that will dissuade you to use LRT as...!* Next, to elicit normative beliefs we asked question: (5) *Please list the individuals who would think you should use LRT as ....* (6) *Please list the individuals who would think you should NOT use LRT as ....* Then, to elicit control beliefs we posed question (7) *Please list any factors, circumstances or things that would make it easy for you to use LRT as....* (8) *Please list any factors, circumstances, or things that would make it difficult for you to use LRT as ....* Lastly, we asked additional question regarding moral awareness issue: (9) *Tell us the positive impacts LRT would bring to the society if people use LRT as ... .*

Since the questionnaire was open-ended, duration of the interview was very sensitive. Interview process lasted approximately 20–25 min including brief introduction of the study and LRT project plan. Most potential respondents had already rejected right after looking at the questionnaire form. In order to ensure valid and complete responses, surveyors needed to explain and make sure that the respondent was already aware of our taking their time prior to giving their consent. Providing compensation could be an optional strategy for such interview in the future.

Obtained data was sorted by classifying all the beliefs that have similar meanings into the same category. Two researchers conducted classification in order to alleviate bias in understanding. We counted the frequency of each category being mentioned in the questionnaire. Salient beliefs were identified as the most commonly reported by majority of the participants.

4 Results and Discussion

4.1 General Characteristics

A total of 150 people participated in the study. Table 2 shows general characteristics of participating respondents. Majority of the respondents were female (69.3%) and were 26 years old on average. More than half (54%) of the participants were students, while those whose occupation such as white collar, entrepreneur, small/medium merchant, and others are evenly distributed around 10%. The participants are mostly from relatively medium-sized household (47.3%) and small-sized household (34%). Minimum monthly income of Medan citizen is around 2.1 million Rupiah ($154), and most of the participants (46.7%) have lower income level. There are only a few respondents (23%) who possess private car, whereas there are 64% who own one or more motorcycle. Lastly, primary modes of transportation of the participants are motorcycle (37%) and public transport such as paratransit "angkot" (34%). There is 20% ride-hailing frequent user among the respondents.

Table 2 Description of participant characteristic (N = 150)

Characteristics	N	%	Characteristics	N	%
Gender			*Monthly income*		
Male	46	30.7	Less than 1 million Rupiah	70	46.7
Female	104	69.3	1–2.5	42	28.0
Age (yrs. old)			2.5–5	21	14.0
16–20	47	31.3	5–10	9	6.0
21–25	51	34.0	10–15	6	4.0
26–35	30	20.0	Greater than 15	2	1.3
36–45	15	10.0	*Car ownership*		
46–55	6	4.0	None	114	76.0
older than 55	1	0.7	1	26	17.3
Occupation			More than 1	10	6.7
Student	82	54.7	*Motorcycle ownership*		
White Collar	16	10.7	None	53	35.3
Entrepreneur	19	12.7	1	57	38.0
Small/medium merchant	18	12.0	2	29	19.3
Other	15	10.0	More than 2	11	7.3
Household size			*Primary mode of transportation*		
Less than 4 persons	51	34.0	Motorcycle	56	37.3
4-6	71	47.3	Car	12	8.0
7-8	22	14.7	Public transport	51	34.0
More than 8 persons	6	4.0	Ride-hailing service	31	20.7

Table 3 shows summary of number of beliefs mentioned by the participants for each belief category. On average, beliefs under the question of disadvantages, repellence aspect, disapproval, and barrier are very low in number. Participants had trouble providing more than one belief regarding negative perception toward LRT. This is most likely because LRT in Medan is still nonexistent, however, people already highly regard the future train system as positive improvement. As we can also see, there are only a few numbers of participants who failed to refer any belief on the questions about positive perception. In addition, there are even 10% (15 people) who stated 5 or more beliefs for advantages, attracting aspects, and positive impacts. An extreme example is given in questions about "Disapproval", where there are only 101 beliefs in total elicited from all the respondents since it is reported that they could not think of anyone who would discourage or be unwilling to use the LRT in the future.

4.2 Behavioral Beliefs

We posed four questions to elicit behavioral beliefs of the respondents. Table 4 presents positive belief regarding advantages and attractiveness. There are five most referred beliefs, which are fairly similar for both questions. The possibility of LRT "*to improve travel time efficiency*" is believed to be popular among our participants. "*Separated from road traffic*", "*comfortable facilities*", and "*affordable fare/travel cost*" are also often mentioned in both categories. LRT's modern image is also believed to be one of the attracting aspects. These are the affirming salient beliefs that will drive future attitude toward using LRT system for daily use in the future.

It is interesting that people believe LRT would improve travel time when they choose to use it for daily commuting. Our participants are mostly people who have daily activities around the proposed stations. It is likely that their current travel time is quite unsatisfactory under traffic congestion, and being in elevated LRT track separated from the road traffic would most probably be a better condition. Travel time can be classified into in-vehicle and out-of-vehicle travel time, of which the former is regarded as better measure to attract more passengers (Dell'Olio et al. 2012).

Table 5 shows negative beliefs underlying instrumental attitudes toward LRT as daily commuting mode. The answers were widely varied, therefore additional time was required to assess each belief and their similarity. Therefore, it was found there are seven beliefs, which are dominantly stated. The most common disadvantages reported by participants were the inability to choose precise location of getting on and getting off the train. Meanwhile, the most mentioned aspect that repels the idea of riding LRT is the possibility of crowding, which is also consequently referred as disadvantage. This belief is most likely derived by the grueling images of jammed crowd exhibited by several train operations in Java Island. It is intriguing to see that there are also many participants who believed that LRT would charge expensive fare. As well as, "*being in queue to wait for their turn*" is regarded as disadvantage. We also

Table 3 Summary of beliefs reference

No.	Beliefs category (*avg. reference per participant*)	% Participants (N = 150) according to number of reference							Total number of reference
		0	1	2	3	4	5	6	
1.	Advantages (2.6)	0.67	2.00	52.67	30.67	8.00	5.33	0.67	393
2.	Disadvantages (1.3)	16.67	41.33	36.00	4.00	2.00	–	–	200
3.	Attraction (2.1)	2.00	20.67	51.33	21.33	2.67	2.00	–	312
4.	Repellence (1.5)	14.67	33.33	44.00	6.67	1.33	–	–	220
5.	Approval (2.4)	1.33	6.67	51.33	32.00	8.67	–	–	360
6.	Disapproval (0.7)	57.33	22.00	17.33	2.67	0.67	–	–	101
7.	Facilitator (1.6)	5.33	43.33	42.00	8.00	0.67	0.67	–	236
8.	Barrier (1.5)	12.67	38.00	38.00	10.67	0.67	0.00	–	223
9.	Positive impact (2.3)	8.00	8.67	39.33	34.00	8.00	1.33	0.67	348

Table 4 Positive behavioral beliefs

No.	A. Advantages	N	%	No.	C. Attractiveness	N	%
1.	Improve travel time efficiency	115	28.5	1.	Secure, clean, and comfortable facilities	72	23.1
2.	Could avoid road traffic congestion	86	21.3	2.	Could improve travel time efficiency	60	19.2
3.	More comfortable experience in travel	47	11.7	3.	Radiates modern image of transportation mode	42	13.5
4.	Affordable fare rate	35	8.9	4.	Traffic is separated from existing road traffic	39	12.5
5.	Easier to use than private vehicle or angkot	33	8.2	5.	Travel cost could be cheaper	21	6.7
	Cumulative		78.6				75.0

Table 5 Negative behavioral beliefs

No.	B. Disadvantages	N	%	No.	D. Repel interest	N	%
1.	Getting on/off the train is strictly limited at station	46	23.0	1.	Overcrowding	45	20.5
2.	Expensive fare	27	13.5	2.	Route and station location might not fulfill my need	37	16.8
3.	Crowded train	22	11.0	3.	Unable to get on/off the train is an inconvenience	30	13.6
4.	More time to queue for entrance, ticket, and getting on	18	9.0	4.	Operation flexibility is low (fixed time schedule and allowable luggage)	26	11.8
5.	Train schedule is very strict	14	7.0	5.	Expensive ticket fare	14	6.4
6.	Noise from the train	13	6.5	6.	Vulnerability to criminal act (pickpocket, sexual harassment)	9	4.1
7.	Distance of station is relatively far from desired place	10	5.0	7.	Content with my own private vehicle	7	3.2
	Cumulative		75.0				76.4

Table 6 Normative belief

No.	E. Approval	N	%	No.	F. Disapproval	N	%
1.	Friends	103	28.6	1.	Friends	28	27.7
2.	Parents	97	26.9	2.	Spouse	26	25.7
3.	Siblings/Relatives	63	17.5	3.	Parents	15	14.9
4.	Spouse	56	15.6	4.	Superior/Senior	9	8.9
	Cumulative		88.6				77.2

found that even though the service is not yet available, majority of the participants already think of being left or forced to arrive early just to catch the train. This could be an important point since higher service level is considered a great recommendation to improve ridership (Currie and Delbosc 2013).

4.3 Normative Beliefs

Since eliciting descriptive normative referents is not possible for our case of study, we solely focus on the injunctive referents of the participants. Descriptive normative referents are those who actually perform the behavior, whereas injunctive referents are those who either approve or disapprove using LRT for daily commuting trip. Table 6 presents the most common referents mentioned by participants.

Large numbers of participants believe their friends would support their using LRT for daily commuting trip. However, "*Friends*" was also mentioned by considerable number of participants as disapproving individuals, followed by spouse with a slight difference in percentage. "*Parents*" could also be regarded as active social referents for the participants since it was mentioned by a lot of respondents in both categories. Meanwhile, spouse is believed to be disapproving referents candidate.

4.4 Perceived Control Behavioral

Control beliefs are originally related to circumstances based on the experience of the participants. In this study, we asked people to relate their past experience using existing public transportation, or similar transport modes to LRT. Given this consideration, control beliefs reported here are most likely logically perceived. Nevertheless, we value and accept such belief to be included.

Among four beliefs about facilitating aspects, easiness of access, and good ticketing system are mentioned by majority of participants, and are best describing the salient beliefs in question. Even though slightly low impact, ticketing has also been found to help improving ridership (Currie and Delbosc 2013). Also, as expected

Table 7 Control belief

No.	G. Facilitator/Ease the behavior	N	%	No.	H. Barrier/Hinder the behavior	N	%
1.	Travel time efficiency	54	22.9	1.	Difficult access to station and tiring	53	23.8
2.	Easily and effortlessly accessible station	51	21.6	2.	Too much queue	46	20.6
3.	Easy and fast payment/ticketing system	47	19.9	3.	Strict train schedule	33	14.8
4.	Cheap fare ticket	25	10.6	4.	Route and station location are not in best interest	29	13.0
				5.	Bad weather problem	10	4.5
	Cumulative		75.0				76.7

most participants reported long queue and bad weather as circumstance that hinder the daily commuting with LRT (Table 7).

4.5 *Additional Construct: Impact Awareness*

In addition to the core construct in TPB, we also wanted to elicit participants' salient beliefs regarding the positive impact LRT would bring to the city in the future. There could be a relation between potential users' future intention and their awareness of positive impact from LRT. We found that majority of the respondents believe LRT would reduce traffic congestions because it would attract more people to shift from private vehicle to public transportation. In other words, people accept the introduction of public transportation in the future as an attempt to improve traffic performance of the city. LRT is also believed as a modern mode of travel, and would boost fresher image of the city. In terms of environmental concern, the result shows that the participants believe LRT generates less emission and could promote cleaner air (Table 8).

An interesting belief elicited was "*cut down road accidents*". This represents how participants think of road safety at current state, and how potential shifting to LRT in the future could save more lives by reducing road accidents.

Table 8 Positive impact beliefs

No.	I. Positive impact	N	%
1	Lessen road congestion	98	28.2
2	Reliable comfort and security	52	14.9
3	Lessen air pollution	47	13.5
4	Increase travel time efficiency	36	10.3
5	Cut down road accidents	23	6.6
6	Emitting image of modern city	23	6.6
7	Free from flood	12	3.4
	Cumulative		83.6

5 Conclusion

In this study, we focus on exploring underpinning beliefs of potential users toward utilizing future LRT for daily commuting trip. We extracted the driving beliefs that generate the core construct of theory of planned behavior, namely behavioral beliefs, normative beliefs, and control beliefs. Impact awareness was also elicited as supplementary. These salient beliefs are useful to develop direct measurement of user's future intention.

In general, findings show that potential users hold more positive beliefs and perceived LRT as a good improvement for the future, especially the possibility of improving travel time efficiency, more comfortable travel experience, reducing congestion, and improving city image. Participating individuals experienced difficulties while discussing any negative aspects, or hindering circumstances regarding the utilization of LRT. Possibly losing time to match the fixed schedule, strict access only at the station, and crowding are the most consistently mentioned negative beliefs by the respondents.

These salient beliefs are helpful to conduct further measurement of future intention using TPB. Nevertheless, the advantage of having such qualitative result can also help policymakers to prepare for intervention before the LRT actually operates. It was previously stated that access to station, whether time or energy consuming, would be a source of disappointment. Another example, long queue is fairly undesirable, whereas good ticketing system would increase the satisfaction and easiness of using the LRT for daily commuting.

This study is part of our research on applying TPB to predict future intention of LRT usage. Therefore, the items developed from this study will be used for direct measurement and more quantitative analysis.

Acknowledgements Authors would like to thank Directorate General of Strengthening for Research and Development, Ministry of Research and Technology Republic of Indonesia for providing research grant under SK No. T/140/E3/RA.00/2019.

References

Acharya SR, MoRichi S (2007) Motorization and role of mass rapid transit in East Asian megacities. IATSS Res 31(2):6–16
Ajzen I (1991) The theory of planned behavior. Organ Behav Hum Dec Process 50(2):179–211
Ajzen I (2006) Constructing a theory of planned behavior questionnaire
Bamberg S, Rölle D, Weber C (2003) Does habitual car use not lead to more resistance to change of travel mode? Transportation 30(1):97–108
Bappenas (2016) Kajian awal prastudi kelayakan Light Rail Transit (LRT) Medan Ministry of National Development Planning/ National Development Planning Agency
Choocharukul K, Fujii S (2007) Psychological factors influencing behavioral intention of private car use in future work trips. J East Asia Soc Transp Stud 7:211–222
Currie G, Delbosc A (2013) Exploring comparative ridership drivers of bus rapid transit and light rail transit routes. J Public Transp 16(2):3
Curtis J, Ham SH, Weiler B (2010) Identifying beliefs underlying visitor behaviour: a comparative elicitation study based on the theory of planned behaviour. Ann Leis Res 13(4):564–589
Dell'Olio L, Ibeas A, Dominguez A, Gonzalez F (2012) Passenger preference analysis: light rail transit or bus versus car. Transport 27(3):276–285
Downs DS, Hausenblas HA (2005) Elicitation studies and the theory of planned behavior: a systematic review of exercise beliefs. Psychol Sport Exerc 6(1):1–31
Francis J, Eccles MP, Johnston M, Walker AE, Grimshaw JM, Foy R, Bonetti D et al (2004) Constructing questionnaires based on the theory of planned behaviour: a manual for health services researchers
Godin G, Kok G (1996) The theory of planned behavior: a review of its applications to health-related behaviors. Am J Health Promot 11(2):87–98
Gomez-Ibanez JA (1985) A dark side to light rail? the experience of three new transit systems. J Am Plan Assoc 51(3):337–351
Joewono TB, Santosa W, Kubota H (2007) The exploration of the sustainability of urban transport in Medan, Indonesia. J East Asia Soc Transp Stud 7:195–210
Joewono TB, San Santoso D, Susilo YO (2015) Paratransit transport in Indonesia: characteristics and user perceptions. J East Asia Soc Transp Stud 11:1346–1361
Lanzini P, Khan SA (2017) Shedding light on the psychological and behavioral determinants of travel mode choice: a meta-analysis. Transp Res Part F Traffic Psychol Behav 48:13–27
Medan Bureau of Transportation (2010) Study of mass transportation (in Bahasa Indonesia)
Muromachi Y (2017) Experiences of past school travel modes by university students and their intention of future car purchase. Transp Res Part A Policy Pract 104:209–220
Nordlund A, Westin K (2013) Influence of values, beliefs, and age on intention to travel by a new railway line under construction in northern Sweden. Transp Res Part A Policy Pract 48:86–95
Panondang CN (2018) Study of operational performance and quality of satisfaction of BRT trans mebidang Corridor 1 & 2 (in Bahasa Indonesia). Master thesis, University of North Sumatera, Medan
Phun VK, Yai T (2016) State of the art of paratransit literatures in Asian developing countries. Asian Transp Stud 4(1):57–77
Pojani D, Stead D (2015) Sustainable urban transport in the developing world: beyond megacities. Sustainability 7(6):7784–7805
Rowe R, Andrews E, Harris PR, Armitage CJ, McKenna FP, Norman P (2016) Identifying beliefs underlying pre-drivers' intentions to take risks: an application of the theory of planned behaviour. Accid Anal Prev 89:49–56
Scherer M (2010) Is light rail more attractive to users than bus transit?: arguments based on cognition and rational choice. Transp Res Rec 2144(1):11–19
Senbil M, Zhang J, Fujiwara A (2007) Motorization in Asia: 14 countries and three metropolitan areas. IATSS Res 31(1):46–58

Sigurdardottir SB, Kaplan S, Møller M, Teasdale TW (2013) Understanding adolescents' intentions to commute by car or bicycle as adults. Transp Res Part D Transp Environ 24:1–9

Tarigan AK, Samsura DAA, Sagala S, Pencawan AV (2017) Medan city: development and governance under the decentralisation era. Cities 71:135–146

Zhang M (2009) Bus versus rail: meta-analysis of cost characteristics, carrying capacities, and land use impacts. Transp Res Rec 2110(1):87–95

A Study of Intended Unsafe Pedestrian Crossing Behaviors at Signalized Intersections in Vietnam

Xuan Can Vuong, Rui-Fang Mou, Trong Thuat Vu, and Thi An Nguyen

Abstract This paper presents a model for pedestrians' intended unsafe crossing behaviors at signalized intersections based on the Theory of Planned Behavior (TPB). The model has been proposed with three factors, including attitude toward unsafe crossing behavior, subject norm, and perceived behavior control that impact on pedestrians' unsafe crossing behavioral intention. Taking Vietnam as an example, through the reliability and validity test of 508 participants, the Structural Equation Model (SEM) of pedestrians' unsafe behavioral intention is constructed to predict the unsafe crossing behavioral intention of pedestrians. Simultaneously, the relationship between influencing factors is obtained. The results show that perceived behavior control is the key determinant of pedestrian crossing intention, followed by attitude toward unsafe crossing behavior, and subject norms. This also shows that the proposed model can effectively explain and predict the unsafe behavioral intention of pedestrians in Vietnam's situation.

Keywords Unsafe behavioral intention · Pedestrians · TPB · SEM

1 Introduction

Pedestrians are vulnerable road users, and pedestrian-related safety is a growing concern all over the world. The World Health Organization (WHO) estimated 1.25 million fatalities worldwide caused by Road Traffic Accidents (RTAs), 22% of RTAs occur among pedestrians in 2013 (WHO 2015), and most pedestrian collisions occur when pedestrians are crossing the road (WHO 2013). The National Highway Traffic Safety Administration (NHTSA) of the US estimated there were 4,743 pedestrian

X. C. Vuong (✉) · R.-F. Mou
School of Transportation and Logistics, Southwest Jiaotong University, Chengdu, China
e-mail: 1323972686@qq.com

X. C. Vuong · T. T. Vu
University of Transport and Communications, Hanoi, Vietnam

T. A. Nguyen
National Economics University, Hanoi, Vietnam

L. T. T. Huong and G. M. Pomeroy (eds.), *AUC 2019*, Advances in 21st Century Human Settlements, https://doi.org/10.1007/978-981-15-5608-1_15

fatalities and 76,000 pedestrian injuries in RTAs. Of these, 20% fatalities occurred at intersections in 2012. (National Highway Traffic Safety Administration 2015). In the European Region, there were 84,589 people who died from road traffic injuries in 2013, about 26% of deaths were pedestrians (Jackisch et al. 2015). In OECD member countries, more than 20,000 pedestrians die from RTAs annually, 33–50% of crashes occur at a pedestrian crossing (OECD 2012). The number of pedestrians' fatalities from 2013 to 2015 in China accounted for about 20% of deaths from RTAs (Traffic Management Bureau-Ministry of Public Security of PRC 2016). Pedestrian safety is also a major and growing problem in Vietnam. According to the National Traffic Safety Committee of Vietnam (NTSC), more than 9,000 people have been killed and thousands have been injured from RTA, including 32% of pedestrian-related injuries in 2012 (NTSC of Vietnam 2012). Risk factors for traffic injuries of pedestrians include higher speed, overdrinking, lack of pedestrian facilities, lack of visibility, mistakes by drivers and pedestrians, and so on (WHO 2013; ECMT 2000), among which drivers and pedestrians' unsafe behaviors are essential causes of road accidents in Vietnam (70–87% (TDSI of Vietnam 2011)), involving pedestrians' accidents. Pedestrians' safety in Vietnam is difficult to assess due to the lack of pedestrians' accident data as well as the lack of effective estimation models for pedestrians. Therefore, understanding the unsafe pedestrian behaviors is an important way to reduce pedestrian violations, promote safety awareness, and prevent pedestrian-related accidents.

Theory of Planned Behavior (TPB) is a widely used framework to better understand the decision-making process of the people who violate traffic rules, and several scholars have implemented the theory to investigate pedestrians' road-crossing behavior. For example, Evans and Norman (1998, 2003) found that the three components of the theory were significant predictors of pedestrians' road-crossing intention, among which perceived behavioral control emerged as the strongest predictor of intentions. Similarly, Barton et al. 2016 also found that attitude and perceived behavioral control predicted intentions to cross streets while under conditions of distraction; Diaz (2002) applied TPB to measure pedestrian risk-taking behaviors found that reported violations, errors, and lapses appeared causally related to the intention to violate regulations of pedestrians, and this in turn with positive attitudes, subjective norm, and perceived behavioral control; Zhou et al. 2009 used TPB to investigate the effect of conformity tendency on pedestrians' road-crossing intentions in China, and they found that pedestrians were greater likelihood in the road-crossing when other pedestrians were crossing the road; Xu et al. 2013 studied jaywalkers and evaluated their influence of past behavior using an extended version of TPB. They found that the past behavior explained 42% of the variance in the intention of pedestrians' illegal road-crossing; Zhou et al. 2016 extended the theory by adding new factors including descriptive norm, perceived risk, and conformity tendency to evaluate their respective impacts on pedestrians' behavior intentions. They showed that conformity tendency was a strong predictor. The effectiveness of the TPB model in predicting pedestrian behavioral intention has also been tested. The TPB model explained 37–49% of the variance in pedestrians' behavioral intentions in (Evans and Norman 1998), predicted 56–65% of the variance in intention to cross in (Carol and Roslyn 2007), and accounted for 30–48,% of the variance in China (Zhou et al. 2009; Zhou and Horrey 2010). However, the application of the TPB to examine pedestrians' violating crossing behavioral intention has not been conducted in Vietnam. Therefore, we aimed to assess whether the TPB explains pedestrians' unsafe crossing behavioral intentions in Vietnam's urban areas.

2 Research Model

The TPB was proposed by Ajzen based on the theory of reasoned action (Ajzen 1991). This theory constructs a theoretical framework for identifying the factors affecting behaviors and their impacts and the relationship between different factors in various fields. The model is mainly used to explain and predict behavior under certain circumstances. The TPB is defined by the attitude toward the behavior, subject norm, and perceived behavior control which constitute three main variables. Attitude toward the behavior is the degree to which performance of the behavior, it is positively or negatively valued, and is determined by the total set of accessible behavioral beliefs linking the behavior to various outcomes and other attributes. Subjective norm refers to the social pressure that an individual perceives when deciding whether to perform a particular behavior, and its main manifestation is the influence of another person or group on individual behavioral decision-making. Perceived behavioral control refers to people's perceptions of their ability to perform a given behavior, and it reflects the actual behavioral control conditions (Ajzen 1991, 2005). The three components are the base of a behavioral intention, which is an indicator of a person's readiness to perform a behavior and is considered as an important antecedent of future behavior. The strength of intention indicates how much people attempt to conduct the behavior. Hence, understanding behavioral intention results in valuable prediction about a given behavior. The model of the pedestrians' unsafe behavioral intention using the TPB is shown in Fig. 1. This study proposes a hypothetical model with the following hypothesis: (H1) Attitude toward unsafe crossing Behavior (AB) will have a positive effect on unsafe crossing Behavioral Intention (BI); (H2) Subjective norms will have a positive effect on the BI; (H3) Perceived Behavioral Control (PBC) will have a positive effect on the BI; (H4) The BI will have a positive effect on Unsafe crossing Behavior (UB); (H5) The PBC will have a positive effect on the UB.

3 Data Collection and Analysis

3.1 Questionnaire Design

According to the constructed model and measured variables and studies related to TPB-based pedestrians' unsafe behavioral intentions, the questionnaire was designed

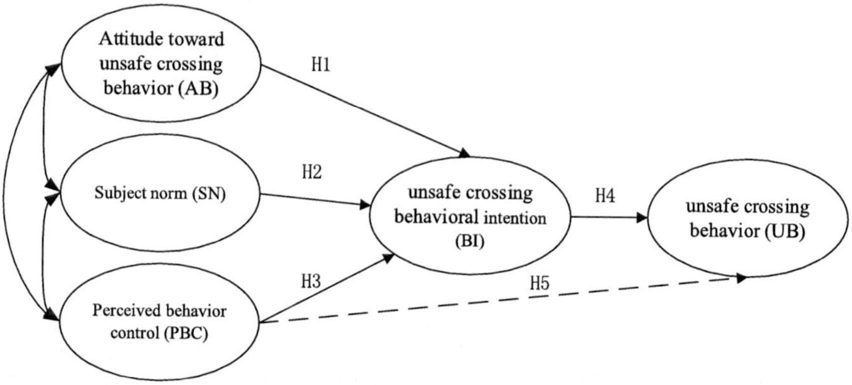

Fig. 1 The model of the pedestrians' unsafe behavioral intention using the TPB

to capture main pedestrians' demographic characteristics and their behavioral intentions, namely attitudes, subject norm, perceived behavior control, perception and experience, the unsafe behavioral intention of pedestrians. The questionnaire was created as a list of items to be rated based on a five-point Likert scale, ranging from "1-strongly agree" to "5-strongly disagree", from "1-never" to "5-always". Additionally, the demographic information of participants wase also incorporated in the questionnaire.

3.2 Data Collection

Questionnaires were randomly distributed at office buildings, schools, and public facilities, the bus stations located along main streets in the center of major cities in Vietnam, including Hanoi, Ho Chi Minh City, Hai Phong, Da Nang, and Can Tho in the period from March to June 2017. Participants completed the questionnaire with the assistance of surveyors. After removing incomplete responses from the initial questionnaires (N = 515), 508 usable samples were obtained for further analysis.

3.3 Data Analysis and Results

Questionnaire Sample Characteristics

After data collection, the information of the valid questionnaires was processed into SPSS 22.0 and AMOS 22.0 for post-processing. The distribution of the questionnaire samples was as follows: 65.6% of male and 34.4% of female with the majority of people aged 20–35 year old (65.6%), followed by those under 20 years old (31.1%) and 3.3% of them over 35 years old; 81.7% of participants was student/pupil, 7.1% of them was official, and 11.2% of them were others occupation. There were 75.8%

of participants having the upper high school degree and 24.2% having the lower high school degree; 13.6% of participants were married and 86.4% of them were unmarried with the average income of below 5 million VND accounts for 83.9% of participants.

Data Reliability and Validity
Cronbach's Alpha (α) is used to test the reliability of all items. Generally, when α is greater than 0.6, the designed scale has good reliability (Hair et al. 1998). Additionally, if "item-total correlation" coefficient of an item is less than 0.3, it is eliminated (Hair et al. 1998). The results showed that the minimum Cronbach's Alpha was 0.603 (> 0.6) and minimum value of "item-total correlation" of all items was 0.415 (> 0.3), which held enough reliability to permit the use of the survey results (see Table 1).

The validity test is carried out to determine whether the scale can effectively reflect the content to be investigated by using the confirmatory factor analysis with the "Principal Components Analysis (PCA)" and the "Varimax" rotation method in the software SPSS. The higher the validity, the more the data can reflect the true characteristics of the set variable content. The results are required to obtain a KMO coefficient for measuring sampling adequacy of 0.5 or greater and a significant level of Bartlett's test less than 0.05 (Hair et al. 2006). Besides, the factor loading coefficient of the items is not less than 0.5, and the percentage of variance higher than 50% (Gerbing and Anderson 1988). PCA identified 5 components accounting for 70.45% of the variance (> 50%). The KMO coefficient of 0.829 with Bartlett's test statistical meaning of 0.00 (< 0.05) was greater than the standard value (0.5). Besides, the factor loading of all items ranges from 0.564 to 0.901 above the lower limit value of 0.5, indicating that the survey results were up to the criteria of convergent validity test.

Results of the Structural Equation Modeling (SEM)
The SEM is employed to explain the relationships among latent hypothetical, unobserved variables and items (observed variables) by using software AMOS. Goodness of fits to validate the model needs to achieve the following criteria: the ratio between chi-square and degree of freedom (df) should be less than 5; Root Mean Square Error of Approximation (RMSEA) should be less than 0.08; Goodness-of-Fit Index (GFI) should be more than 0.8; Comparative Fit Index (CFI), Incremental Fit Index (IFI), and Tucker & Lewis Index (TLI) should be more than 0.9. The result of the model assessment is shown in Fig. 2. The result shows that these estimated models fit the data reasonably well. Chi-square/df = 3.512, which is lower than the recommended value of 1.488. TLI = 0.912, CFI = 0.926, GFI = 0.901, IFI = 0.926 are more than the cut-off value of 0.90; RMSEA value of 0.07 less than the upper limit value of 0.08 for reasonable fit.

The standardized weights of items range from 0.502 to 0.905 above the lower limit value of 0.5, indicating that the model is well convergent. The standardized weights of latent variables are shown in Table 2. From the standardized weights, we can find that the impact of all kinds of factors on unsafe crossing behavioral intention and unsafe crossing behavior is different, and the estimated value range from 0.118 to 0.611.

Table 1 List of items by questionnaire survey

Latent variables	Items	Statement	M	S. D.
Attitude toward unsafe crossing behavior (AB) ($\alpha = 0.872$)	AB1	When you cross the road during a red signal, there will be an accident.	1.63	0.79
	AB2	The accident will be serious if you cross the road at red light	1.73	0.79
	AB3	When you cross without paying attention to traffic, there will be an accident	1.77	0.73
	AB4	The accident will be serious if you cross without paying attention to traffic	1.62	0.74
Subject norm (SN) ($\alpha = 0.908$)	SN1	My family advises I don't cross roads during a red signal	2.19	0.94
	SN2	My family advises I cross with paying attention	2.05	0.88
	SN3	My friends advise I don't cross roads during a red signal	2.24	0.96
	SN4	My friends advise I cross with paying attention	2.17	0.91
Perceived behavior control (PBC) ($\alpha = 0.603$)	PBC1	It is not easy for me to take a violating crossing behavior	2.39	0.97
	PBC2	I have the ability to cross roads during a red signal when there is little traffic on the road (R)	2.00	1.01
	PBC3	It is easy for me to control a violating crossing behavior	2.30	0.98
Unsafe crossing behavioral intention (BI) ($\alpha = 0.869$)	BI1	I'll take risk crossing the road to save time	1.57	0.87
	BI2	I'll cross outside the pedestrian crossing	1.40	0.73
	BI3	I'll cross roads during a red signal when there is little traffic on the road	1.61	0.82
	BI4	I'll cross without paying attention to traffic lights	1.31	0.68
Unsafe crossing behavior (UB) ($\alpha = 0.815$)	UB1	I cross roads during a red signal to save time and be convenient	1.61	0.99
	UB2	I cross roads during a red signal when there is no oncoming traffic	1.64	0.90
	UB3	To be convenient, I cross without paying attention to traffic lights	1.46	0.93
	UB4	I cross roads outside designated locations to be convenient	1.97	1.12

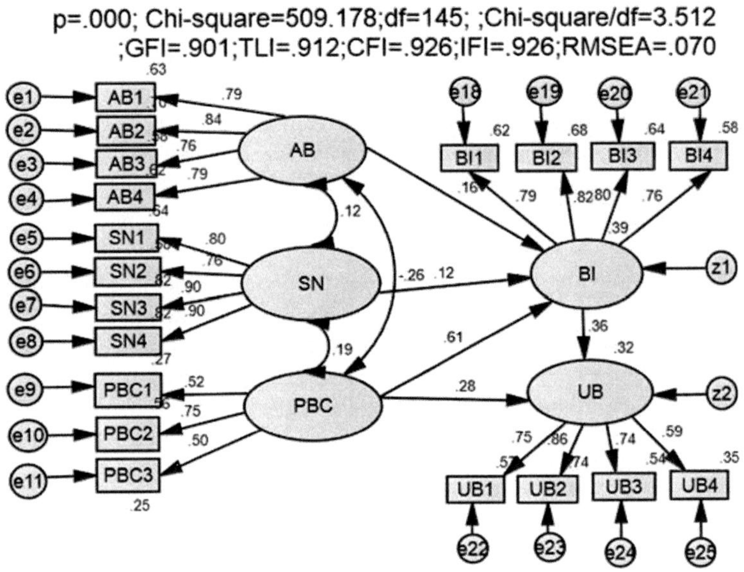

Fig. 2 The result of the model assessment

Table 2 Standardized weights of SEM analysis

Path	Std. weights	S.E.	C.R.	P-value	Accepted hypothesis
BI < – SN	0.118	0.045	2.37	0.018	H1
BI < – AB	0.158	0.056	3.057	0.002	H2
BI < – PBC	0.611	0.098	8.551	***	H3
UB < – BI	0.358	0.074	5.482	***	H4
UB < – PBC	0.276	0.116	3.718	***	H5

Note "***"stands for a significant difference $P \leqq 0.001$

The largest influence on unsafe crossing behavioral intention is perceived behavior control (0.611), followed by attitude toward unsafe crossing behavior (0.158), and subject norm (0.118). For unsafe crossing behavior, the impact of unsafe crossing behavioral is more than perceived behavior control. Besides, the initial hypothesis is accepted. The R-square for behavioral intentions, unsafe behaviors are 0.394 and 0.322, respectively. This proves that the assessment results of the model are also consistent with previous studies.

4 Conclusion

This study applied TPB and SEM to construct an assessment model for the pedestrians' unsafe crossing behavioral intention in Vietnam. In the model, the psychological behaviors of pedestrians such as human attitude and self-control ability are quantified, so that the quantitative analysis becomes qualitative analysis. Analytical results show that applying TPB and SEM to the assessment of pedestrians' unsafe crossing behavioral intention in Vietnam is feasible. Of the three main dimensions, including Perceived Behavior Control (PBC), Attitude toward unsafe crossing Behavior (AB), and Subject Norm (SN), the PBC had the greatest impact on the pedestrians' unsafe crossing behavioral intention. This shows that pedestrians' unsafe crossing behaviors are mostly influenced by pedestrians' perceptions of their ability to perform a behavior. However, to more understanding the pedestrians' unsafe crossing behavioral intention, other factors such as past behaviors, perceived risk, etc. should pay attention. Some factors will be examined in the next steps. The results from this research are expected to be very useful in understanding of pedestrian crossing behavioral intention at signalized intersections in Vietnam's urban areas, and may assist to design more effective interventions and safety campaigns, such as changing people's attitude toward violating behaviors, increasing their safety awareness, etc. in order to reduce pedestrians' unsafe crossing behaviors.

Acknowledgements We appreciate the support of this work from the MOET Science & Technology Program of Vietnam (Grant No. CT.2019.05.04), the MOT Science & Technology Program of Vietnam (Grant No. DT174021), and the National Key R&D Program of China (Grant No. 2016YFC0802209).

References

Ajzen I (1991) The theory of planned behavior. Organ Behav Hum Decis Processes 50(2):179–211

Ajzen I (2005) Attitude, personality and behavior-2nd Edn. McGraw-Hill Education (UK)

Barton BK, Kologi SM, Siron A (2016) Distracted pedestrians in crosswalks: an application of the theory of planned behavior. Transpo Res Part F Traffic Psycho Behav 37(s 4–5):129–137

Carol H, Roslyn H (2007) The effect of age, gender and driver status on pedestrians' intentions to cross the road in risky situations. Accid Anal Prev 39(2):224–237

Dıaz EM (2002) Theory of planned behavior and pedestrians' intentions to violate traffic regulations. Transp Res Part F: Traffic Psycho Behav 5(3):169–175

ECMT (2000) Safety in road traffic safety in road traffic. OECD Publications Service

Evans D, Norman P (1998) Understanding pedestrians' road crossing decisions: an application of the theory of planned behaviour. Health Educ Res 13(4):481–489

Evans D, Norman P (2003) Predicting adolescent pedestrians' road-crossing intentions: an application and extension of the theory of planned behaviour. Health Educ Res 18(3):267–277

Gerbing DW, Anderson JC (1988) An updated paradigm for scale development incorporating unidimensionality and its assessment. J Mark Res 25(2):186–192

Hair JF, Anderson RE, Tatham RL, Black WC (1998) Multivariate data analysis. Prentice Hall, New Jersey

Hair JF, Black WC, Babin BJ, Anderson RE, Tatham RL (2006) Multivariate data analysis, vol 6. Pearson Prentice Hall, Upper Saddle River, NJ
Jackisch J, Sethi D, Mitis F, Szymański T, Arra I (2015) European facts and the global status report on road safety 2015. WHO, Regional Office for Europe
National Highway Traffic Safety Administration (2015) Traffic safety facts, 2012 data: pedestrians. Ann Emerg Med 65(4):452
NTSC of Vietnam (2012) Traffic safety annual reports. Hanoi, Vietnam
OECD (2012) Pedestrian safety, urban space and health-research report. International transport forum
TDSI of Vietnam (2011) National road traffic Safety strategy to 2020 and Vision to 2030-Final Report
Traffic Management Bureau-Ministry of Public Security of PRC (2016) Statistics annual report of road traffic accident in China
WHO (2013) Pedestrian safety—a road safety manual for decision-makers and practitioners
WHO (2015) Global status report on road safety 2015. WHO Press, Geneva
Xu Y, Li Y, Zhang F (2013) Pedestrians' intention to jaywalk: automatic or planned? A study based on a dual-process model in China. Accid Anal Prev 50:811–819
Zhou R, Horrey WJ (2010) Predicting adolescent pedestrians' behavioral intentions to follow the masses in risky crossing situations. Transp Res Part F Traffic Psychol Behav 13(3):153–163
Zhou R, Horrey WJ, Yu R (2009) The effect of conformity tendency on pedestrians' road-crossing intentions in China: an application of the theory of planned behavior. Accid Anal Prev 41(3):491–497
Zhou H, Romero SB, Qin X (2016) An extension of the theory of planned behavior to predict pedestrians' violating crossing behavior using structural equation modeling. Accid Anal Prev 95:417–424

The Impact of E-Mobility on Greenhouse Gas Mitigation in Vietnam

An Minh Ngoc and Khuat Viet Hung

Abstract This paper aims to develop possible scenarios of electric bus introduction and their impact on the Vietnam Greenhouse Gas (GHG) mitigation. Give the foreseen increased number of bus until the year 2030, three different scenarios in terms of penetration of electric vehicles have been considered, namely 10, 20, and 40% of new vehicle sales. A detailed bottom-up energy model of the Vietnamese bus fleet has been developed. The vehicle fleet is analyzed in terms of energy consumption and carbon emissions. The analyzed scenarios suggest that the introduction of electric vehicles would always lead to primary energy savings. In particular, the increase of the penetration corresponds to a decrease in primary energy consumption and carbon emissions.

Keywords E-Mobility · Electric bus · GHG

1 Introduction

Climate change is one of the most serious challenges to people nowadays. Each country needs to make a specific contribution to climate change response in order to protect the Earth's climate system for the current and next generations. In Vietnam, transport is becoming a large and growing contributor to total Greenhouse Gas (GHG) emissions, accounting for 18% of total CO_2 equivalent in 2014. The primary GHGs produced by the transport sector are carbon dioxide (CO_2), methane (CH_4), nitrous oxide (N_2O), hydrofluorocarbons (HFC), sulfur hexafluoride (SF_6), and trifluoride nitrogen (NF_3) (WHO 2006). Of these, CO_2, CH_4, and N_2O are considered as the major contributors. However, N_2O emissions are marginal, and therefore only CO_2 and CH_4 emissions are included (EEA 2018). CO_2 emissions are determined based

A. M. Ngoc (✉)
University of Transport and Communications, 3 Cau Giay, Hanoi, Vietnam
e-mail: anminhngoc@utc.edu.com

K. V. Hung
National Traffic Safety Committee, 80B Tran Hung Dao, Hanoi, Vietnam
e-mail: kviethung@mt.gov.vn

L. T. T. Huong and G. M. Pomeroy (eds.), *AUC 2019*, Advances in 21st Century Human Settlements, https://doi.org/10.1007/978-981-15-5608-1_16

on the energy consumption according to the IPCC methodology (2006). Direct as well as indirect GHG emissions of CNG buses also include methane slip. Leakage of unburnt methane is important due to the high Global Warming Potential (GWP) of CH_4. Methane slip is caused within the vehicle in the crankcase and the exhaust pipe and "upstream" due to leaks in the gas pumps and wells.

Transport GHG emissions account for 30.55 $MtCO_2e$ in 2014 in Vietnam. Recently, transport GHG emissions have been growing steadily. From 2010 to 2014 transport GHG emissions increased by 10%. Vietnam's increasing demand for greater mobility and the high motorization rate pose significant challenges for Vietnam's potential for reducing greenhouse gas emissions in this sector. In 2014, emissions from on-road vehicles accounted for 89.7% of transport GHG emissions. Road transport, which contributed 94% of passengers and 76% of freight tons, represents one of the reasons for this large share of emissions. Emissions are projected to increase in a Business-As-Usual (BAU) scenario, at an annual average rate of 6–7%, reaching 89.1 million tons in 2030.

With the rise of emission issue in transport, the interest in renewable energy is growing rapidly as they could replace polluting combustion engines. In light of this, the electrification of the transport sector may offer important opportunities in order to increase its sustainability, considering the increasing rate of electricity generated by renewable. Electric vehicles are a key element in transport sustainable development as they could reduce its energy and carbon intensity, as well as the associated pollutant emission.

The impact on GHG reductions will be far higher by deploying commercial electric vehicles instead of private units due to higher fuel usage, higher mileage, and longer life span of commercial vehicles. Replacing one urban diesel bus with an electric unit has the same impact as replacing 35 fossil fuel passenger cars or 300 motorcycles. Therefore, a focus on commercial vehicles maximizes the emission impact.

Currently, Vietnam Government has pursued a series of policies to encourage its subordinates to promote clean and energy-efficient road transport vehicles by building a framework of incentives and tax reduction as well as loan supporting for low emission vehicles including electric ones.

The paper aims at estimating to what extent is the primary energy consumption reduced in Vietnam if e-buses are introduced, how much carbon emissions can be saved.

A bottom-up model is implemented for assessing the forecast of the vehicle stock, analyzing the necessary electricity supply, and checking the consequence emissions. In this way, it is possible to test different scenarios in terms of bus fleet development and specific consumption of the vehicles.

2 Case Study: Vietnam Vehicle Fleet

The Vietnam bus fleet is composed of ~11 thousand of vehicles, and data shows that an increase in the number of vehicles was steadily registered in the period 2010–2017. In fact, the Vietnam bus fleet has been increased due to the pressure of the lack of infrastructure system leading to serious traffic problems such as traffic congestion, traffic accidents, and government realized that public transport should be the key pillar to deal with these issues.

In terms of compliance with European Emission regulations, almost vehicles are compliant only with Euro 2 (98%) regulations, which are the oldest ones. This means that, on average, the circulating buses in Vietnam are characterized by higher pollutant emissions.

In terms of fuel, must of the bus fleets in Vietnam are fueled with diesel. Recently, a relevant increase in the use of CNG buses has been observed, especially in big cities such as in Hanoi and Ho Chi Minh City.

Approximately half of the fleet is represented by medium buses, 46.7%, whereas the other relevant share is represented by middle buses, 31.1%, finally large buses represent only a marginal share, 22.2%. It can be observed that the average dimension of the large buses tends to decrease due to a fall of the total carrying capacity of buses and the lack of adequate road infrastructure (Fig. 1).

3 Methodology

A bottom-up model has been developed to determine the impact of the introduction of electric buses on emission mitigation. Firstly, an estimation of the amount of evolution of buses in Vietnam is led and the fuel consumption associated is calculated. Then,

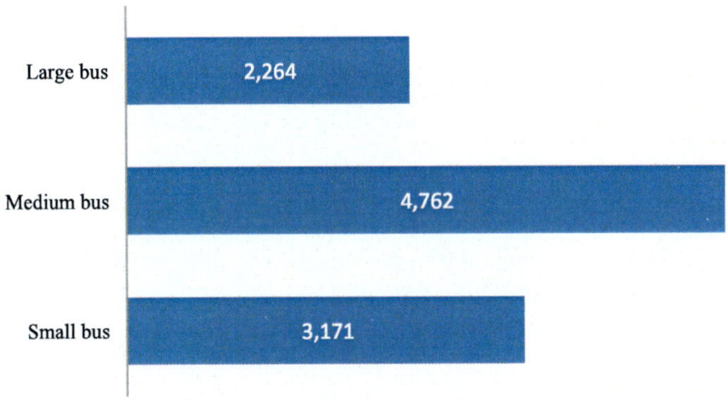

Fig. 1 Bus fleet, 2017. *Source* Ministry of Transport (2017)

the increasing introduction of electric busses is analyzed determining the related primary energy consumption and carbon emissions.

The first input data were the historical number and typologies of buses circulating in Vietnam.

Bus market study includes four steps:

- **Step 1**: Forecast of public transport dynamics: current state and trends in passenger turnover and number of buses are analyzed at this step.

  Input: Data and statistics from General Statistics Office (2010–2017), Ministry of Transport (2010–2017), Ministry of Natural Resources and Environment (2010–2019) and relevant publications.
  Output: Forecast of transportation volume, number of vehicles by types until the year 2030.

- **Step 2**: Forecast of penetration rate: potential number of electric buses is forecasted at this step.

  Input: Forecast of bus fleet until the year 2030, survey data from market agents and policy analysis.
  Output: Forecasted number of electric buses by 4 scenarios (10, 20, 30, and 40%) until the year 2030.

- **Step 3**: Forecast of demand for fuel consumption: fuel required for the bus fleet is forecasted at this step.

  Input: Number of electric buses until the year 2030, assumptions on power consumption.
  Output: Power demand from electric buses.

- **Step 4**: Estimation of the carbon emissions. Carbon emissions are estimated by considering the appropriate fuel emissions factors. Emissions included in this research are greenhouse gas emissions and local pollutants including Particle Matter ($PM_{2.5}$), NO_x, and SO_2 (Table 1).

4 Results

Three scenarios of electric buses penetration are considered, namely 10, 20, and 40% for new vehicle sales and compared to the Business-As-Usual (BAU) scenario. In all scenarios, the share of electric bus type will remain as diesel buses, it is therefore assumed that 30% of electric buses belong to the small category and the 50% to the medium on, and the rest is for the large one. The aim of testing these scenarios is to estimate the maximum potential for energy savings and emission reductions, as well as a realistic estimation of achievable results in short-medium period.

BAU scenario is projected to increase with growing mobility demand and motorization. Based on population and economic growth, Passenger-Km Traveled (PKT)

Table 1 Default calculation values

Description	Value	Source
Net calorific value of diesel	43.0 MJ/kg	IPCC (2006), Table 1.2
Net calorific value of CNG	48.0 MJ/kg	IPCC (2006), Table 1.2
CO_2 emission factor of diesel	74.1 gCO_2/MJ	IPCC (2006), Table 1.4
CO_2 emission factor of CNG	56.1 gCO_2/MJ	IPCC (2006), Table 1.4
Global warming potential of BC	900	IPCC (2013), Table 8.A.6
Global warming potential of methane	28	IPCC (2013), Table 8.A
Well-to-tank mark-up factor for diesel and CNG	Diesel: 23% CNG: 18%	UNFCCC (2014), Table 3
Methane slip as % of CNG consumption TTW and WTW	TTW: 1.1% WTW: 2.3%	ICCT (2015), ICCT (2018), Table 4
Carbon electricity grid factor Vietnam for the year 2016	0.500 kg CO2/kWh	OECD/IEA for CO_2 emissions and IEA for electricity production minus losses

is projected to increase at an annual average growth rate of 5.6% from 2017 to 2030. Following this, the total bus fleet is projected to increase at an annual average rate of 5.2% during the same period (Fig. 2).

Without any electric energy policies and measures to reduce GHG emissions, the current composition of the energy sources is estimated to remain largely, with heavy reliance on diesel fuels, as depicted in Table 2.

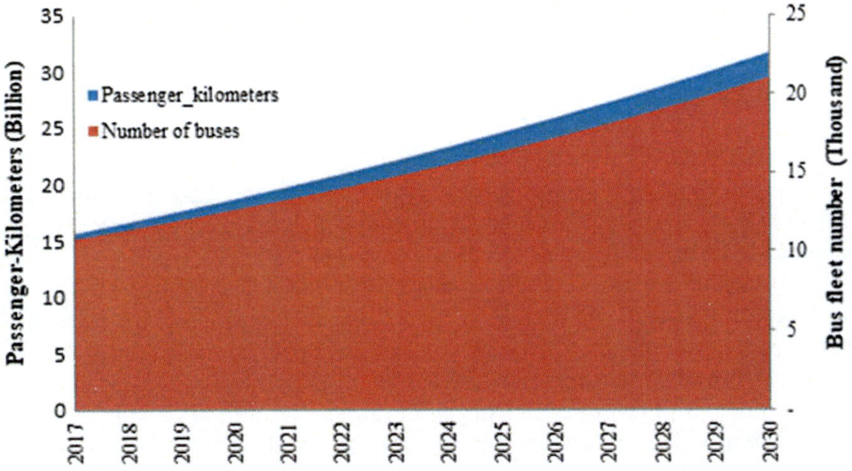

Fig. 2 Projection of passenger-kilometers traveled and vehicle numbers under BAU. *Source* Ministry of Transport (2017)

Table 2 Projected energy consumption by source under BAU

Fuel type	Unit	2017	2020	2025	2030
Diesel	Gg	126.6	149.2	194.8	253.2

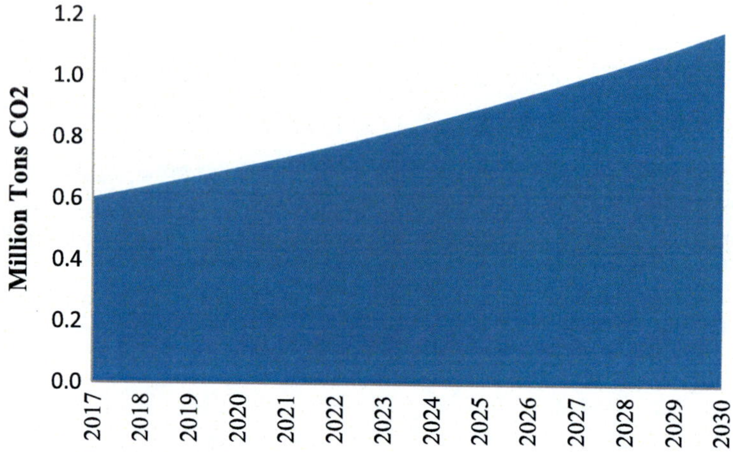

Fig. 3 Projected CO_2 emissions by bus system under BAU. *Source* Ministry of Transport (2017)

Consequently, emissions are projected to increase at an average of 5.1% per year, reaching nearly 1.2 million tons of CO_2 emissions in 2030 (Fig. 3).

Based on BAU result, the impact of different scenarios is analyzed in terms of primary energy consumption and CO_2 emissions.

Scenario 1 includes the shift of 10% of new purchased buses to electricity from 2020. This would result in a slight reduction in projected fuel consumption compared to under BAU, leading to a reduction in CO_2 emissions compared to BAU. The diesel consumption would drop by 7.2% in 2025 and by 11.2% in 2030. The CO_2 emissions show a reduction of 76 thousand tons in 2030.

If the percent of new vehicle sales increases to 20%, the same phenomena are observed: there is a decrease in energy consumption equal to 57 Gg up to 2030, a reduction of carbon emission of 153 thousand tons in 2030.

With scenario 3, diesel consumption would be lower than Business-As-Usual (BAU) by 29% in 2025 and by 45% in 2030. Compared to BAU and the two above scenarios, steeper emissions reduction would occur in public transport activity. Up to 40% of electric buses, it reveals that 113 Gg of primary energy and 0.31 Mt of carbon emissions can be saved. Saving results are shown in Tables 2 and 3.

As shown in Table 4, investments or policies toward fuel shift from diesel to electricity in public transport would result in much greater emissions reduction over a longer period.

Figure 4 depicts their contributions to the overall emissions reduction against the Business-As-Usual (BAU) scenario.

Table 3 Projected energy consumption under BAU and scenario

	Unit	2017	2020	2025	2030
BAU	Gg	126.6	149.2	194.8	253.2
Scenario 1	Gg	126.6	147.0	180.8	225.0
Scenario 2	Gg	126.6	144.9	166.8	196.7
Scenario 3	Gg	126.6	140.6	138.7	140.1

Table 4 Bus CO_2 emissions under BAU and scenario

	Unit	2017	2020	2025	2030
BAU	Mil.tons	0.61	0.70	0.90	1.16
Scenario 1	Mil.tons	0.61	0.7	0.87	1.08
Scenario 2	Mil.tons	0.61	0.69	0.83	1.01
Scenario 3	Mil.tons	0.61	0.68	0.75	0.86

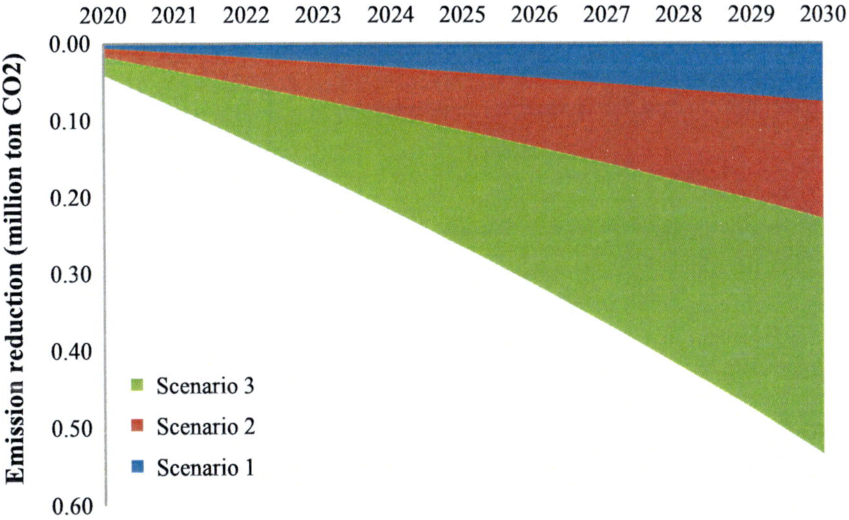

Fig. 4 Reduction of CO_2 emissions by each scenario

Among the three scenarios, the implementation of 40% of new electric buses is considered the most effective GHG emissions mitigation action. Under the impact of this scenario, CO_2 emissions started to largely decline from 2022. Looking ahead, CO_2 emissions are projected to reduce by 9% (2022), 17% (2025), and 26% (2030) in the comparison of BAU.

5 Discussion

Using electric buses shows a significant reduction of GHG emissions in transport sector. In particular, electric buses show 50% lower GHG emissions than diesel Euro IV buses and 65% lower emissions than EEV CNG buses. They are therefore clearly a very effective mean to reduce GHG emissions from the transport sector. In addition, the above analysis shows that public transport activities in Vietnam could achieve a 26% reduction in CO_2 emissions by deploying the measures of electric bus development.

Vietnam can adopt and implement economically feasible mitigation measures to achieve significant emissions reductions in the rapidly growing transport sector. The emissions reduction would help Vietnam meet its Nationally Determined Contribution (NDC) targets, along with the policies and investments in other sectors, especially in power generation.

The electrification of transport is no longer in doubt, but an ongoing trend. Each country must choose the right path to enable electrified transport and help prevent and arrest increasing pollution and higher greenhouse gas emissions. This requires concerted efforts and strategic decision-making at multiple levels of government and industry that are involved in the policies and manufacturing of electric vehicles, the construction of charging stations, the provision of mobility services, the mix of the energy supply, and interactions with financial institutions.

6 Conclusions

The paper investigates the impact of the introduction of electric buses to Vietnam vehicle fleet and its impact on the energy system.

A specific bottom-up model has been implemented in order to estimate the future development of the bus fleet till 2030, the penetration of electric buses, and the corresponding effects in terms of energy consumption and carbon emissions.

Furthermore, three scenarios have been analyzed considering penetration levels of electric buses equal to 10, 20, and 40% of new vehicle sales. The analysis of the scenarios highlights that the introduction of electric buses determines positive effects on Vietnam energy system.

Up to 2030, the implementation of new electric buses contributes greatly to the reduction of GHG emissions in the public transport activity, cumulatively accounting for up to 0.31 million tons CO_2 emissions reduction during 2020–2030. With the above measures, the public transport activity can reduce CO_2 emissions by 26.0% compared to BAU in 2030.

Electrification of mobility solutions represents an area with an expectation of notable technology innovation, and market maturation should be encouraged in the near future because it will be the highest contributor to emissions reduction.

References

EEA (2018) Air pollutant emission inventory guidebook Version 2016 update December 2018

GSO (General Statistics Office of Vietnam) (2016) Statistical yearbook of Vietnam 2016. Government of Vietnam, Hanoi. https://www.gso.gov.vn/default_en.aspx?tabid=515&idmid=5&ItemID=18533

ICCT (2018) Effects of battery manufacturing on electric vehicle life-cycle greenhouse gas emissions. https://www.theicct.org/publications/EV-battery-manufacturing-emissions

ICCT (2015) Assessment of heavy duty natural gas vehicles emissions: implications and policy recommendations

IPCC (2013) Fifth assessment report (AR5). https://www.ipcc.ch/report/ar5/

IPCC (2006) IPCC guidelines for national greenhouse gas inventories

Kontgis C et al (2014) Monitoring peri-urbanization in the greater Ho Chi Minh City metropolitan area. Appl Geogr 53(2014):377–388

MoNRE (Ministry of Natural Resources and Environment) (2010) Vietnam's second national communication to the united nations framework convention of climate change. Ministry of Natural Resources and Environment, Vietnam, Hanoi. https://unfccc.int/sites/default/files/resource/Vietnam%20second%20national%20communication.pdf

MoNRE (Ministry of Natural Resources and Environment) (2016) Intended nationally determined contribution of Vietnam. Technical Report. Ministry of Natural Resources and Environment, Vietnam, Hanoi. https://www4.unfccc.int/sites/ndcstaging/PublishedDocuments/Viet%20Nam%20First/VIETNAM%27S%20INDC.pdf

MoNRE (Ministry of Natural Resources and Environment) (2019) Third national communication of Vietnam to the united nations framework convention of climate change. Revised April 20, 2019. Ministry of Natural Resources and Environment, Vietnam, Hanoi. https://unfccc.int/sites/default/files/resource/Viet%20Nam%20%20NC3%20resubmission%2020%2004%202019_0.pdf

OECD/IEA (2018) CO_2 emissions from fuel combustion

UITP (2017) An electric solution for urban bus networks

UNFCCC (2014) CDM methodological tool: upstream leakage emissions associated with fossil fuel usage, version 02.0

WHO (2006) WHO air quality guidelines for particulate matter, ozone, nitrogen dioxide and sulfur dioxide

Policies and Measures to Create Efficient and Low-Carbon Transport in Urban Area: Case Study in Hochiminh City

Vu Trong Tich and An Minh Ngoc

Abstract The transport sector contributes a significant source of greenhouse gasses and other pollutants that result in climate change. Therefore, it is crucial to change in the direction of policies and measures that lead to sustainable development by diminishing the environmental consequences and other negative impacts of transport systems. Any proposed solutions should be comprehensive and encompass the transport framework and the operation of individual modes of transport, fostering low-carbon transport that furthers sustainable development in urban areas in Vietnam. This paper looks at the current state of transport development in Hochiminh City, and propose a way to create efficient and low-carbon transport on it.

Keywords Urban transport · Climate smart city · Policy

1 Introduction

Environmental aspects have been paid attention in Vietnam since two decades. Energy issues were first mentioned in 1993, with the Law No. 29-L/CTN, providing for environmental protection. This law was set as the first milestone that marked the Vietnamese government's concern about GHG emissions.

GHG inventory in 2014, revealed that the transport sector created 30.55 $MtCO_2e$ or 10.8% of total CO_2e emissions (World Bank 2019). An explosion of urbanization and economic growth significantly contributes to mobility, and in turn leads to a huge volume of GHG emissions in transport sector.

Many traffic management measures can help to achieve the sustainable urban transport objectives. These include developing public transport, organizing traffic management, vehicle control, vehicle emission control, and managing fuel quality. However, a number of constraints affect the optimal utilization of these solutions.

V. T. Tich · A. M. Ngoc (✉)
University of Transport and Communications, 3 Cau Giay, Hanoi, Vietnam
e-mail: anminhngoc@utc.edu.vn

V. T. Tich
e-mail: vttich@utc.edu.vn

L. T. T. Huong and G. M. Pomeroy (eds.), *AUC 2019*, Advances in 21st Century Human Settlements, https://doi.org/10.1007/978-981-15-5608-1_17

Based on a case study from Hochiminh City (HCMC), this paper aims to provide guidance for possible policies and measures that can be implemented to create efficient and low-carbon transport in urban areas.

Located in southeast Vietnam, HCMC consists of 24 districts and 259 wards with a total area of 2,095.39 km^2. In 2017, the average population of HCMC has reached over 8.64 million, where 80.9% lives in the urban area (Hochiminh City Statistics Office 2017). Representing 9% of the total population of Vietnam, it is the most populous and dense province in the country, followed by Hanoi with a total population of 7.4 million (Hochiminh City Statistics Office 2017). Over the past 10 years, the population growth rates range from 1 to 1.3% annually (Hochiminh City Statistics Office 2017).

Lately, transport is becoming one of the major sources of Greenhouse gas (GHG) emissions in HCMC, where 16.68% of the GHG emissions in 2013, came from the road transport sector (JICA 2013), as a result of gasoline and diesel combustion vehicles. The high growth of motorization also created many other negative impacts, such as low air quality, accidents, and fatalities, where in 2015, 9% of the traffic accidents involved children (National Transport Safety Committee 2015).

With private vehicle trips make the majority of the people's trip in HCMC, it is clear that a policy to reduce private vehicle is needed, complemented with the provision of high-quality public transport service and good access to public transport.

The study provides an analysis of the current mobility condition and plan made by the government to improve mobility in HCMC. It also shows a range of possible policies and measures that should be implemented to create low-carbon transport, and the recommended action plan and roadmap to achieve the goal of climate-smart city for HCMC.

2 Methodology

In order to achieve the main objective of this paper, a specific methodology was applied (Fig. 1).

Step 1: Literature

One of the main objectives of this paper is to specifically address the policies and measures for low-carbon transport in urban areas. In order to gain this objective, the paper starts with reviewing the state-of-art of the low-carbon transport measures in

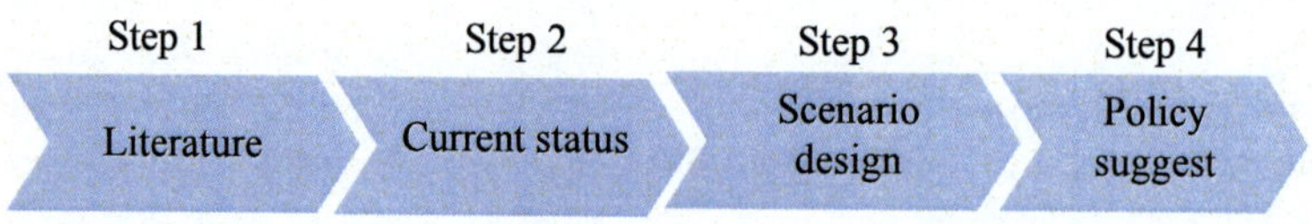

Fig. 1 Methodology

order to found the theoretical background for the low-carbon transport concept in urban areas in general and in HCMC in particular.

Step 2: Current Status

This section defines the mobility conditions and analyzes the environmental aspects of HCMC, in order to identify the impact of transport to the environment.

Step 3: Scenario Design

The goal of scenario design is to establish the options of strategies and traffic management measures and to estimate the mobility impact that would reduce emissions.

Step 4: Policy suggest

As the final step, the results from step 3 are summarized and discussed in order to suggest some key policies and measures.

3 Literature Review

Low-carbon transport measures, by using Avoid-Shift-Improve approaches, can reduce transport GHG emissions by pushing a series of actions such as modal shift, fuel economy, eliminating private transport modes.

Measures for Avoiding Traffic Demand focus on the reduction of traffic demand in heavy areas. These measures aim to make a combination, substitution, and modification of travel demand (Manfred Boltze 2003).

Measures for Shifting Traffic Demand focus on changing travel behaviors by time, mode, and location (Manfred Boltze 2003). The increase in public transport use is a performance indicator of this solution.

Measures for Improving Traffic Demand focus on the efficiency of the vehicle fleet and use.

There are different ways to categorize low-carbon transport measures. The reviewed literature shows that these above measures can be categorized into measures of each transport mode (Fig. 2).

Measures for Public transport (PT) aim to enhance the attraction of public transport services, in order to, encourage citizens to use public transport services (Speer 1993; ITE et al. 1993; Littman 2002).

Measures for Nonmotorized transport (NMT) aim to create safe and comfortable facilities for pedestrian and bicycle traffic (Berman 1995; Littman 2002).

Measures for Individual motorized vehicle (IMV) aim to increase the efficiency of these modes. These measures mainly related to vehicle technical issues such as emissions standards and fuel economy.

Measures for Intermodal transport mostly aim to encourage the integration of public–private modes. For example, the facility for park and ride allows creating an auto-transit interface (Vuchic 2005).

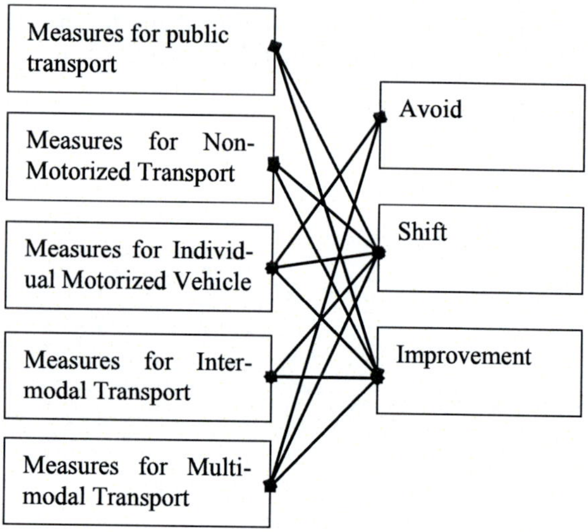

Fig. 2 Categorization of low-carbon transport measures

Measure for Multimodal transport aims to encourage all sustainable transport modes that are environmentally friendly, and ensure sustainable mobility to the citizen (Berman 1995).

4 Current Status

4.1 A Review of HCMC Mobility

Private vehicles, especially motorcycles, dominate people's trip in HCMC. Motorcycle accounts for around 80% of the trips, and private cars take around 10% of the mode share. Despite being highly visible on the road, public transport only plays a small role in serving the mobility in HCMC, that fulfills only 6% of travel demand, in the form of regular busses (Fig. 3).

In 2017, there were 7.36 million motorcycle and 640,000 private cars registered in HCMC with a growth rate of 16% per annum (cars) and 8% per annum (motorcycle). This has brought the car motorization rate from 60 cars per 1,000 inhabitants in 2008 (Emberger 2016), to 76 cars per 1,000 inhabitants in 2017, and from 390 motorcycles per 1,000 inhabitants to 871 motorcycles per 1,000 inhabitants between 2008 and 2017. Another big city in Southeast Asia, Jakarta, has an ownership rate of 494 motorcycles per 1,000 inhabitants in 2014 (ITDP 2019), while the rate in Vietnam, is approximately 490 motorcycles per 1,000 inhabitants in 2016 (National Transport Safety Committee 2015). This condition creates major transportation issues in HCMC such as congested roads, noise, and air pollution, as well as high traffic accidents.

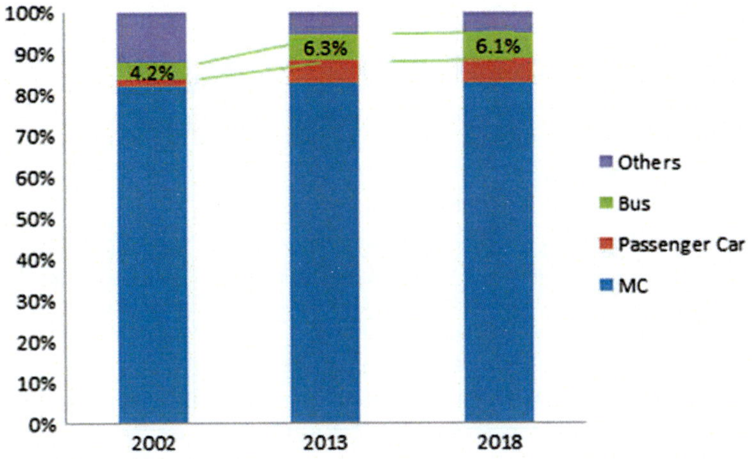

Fig. 3 Mode share in HCMC. Source: TEDI (2018)

Compared to motorcycle, bus is very unattractive. Although the fare level is low (USD 0.25–0.3), there are many aspects such as speed, operational system, supporting infrastructure, and policies of the current bus system, which makes it unreliable and difficult to access. The bus also takes much longer travel time compared to the motorcycle.

The rapid growth of motorization is still ongoing, and if such condition is not tackled with any effort to restrict private vehicle use and improve public transport, it would lead to loss of city's productivity and deteriorate the quality of life in HCMC. Additionally, the new ride-hailing App service such as Grab or Go-Viet make people in HCMC to be more motorcycle-dependent society. If not managed properly, public transport share might be close to zero over the next decade.

4.2 Policy Context in Vietnam and Hochiminh City

With the rise of emission issues, Vietnam has pursued a series of policies in the efforts toward mitigating emissions. Nationally Determined Contributions (NDC) of Vietnam, specifies the commitment to reduce 8% of its GHG emissions below business-as-usual in 2030, or equivalent to about 62 million tons of CO_2, in the sectors of energy, transport, agriculture, and waste management.

Among the policies, public policy tools available to bring GHG emissions down from the transportation sector should be reviewed and analyzed. Review of transport policies in one hand will help to assess the GHG effects, in another hand, to increase awareness of policymakers and other decision makers on the environmental impacts of transport policies and measure, pushing them to promote the effective strategies for managing and reducing GHG emissions through a better understanding of the emissions impact of policies and actions.

Transport policies provided co-benefits, not only for addressing climate change such as reduction of emissions and congestion but also other social and economic benefits. The following lists the policies in accordance with avoid (A), shift (S), and improve (I) categorization.

It appears that most of the public transport measures and individual motorized vehicle measures are being implemented. In accordance with the policy of fuel economy, it reveals that mandatory labeling for a new car (up to 7 seats) is in place. It appears although some incentives for energy-efficient vehicles exist but no policy discussion on CO_2-based vehicle taxation.

As listed in Table 1, the measures for promoting public transport service in HCMC are seen obviously in terms of route, scheduling, accessibility, and passenger information, especially for the bus system. Other efforts to increase the market share of public transport are expected from the development of Bus Rapid Transit (BRT) and Urban Mass Rapid Transit (UMRT). In the transport master plan, city schedules to operate six BRT lines and eight UMRT lines until 2030 (Table 2). However, this is an ambitious plan because the implementation of the first BRT line and the first UMRT line has encountered many obstacles so far.

In parallel with public transport measures, measures for eliminating the growth of private vehicles have been paid enough attention to HCMC recently. A series of policies has been implemented regarding parking management, vehicle road fees, etc. They contribute to the increase of the public budget but less unlikely to succeed in reduction of individual traffic demand.

Measures for Nonmotorized Transport (NMV) are not yet fully implemented. Pedestrian zones have been established somewhere in HCMC but often illegal occupied by private vehicles from street-side shops. Cycling infrastructure is completely lacking, cyclists are riding in mixed traffic flows and competing with large engine vehicles such as cars, buses, and trucks.

When reviewing the progress of work done by the government to implement the mass transit so far, it is clear that the implementation mass transit system is still far from the goal. So far, out of 8 UMRT lines in HCMC, only UMRT Corridor 1 and UMRT Corridor 2 are moving with construction and preparation. Whereas the implementation for the remaining corridors remains unclear.

To meet the goals, HCMC needs to find a way to solve these problems and to speed up the implementation process by finding the most efficient and economic mode of public transport.

5 Development of GHG Emissions Scenarios

In order to suggest the appropriate policies to create efficient and low-carbon transport, it is necessary to have an evaluation of policy impacts to reduce CO_2 emission.

Table 1 Policy and measures

No	Measure	A/S/I	Current Status
PT	**PT measure**		
1	PT route improvement	S	●
2	PT scheduling improvement	S	●
3	PT accessibility improvement	S	⊕
4	PT ROW prioritization	S	○
5	PT user incentive	S/I	●
6	PT information service	S	●
7	PT management center	I	●
8	Smart ticketing system	S/I	○
9	Clean energy vehicles	I	⊕
NMT	**NMT measure**		
1	Sidewalks and crossing facilities	S/I	●
2	Bicycle land and facilities	S/I	○
3	Non-motorized transport zone	S/I	●
4	NMT traffic information service	S/I	○
5	Electric bicycle	I	○
IMV	**IMV measure**		
1	Road pricing for private car	A/S	○
2	Parking pricing system	A/S	●
3	Parking information service	S	●
4	Tax and fees	S	●
5	Car/MC labeling	I	●
6	Fuel efficiency standard	I	○
7	Electric car/MC	I	○
MIM	**Multimodal and intermodal transport measure**		
1	Ring road	A	●
2	Signalization of intersection control	I	●
3	Speed reduction	I	●
4	Land use change	A	●
5	Park and ride facility	S	○

Note: A-Avoid; S-Shift; I-Improve
● Implemented; ⊕ Partly implemented
○ Not implemented;

Two scenarios have been studied based on the projected demand for travel considering the passenger-kilometers of travel. Scenarios have been generated to study the amount of CO_2 emission in 2030, which are as follows:

- Business-As-Usual (BAU): This scenario considers the natural development of transport activities without any added transport policies and measures after base year that aim to mitigate GHG emissions.

Table 2 Policy for public transport development

	Legal documents	2020–2030
BRT development	Decision 568/QD-TTg on Transportation Development Planning in HCMC; Decision 280/QD-TTg dated 8th March 2012 on Program of Development of Bus system in period 2012–2020	6 BRT lines
Metro development		8 UMRT lines
Bus development		Public Transport in HCMC reaches a market share of 20–25%

Table 3 List of mitigation actions

No	Measure name	Scenario 1
1	Development of BRT system	In 2021, BRT line 1 will operate; In 2025, two new BRT lines will operate
2	Development of metro system	2022: UMRT line 1 will operate
3	Promotion of electric motorcycle	It reveals that E-motorcycle accounts for 5% of the new vehicle fleet in 2016, and increase dramatically. Therefore we assume that E-motorcycle make up 14% of new fleet annual from 2019
4	Promotion of electric cars	We assume that electric cars account for 5% of the new car fleet annual from 2025
5	Promotion of electric busses	We assume that E-bus will operate since 2025

– Scenario of mitigation actions: The list of plans, that was mentioned in the transport master plan, already reviewed, and other assumptions are built to seek appropriate and feasible measures in mitigation (Table 3).

5.1 *Business-as-Usual Reference Scenario*

Under the BAU scenario, a simple regression model is applied to estimate the transport activities. The growth of passenger-km traveled (PKT) and Freight ton-km traveled (FTKT) are projected in relation to urban population, passenger volume, freight volume, load factor, and historical vehicle fleet. The result shows an impressive increase in demand. From 2014 to 2030, transport activities in terms of passenger-kilometer and ton-kilometer increased by 3.86–9%, respectively. This growth is traveled with motorization and increasing heavy-vehicle traffic. Consequently, infrastructure development will not be unable to address with such growth.

In this trend, road transport contributes to almost passenger transport, i.e., 97.26% in terms of passenger-kilometers. In freight transport, the road sector shares 17% in terms of ton-kilometers. Inland waterway shows a major share in freight transport with 82.7% in terms of ton-kilometers (Fig. 4).

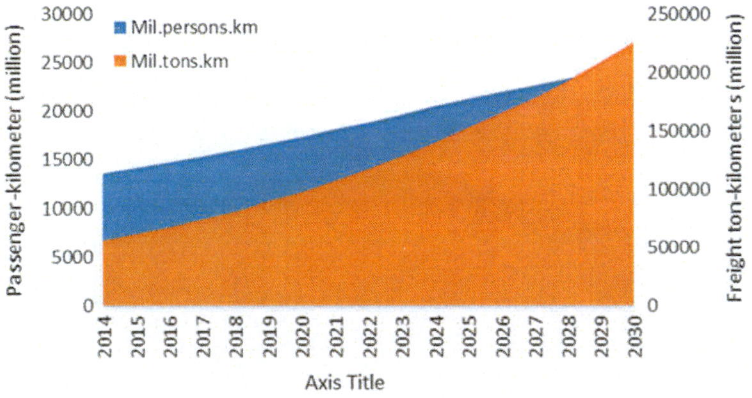

Fig. 4 Projection of PKT and FTKT under BAU

Due to the vast majority of public funding allocation for road infrastructure in HCMC, transport activities almost use road infrastructure for passenger transport. The number of motorized vehicles is projected to rapidly increase, with a high share of motorcycles (77.3% by 2030), following by passenger cars with 21.4% in 2030 (Fig. 5).

Without additional development of actions for mitigating negative impacts on the environment from traffic activities after 2014, an increase of less sustainable transport mode and reliance on fossil fuels will have significant environmental consequences. The composition of the energy sources is estimated to reach 695 kTOE. As a result, emissions are estimated to increase at an annual average of 5.5%, reaching nearly 12.81 million tons of CO_2 emissions in 2030 (Fig. 6).

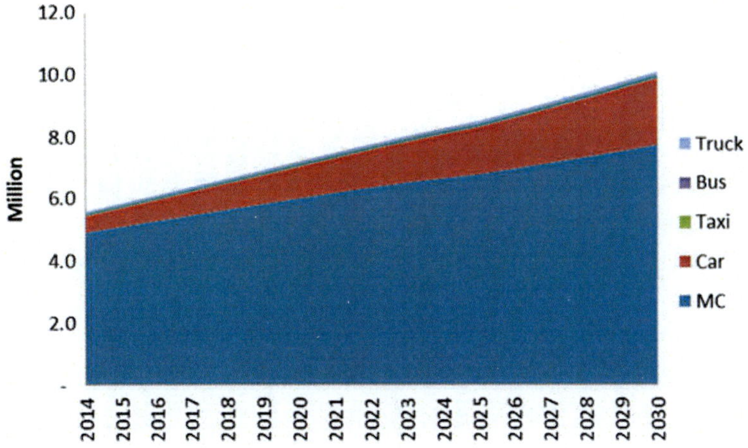

Fig. 5 Projection of vehicle numbers by vehicle type

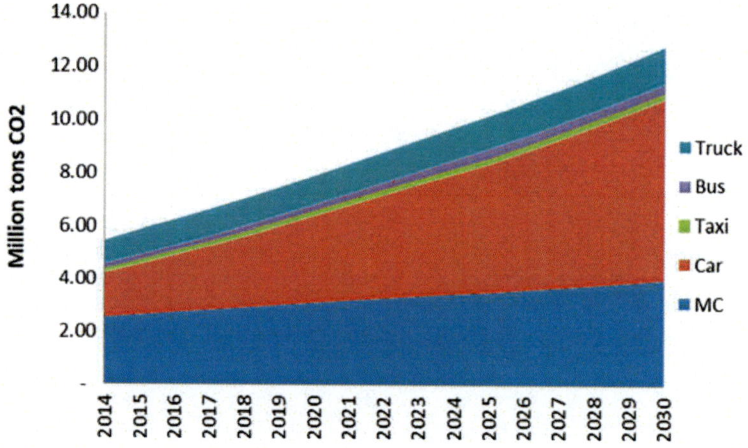

Fig. 6 Projected CO_2 emissions by transport under BAU

5.2 *Scenario—with Domestic and International Resources*

The possibility of policy will significantly depend on support from domestic and international resources. On the contrary, the government's commitment and actions to reduce emissions will create a positive signal to receive these allowances.

As mentioned in the previous section, five mitigation options will be invited to examine their impact on emissions reduction.

By using VISUM in four-steps model, the development of BRT and Metro system will result in significant shifting from private to public transport as illustrated in Fig. 7.

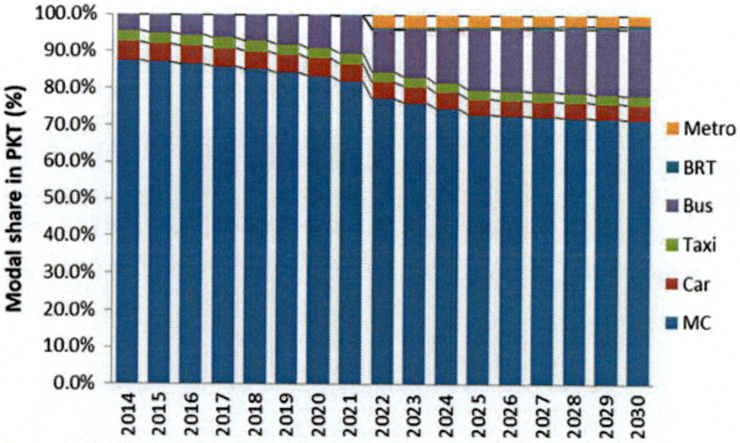

Fig. 7 Projection of modal share in PKT

The above measures would donate a reduction in CO_2 emissions compared to BAU due to a significant decrease in fuel consumption. A part of this contribution comes from the shift from private traffic demand to public transport. The shift from fossil fuel vehicles to electric vehicles (electric motorcycle, electric cars, and electric busses) also results in a reduction in gasoline consumption, leading to a reduction in CO_2 emissions compared to BAU. The result from mitigation actions shows a reduction of 6.9% of CO_2 emission compared to BAU (Fig. 8).

In a combination of these five mitigation options, investments or policies towards electric motorcycles would lead the highest proportion emissions reduction during the 2014 to 2030 period. Investment of the UMRT system would result in gradual emissions reduction over a long period, but the modal shift from private to public transport would clearly realize (Fig. 9).

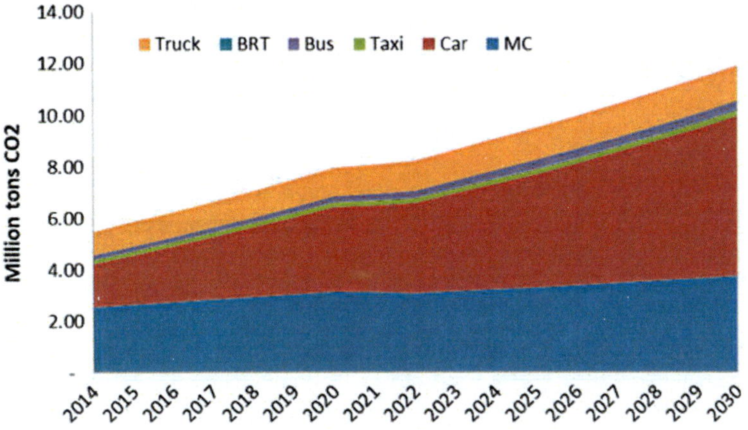

Fig. 8 CO_2 emissions by road transport under scenario

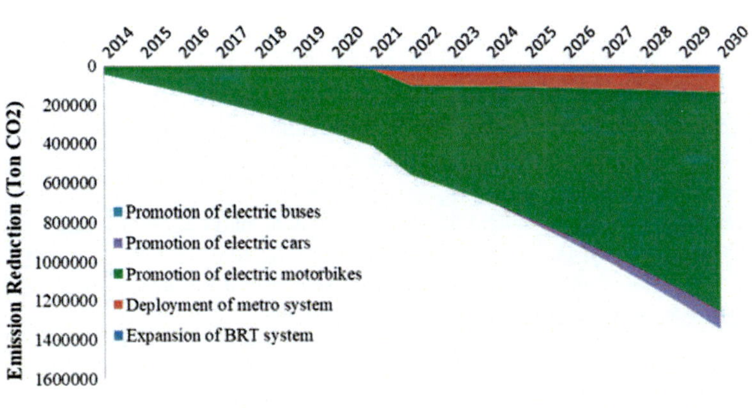

Fig. 9 CO_2 emissions reduction by each mitigation measure

In summary, the selected policies are recognized to have significant impacts on reducing GHG emissions as follows:

Modal shift—transfer passenger transport from motorcycle to UMRT, BRT, and bus: 139,246 tons CO_2 e.

Fuel shift—promotion of electric vehicles: 1.2 million tons CO_2e.

6 Policy Orientation to Create Efficient and Low-Carbon Transport in Urban Areas

The above results show that fuel shift is the best option to reduce the GHG emissions, following by modal split. Therefore, from the policy's respective, it should encourage electric vehicles (EVs) and the development of public transport.

6.1 EVs

EVs policies focus on two groups: financial and nonfinancial incentives. Financial incentives are given under direct subsidies or indirect subsidies to promote the use of electric vehicles. Reduction of tax or fiscal incentive is seen as a kind of direct subsidies, while indirect subsidies include the build of charging infrastructure or reduction of energy costs.

Development of public charging infrastructure should be considered as specific encouragement because it helps to promote EVs dramatically. Travelers often hesitate in using EV due to the inadequacy of charging stations, while their competitors are not able to set up their own charging stations.

Second EVs specific incentives should focus on commercial transport activities. A successful growth of the EVs market realized in many cities such as Chinese cities, European cities is a positive signal to the market in HCMC, where awareness of environmental pollution has changed significantly recently.

Nonfinancial factors should be considered to support development of EV fleets. Development of a zero-emission zone at the city center is an example of nonfinancial incentives. Others include the setting of user-friendly charging stations, free road fees for EVs, and free parking fees.

In the long term, national policy should include banning fossil fuel vehicles. Looking outside countries, Norway, is the first country in giving regulation that bans fossil fuel vehicles, following by more than 14 other countries, including China, with the targeting year from 2025 (Kim and Grütter 2019).

6.2 *Public Transport Development*

To accelerate the implementation progress of mass transit, instead of waiting to implement the MRT, the BRT system can be prioritized to be first implemented on the MRT corridor. The BRT which has cheaper, quicker, and the same carrying capacity as the MRT will have a similar benefit to carry the same passenger demand as the MRT.

HCMC should also build good cooperation with all stakeholders. The government should always include bus operators during the whole improvement process of the bus system because the program would have not been possible to succeed without their support and participation.

In order to manage traffic demand, there has to be good policy in place, and the use of the electronic and online system for the operation could help the government operate the program easier and reduce any possible failure on the implementation. The government should have strict law enforcement and fining mechanism.

7 Conclusions

Ho Chi Minh City will increasingly face severe transport problems due to the increase in individual travel demand, resulting in the deteriorating environmental quality. Transport is the second largest source of GHG, it contributes to the rate of 23.1% of total GHG emissions in HCMC.

Without mitigation options, transport activities occur at a natural level. Under this condition, GHG emissions from transport activities have reached at 12.81 million tons by 2030. In case some mitigation scenarios are added such as fuel shift or modal share, GHG emissions can be reduced by 6.9% compared to BAU. This is an impressive signal given that NDC commitment at the national level expects the target of 8% lower in all areas including energy, transportation, agriculture, and waste management. In which, promotion of EV contribute a reduction of 1.2 million tons of GHG emissions in HCMC.

In order to achieve this target, policies need to be seriously complemented to encourage the use of electric vehicles. The current policies obviously show the lack of regulations regarding the promotion of electric vehicle use. Especially, there are no coherent financial incentives to encourage both EVs supply and demand.

The shift from private transport to public transport will give a reduction of 0.14 million tons of GHG emissions over the next 10 years. It can be seen that the transition from individual to public transport will have a gradual impact on emissions reductions. A part of this reason that people will change their travel behavior gradually since the mass rapid transit have operated. However, the assumption is only established with one UMRT line while the metro system covers 8 lines in HCMC. In case of completing whole metro system, it therefore will be enough to create drastic change in travel behavior.

The paper aims at providing evidence through the calculation of GHG emissions under different scenarios. The impacts of each group of policies or measures on GHG reduction have been explicit over a period of time. These results facilitate the implementation of environmentally sustainable transport in Vietnamese urban areas in general and in HCMC in particular.

References

Berman (1995) Congestion control and demand management. Public Roads Online 58(3)
Emberger G (2016) Urban transport in Ho Chi Minh City, Vietnam
Hochiminh City Statistics Office (2017) Statistics yearbook
ITDP (2019) Count Survey in HCMC downtown
ITE, Georgia IT, K.T. Analytics Inc. et al. (1993) Implementing effective travel demand management measures. Institute of Transportation Engineers, Washington D.C
JICA GHG (2013) Inventory of Ho Chi Minh City
Kim K-J, Grütter J (2019) E-mobility options for ADB developing member countries. https://www.adb.org/publications/e-mobility-adb-developing-member-countries
Littman (2002). Mobility management, sustainable transport: a source book for policy-maker in developing cities. Deutsche gesellschaft fur technische zusammenrabeit GmbH, TZ Verlagsgesellschaft mbH, Eschborn
Manfred Boltze (2003) Intermodales Verkehrsmanagement: Standortbestimmung und Hinweise zur Weiteren Entwicklung, Deutsche Gesellschaf für Ortung und Navigation e.V
National Transport Safety Committee (2015) Report of national traffic safety strategies
Speer et al. (1993) Frankfurt urban integrated traffic management. Dezernat Bau and Strabenverkehrsbehorde, Stadt Frankfurt and Main
TEDI (2018) Survey report of Hochimin City transport
Vuchic V (2005) Urban transit: operations, planning, and economic. Wiley
World Bank (2019) Addressing climate change in transport, vol 1, Pathway to Low-Carbon Transport. http://documents.worldbank.org/curated/en/581131568121810607/Volume-1-Pathway-to-Low-Carbon-Transport. Accessed 11, 2019

Motorcycles in a Long-Term Perspective: Case of Ho Chi Minh City

Thanh Tu Nguyen

Abstract Currently, motorcycles are the dominant mode of transport not only in Ho Chi Minh City (HCMC) but also in the major cities in Vietnam. How did motorcycle pass from the ego of the bicycle in the 1990s, and can we avoid an automobile boom that is experienced by the industrialized countries? Demography is a powerful explanatory factor for long-term changes in mobility. Indeed, the distance traveled per person per day (both overall and in the dominant mode) follows a bell curve during the life cycle, which peaks when people are at the age of forty. The demographic transition (slowdown in population growth and aging) should lead to a slowdown in overall mobility. The aim of this paper is to understand the impact of demographic factors on long-term travel demand. By using a demographic-based model, the author estimated the motorcycle trend in HCMC, in particular, estimated the share of the motorcycle in the total mode of transport; the total travel distance, and the trip rate by motorcycle.

Keywords Motorcycle · Mobility · Age-cohort model

1 Introduction

Currently, the motorcycle is the dominant travel mode not only in HCMC but also in most of the cities in Vietnam. The history of the motorcycle in Vietnam is fascinating and tells us a lot about the country's development history. From a luxury product until the 1990s, it is now very popular for all travel patterns. According to the Department of Transport, there were about 7 million and 30 thousand new motorcycles adding more for each month in HCMC in 2017. Given this growth, the city authorities expected more than eight million motorcycles in 2020.

The analysis is undertaken at HCMC, the biggest city in Vietnam. The HCMC's population in 2018 was approximately 9 million inhabitants and was predicted to reach 13.9 million in 2025. HCMC plays an important role in

T. T. Nguyen (✉)
Faculty of Transport and Economics, University of Transport and Communications, Hanoi, Vietnam
e-mail: ngthanhtu@utc.edu.vn

L. T. T. Huong and G. M. Pomeroy (eds.), *AUC 2019*, Advances in 21st Century Human Settlements, https://doi.org/10.1007/978-981-15-5608-1_18

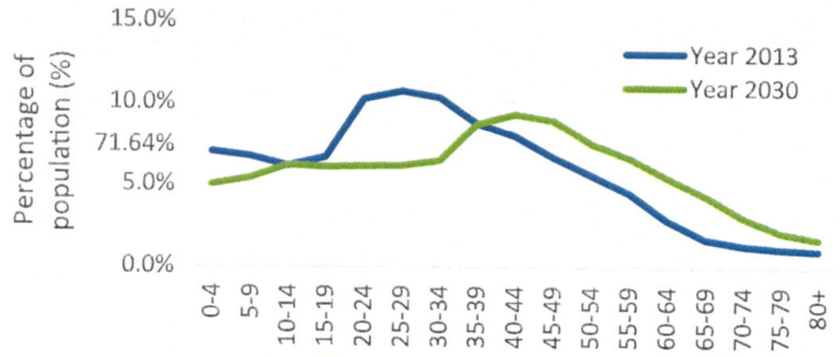

Fig. 1 Aging in Ho Chi Minh City

Table 1 Demographic change in Ho Chi Minh City

Year	2012	2018	2030*
Population	7833	8441	14401
Male	3995	4288	7236
Female	3838	4152	7165
Average age	32.1	33.8	38.3

*The population was estimated by author, included migrants arrived HCMC from other provinces

the economic development of Vietnam. The road transport system includes bikes, motorcycles, cars, busses, taxis, and in which, the motorcycle is the most popular vehicle in the city.

With life expectancy on the trend of rising, HCMC and other cities in Vietnam, are facing challenges of an aging population regarding employment, labor, health care, and social security. Therefore, many solutions are needed to improve elderly life quality (Fig. 1 and Table 1).

Travel behavior is influenced by demographic factors and urbanization. To describe this impact, we use a model for measuring travel characteristic such as trip distance, trip rate, and motorcycle share in the period 2012–2030 in HCMC. This model, based on the population analysis, describes the mathematical relationship between the travel behavior and the age, the period in life, and the cohort (birth year), and we call Age-Cohort-Period or APC model. Three main parameters of the population are explained as follows:

- Age: shows the effect of age on travel behavior. By analyzing mobility data, we can see a characteristics curve of mobility changes that could be related to age.
- Cohort (year of birth): which considers the travel behavior of a group of people born in the same period, who have a common experience. We can forecast the travel behavior in a long-term perspective by analyzing the gap effect between the cohorts.
- Period (year): the effect of changes occurred at a particular time on travel.

In this study, the period effect is not considered because the gap between two periods is 7 years only, and there are not so many changes in socioeconomic during this time.

2 Travel Demand Modeling

Household travel surveys aim to collect the mobility data in an urban area. It is useful for analyzing the daily mobility and transport planning (Krakutovski 2004). But there are cross-sectional surveys, which describe the picture of mobility in urban area of a single day. Research of (Gallez 1994; Madre 1989), shows that they have limitations for long-term travel demand forecasting. The study of (Grimal 2011), indicates that the projection of car ownership cannot be estimated from cross-sectional survey. The model based on the panel data offer much more advantages than conventional cross-sectional approaches, and it is appropriate for travel behavior analysis and demand forecasting (Kitamura 1990).

This study tried to explain the relationship between the travel behavior and the age factors (which is presented by the components of life cycle, generation, and period mentioned above Dejoux et al. 2009; Krakutovski 2004; Madre and Armoogum 1996).

2.1 Age—Cohort Model in Long-Term Travel Demand Forecasting

Age period cohort (APC) analysis is often used in the research of changes overtime in epidemiology. In transportation planning, there are some researches of (Bussière 1996; Dejoux et al. 2009; Krakutovski 2004; Madre et al. 2000; Tapia-Villarreal 2014; Nguyen and Madre 2016) which applied this model for travel demand forecasting, not only in developed countries but also in developing countries. In fact, the Time, Age, and Year of birth affect travel demand.

In order to estimate these effects (age, generation, and period) on travel behavior, and to project the future mobility, a demographic approach was developed by the INRETS researchers in the 1980s. This approach mainly based on the study of (Madre 1989; Madre and Armoogum 1996) first for car ownership and then for travel behavior (see Fig. 2). They have shown that data from panel surveys can be used for this model.

In most studies, the regression parameters show that the travel characteristics were estimated approximately by the Age-Cohort model and by analysis criteria such as age, gender, zone residential, etc. Regarding the precision in the forecast (Dejoux et al. 2009), shows that the errors are about 10–15% when applying the Age-Cohort model to Montreal and Paris. But with a small size of the sample, this error can be higher, from 30 to 40%. Besides the effects of age and cohort (Tapia-Villarreal

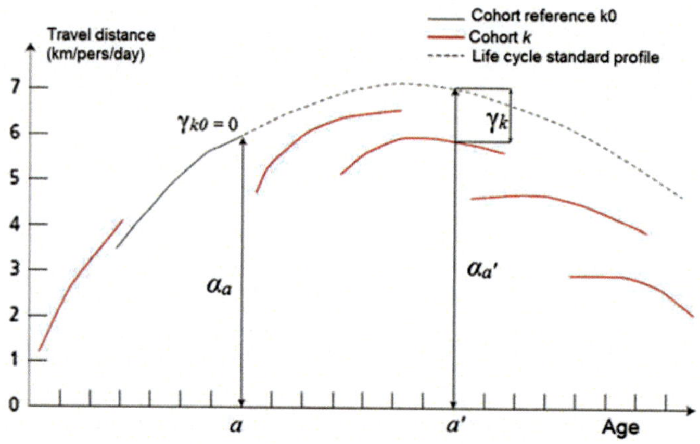

Fig. 2 Travel behavior by a group of age, based on (Gallez 1994)

2014), tried to investigate the effect of urban form on travel behavior by using the same model. He has shown that the person living in a lower population density area has longer travel distance than the one living in the city center.

The travel demand forecasting model is referenced from (Dejoux et al. 2009). This model is mainly based on the Age-Cohort model, which estimates the impact of age and cohort on travel behavior. The variance analysis model is as follow:

$$\mu_{a,g} = \sum_{a \varepsilon A} \alpha_a I_a + \sum_{a \varepsilon A} \gamma_g I_g + \varepsilon_{a,g}$$

In which:

- $\mu_{a,g}$: measurement of travel criteria (trip rate, total trip distance per day and motorcycle share)
- a is the age group of respondent and g is their generation or year of birth;
- α_a measurement of travel criteria of a preference generation at the age group a.
- I_a: dummy variable of the age group a;
- γ_g: measurement of the difference between the generation g and preference generation $\gamma_g 0$;
- I_g: dummy variable of the generation g.
- $\varepsilon_{a,g:}$ model errors.

2.2 *The Model for Travel Demand Forecasting*

The travel demand characteristics (trip rate, daily distance traveled, and motorcycle share) of a person in zone residential (z), gender (s) at the date (t) is calculated by the formula:

$$\mu_{a,g}^{z,s} = \alpha_a^{z,s} + \gamma_g^{z,s}$$

In which:

- $t = a + g$ (a is the age group of respondent and g is their generation or year of birth);
- $\alpha_a^{z,s}$: measurement of travel criteria of a preference generation at the age group a, of gender s at residential z;
- $\gamma_g^{z,s}$ measurement of the difference between the generation g and preference generation of gender s at residential z;

The total travel demand of population at the date t is calculated by the formula:

$$N_t = \frac{\sum_{s=1}^{2}\left(P_{a,t}^{s} * \pi_{a,k=t-a}^{s}\right)}{\sum_{s=1}^{2} P_{a,t}^{s}}$$

where:

$P_{a,t}^{s}$ is the projected population at the date t of gender s.

3 Application of the Age-Cohort Model in Travel Demand Modeling in Ho Chi Minh City

The measurement is calculated for "five-year" standard cohort in demographic analysis. The age groups with the limited sample are needed to be aggregated. People who are more than 65 years old were arranged in the same group '65 and above'. People who were born between 1976 and 1980 were grouped with the same generation "1976–1980" and this generation is selected as the generation of preference in the model.

In order to be able to analyze the effects of Age and Generation on travel behavior, Age-Cohort model calibration which based on the variance analysis needs mobility data collected from at least two surveys. It is better to have more two surveys to get a residual term from the factors not included in the model (i.e., economic factors). But in HCMC, there are only two mobility data sources with enough sample size:

- First, the survey conducted in 2012, in the Data Collection Survey for Railway for major cities in the Vietnam project with a sample size of 20,000 households, thanks to the support of Japan International Cooperation Agency (JICA).
- Second, the household travel survey in 2018, as a part of the Urban Traffic Modeling Project in HCMC with the participation of 9000 households.

Two models of variance analysis for two genders were run. Travel behavior is measured by three variables:

- Total number of trips undertaken by motorcycle (average trip number per day and per person);

- Daily trip distance undertaken by motorcycle;
- Motorcycle modal share (the number of trips among total trips undertaken by all modes of transport).

3.1 Model Age-Cohort of Daily Distance Traveled by Motorcycle

The Age-Cohort model shows that the age group from 40 to 49 for men and 50–54 years for women, the daily distance traveled by motorcycle is the highest. In term of cohort effects, people born earlier has less travel distance than those born after. Specifically, those born before 1930, will have the least distance traveled. Concerning gender, cohort effects for women are larger than for men (Table 2).

3.2 Model Age-Cohort of Motorcycle Trip Rate

The model parameters are shown in the following Table 3.

The age-cohort model shows that the motorcycle share is highest for men at age 35–39, with 84.2% of total trips. This maximum trip rate by motorcycle is 79.9% for women from 25 to 29 years old. When comparing the sexes, the trip rate of the motorcycle of women is often lower than men at the same age.

3.3 Validation of Model

As the model only gives results every five years, we used linear interpolation to estimate the results of the two surveys. In order to validate the model, it is necessary to compare the data collected with model results. Thus, we used a regression between these values at the most detailed level, i.e., by crossing the variables: 2 genders; 17 age groups, and 2 survey dates, so there are $2 \times 2 \times 17 = 68$ points for the regression. The model is validated if R-square and the model parameter of the regression is approximatively 1, the model constant is not significantly different from zero.

The results in Table 4, show that the travel demand forecasting using an Age-Cohort model seems to be quite appropriate.

Table 2 Result of the distance traveled by motorcycle based on the Age-Cohort model

Age	Men		Women	
	Estimation of parameter	Err. standard	Estimation of parameter	Err. standard
5–9	9.94	0.50	8.76	0.53
10–14	10.40	0.47	8.09	0.49
11–19	14.24	0.50	12.06	0.49
20–24	19.41	0.46	15.76	0.45
25–29	18.12	0.41	14.55	0.38
30–34	19.03	0.29	15.05	0.27
35–39	19.06	0.29	15.50	0.25
40–44	19.29	0.40	15.99	0.35
45–49	19.60	0.51	17.53	0.45
50–54	19.05	0.61	18.17	0.55
55–59	14.96	0.72	15.91	0.67
60–64	14.53	0.86	15.69	0.80
65–69	14.15	1.15	14.89	1.05
70–74	12.77	1.63	14.93	1.42
75–79	12.74	2.08	13.71	1.96
80+	12.98	2.88	12.49	2.41
Cohort	Men		Women	
	Estimation of parameter	Err. standard	Estimation of parameter	Err. standard
<1930	−6.10	3.19	−11.01	2.70
1931–1935	−5.53	2.54	−10.08	2.27
1936–1940	−5.25	1.88	−8.53	1.70
1941–1945	−5.17	1.41	−7.90	1.30
1946–1950	−5.18	1.02	−8.19	0.95
1951–1955	−4.79	0.81	−7.50	0.76
1956–1960	−4.26	0.68	−7.09	0.63
1961–1965	−3.36	0.59	−6.04	0.53
1966–1970	−2.16	0.49	−3.54	0.43
1971–1975	−1.12	0.38	−1.26	0.34
1976–1980	0	0	0	0
1981–1985	1.07	0.37	2.00	0.34
1986–1990	2.17	0.47	3.70	0.44
1991–1995	2.88	0.55	5.95	0.53
1996–2000	3.61	0.64	6.13	0.63
2000–2005	2.91	0.75	4.04	0.75
2006–2010	3.48	0.86	4.47	0.90
2010+	4.66	1.16	4.58	1.28

Table 3 Result of motorcycle share based on the age-cohort model

Age	Men		Women	
	Estimation of parameter	Err. standard	Estimation of parameter	Err. standard
5–9	0.648	0.002	0.673	0.002
10–14	0.479	0.001	0.479	0.002
11–19	0.487	0.001	0.473	0.001
20–24	0.801	0.001	0.769	0.001
25–29	0.836	0.001	0.799	0.001
30–34	0.835	0.001	0.786	0.001
35–39	0.842	0.001	0.788	0.001
40–44	0.831	0.001	0.756	0.001
45–49	0.823	0.001	0.719	0.001
50–54	0.782	0.001	0.628	0.001
55–59	0.741	0.001	0.551	0.002
60–64	0.663	0.002	0.472	0.002
65–69	0.588	0.002	0.459	0.003
70–74	0.402	0.004	0.333	0.004
75–79	0.422	0.005	0.373	0.006
80+	0.413	0.008	0.272	0.007
Cohort	Men		Women	
	Estimation of parameter	Err. standard	Estimation of parameter	Err. standard
<1930	−0.034	0.002	−0.034	0.002
1931–1935	−0.049	0.002	−0.049	0.002
1936–1940	−0.021	0.001	−0.021	0.001
1941–1945	−0.004	0.001	−0.004	0.001
1946–1950	−0.007	0.001	−0.007	0.001
1951–1955	−0.004	0.001	−0.004	0.001
1956–1960	0.006	0.001	0.006	0.001
1961–1965	0.009	0.001	0.009	0.001
1966–1970	−0.017	0.001	−0.017	0.001
1971–1975	−0.095	0.001	−0.095	0.001
1976–1980	0	0	0	0
1981–1985	−0.061	0.002	−0.061	0.002
1986–1990	−0.017	0.002	−0.017	0.002
1991–1995	0.003	0.003	0.003	0.003
1996–2000	−0.034	0.002	−0.034	0.002
2000–2005	−0.049	0.002	−0.049	0.002
2006–2010	−0.021	0.001	−0.021	0.001
2010+	−0.004	0.001	−0.004	0.001

Table 4 Regression result of model validation

Model	R^2	Gradient		Constant	
		Estimation of param.	*t*-value	Estimation of param.	*t*-value
Distance traveled	0.99	0.96	61.4	0.5	3.42
Motorcycle share	0.99	1.005	101.1	−0.005	−0.85

Table 5 The daily travel distance by motorcycle

Year	2012	2018	2030
Men	14.6	16.7	17.2
Women	12.3	14.9	16.3
All	13.5	15.9	16.8
Total number of trip per day			
Men	2.89	3.13	3.14
Women	2.86	3.12	3.11
Percentage of trip by motorcycle			
Men	71.2%	71.0%	69.7%
Women	64.3%	62.8%	61.9%

4 Long-Term Forecasting of Motorcycle in Ho Chi Minh City

4.1 Estimation of Daily Travel Distance by Motorcycle

Based on the result of the Age-Cohort model, we predicted the travel distance by motorcycle in 2030, in HCMC.

The total distance traveled in Ho Chi Minh City is in the increasing trend. The reason for this trend is the rapid urbanization in the suburban area and rural area from the year 2010 so that the average distance traveled per person per day increased from 13.5 to 15.9 km between 2012 and 2018, and could reach 16.8 km in 2030. The travel distance of men is greater than women, but the gap between them is decreasing in 2030 (Table 5 and Fig. 3).

4.2 Estimation of Trip Rate by Motorcycle

The daily number of trips traveled increased between 2012 and 2018, from 2.89 trips to 3.14 for men and 2.86 to 3.11 for women. By 2030, the average number of trips seems to be stable compared to 2018. This phenomenon can be explained by the increase in mobility among people over 40 years, which can compensate for the decrease of travel demand due to the aging of the population.

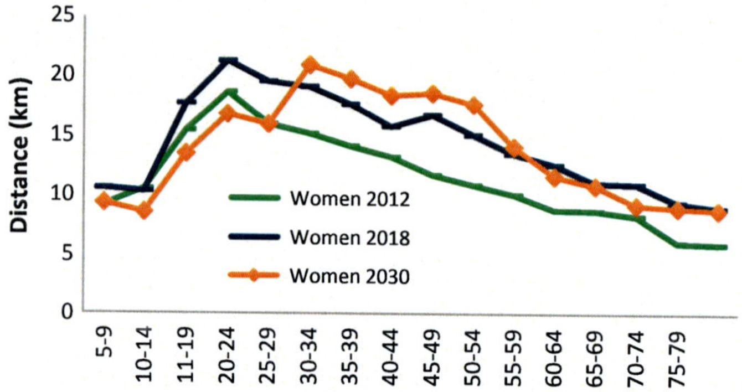

Fig. 3 Distribution of daily distance of women by age group

In recent years, Vietnam is experiencing a rapid growth in car use. This is the reason why the modal share of the motorcycle in the big cities is decreasing by 2030. Specifically, for men, this rate decreases from 71.2% in 2012 to 71.0% in 2018, with a projection at 2030 of 69.7%. This rate is in the same way for women.

5 Conclusion

This study used the Age-Cohort model in order to predict the travel demand in HCMC, in which the travel behavior (number of trips, travel distance, motorbike-use rate) is the dependent variable and demography characteristics (age, sex) are independent variables. Socioeconomic factors (income, economic growth, etc.) are not considered in this model. The study shows that the travel behavior in HCMC will be influenced by demographic and aging factors. The elderly percentage is highly increasing because of the longevity and reduced fertility rate. In 2030, the built-up area in HCMC increase rapidly and could continue to 2050. By 2030, more than 60% of HCMC land is urban area and about 50% of population living in the suburban.

The major changes in future travel demand were predicted by an Age-Cohort model. We see the difference in travel behavior between men and women. Women have lower mobility than men in both trip frequency and trip distance, but this difference is more and more decreasing, like in most cities over the world. The results of model show that the rate of motorcycle in the total mode of transport in HCMC in 2030 decreased from 71.2% in 2012 to 69.7% for men. This rate for women is predicted in deacreasing trend from 64.3 to 61.9%. The rate of using motorcycles is quite high and motorcycles will be still the main mode of transport in the urban area. However, with the economic development, cars use is growing rapidly in HCMC.

References

Bussière Y (1996) Urban growth, motorization, and pollution in mid-sized third world cities—future perspectives derived from two case studies: Marrakech and Puebla. In: CODATU-VII. New-Delhi, India

Dejoux VB, Madre Y, Armoogum J (2009) Projection of the daily travel of an ageing population: the Paris and Montreal case, 1975–2020. Transp Rev 30(4):495–515

Gallez C (1994) Identifying the long term dynamics of car ownership: a demographic approach. Transp Rev Transnatl J 14(1):83–102

Grimal R (2011) A disagregate model of car ownership based on French national transport survey. In: European transport conference 2011. Association for European Transport, Glasgow

Kitamura R (1990) Panel analysis in transportation planning: an overview. Transport Res A Gen 24(6):401–415

Krakutovski Z (2004) Améliorations de l'approche démographique pour la prévision à long terme de la mobilité urbaine. In: Thèse de doctorat en Transport

Madre JL (1989) Projection du trafic automobile sur les routes nationales et les autoroutes françaises. In: Fifth world conference on transport research. Yokohama, Japan

Madre, Armoogum J (1996) Motorisation et mobilité des franciliens aux horizons 2010–2020. INRETS

Madre J-L, Hubert J-P, Armoogum J, Strambi O, Bussière Y (2000) Forecasting pollutant emissions by automobiles in three large Metropolitan Areas: Sao Paulo, Montreal and Paris. CODATU IX, Mexico

Nguyen TT, Madre J-L (2016) Projection of the daily mobility and CO_2 emission in Hanoi using the demographic-based model. In: ATRANS symposium 9th, Bangkok, Thailand

Tapia-Villarreal (2014) Urban form, demography and daily mobility forecasts: comparative analysis France-Mexico. In: Doctoral dissertation. Univ. Paris 1 Panthéon-Sorbonne and IFSTTAR, p 382

Transformations in Land Management

GIS and Fuzzy Logic Approach for Providing Land Value Information: A Case Study in Hanoi City

Quang Cuong Doan

Abstract Land value plays a vital role in the success of land distribution processes. Land valuation is a process of assessing and synthesizing elements of land related to land prices, which help the government to plan tax and real estate market policies. However, current land prices are issued by the People's Committees of Hanoi city, mainly based on location and route factors, and not on a combination of factors affecting land values. Therefore, in this study, nine factors impacting land values were generalized into three groups including (1) public services factors, (2) environmental factors, and (3) socio-cultural factors. Thereafter, a case study using geography information system and Fuzzy logic was implemented with a view to building thematic raster classes and finding the land value. The results show the distributions of land value zones in the study area, which reflect reality. Based on these findings, managers can make policy plans for land use to ensure sustainable development. Each land value is assigned to a definite plot of land and will be available to the public via e-government.

Keywords GIS · Fuzzy logic · Land value · Hanoi

1 Introduction

One of the vital content of the government to achieve sustainable development is effective management land information (Bovkir and Aydinoglu 2018). Determining the value of the real estate has become an important and indispensable element in real estate management (Holden and Bezu 2016). Land value is the main factor in the success of land allocation processes (FAO 2003). It also plays an important role , because there are many problems that involve the socioeconomic life of developed and developing countries (Johnstone 2004). This is also a process to reflect the quality of land by using a dimensionless score for land plots of an area (Demetriou 2016). Based on land value information, the government can adjust the legal framework to

Q. C. Doan (✉)
Faculty of Geography, VNU - University of Science, Vietnam National University, 334 Nguyen Trai, Thanh Xuan, Hanoi, Vietnam
e-mail: doanquangcuong_t58@hus.edu.vn

L. T. T. Huong and G. M. Pomeroy (eds.), *AUC 2019*, Advances in 21st Century Human Settlements, https://doi.org/10.1007/978-981-15-5608-1_19

plan policies on taxes, market capitalization, and real estate market management. Based on land value information, the government can adjust the legal framework to plan policies on taxes, market capitalization, and real estate market management. One of the tools to support land valuation is based on geospatial data and GIS technology. Geographic information systems (GIS) are computer programs that allow the processing of geographic information and location. Currently, GIS world has been widely used in many cities to develop plans. The spatial data on land can be defined at specific locations through geocoding (Samuels-Kalow and Camargo 2019). Maps can be layered to show the relationship between different types of data (such as public service quality and environmental quality).

Land is a valuable asset and land values are influenced by many factors. Many authors have made great achievements in determining land value, by combining factors affecting land prices; Bovkir and Aydinoglu (2018) built up a system to provide land value information based on national geographic data infrastructure (Bovkir and Aydinoglu 2018). Studies have shown that GIS is an effective tool in determining land value (McCluskey et al. 1997). In determining the value of land, the indicators affecting the value of land are very diverse. The degree of influence of these indicators is implemented by methods such as AHP (hierarchical analysis process), multiple regression, neural network, and fuzzy logic (Shenavr and Hosseini 2014; Demetriou 2016; Nghiep and Cripps 2001). In particular, fuzzy logic method is highly effective in land valuation (Sarip and Hafez 2015; Bogataj et al. 2011; Hajnal 2014).

Quoc Oai is a suburban district located in the west of Hanoi capital. According to the general planning of Hanoi's capital to 2030–2050, Quoc Oai district will develop towards becoming an eco-town, and many projects are planned for development. Therefore, the provision of land value information will serve as a basis for planning future land use policies and real estate market development.

Along with the rapid development of cities, land values are increasing. One of the factors directly affecting public utilities such as urban railways, airports, urban areas. Through the study of Lin and Hwang, 317 residential areas within a 400 m radius of the railway see their bottom price increase by 480 USD/m^2 from 1993 to 1995 (Lin and Hwang 2004). Land prices increased by 16% for residential areas within 1 km of the railway, and this value was inversely proportional to the distance to the station (Anantsuksomsri and Tontisirin 2015; Sun et al. 2016). In addition, the value of land will increase significantly if an ecological value is provided. For example, parks have a significant impact on transfer prices and rental prices (Hammer et al. 1974; Kim et al. 2019). In the city of Philadelphia, land with a distance of fewer than 90 m from a park will have a rent that is twice as high as that with a distance of 200 m (Hammer et al. 1974). At the macro level, land values in Israel, are influenced by economic, demographic, and environmental factors, in which factors closely related to land values include: immigration, income average, lending rates (Kheir and Portnov 2016). In addition, utilities serving daily life such as schools and hospitals also significantly impact land prices. Based on the geographically weighted regression (GWR) model, Nakamura points out that the quality of living environment and business ability has a positive impact on land values (Nakamura 2019). In Vietnam, land use right certificates are a legal basis in real estate transactions, so legal status has a significant impact on land value. Besides legal factors, transportation has an important meaning in economic and cultural development. The presence of transportation systems increases asset value by improving accessibility. The transport network facilitates the circulation of goods, reducing transportation and investment development costs. According to Demetrious, the transport network (expressway, national highway) has a greatly influences on the land value. In areas with national highways and railway stations, land values are higher from 9.6 to 10.1% compared to areas without roads and railways passing through (Malaitham et al. 2018). According to research by Qisheng Pan, in Shanghai-China city, the distance to city centers increases by 1 km, the land price will decrease by 4% (Pan et al. 2014). On the other hand, activities from shopping centers, bank branches, hospitals, schools, airports, green spaces are factors that change land values, specifically shopping centers office make rental rates higher (Bollinger et al. 1998). Moreover, the quality of schools, security, crime rates, and pollution levels also have implications for land values, especially in densely populated areas such as industrial parks and streams; immigrants pushed up the demand for land use, so the value of land increased sharply.

2 Method

2.1 Spatial Analysis Via GIS

Geographic information system (GIS) is used as a powerful and integrated tool with a unique ability to store, manipulate, analyze, and visualize environmental data referenced geographically (Jelokhani-Niaraki et al. 2018). Therefore, GIS is used to analyze the spatial relationship between objects, assess the number of spatial indicators, and extract secondary information. In the study, some spatial analyses used were.

The Euclidean Distance is the distance of the straight line (distance calculated by the fly) between two points on a plane. In the Euclidean distance analysis method, the distance is calculated as the distance from the center of the source cell to the center of the surrounding cells. The algorithm to determine this Euclidean distance is as follows: for each cell, the distance to source cells is determined by calculating the length of the hypotenuse based on two values x_max and y_max are the two values of the sides.

Overlaying information layers (Overlay).

In this paper, overlaying the raster information layer will be used to create a common raster layer of component criteria. With overlaying raster data layers, each raster layer is a matrix of the pixel which has the same size. The value at each position on a matrix is combined with the value of the corresponding position on another matrix to derive the new value (Fig. 1).

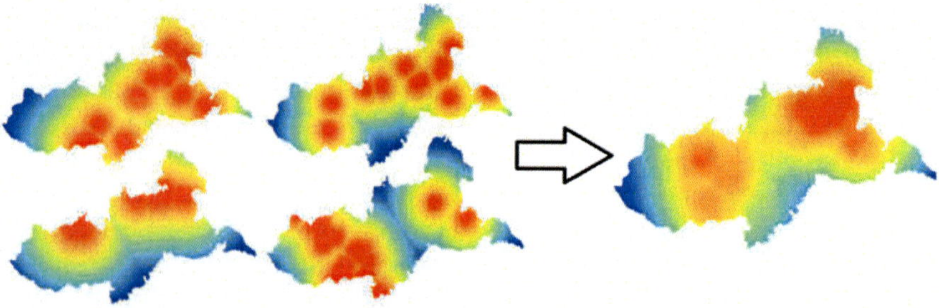

Fig. 1 Overlay function in ArcGIS

2.2 *Fuzzy Theory*

Land values are influenced by many different factors. Therefore, when determining land value, it will be more effective if supported by statistical methods (YaĞMahan and Gülgen 2018), such as AHP multi-criteria analysis, artificial neural, geographical weight (Bogataj et al. 2011; Hajnal 2014). In the statistical process, the relationship between land values and influencing factors is expressed in the form of mathematical models, in which mathematical operators in Fuzzy models help to bring about quick results during processing. Fuzzy theory is a method that allows evaluating many different levels of an element's ability to belong to a set. Fuzzy logic suggests that each element in a fuzzy set can assume a value from 0 to 1. Consider X as a set of x elements. Fuzzy set is a set expressed as a member function μM (x) representing the ability to x in the M. Set the value representing the membership level of an element x ranges from 0 to 1. If $\mu M(x) = 0$ means that element x is completely non-part M and vice versa, if μM $(x) = 1$ means that element x is completely part of M (Fig. 2).

There are many fuzzy functions, but the trapezoidal (1) and triangular (2) forms are most commonly used.

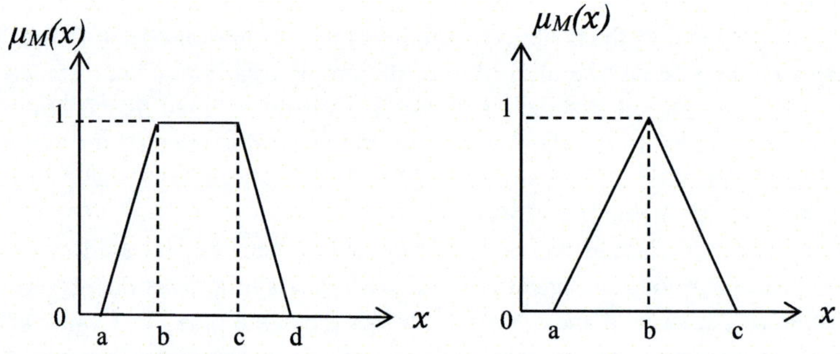

Fig. 2 The main fuzzy function

$$\mu_{M(x)} = \begin{cases} 0 & x < a, x > d \\ \frac{x-a}{b-a}, & a \le x \le b \\ 1 & b < x < c \\ \frac{d-x}{d-c}, & c \le x \le d \end{cases} \quad (1)$$

$$\mu_{M(x)} = \begin{cases} 0 & x < a \\ \frac{x-a}{b-a}, & a \le x \le b \\ \frac{c-x}{c-b}, & b \le x \le c \\ 0 & x > c \end{cases} \quad (2)$$

3 Study Area

3.1 Study Area and Data

Quoc Oai district is a suburban district of Hanoi Capital, including 21 administrative units (1 town and 23 communes) with a natural area of 147.01 km^2. Quoc Oai district is about 20 km from the city center. The Quoc Oai district's boundary is bordered to the east by Dan Phuong and Hoai Duc districts; Hoa Binh province to the west; the Chuong My district to the south; and Thach That and Phuc Tho district to the north. Quoc Oai is a district located in the transition zone between mountainous and plain areas. There are two key transportation routes passing through Lang-Hoa Lac and Ho Chi Minh roads, so there are many advantages for urban and industrial development. In addition, Quoc Oai district also has a complex of Thay pagoda with many unique cultural features (Fig. 3).

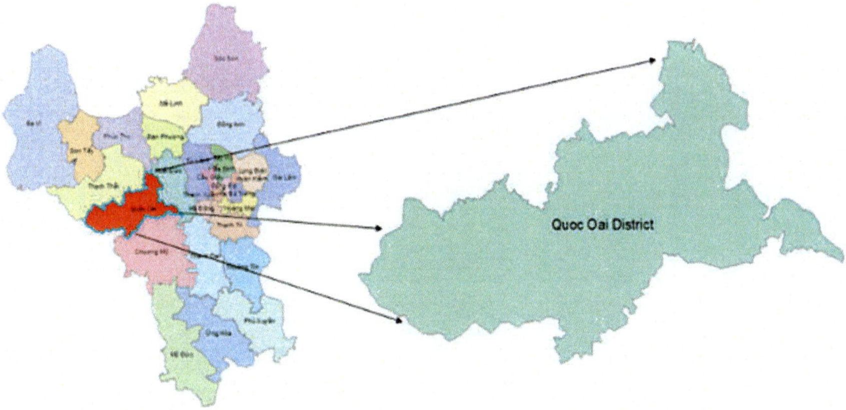

Fig. 3 Study area

3.2 *Select Factors that Affect Land Values*

Based on the current situation of natural, socioeconomic, and data sources. The study of determinants of land value factors includes the following groups of factors: (1) public service factors, (2) environmental factors, (3) sociocultural factors. The details are shown in the Table 1.

4 Results and Discussion

To build fuzzy functions for influencing factors, research data collection of public service groups, aboveground features, and cultural and social factors. For each data set, the raster layer is set using a distance measuring tool (Euclidean distance). The distance between cells on the surface is calculated by the Pythagorean formula. If cell A (p1; p2) and cell B (q1; q2), then the distance between the two cells is determined by

$$d(A, B) = \sqrt{(q_1 - p_1)^2 + (q_2 - p_2)^2} \tag{3}$$

Fuzzification of Public Service factors
In the group of public service factors, data on national highways, utilities such as hospitals, health care centers, education centers, and airports are collected as shapefiles. The ideal distance for main roads is 100 m, the distance to roads more than 1000 m will not make much sense. For facilities such as hospitals and schools, the ideal distance is defined as 500 m, due to good accessibility and the benefit of encouraging people to use public transport or walking (Construction 2008). From 500 m upwards, the value of member function decreases. The ideal distance to an airport

Table 1 Factors affecting land prices in Quoc Oai district

No	Valuation criterions	Groups of factors
1	Highway	Public service factors
2	Hospitals, health care centers	
3	Education	
4	Markets	
5	Urban areas	
6	Industries area	Environmental factors
7	Green space	
8	Waste	
9	Historical	Sociocultural factors
10	Administrative center	

Table 2 Fuzzy member in the public service group

Valuation criterions	Function members
Highway	$f(x) = \begin{cases} 1, 0 \le x \le 100 \\ \frac{1000-x}{1000-100}, 100 < x < 1000 \\ 0, x \ge 1000 \end{cases}$
Hospital, Heath care center, Education, Market	$f(x) = \begin{cases} 1, 0 \le x \le 500 \\ \frac{1500-x}{1500-500}, 500 < x \le 1500 \\ 0, x \ge 1500 \end{cases}$
Urban area	$f(x) = \begin{cases} 1, x \le 1000 \\ \frac{2000-x}{2000-1000}, 1000 < x \le 2000 \\ 0, x > 2000 \end{cases}$

is 3000 m, which has not only many facilities but also has comprehensive planning, potentially brings more value, from 4000 m or more, the impact of the airport to the price land no longer has much meaning (Table 2).

Fuzzification of Environmental factors

In the group of environmental factors, industrial parks attract a large number of workers to work, so the demand for housing and using public utilities increases fast, making land values rise. The ideal distance to the industrial zone is 800 m, from 2000 m or more, the land value gradually decreases due to the more difficult access to the industrial zone.

Urban green space has a positive impact on human health (Cole et al. 2019). Moreover, the urban space is the attraction of many entertainment and entertainment areas. Therefore, the value of the land will be higher. According to the survey, the ideal distance to green spaces is 600 m. Distance less than 1000 m to the park facilitates sport and walking. Besides, waste factors have a negative influence on the environment; especially, the impacts are most significant around 500 m. It may cause pollution of water, air, and soil environment. Around a radius of 1000 m, the impact of pollution is a significant decrease (Table 3).

Fuzzification of Socio-cultural factors

In the sociocultural factor group, Quoc Oai district administrative center is located in the central area. It has its management campus connected to other agency headquarters and plays an important role in the central area. The ideal distance for this factor is less than 500 m, from 500 to 1000 m, the impact on the land price decreases. In the Quoc Oai district, the complex of Thay pagoda is a unique relic with many cultural and historical values. Therefore, the surrounding area's land is quite expensive. Under 800 m, the influence of this monument on land value is most significant. From 800 to 1500 m this effect decreases (Figs. 4, 5 and Table 4).

Table 3 Fuzzy member in aboveground feature group

Valuation criterions	Function members
Industries area	$f(x) = \begin{cases} 1, x \leq 800 \\ \frac{2000-x}{2000-800}, 800 < x \leq 2000 \\ 0, x > 2000 \end{cases}$
Green space	$f(x) = \begin{cases} 1, x \leq 600 \\ \frac{1000-x}{1000-600}, 600 < x \leq 1000 \\ 0, x > 1000 \end{cases}$
Waste	$f(x) = \begin{cases} 0, x \leq 500 \\ \frac{x}{1000}, 500 < x \leq 1000 \\ 1, x > 1000 \end{cases}$

4.1 *Overall Assessment*

Composite raster classes are implemented based on the raster of member functions. Each raster layer of member functions affects the land value differently. Therefore, the weights of the member functions are determined by averaging raster values of each element in the zoning as shown in the Table 5.

We have formula: Land value = "Highway" * 0.14 + "Health care factor" * 0.06 + "Education" * 0.15 + "Markets" * 0.11 + "Urban areas" * 0.06 + "Green space" * 0.04 + "Industries" * 0.16 + "Waste factor" * 0.04 + "Administrative center" * 0.12 + "Historical" * 0.12.

From the synthetic raster class, it can be seen that the highest value areas belong to the center of Quoc Oai town; Sai Son area—where the complex of Thay Pagoda is located. Besides, the area along the Thang Long highway also has a high land value. In the area adjacent to Universities (Hanoi National University, Forestry University), the land value is high. These areas, in the future, have great potential to develop into urban areas, connecting the Quoc Oai district with Xuan Mai urban area in Chuong My district and Hoa Lac hi-tech park in Thach That district. However, the development of urban areas in this area should pay attention to the function of urban ecological planning until 2030. By contrast, the areas of midland and mountainous areas in the South West have limited access and lower land values. In some areas, although the land value is not high, it is possible to develop eco-economic agriculture to create economic efficiency (Fig. 6).

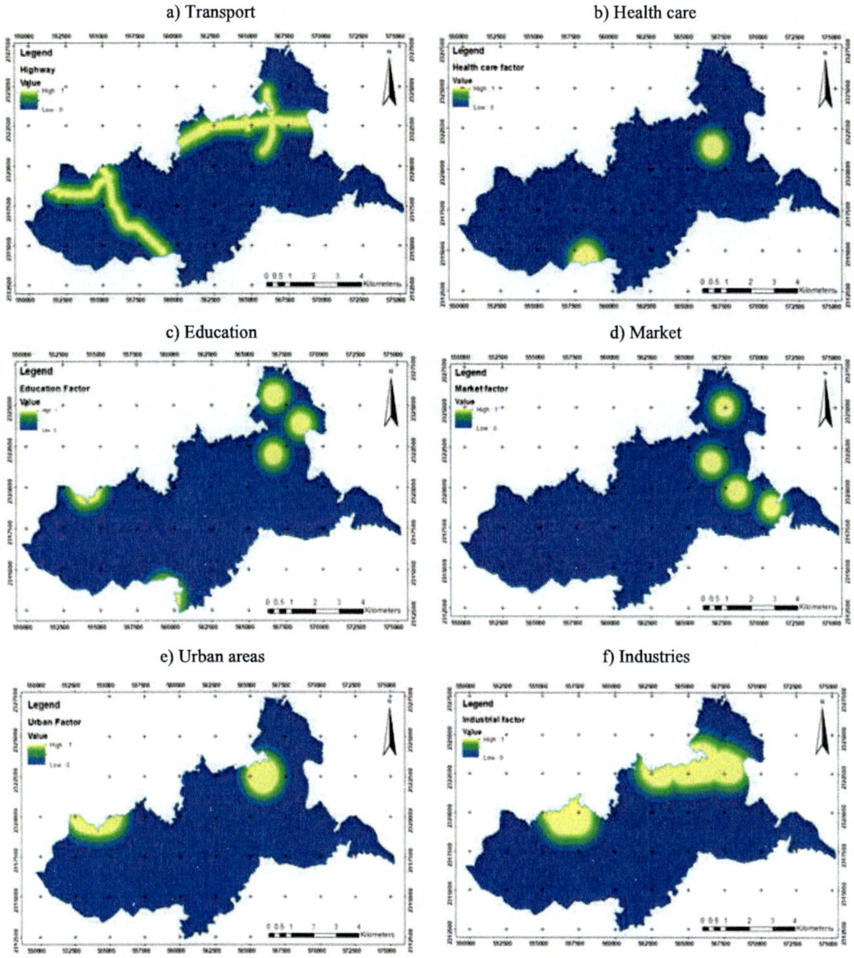

Fig. 4 Fuzzy membership surfaces of 6 criterions

5 Conclusion

Providing land value information plays an important role in financial planning and land policy. To accurately assess the potential, land values need to be evaluated based on the composition of factors affecting the land treatment. Based on the strength of spatial analysis of GIS and the handling of Fuzzy's complex problems, the information on land values provided contributes to effective and sustainable land use. For example, Quoc Oai district, land values were made to provide real estate-related information based on a combination of 10 indicators. Besides, value indexes according to each criterion are included in each parcel of land to complete and diversify information on land, contributing to the development of the real estate market.

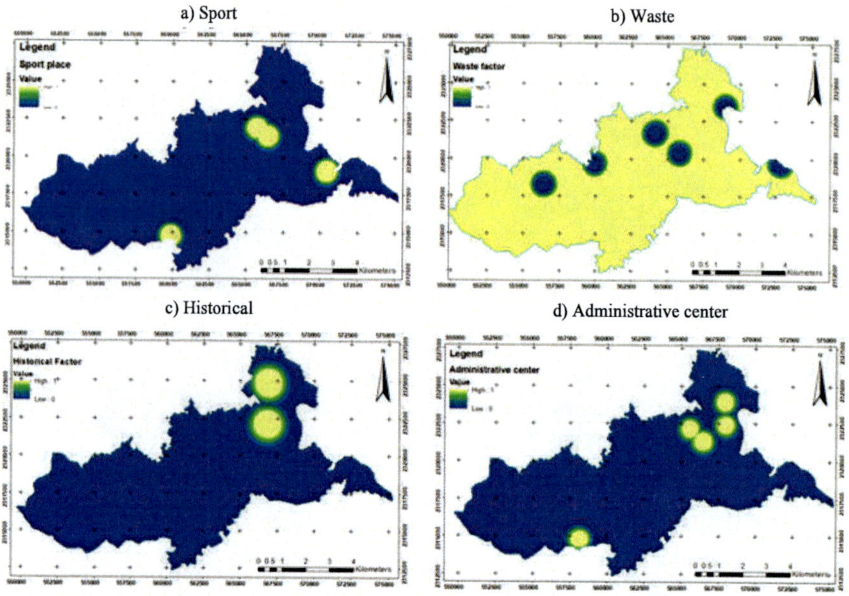

Fig. 5 Fuzzy membership surfaces of 4 criterions

Table 4 Fuzzy member in the sociocultural group

Valuation criterions	Function members
Cultural relics	$f(x) = \begin{cases} 1, x \leq 800 \\ \frac{1500-x}{1500-800}, 800 < x \leq 1500 \\ 0, x > 1500 \end{cases}$
Administrative center	$f(x) = \begin{cases} 1, x \leq 500 \\ \frac{1000-x}{1000-500}, 500 < x \leq 1000 \\ 0, x > 1000 \end{cases}$

Table 5 Thematic weights

No.	Valuation criterions	Weight
1	Highway	0.14
2	Hospitals, health care centers	0.06
3	Education	0.15
4	Markets	0.11
5	Urban areas	0.06
6	Industries area	0.04
7	Green space	0.16
8	Waste	0.04
9	Historical	0.12
10	Administrative center	0.12

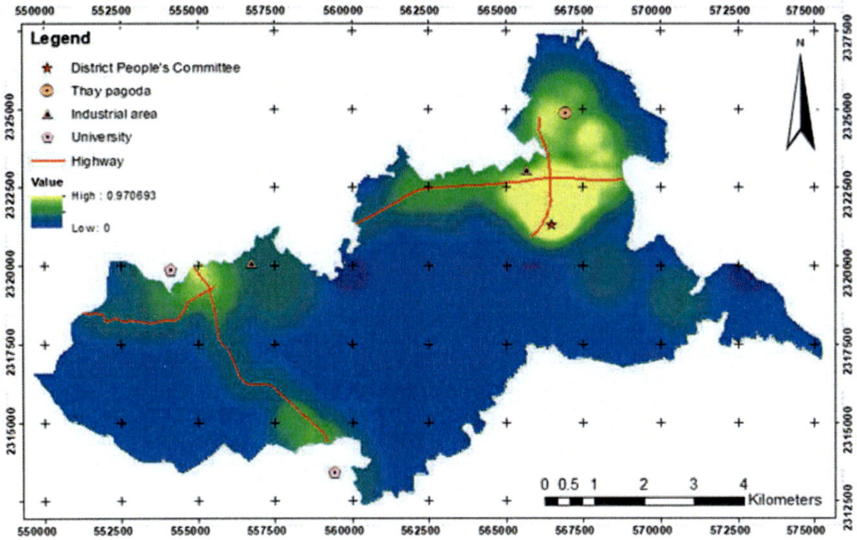

Fig. 6 Weighted fuzzy overlay map

References

Anantsuksomsri S, Tontisirin N (2015) The impacts of mass transit improvements on residential land development values: evidence from the Bangkok Metropolitan region. Urban Policy Res 33(2):195–216

Bogataj M, Tuljak Suban D, Drobne S (2011) Regression-fuzzy approach to land valuation. Central Eur J Oper Res 19(3):253–265

Bollinger CR, Ihlanfeldt KR, Bowes DR (1998) Spatial variation in office rents within the Atlanta Region. Urban Stud 35(7):1097–1118

Bovkir R, Aydinoglu AC (2018) Providing land value information from geographic data infrastructure by using fuzzy logic analysis approach. Land Use Policy 78:46–60

Cole HVS, Triguero-Mas M, Connolly JJT, Anguelovski I (2019) Determining the health benefits of green space: does gentrification matter?. Health Place 57:1–11

Construction MO (2008) QCXDVN 01: 2008/BXD, Vietnam construction standards—construction planning

Demetriou D (2016) The assessment of land valuation in land consolidation schemes: the need for a new land valuation framework. Land Use Policy 54:487–498

FAO (2003) Overview of land value conditions. Rome

Hajnal I (2014) Continuous valuation model for work-in-progress investments with fuzzy logic method. Procedia Eng 85:206–213

Hammer TR, Coughlin RE, Horn ET (1974) The effect of a large urban park on real estate value. J Am Inst Plann 40(4):274–277

Holden ST, Bezu S (2016) Preferences for land sales legalization and land values in Ethiopia. Land Use Policy 52:410–421

Jelokhani-Niaraki M, Sadeghi-Niaraki A, Choi S-M (2018) Semantic interoperability of GIS and MCDA tools for environmental assessment and decision making. Environ Model Softw 100:104–122

Johnstone (2004) Guidelines for the use of automated valuation models for UK. RMBS Trans Standard Poor

Kheir N, Portnov BA (2016) Economic, demographic and environmental factors affecting urban land prices in the Arab sector in Israel. Land Use Policy 50:518–527

Kim H-S, Lee G-E, Lee J-S, Choi Y (2019) Understanding the local impact of urban park plans and park typology on housing price: a case study of the Busan metropolitan region, Korea. Landscape Urban Plann 184:1–11

Lin J-J, Hwang C-H (2004) Analysis of property prices before and after the opening of the Taipei subway system. Reg Sci 687–704

Malaitham S, Fukuda A, Vichiensan V, Wasuntarasook V (2018) Hedonic pricing model of assessed and market land values: a case study in Bangkok metropolitan area, Thailand. Case Stud Transp Policy

McCluskey W, Deddis W, Mannis A, McBurney D, Borst R (1997) Interactive application of computer assisted mass appraisal and geographic information systems. J Prop Valuat Invest 15(5):448–465

Nakamura H (2019) Relationship among land price, entrepreneurship, the environment, economics, and social factors in the value assessment of Japanese cities. J Clean Prod 217:144–152

Nghiep N, Cripps A (2001) Predicting housing value: a comparison of multiple regression analysis and artificial neural networks. Real Estate Res 22(3):313–336

Pan Q, Pan H, Zhang M, Zhong B (2014) Effects of rail transit on residential property values. Transp Res Rec J Transp Res Board 2453(1):118–127

Samuels-Kalow ME, Camargo CA (2019) The use of geographic data to improve asthma care delivery and population health. Clin Chest Med 40(1):209–225

Sarip AG, Hafez MB (2015) Fuzzy logic application for House price prediction. Int J Prop Sci 5

Shenavr B, Hosseini SM (2014) Comparison of multi-criteria evaluation (AHP and WLC approaches) for Land capability assessment of Urban development in GIS. Int J Geomat Geosci 4(3)

Sun H, Wang Y, Li Q (2016) The impact of subway lines on residential property values in Tianjin: an empirical study based on hedonic pricing model. Discrete Dyn Nat Soc 1–10

YaĞMahan G, Gülgen F (2018) Taşınmaz değerlemede Coğrafi Bilgi Sistemlerinin kullanılması: Mardin Midyat örneği

A Study on the Ecological Balance Capacity of Hanoi Green Corridor

Quynh Chi Le, Dinh Viet Hoang, Van Tuyen Nguyen, and Quoc Thai Tran

Abstract The "Hanoi Capital Construction Master Plan to 2030 and Vision to 2050" has been approved and implemented by the Vietnamese government since 2011 with the goal of channeling city development to become a sustainable capital in Asia. To achieve that goal, the green corridor has been established with four crucial roles, including: (1) to control the urban sprawl (2) to preserve the landscape and natural value (3) to be a logistics area for the central city (4) to consolidate the environment balance. However, different from the city green corridor in other countries, Hanoi green corridor embraces various high-density residential areas and new development projects, which has resulted in land use transformation. Therefore, the paper has two objectives, including (1) evaluating the ecological balance capacity of Hanoi green corridor for the whole city and (2) assessing the change of the afore-mentioned capacity since establishment due to city transition. The research methodology is to use the plant biomass calculation tool, combining with high-resolution satellite images in different years. The research result will quantify the ability to balance CO_2 in the air of Hanoi green corridor and its change, providing a scientific basis for adjustment of land use planning toward a sustainable city development.

Keywords Biomass · CO_2 absorption · Air balance capacity · Green corridor · Hanoi

1 Introduction

Cities nowadays are facing with environmental and ecological problems due to the industrial development and modernism process, especially in the developing countries where the urbanization has been boosted up recently. Ecological sustainability of cities seems to have global priority and being placed on a green agenda. Protection

Q. C. Le (✉) · V. T. Nguyen · Q. T. Tran
National University of Civil Engineering, Hanoi, Vietnam
e-mail: chilq@nuce.edu.vn

D. V. Hoang
Vietnam Japan University, Hanoi, Vietnam

L. T. T. Huong and G. M. Pomeroy (eds.), *AUC 2019*, Advances in 21st Century Human Settlements, https://doi.org/10.1007/978-981-15-5608-1_20

of urban green spaces, making them and sustainable design, are placed at the subset; in which urban green corridor (or urban greenway corridors) are the most important component (Sara et al. 2015).

The origin of the green corridor planning approach, introduced with the purpose of preserving and providing the continuity of urban open spaces, are based on Olmsted's "Parkways" concept in America, and the Garden City concept of Ebenezer Howard, during the twentieth century in England (Toccolini et al. 2006). The notion of the green corridor has been involved for more than one hundred years, which is largely dependent on the context of urban development and urban issues. Nowadays, indicators of green corridors are "Providing green spaces, Pollutions and pollutants, Connectivity and continuity, Visions and Perspectives, Multi-functional structure, Tourism and income, Access ability and transportation, Social relationship" which are used as factors to promote sustainable development in the urban landscape, a key part for sustainable development. Besides economic benefits, recreational benefits, transportation benefits, the green corridor bring environmental benefits, including flood hazard reduction, reduction of bank erosion and downstream sedimentation, riparian habitat enhancement and biodiversity, protect the natural environment, preservation of habitat, provide wildlife corridor and wildlife migration, microclimate enhancement, increase air quality, enhance environmental quality (Sara et al. 2015).

In Vietnam, the green corridor planning approach has been applied first in the master plan of Hanoi capital city since 2011. Vietnam is currently experiencing one of the most intensive urban transitions in the world with an average growth rate of cities and towns of 6% per year, increasing the national share of Vietnam's urban population from one-third to one-half (MoC 2009). As the nation's capital city and second largest agglomeration after Ho Chi Minh City, Hanoi is one of the key sites of this urban transition. Hanoi's administrative boundaries were enlarged, increasing the area from 92,739 km^2 in 1991 to 33,446 km^2 in 2008, to include Ha Tay, Me Linh district (Vinh Phuc), Luong Son district (Hoa Binh) (HUPI 2014). The official population of the capital city is expected to increase from 6.4 million in 2010 to 9.0–9.2 million in 2030, 10.8 million in 2050, according to the Decision 1259/QĐ-TTg, which was issued by Prime Minister in 2011. To adapt with new development context, "Hanoi General Construction Master Plan to 2030, vision 2050" has been approved and implemented since 2011; in which, one of the main objectives of the master plan is to protect the environment through maintaining the natural environment; it leads to the establishment of a wide range of Green Space networks in the city, including Green Corridor, Greenbelt along Nhue river, green wedges, and urban parks. The Green Corridor including rural areas, systems of rivers, lakes, natural forests, agricultural areas,... is strictly protected to become an urban logistics area, preserving the landscape and ensuring the living environment. Hanoi Green Corridor (GC) has been set up with targets of controlling urban sprawl, protection fertilized agriculture land, reducing flood hazard, conservation cultural heritage, encouraging the friendly environment activities, maintaining and improving the traditional villages, supporting eco-tourism development, facilitating the transportation connect central city and satellite towns (VIUP 2011) (Fig. 1).

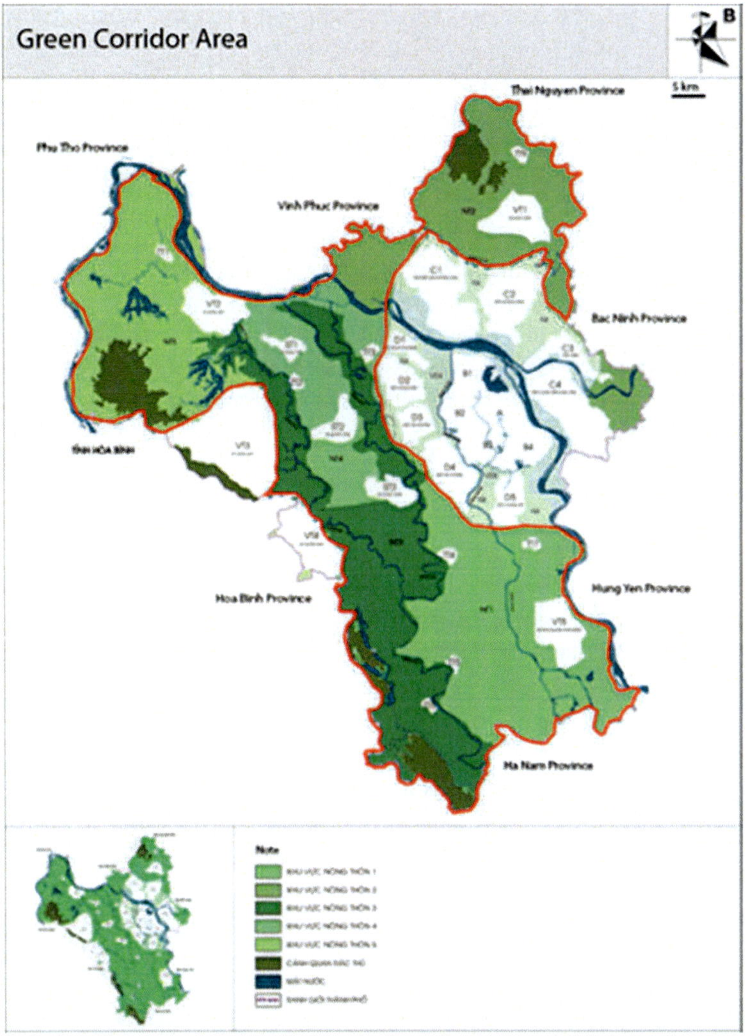

Fig. 1 Hanoi green corridor boundary (VIUP 2011)

Together with the urbanization process, the rapid spread of urban areas and changes in land use have led to significant and remarkable urban heat island (UHI) (Cui and Shi 2012). In Hanoi, there is a difference of 6 °C that exist between the low surface urban heat island (SUHI) intensity area in rural areas and high-intensity areas in the metropolitan core area; urbanization has created impacts on the development of SUHI in Hanoi during the period 1999–2016, the atmospheric UHI phenomenon is accelerated in the city (Thi Mai et al. 2019). The vegetation in the Hanoi Green Corridor is a crucial factor regarding the fact that trees play an important role in reducing UHI by reducing the amount of CO_2 in urban environments (McHale et al.

2009). The impacts of Green Corridor to minimize the negative impacts of the UHI has not been scientifically evaluated in the overall planning scheme (Andhang 2015), especially the equilibrium role in the air of the Green Corridor. The paper's objective is to quantify the ability to absorb CO_2 in the air through plant biomass tool and balance capacity in the air of Hanoi Green Corridor. Moreover, the difference with the Green Corridor in developed countries which is largely constituted by forest and park, the Hanoi Green Corridor embraces a large number of existing urban projects, village population system with high-density and many other forms of ongoing activities. In a surveyed sample plot of 480 ha of land in the Hanoi Green Corridor, there exist various functions including urban development projects (46%), villages (30%), relics, agricultural land (11%), and river (9%). Therefore, it is in need to grasp the correlation between air balance capacity and land use transformation. These results will provide a scientific basis and facilitate the adjustment of land use planning in the future.

2 Methods

2.1 *Biomass Determination*

Biomass is defined as the aboveground standing dry mass of live or dead matter from tree or shrub (woody plant) life forms, expressed as a mass or mass per unit area (CEOS 2019). Biomass is the sequestration of carbon from the atmosphere being absorbed into plant tissue through photosynthesis and stored (sometimes for centuries) in woody tissues. The sequestration of carbon from the atmosphere is an important ecosystem service provided by urban vegetation (Nowak 1994; Rogers et al. 2015) (Fig. 2).

To estimate aboveground biomass beyond the tree level, inventory techniques and remote sensing are used. Inventory techniques are applied in both traditional forestry and urban studies (Nowak and Crane 2002; Davies et al. 2011) where a representative sample of trees is measured. However, data acquisition for field inventory can be expensive, time-consuming, and is often incomplete (Wilkes et al. 2018). Remote sensing presents an opportunity to capture synoptic, fine spatial resolution data. This has been already widely applied to estimate the aboveground biomass, across a range of scales, using both active and passive sensors from space-based and aerial platforms (Baccini et al. 2008; Saatchi et al. 2011; Asner et al. 2010). Remote sensing methods for estimating the aboveground biomass can be categorized into (i) area-based and (ii) individual tree detection (ITD) methods (Coomes et al. 2017; Duncanson et al. 2014). Area-based methods use summary statistics of canopy structure to develop a statistical association with field inventory data, whereas ITD methods measure crown scale metrics to be used directly with allometry (Wilkes et al. 2018). In this paper, area-based methods have been achieved by combining Landsat and LiDAR to increase the reliability of the calculation result; filed inventory data has not been

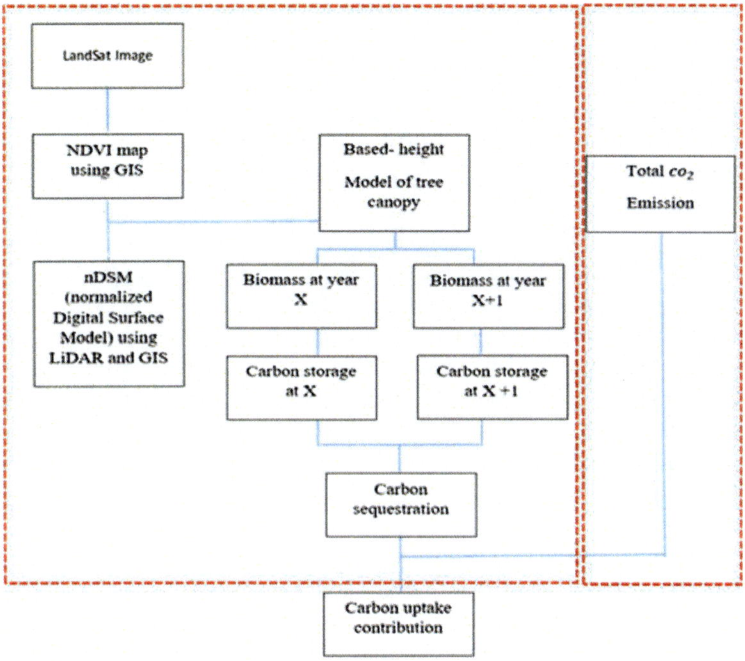

Fig. 2 Methodology chart

collected due to too large and complex research area; instead, a suitable estimation model with algorithms developed by Raciti et al. (2014), is applied.

Research using satellite images to identify plant biomass in the world has been applied for about 30 years. However, most of the studies were conducted in temperate regions because the vegetation in this region has a relatively simple tree structure and composition, the uniformity of the crown is quite high (Trotter et al. 1997; Wu et al. 1994). In the humid tropics, due to the complex structure and diverse tree species composition makes it difficult to determine plant biomass from images, resulting in high biomass/volume determination (Lu 2006). The remote sensing-based biomass estimation methods assume that vegetation information captured by sensors is highly correlated with the aboveground biomass. According to this assumption, the key for biomass estimation is to use appropriate variables and to develop suitable estimation models if sufficient sample plots are available.

Image Selection and Potential Variables on Satellite Imagery Related to Plant biomass

There are several optical satellite imagery types including Landsat, SPOT, ASTER, CBERS, QuickBird, MODIS, AVHRR, etc., which can be used to determine biomass (Luther et al. 2006). In this paper, satellite image data used includes Landsat 8 satellite images of the GC area at 3 points time (the year of 2015, 2016, 2019), combined from United States Geological Survey (USGS), which is a popular open-source that provides remote sensing images with a resolution of 30 m. After downloading the image, the image are fed into the WGS84 coordinate system in ArcGIS, in accordance with the Vietnamese coordinate system. Remote sensing images in the GC area are defined in the Autumn and Spring, with the selected cloud conditions being less than 10% or 20%.

LiDAR images can provide information on height, canopy diameter, trunk diameter, etc., leading to better biomass/forest volume determination than optical or radar images (Bergen et al. 2009). The LiDAR data product, which is obtained from the United States Geological Survey website (USGS), and taken in July 2014, is separated into three geospatial layers including "first return", "last return", and "bare earth model". Accuracy is defined as 100 cm in horizontal and 30 cm planes in a vertical orientation. Due to limited conditions in terms of LiDAR data, 2014 data, will be used to overlap with normalized difference vegetation index (NDVI) calculation results for all 3 years of 2015, 2016, and 2019.

Landsat images will be extracted through NDVI to locate the vegetation, then combined with LiDAR data to identify the plants at each location. NDVI is calculated based on the difference of reflected near-infrared light and red light on plants. Because the leaves reflect strongly with near-infrared radiation (NIR), the leaves' chlorophyll strongly absorbs the red light of the radiation in the visible region. NDVI is often used to estimate primary productivity, as well as plant biomass, monitoring forests and plants. The higher the value of NDVI (from -1 to 1), the stronger the photosynthetic activity (Gamon et al. 1995).

For Landsat 8, NDVI is calculated as the differences in surface reflectivity between NIR (band 4) and red band (band 3), divided by the sum of the NIR band and the red band, expressed through the formula

$$\text{NDVI} = \frac{\rho(\text{NIR}) - \rho(\text{RED})}{\rho(\text{NIR}) + \rho(\text{RED})} \tag{1}$$

ArcGIS has been used as an effective tool to calculate NDVI and export the NDVI map. For all pixel images with NDVI values greater than 0.1 will be used to estimate plant biomass, pixels with NDVI values of less than 0.1 will be excluded because there is no significant biomass value (Raciti et al. 2014).

The NDVI map is overlapped with LiDAR-derived normalized digital surface model (ndsm). In this stage, each pixel of the image selected to calculate the biomass will be arranged in one of four height groups (from 1 m to the maximum height of the canopy), including herbaceous vegetation (grass and not woody plants, $H \leq$ 1 m), shrub (shrubs and trees of $H \leq 2$ m height), tall shrubs (trees with common height H from 2 to 5 m), woody plants (height $H > 5$ m). This classification system has been selected to estimate the biomass for large areas and has diverse vegetation, particularly optimal for areas with high plant densities (Mette, Hajnsek 2003).

Identify Suitable Algorithms for Biomass Determination Model

Raciti et al. (2014), has developed a model of canopy height estimation in the city of Boston. This model was constructed from the linear regression method by the trunk diameter method at breast height of Jenkins and partners in 2003, combined with the canopy segment results based on height, and LiDAR data. This is the most specific equivalent equation that can be applied in all types of terrain. Received biomass results were compared with the results got from the method using standard plots and field measurements; the correlation is of 0.87 and the accuracy is 85%. The advantage of this model is that it is possible to estimate biomass for all types of areas, giving reliable results. Formula (2) shows the estimated model of Raciti in 2014.

$$B = (2.1015 * H + 0.8455) * S \quad (2)$$

B is the biomass calculated in kg. H is the height of each 1 m^2 type of tree classified, S is the total area of trees in height H.

2.2 CO_2 Emissions Data

Since 2015, Hanoi has started greenhouse gas inventory and reports every 2 years; the method is in accordance with the guidance of the Intergovernmental Panel on Climate Change (IPCC). The report "The situation of implementing the tasks of climate change in 2017, and the direction of the tasks in 2018, in Hanoi area" prepared by the Ministry of Natural Resources and Environment (MNRE), shows that, from 2015 to 2017 (in 2 years), Hanoi discharged 5.276 million tons of CO_2 equivalent, of which CO_2 alone is accounted for 51.5% (2.7174 million tons). Therefore, in 2015, Hanoi discharged 1.35857 million tons of CO_2, in which a high proportion of emissions are coming from domestic waste treatment and transport sectors.

2.3 Land Use/Land Cover (LULC)

In the 1970s, the tool applied to classify land cover from Landsat image was a direct analysis, and the use of pixel-based classification methods was also used. In addition, the maximum likelihood classifier (MLC) has become the most commonly used classification method for Landsat images for more than 40 years. Therefore, in this paper, MLC is used to develop LULC maps with 8 types of land (IPCC 2006). In addition, changes in LULC over the years will also be shown through charts (Table 1).

Table 1 Characteristics of land types classified by IPCC 2006

	Classes	Include
1	Residential	Rural residential and urban residential
2	Industrial, commercial	Industrial parks, export processing zones; trading and service
3	Active agriculture	Land for the cultivation of perennial trees, land for cultivation of annual crops, paddy land
4	Inactive agriculture (Wetland and Dry land)	Wetland land: pump water in the field to prepare for farming works Dry Land for the cultivation of perennial trees, land for cultivation of annual crops
5	Water	Land with rivers, streams, canals, springs (located in rivers, natural lakes)
6	Tree	Land for special-use forests Land for protection forests Land for production forests
7	Sandbars	Riparian land
8	Bare soil	Unused land and unused mountain land

Source IPCC (2006)

3 Finding

3.1 Results of Estimated Plant Biomass in Hanoi Green Corridor

After applying ArcGIS to calculate NDVI and export NDVI with the formula (1), the following maps in 3 point times were achieved (Fig. 3).

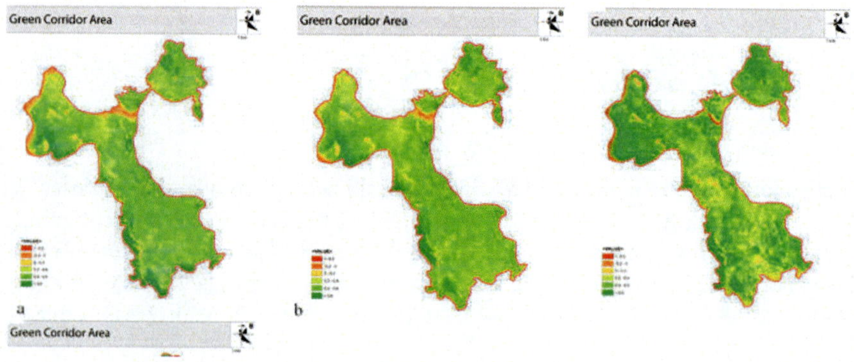

Fig. 3 NDVI maps in 2015, 2016, 2019

The NDVI maps then are overlapped with nDSM obtained from LiDAR data; each pixel of the image selected for biomass calculation will be arranged in one of four height groups. From the data on total pixels for each type of tree in the Green Corridor, the area for each category is derived as follows (Fig. 4).

Aboveground biomass estimation results shows the change in the amount of carbon stored in green trees in the Green Corridor area by years. There is a significant amount of plant biomass stored in the Green Corridor. Analyzing the satellite image taken on September 15, 2015, shows that there are 1029.8 km^2 natural land out of a total area of 2273.2 km^2 Hanoi Green Corridor, this value for 2016 is 1009.6 km^2 and for 2019 is 976.2 km^2.

The amount of biomass calculated by formula (2) in 2015 is 6,449,641.67 tons, equivalent to 9,459,474.45 tons CO_2. Estimated results for 2016 and 2019 are 6,457,686.85 tons and 6,462,306.83 tons, respectively, equivalent to 9,471,274.05 tons CO_2 and 9,478,050.01 tons CO_2 (Fig. 5).

From 2015 to 2016, the Hanoi Green Corridor's biomass increased by an amount of 8,045.18 tons. This figure rose in the period of 2016–2019 by 4,619.97 tons.

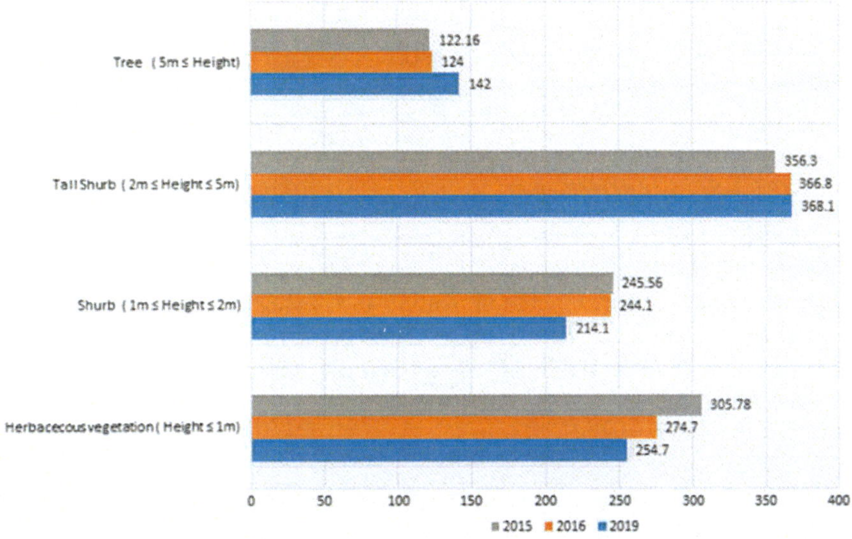

Fig. 4 Statistic of area for each type of tree in the Green Corridor in 2015, 2016, 2019 (km^2)

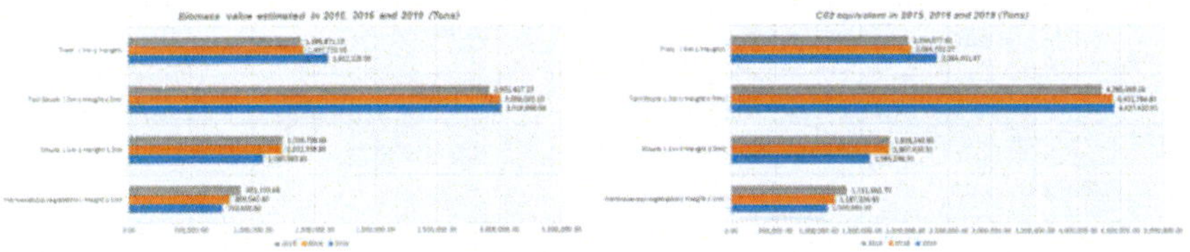

Fig. 5 Biomass and CO_2 equivalent in Hanoi Green Corridor in 2015, 2916, 2019 (tons)

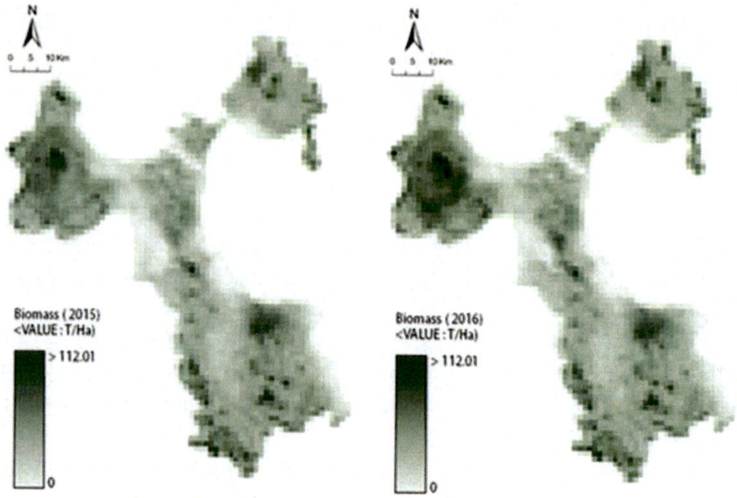

Fig. 6 Biomass map of Hanoi's Green Corridor in 2015, 2016

The annual average absorption value is only 1,355.19 tons/year from 2016 to 2019 (Fig. 6).

3.2 *Change in LULC of Hanoi's Green Corridor*

In general, the classification results show that agricultural land area, including active and inactive, has remained a high proportion and is allocated throughout the GC (more than 30% in) (Fig. 7 and Table 2). Trees occupy a considerable area, most concentrated in the West and Southwest. In addition, high-density residential areas are located along the expressway connecting the city center and satellite towns, especially in the year 2019.

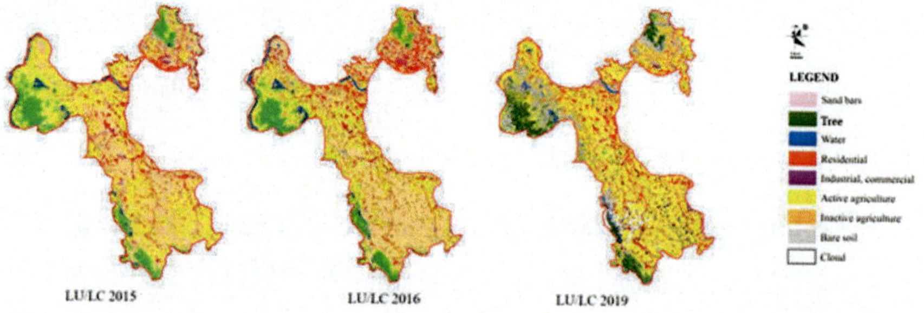

Fig. 7 Transformation of LULC in Hanoi Green Corridor by 2015, 2016, 2019

Table 2 LULC classification results 2015, 2016, 2019 (%)

		2015	2016	2019
1	Residential	12.23	12.35	14.12
2	Industrial	5.18	5.18	7.37
3	Active agriculture	35.12	36.21	32.97
4	Inactive agriculture	9.34	9.47	11.12
5	Water	7.26	7.37	6.98
6	Tree	29.01	28.58	26.74
7	Sandbars	0.56	0.41	0.43
8	Bare soil	1.3	0.43	0.27

Eight types of land use are classified according to IPCC 2006. A scrutiny study on the map and data table shows that the land use types have changed significantly in 2016 and 2019, the biggest change belongs to residential land, tree land, and industrial land. In 2015, the total proportion of residential and industrial areas was 12.23 and 5.18%, rising to 14.12 and 7.37% in 2019. Therefore, from 2015 to 2019, a significant increase in construction land has been increased, to be detailed, a raise of 1.89% in residential areas and 1.82% in industrial parks. Meanwhile, the tree land area has decreased from 29.01% in 2015 to 28.58% in 2016 and 26.74% in 2019. In the urbanization process, Hanoi GC's LULC has changed significantly, it is attributed to urban development, an increase in residential or industrial areas. Population growth, economic development, and policy reform have played an important role in promoting these changes. This is a potential risk for invasion of trees in the GC and reduces the ability of the GC to support for urban ecological balance.

4 Discussion

4.1 Assess the Balance Capacity in the Air of Hanoi Green Corridor

Results of estimating the absorption capacity of GC compared to the total of Hanoi emission.

The difference in carbon stocks in the Green Corridor plant is the amount of CO_2 that trees absorb between the two point times. From the estimated biomass results in 2015 and 2016, 11,799.59 tons of CO_2 has been removed from the air by the Green Corridor in Hanoi. Meanwhile, according to the report of the Ministry of Natural Resources and Environment (MNRE), Hanoi City discharged 1.35857 million tons of CO_2 gas into the air. The Hanoi Green Corridor has met 0.87% of the city's demand for environmental services.

Table 3 CO_2 absorption capacity in Hanoi's GC, Seoul's green belt, and Dakota's green space

	City	Biomass calculation area	Uptake %
1	Hanoi green corridor	2,273.2 km^2	0.87
2	Seoul green belt	606 km^2	1.5
3	Dakota green space	856 km^2	7.35

Comparison of absorption capacity of Hanoi GC with similar models in the world.

When considering the results of Jo (2002), the CO_2 absorption capacity of Hanoi's GC is only 50% of the CO_2 absorption capacity of Seoul's Green Belt (1.5%). For the results of Chang Zhao (2015), the environmental service capacity of Dakota's Green Space is significantly greater than that of Hanoi's GC (7.3%). Below is a table comparing the ability to absorb CO_2 in the air of the Hanoi's GC, the Seoul's Green Belt and the Dakota's Green Space (Table 3).

4.2 *Enhance the Ecological Balance Ability of the Green Corridor in Hanoi*

In terms of CO_2 absorption capacity, Hanoi Green Corridor only meets 0.87% of the ecological environment needs for the whole city. The decrease in trees area and rapid increase of construction land are the main culprits for the GC to meet only a small amount of ecological needs of Hanoi city.

In order to enhance the role of the GC, it is necessary to consider the proportion of green areas in the GC. If considering the model of Seoul's Green Belt as a good lesson of an effective green space planning outside the city center, Hanoi's GC needs to meet 1.5% to 2% of CO_2 balance demand. If only consider changing the value of tree land and keep the value of other types of land, the Hanoi's GC needs to have a minimum area of tree of 49.2% compared to 28.58% in 2016. If we remain the tree land ratio of the GC is 28.58% (2016), changing plant characteristic is essential. Considering data in 2016, the proportion of trees under 1 m tall (with negligible biomass) is 274.7 km^2; if it is replaced by woody plants (with high biomass), the CO_2 absorption capacity of the GC will be significantly increased. In addition, restriction of the area of construction area (including residential areas and industrial zones) is very important because residential areas and industrial parks are the main causes of greenhouse gas emissions and do not contribute to ground biomass (IPCC 2006).

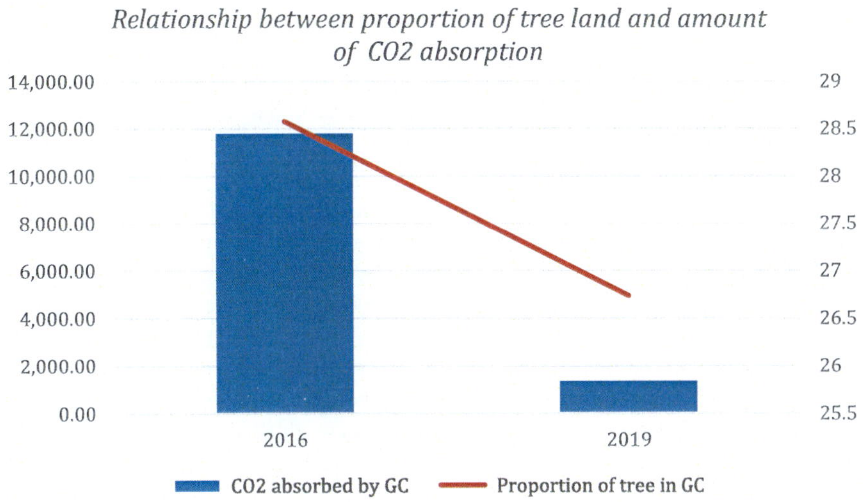

Fig. 8 Relationship between the proportion of tree land and amount of CO_2 absorption

4.3 Assess the Ecological Balance of the Green Corridor in the Future

According to the results of biomass estimation at three time points including 2015, 2016, and 2019, the amount of biomass stored in plants of the GC in Hanoi, has gradually increased. This reflects the development of plants. However, compared to the year 2016, the GC absorbed 11,799.59 tons CO_2, in 2019, only 1.355.19 tons CO_2/year are absorbed. Thus, we are able to see a marked decrease in CO_2 balance over 5 years (9 times reduction).

The following chart shows the relationship between the trees area of GC and the amount of CO_2 absorption in the period of 2016 and 2019. The rapid reduction of trees' land has led to a decline in the CO_2 absorption capacity of the GC. The trend of the graph goes down, and in the near future, carbon stocks will change seriously according to the negative side. If there is no control of the development of urban function in GC, the urban heat island will be much worse in the future (Fig. 8).

5 Conclusion

Through the process of calculating biomass at three point times (2015, 2016, 2019), there is a large amount of biomass stored by plants in the Green Corridor. However, the CO_2 absorption capacity of plants in the Green Corridor is quite small, only reaching 0.87% of total CO_2 emissions in Hanoi, in 2015 (in comparison with this figure in Seoul's Green Belt reaching 1.5%). Through analysis of land use maps, the change in the largest area ratio in three types of land: tree land, industrial land, and residential land. This change has a close relationship with CO_2 absorption capacity of the Green Corridor Hanoi. While the area of tree land decreases sharply, the construction area increases rapidly from 2015 to 2019. The reduction of the green area means that CO_2 absorption will decrease rapidly and the amount of CO_2 released into the environment will increase.

The results of biomass estimation and assessment of CO_2 balance in the air of the Hanoi's GC would be a database to serve the planning and management of the GC. In addition, the paper recommends that there should be studies to assess the impact of species composition in the flora of Hanoi, in general, and the GC in particular, in order to enhance the ability to meet the ecological needs of the city. Assessments of the change in LULC of the Green Corridor area in Hanoi, contributing to the Urban Heat Island (UHI) phenomenon is also one of the researches that need attention in the future. Greenhouse gas inventory is a very important activity. Inventory activity data not only provides information for the city's GHG mitigation plans, but also an important database for scientific research. In Hanoi, the City has been conducting a greenhouse gas inventory since 2015. However, this activity is often interrupted and limited in providing data regularly. We recommend the General Statistics Office, Ministry of Natural Resources and Environment (MNRE), to conduct the annual GHG inventory.

References

Asner GP, Powell GV, Mascaro J, Knapp DE, Clark JK, Jacobson J, Kennedy-Bowdoin T, Balaji A, Paez-Acosta G, Victoria E, Secada L, Valqui M, Hughes RF (2010) High-resolution forest carbon stocks and emissions in the Amazon. Proc Natl Acad Sci USA 107(38):16738–16742. https://doi.org/10.1073/pnas.10048

Baccini A, Lapor te NTT, Goetz SJ, Sun M, Dong H (2008) A first map of tropical Africa's above-ground biomass derived from satellite imagery. Environ Res Lett 3(4):1–9. https://doi.org/10.1088/1748-9326/3/4/045011

Bergen KM, Goetz SJ, Dubayah RO, Henebry GM, Hunsaker CT, Imhoff ML, Nelson RF, Parker GG, Radeloff VC (2009) Remote sensing of vegetation 3-D structure for biodiversity and habitat: review and implications for lidar and radar spaceborne missions

CEOS land product validation subgroup (2019) https://lpvs.gsfc.nasa.gov/Biomass/AGB_home.html. Accessed 26 Sept 2019

Coomes DA, Dalponte M, Jucker T, Asner GP, Banin LF, Burslem DFRP, Lewis SL, Nilus R, Phillips OL, Phua MH, Qie L (2017) Area-based vs tree-centric approaches to mapping forest carbon in Southeast Asian forests from airborne laser scanning data. Remote Sens Environ 194:77–88. https://doi.org/10.1016/j.rse.2017.03.017

Cui L, Shi J (2012) Urbanization and its environmental effects in Shanghai, China. Urban Clim 2:1–15

Davies ZG, Edmondson JL, Heinemeyer A, Leake JR, Gaston KJ (2011) Mapping an urban ecosystem service: quantifying above-ground carbon storage at a city-wide scale. J Appl Ecol 48(5):1125–1134

Duncanson LI, Cook BD, Hurtt GC, Dubayah RO (2014) An efficient, multi-layered crown delineation algorithm for mapping individual tree structure across multiple ecosystems. Remote Sens Environ 154:378–386. https://doi.org/10.1016/j.rse.2013.07.044

Gamon JA, Field CB, Goulden ML, Griffin KL, Hartley AE, Joel G, Penuelas J, Valentini AR (1995) Relationships between Ndvi, canopy structure, and photosynthesis in three californian vegetation types

Lu D (2006) The potential and challenge of remote sensing-based biomass estimation

Luther JE, Fournier RA, Piercey DE, Guindon L, Hall RJ (2006) Biomass mapping using forest type and structure derived from landsat TM imagery

McHale MR, Burke IC, Lefsky MA, Peper PJ, McPherson EG (2009) Urban forest biomass estimates: is it important to use allometric relationships developed specifically for urban trees. Urban Ecosyst 12:95–113. https://doi.org/10.1007/s11252-009-0081-3

Nowak DJ (1994) Atmospheric carbon dioxide reduction by Chicago's urban forest. In: Mcpherson EG, Nowak DJ, Rowntree RA (eds) Chicago's urban forest ecosystem: results of the Chicago urban forest climate project. USDA Forest Service General Technical Report NE-18, Radnor, pp 83–94

Nowak DJ, Crane DE (2002) Carbon storage and sequestration by urban trees in the USA. Sage Publications, Thousand Oaks

Raciti SM, Hutyra LR, Newell JD (2014) Mapping carbon storage in urban trees with multi-source remote sensing data: relationships between biomass, land use, and demographics in Boston neighborhoods

Rogers K, Sacre K, Goodenough J, Doick K (2015) Valuing London's urban forest. Results of the London i-Tree eco project. Treeconomics, London, p 82

Saatchi SS, Harris NL, Brown S, Lefsky M, Mitchard ETA, Salas W, Zutta BR, Buermann W, Lewis SL, Hagen S, Petrova S, White L, Silman M, Morel A (2011) Benchmark map of forest carbon stocks in tropical regions across three continents. Proc Natl Acad Sci 108(24):9899–9904. https://doi.org/10.1073/pnas.10195

Sara GE, Mona M, Saeed G (2015) Studying the relationship between urban green corridors and sustainable urban landscape. Int J Sci Technol Soc 3(2–1):36–40. https://doi.org/10.11648/j.ijsts.s.2015030201.18

Thi Mai N, Lin T-H, Chan H-P (2019) The environment effects of urban development in Hanoi, Vietnam from satellite and meteorological observation from 1999–2016. Sustainability 11(6):1768. https://doi.org/10.3390/su11061768

Toccolini A, Fumagalli N, Senes G (2006) Greenways planning in Italy: the Lambro river valley greenways system. Landsc Urban Plann 76:98–111

Wilkes P, Disney M, Boni Vicari M, Calders K, Burt A (2018) Estimating urban above ground biomass with multi-scale LiDAR. Carbon Balance Manage 13:10. https://doi.org/10.1186/s13021-018-0098-0

Livable and Smart Cities

Digital Addressing of Historical City Morphology: The Case of Lalitpur City in Kathmandu Valley, Nepal

Krishna Prasad Poudel and Suresh Shrestha

Abstract Lalitpur Metropolitan City (LMC) covers 36 km^2 area, 68,353 households, and 276,479 total populations constituted as a metropolitan in 2016, by the Government of Nepal. The core area of the city covers approximately 8 km^2 and comprised of 70% households, built in around 3rd century B.C. Digital metric addressing system is being implemented for the historical city morphology of Lalitpur City. In this process, a detailed base-map of 1:2500 scale had been prepared from the high resolution stereo satellite imageries (0.32 m resolution) for the whole municipal area and 1:500 scale for the core city area with the large-scale photographs acquired by unmanned aviation vehicle (UAV). The differential global positioning system (DGPS) and traverse leveling survey had been administered to attain positional accuracy to construct accurate base-map through the photogrammetric procedures. Base-map depicts detailed surface features including the contour of one-meter interval, land use, and road network, building footprint, major social and cultural landscapes. From the detailed field survey, every building unit with the main doorway (entrance gate) was marked in the base-map. Distance from the road junction to the building doorway was computed and assigned each building/house number corresponding to the distance value in meter. Those numbers were designed in a system that never repeats or duplicates to another house number. The whole process of exercise of digital metric addressing of historical city morphology provides an intensive practical lesson on recently developed space-place hybrid data integration, real ground-based large-scale mapping, and diverse city landscape applications. The shared experience may be referenced for the morphological studies in similar other historical Asian cities.

Keywords Urbanization · Addressing · City morphology · Base-map

K. P. Poudel (✉)
Central Department of Education, Tribhuvan University, Kathmandu, Nepal
e-mail: kpoudel.Pokhara@gmail.com

S. Shrestha
GeoSpatial Systems Pvt. Ltd, Kathmandu, Nepal
e-mail: geosp@geosp.com

L. T. T. Huong and G. M. Pomeroy (eds.), *AUC 2019*, Advances in 21st Century Human Settlements, https://doi.org/10.1007/978-981-15-5608-1_21

1 Introduction

Cities are the agglomeration of a large number of population with high density, diversity in functions and amenities, and services. The United Nations (UN), defines the cities as "large numbers of people live and work; they are hubs of government, commerce and transportation" (UN 2018:1). However, the delineation of a city boundary always has some differences. So far, not a single standardized international criterion exists for determining the boundaries of a city and often multiple boundary definitions are available for any given city (UN 2018:1). The common characteristics of cities have a large concentration of people in different administrative units, functional zones, and land-use types. Cities have a concentration of a large number of population within a certain extent of the geographical territorial unit either in a contiguous functional, administrative or formal delineated boundary. In terms of the world population distribution, 23% of the world's population lived in a city with at least 1 million inhabitants in 2018. The projected figure for 2030 shows 28% of people worldwide will be concentrated in cities with at least 1 million inhabitants (UN, 2018:3). This shows that the share of the global population living in cities is quite high and increasing tremendously. In a statement of Undersecretary General and Executive Director of UN-Habitat in April 2019, mentioned that "we live in an urban world: more than 55% of the world population lives in urban areas today; this likely grow to 68% by 2050. Human settlements are the local, national, and global drivers of economic prosperity and development, with up to 80% of the global Gross Domestic Product (GDP) generated in our towns, cities, and metropolises. Cities are also the hubs for governance and state institutions, for ideas, commerce, culture, science, and for social, human, and individual development" (UN-Habitat, 2019:v). The statement clearly hints the significance of the urban centers in the contemporary global context and also the forecast for 2050. In 1992, the World Bank had pointed out the world urban managers and policymakers require making strategic decisions on environmental protection, infrastructure development and maintenance, land development, and land administration and taxation. In many cities in the developing countries, they also had to respond to a specific pressure, such as rapid population growth and an expansion in urban areas. Rarely do they have access to up-to-date base-maps and systematic information on the extent of settlements, land-use patterns, environmental problems, and infrastructural facilities (World Bank 1992:v). Having considering the expansion of global urbanization patterns, in 2005, the World Bank has reported that many cities in the developing world have experienced extremely rapid growth. Because of such rapid growth, inadequate identification systems have created a worrisome predicament for urban services (World Bank 2005:2). The document clearly hints to develop the system of street coordinates, find ways around a constantly growing city, make easy on dispatching ambulances, firemen, or law enforcement personnel quickly, deliver the mail and message to private houses promptly, efficient delivery of municipal services to the public, pinpoint of the breakdowns in water, electricity, and telephone systems, and also the set up an efficient tax collection systems from the city dwellers. For all these activities the report has given the way out through the street addressing systems (World Bank 2005) and implemented in different countries by using the geo-information systems (Jezzini 2017; Demir and Raskar 2018; Demir et al. 2018). The addressing was also practiced in Kathmandu Metropolitan City, in early 2000 (KMC and EU 2004). However, it was not fully integrated with digital base-map and was also prepared on the base of a paper-based cadastral map dated the early 1970s, and addresses were given on measuring the doorway location with measuring tape on the ground. That also has the problem of implementation at the public utilization and continuation due to its unorganized database. Despite that, some of the lessons learned from the Kathmandu Valley Mapping Project (KVMP), have been more fruitful in the present context, too.

Upon this ground, the addressing urban houses based on the street network systems seems to be widely accepted techniques for the efficient and scientific urban governance system. However, adapting the systems practiced elsewhere in modern town and cities contains several practical, preparation, and implementation challenges' in the case of old historical cities. Specifically, old cities built as of the centuries-old architectural design, cultural settings, security provisions, economic production patterns and living standards, and the magnitude of urban services, amenities, and infrastructures have different scenarios. Therefore, the present paper has been highlighting the process of digital addressing system with the application of space-based imaging systems to place-based technological and knowledge integration in the historical Lalitpur City, located in the Kathmandu Valley, Nepal. The main aim of the research was to establish practical and scientific addressing system in Lalitpur City, to support efficient and good governance practices, better emergency planning, pro-poor policies, and equitable resource mobilization and balance development within the city to attain the sustainable cities and communities envisioned by the sustainable development goals (SDGs).

2 Lalitpur City

The Lalitpur City is located in the central part of the country, and it's one of the oldest cities in the Kathmandu Valley (Fig. 1).

Lalitpur has been declared "Metropolitan City" by the Government of Nepal (GoN), in 2016; it covers 36 km^2 area, 68,353 households, and 276,479 total populations (CBS, 2012). The core city covers approximately 8 km^2 area and comprises 70% households, built in around 3rd century B.C. (http://lalitpurmun.gov.np/en/node/4) along with the Kathmandu Valley Civilization (https://whc.unesco.org/en/list/121/). For many centuries, Lalitpur was a sovereign city-state. It has been developed as one of the most artistic and colorful cities within the Kathmandu Valley. This City is also inscribed within the UNESCO World Heritage Site. The ancient city center is highly complex; several courtyards are located inside the small and narrow entrance. Even some of them are inside the individual house unit. Water resources, open space, and public spaces were built and managed properly for a long historical time. City morphology is highly complex. In such historical city morphology, it has a great

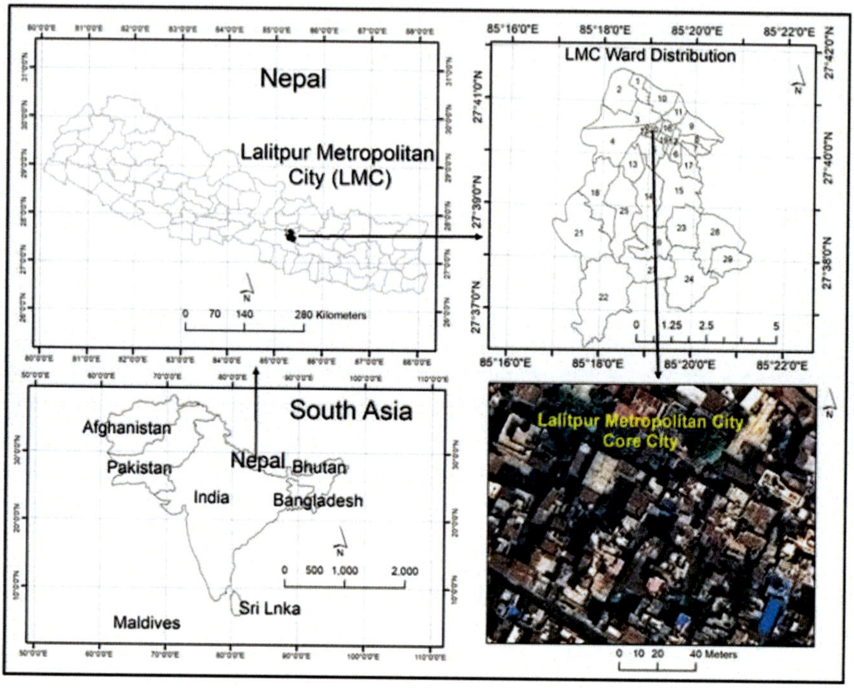

Fig. 1 Location map of Lalitpur City

challenge in managing the current growth of urban population, modern infrastructure development, city management, disaster risk reduction, and management activities. From the implementation of urban governance systems, it seems a herculean task.

The Lalitpur Metropolitan City (LMC), has different morphological distributions within its extension. The core city has very complex morphology. It has a distinct boundary confined within the close vicinity of Old Durbar Square named "Patan Durbar Square". "Core City" has several interior squares or courtyards called *"bahas"* and *"bahis"* (Locke 1985). The core city was gradually expanded on its outskirts in the course of time. The expansion was at its rapid speed after the construction of Ring Road, in the late 1970s. The gradual expansion of the city caused due to the in-migration of the people to Kathmandu Valley, from the other parts of the country with various purposes (Thapa 2009). Within the LMC there are some satellite old inhabitant centers where local residents used to follow agriculture and small local and domestic craft industries like carpentry, sculpturing, cloth weaving, food processing, jewelry, and so on mainly located in Khokana, Bungamati, Harisiddhi, Thecho, Dhapakhel, etc. Gradually those centers were also converting to satellite hub after building road connectivity with Ring Road and Core City (Thapa and Murayama, 2009). The Valley has been taken place the urbanization processes with an alarming trend after the expansion of connectivity (Poudel 2012). The fast increasing urbanization process exerts impact over different aspects. Addressing the

Table 1 Data types and sources

Data type	Source	Resolution/scale	Date of acquisition
Stereo satellite image	WorldView 3 3 digital globe	0.32 m pixel size	25th March 2019
Cadastral map	Land revenue department, Government of Nepal	1:500	Regularly updated since 1975 to current
Unmanned Aviation Vehicle (UAV) or drone images	GeoSpatial systems Pvt. Ltd.	Taken from 50–100 m elevation	May–July 2019
Ground Control Points (GCPs)	GeoSpatial systems Pvt. Ltd.	Less than 30 cm accuracy level	May–July 2019
Leveling Survey	GeoSpatial systems Pvt. Ltd.	Less than 20 cm accuracy level	May–July 2019
Field survey	GeoSpatial systems Pvt. Ltd.	Household level	May–Dec 2019

morphology thus provides factual evidence about the city expansion and provides foundations for its better management. In the present context, a general overview of the entire metropolitan city has been made but the detailed digital addressing has been confined within the historical core city.

3 Data Types and Sources

Being a native researcher and one of the authors born and grown-up within the historical core city of Lalitpur, long personal experiences, observations, technical innovation, and requirements have been assessed for a long time. Well and up-to-date with the city morphology, identification of each and every street, building blocks, courtyards, and traditional resource centers, infrastructure building processes, urban amenities supply, delivery systems, and governance systems have been assessed for a long time. Upon these assets of prior background knowledge about the city, different data types have been acquired from various sources (Table 1).

4 Data Processing

The collected data had been processed to attain the stated goals. The base-map of the city was prepared by integrating the stereo satellite image, Drone image, Differential Global Positioning Systems (DGPS), results of the instrumental Travers Leveling Survey, and other ancillary information and field survey. The positional accuracy of the area has been affined with the DGPS survey (coordinates were within 0.008 m

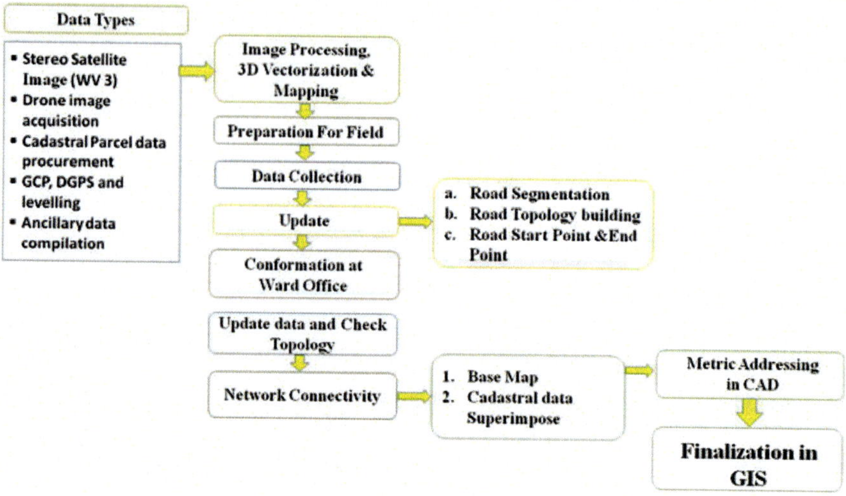

Fig. 2 Data types and processing procedures

in X and Y and 0.05 m in Z) and leveling survey points and produced accurate (closing error of accuracy level were <20 cm for all routes). The data source and the processing procedure are given in the Fig. 2.

Contour lines of 2 m intervals were generated from the stereo satellite image in outside and 1 m in core area from the Drone image. Street network, land use, building footprints, and other spatial referenced objects were vectorized from the 3-D imageries with the help of photogrammetric process and displayed in the base-map. A constructed street network was used for addressing purposes.

The whole metropolitan city area was divided into five addressing zones, and also kept the Municipal Boundary and Ward Boundary as it is with a purpose to maintain the municipal governance systems as they used to follow. Triangulation of positional accuracy of the features like street and building depicted in the base-map was made by superimposing the vector layer of Cadastral parcels. Although the cadastral parcel data were prepared in the early 1970s with plain table survey, they were not exactly superimposed with the current building footprint but gave ideas on identifying correspondence parcel with existing building footprints and road networks.

The distance from the start node of the street (junction or cross-road) to the main entrance or doorways of the building/compound was computed after fixing the start node and end node of each road segment's centerline. The whole process of building number addressing was done by using an interface developed in AutoCAD Map on the base of defined street code/road name in the base-map. In parallel, the building numbers were addressed in AutoCAD, and the data structure was transferred to ArcGIS Environment. Street signage and house number plate were designed. Each building with their metric number again was verified in the field and given the number to each building owner by quoting their *Tole* name, addressing zone number, ward number, and street name. Those signage plates and house number formats were finally handed over to the metropolitan office for their further execution. After the completion of referring house numbering in digital format, those were managed in the digital database of LMC, and also developed the webGIS systems for Metropolitan's wider application.

5 Procedures for the Addressing

In the core area, several building blocks are adjoined with each other. Those buildings were constructed in the past. However, the major layout of the city has not much changed but the rest of the factors, such as the number of storeys, building density; urban services, etc., have been changed alarmingly. With a view to address the urban landscape, it had been adapted different techniques, integrating various tools and viewed the factual situations with different perspectives. High resolution stereo satellite imageries had been used to produce the base-map with 3D vectorization of all features and produced Orthophoto of the city landscape. It had been geo-referenced with the benchmark of global position as per the benchmark established at Kathmandu Valley, by the Department of Survey of Government of Nepal. Newly established 10 GCP stations in a triangulated network have been established by using the DGPS survey and leveling within the LMC. The vertical height of each point had also been triangulated by the traverse leveling survey. Being a vertical expansion of the city landscape due to the multistory building, so far the high resolution Digital Globe Imageries (0.32 m) had experienced a little comfortable even though it has been properly preprocessed and enhanced the quality through photogrammetry process available in the machine reading software. Thus, the vertical and near-vertical imageries of the historical city core area had been acquired by UAV from the height of 50 m to 100 m as per permission granted by the government authorities. Image resolution and Orthophoto of the both Digital Globe and UAV have observed with remarkable differences (Fig. 3a and b).

The conceptual basis for digital addressing follows the streets or paths in a systematic pattern. The street-based addressing has been adapted that each dwelling unit is connected by at least one path and the dwelling unit has one main doorway connected in a single path. The street can be numbered according to their segment which can be separated by a junction. Following the distance of a segment of the path from its junction to the building, the doorway helps to track the building location. So, it requires the street code number (Road ID), name of the street, and junction identification number (ID). Tracking street ID made easier by the identification of *Tole* name (geographical entity of local unit of conventional addressing based on the identified community) and *Ward* number of the Metropolitan City. The digital array of the number follows the province, district, municipalities, addressing zone, ward number, street ID, and house number in a sequential order by which not a single building number never repeat and duplicate, and remains identical (Fig. 4).

Fig. 3 Orthophoto of the core city area, **a** Digital Globe and **b** UAV

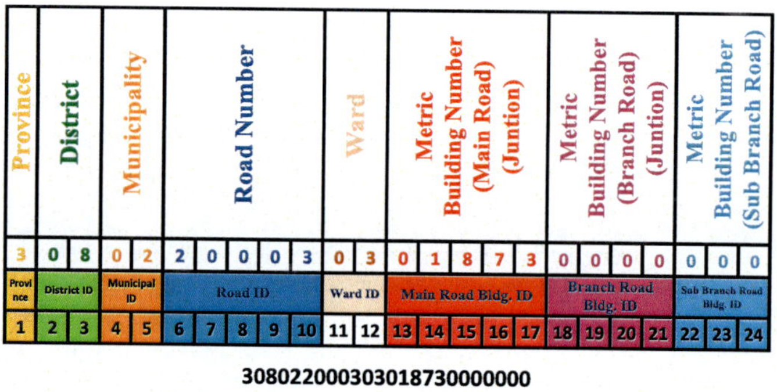

Fig. 4 Digital array of the number to each parameter

Therefore, the street signifies the network within the system and each street segment with their number provides the base for the house numbering. Following this conceptual ground, a detailed street map of the city had been prepared prior to the digital addressing (Fig. 5).

Street networks have been classified based on two different bases, i.e., the width of the street and the surface types of the street. On the basis of width, there are five types which are *path* (=>14 m width with sidewalk track), *sadak* (=>8 to 14 m width with or without sidewalk track), *marga* (=>3 m to 8 m width without sidewalk track), *galli* (<3 m width), and inside *Baha* and *Bahi* or *chowk* or *courtyard* streets. Based on surface types there are blacktopped, gravel, and earthen. All the street segments had been presented by a different symbol, labeled with identification number, name, and marked with junctions.

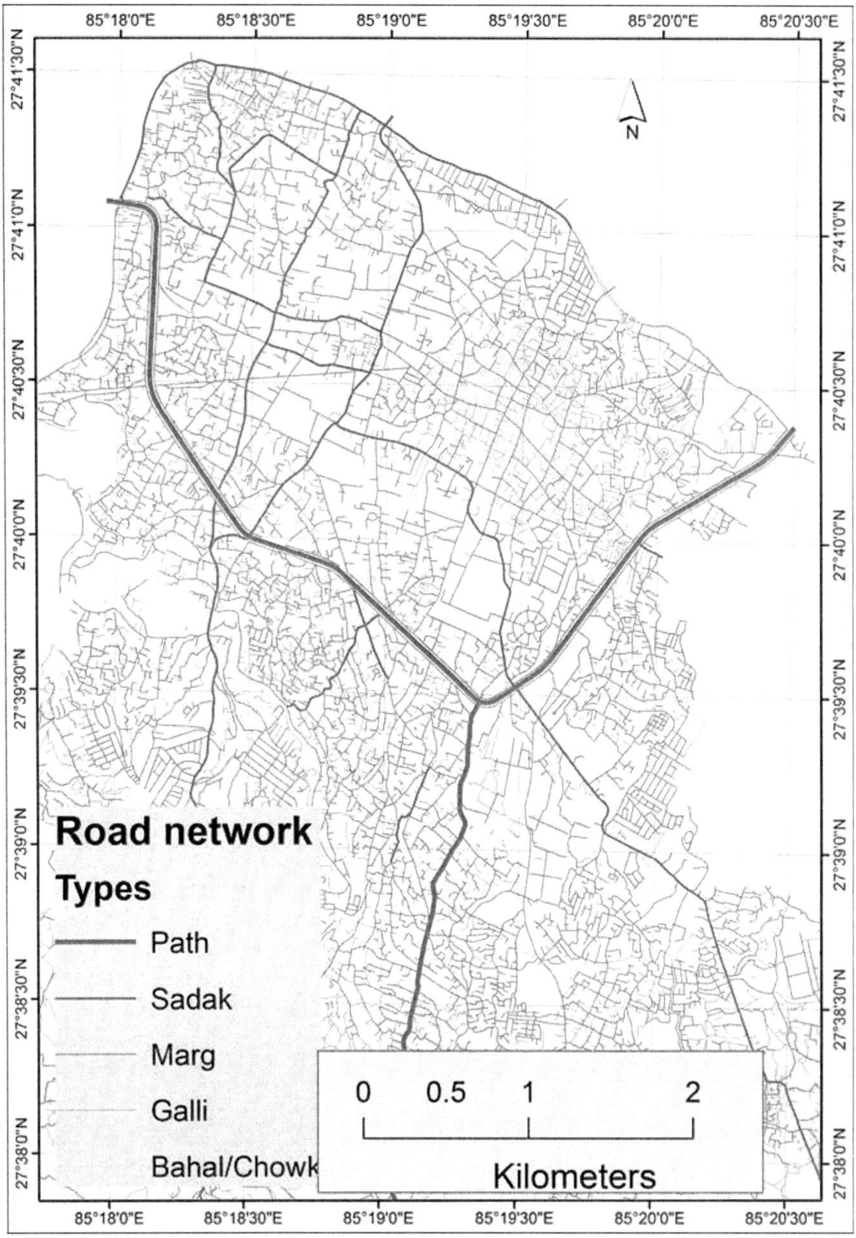

Fig. 5 Street network of the city

Being the landscape is very complex, some of the streets and underpasses were not even visible on a 3-D stereo plotter as well in Orthophoto. Similarly, several narrow streets were covered by building rooftops. Again some streets were through-fare under the private building structure and reach to the courtyard inside the building rows. To make it more accurate, mapping the digital layer of Cadastral Parcel was superimposed over the base-map and Orthophoto. Extensive field verification was carried out on such complexities using tablets with the background of satellite and Drone Ortho maps. In parallel, all maps were printed in an enlarged scale of 1:250 and obtaining all features information needed for the base-map and addressing system from the field. Street centerline, street width, and all street network extension were plotted precisely for the purpose of obtaining the base-map more accurately. Finally, following all those processes and with intensive field verification, the base-map of 1:500 scale in the core city area and 1:2500 scale in the outer side had been prepared (Fig. 6).

The Base-map of the core city area shows the internal structure of the city has very complex agglomeration of the buildings. There are few wider roads (*sadak*) outside, but several small and narrow streets (*galli*) are entering inside the complex agglomeration of the building. There are several small and large courtyards (*bahas* and *bahis* or *chowk*) inside surrounded by buildings. Those courtyards (*bahas* and *bahis* or *chowk*) relate to the wider road network by the narrow streets. Some of the narrow streets are through-fare even beneath the individual buildings. Some of such narrow and small streets have low height and doors to close for the security

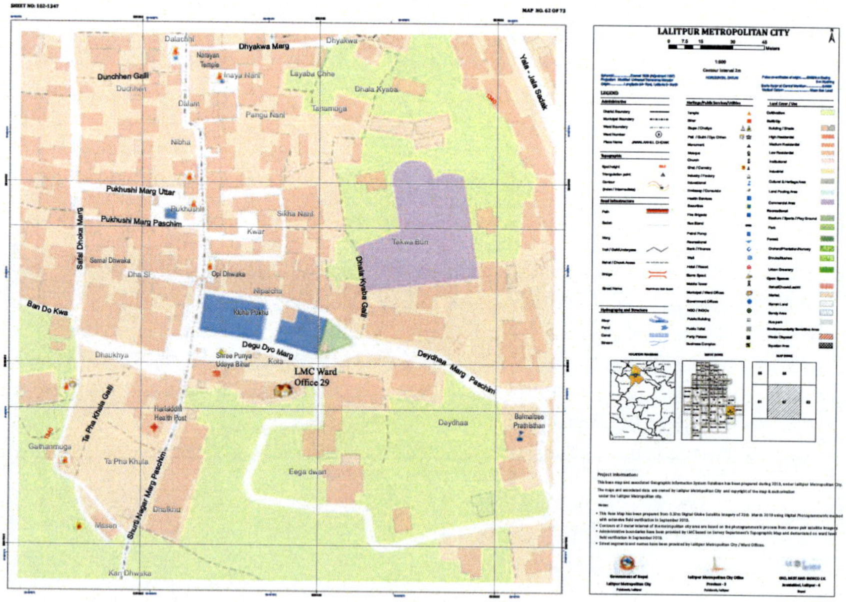

Fig. 6 Base-map

purpose in the night or some possibility of outside threats. Within the premises of the courtyard, there are deep wells, some places even have stone spouts and ponds for drinking water, Stupas and temples to pray and worship, rest places (phalcha) for sitting and gossips. Most of the buildings have the main door facing towards the courtyard sides and towards the main road fully blocked by the wall. Thus, the core city has a typical archeological value system followed as per the need of the time in the past. Addressing such a complex city structure in the present context has put forth great challenges. For such challenges, deferent rules had been established prior to the execution (Fig. 7a, b, c.).

The systematic rules have been followed to assign the numeric value of the distance from the main junction of the path to the main doorway of a building. It has an order of odd number on the left direction and even number on the right direction of the path from the start point of start junction towards the end junction of the street. Thus, everyone could easily manage to find the building according to the ascending and descending array of the number from the road identification number or junction. In database management, it gives a systematic procedure and helps to arrange data of each road type, code ID, junction, and building number in an easily accessible and retrieval mode.

6 Results

Digital metric addressing of highly complex core city areas faced number of challenges. Directions of streets are not clearly recognizable. The junctions are not easily distinguishable. People use different streets as of their convenience, thus, it is not easy to identify the main doorway. Narrow streets and even streets beneath the private buildings are sometimes difficult to identify as through-fare or dead-end. Some house units have a single door and shared by more than one family. Inner courtyards are used for many different purposes, like through-fare, meeting and drying their cereals, and for the source of sunlight during the winter months. They are also used for gathering during special events. Thus, the houses have the main door towards the inner courtyard. In such cases addressing based on the street network system was not an easy task. Despite several such challenges, digital addressing of the street, building, public structures, cultural monuments, and even drinking water sources have been marked in a systematic way (Fig. 8).

For the whole process, the prepared detailed street network map of the area was used. The linear alignments of the vectorized streets were verified by the superimposed of the Cadastral layers. Each street segment was numbered with a specific identifier. Those IDs were assigned in a systematic way following the direction of from node to node. Junctions of the street segment were usually marked at the start and the end node. That has usually given the direction. Following the procedures, each street was marked with signage plate (Fig. 9).

Based on that direction, each housing unit follows an array of a systematic number corresponding to the distance of road junction to the main doorway of the housing

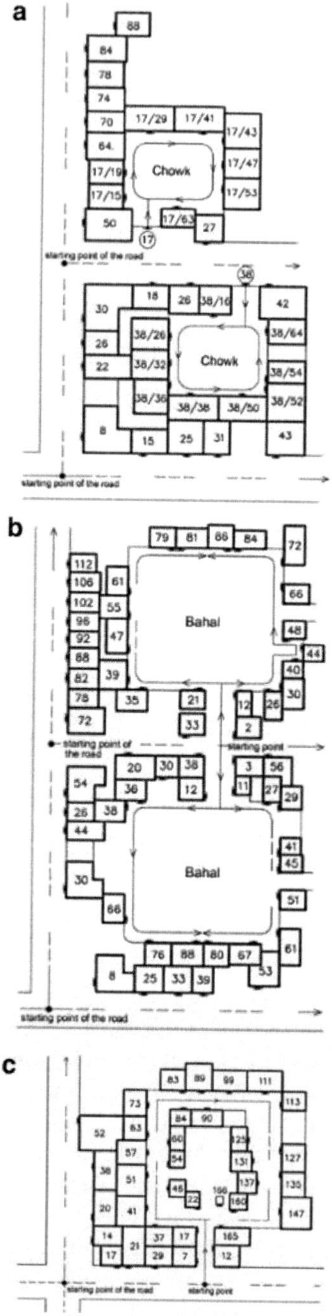

Fig. 7 **a** Addressing system follows the path direction from the main junction and the small courtyard (*chowk*) both side of the path with numeric value as of the distance from the main junction. **b** Addressing system follows the path direction from the main junction and the large courtyard with numeric value as of the distance from the main junction. **c** Addressing system follows the path direction from the main junction and the large courtyard with building inside the large courtyard given numeric value as of the distance from the main junction

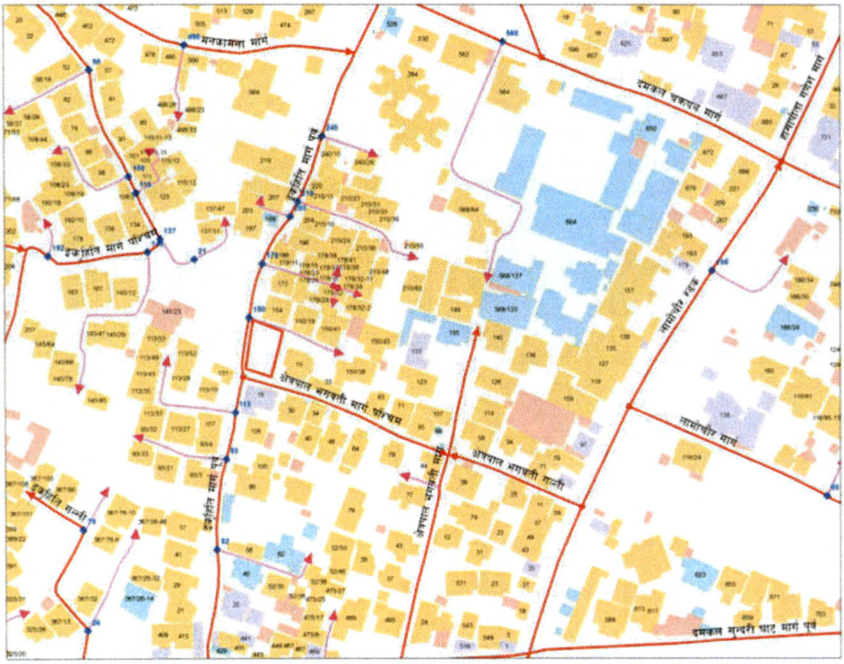

Fig. 8 An example of house number (need to change with house numbered one)

Fig. 9 An example of a street signage plate where it shows the street name and the Identifier number in both english and Nepali language

unit. The process has given all the building numbers according to the road alignment. For its public recognition, each building/house/compound number has been tailored with *Tole* name, street name, number with addressing zone, and Metropolitan *Ward* number (Fig. 10).

From the real ground exercise, each building/house unit marked with a systematic array of addressing. Those databases were digitally coded and maps were printed in

Fig. 10 An example of building/house number plate

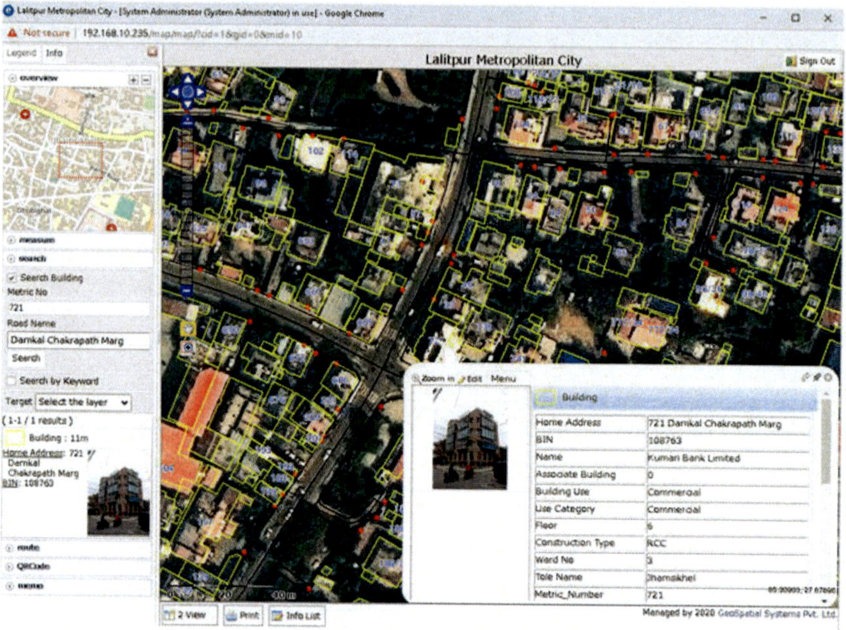

Fig. 11 Prototype of webGIS platform

hardcopy for the public and office use. For the Metropolitan office, all the data were managed in GIS/CAD format and stored in the Server. For the wider circulation and day to day use all the data were maintained in the webGIS platform (Fig. 11).

WebGIS platform had been developed within the open-source access mode.

WebGIS system uses three major components to read spatial data (Vector and Raster). It uses a map engine to read spatial data such as Shapefiles, KML files, etc., along with Spatial database engine to read spatial information integrated with spatial data. It displays maps and layers with the most prolific ways on the web by using web map services like Open Layers, Leaflet, and so on. However, the dynamic changes of urban phenomena require a regular update from the authorized level, thus the database access has been given to the Metropolitan Engineering and GIS Department.

7 Conclusion

Historical urban areas are the agglomeration of several urbanization processes during the time since their establishment. Lalitpur City, also falls within such categories where several urban structures were built before the Christian era. Digital addressing is a recently developed scientific technique that follows clear methodological procedures. In this exercise space-place information integration approach has been followed for the digital addressing of historical core city area of Lalitpur. In this approach, high resolution stereo satellite imageries, very high resolution Drone imageries, DGPS, traverse ground leveling survey, integration of Cadastral parcels, and intensive field survey and verification have been used. At the end of the whole exercise detail base-map with landscape characteristics, land use, each street with their positional accuracy, surface types, width level, and junctions were marked. Based on the highly accurate street addressing, all the individual building units, social and cultural monuments, and landmarks were addressed with a specific identifier. Those positional referenced identifiers guide the urban planners, service providers, and rescue operators to identify the specific location for the services. In the course of time, the landmarks and identifiers will be the reference units for the future expansion of the city. Therefore, digital addressing of the current case of the historical core city of Lalitpur, will lead its benchmark for the urbanization management within the country and abroad on geographical contexts.

Acknowledgements This paper is based on the on-going Project conducted under the financial support of Lalitpur Metropolitan City (LMC) to the GeoSpatial Systems Pvt. Ltd. Authors sincerely acknowledge LMC and all collaborative organizations.

References

Bengt P (1992) Urban applications of satellite remote sensing and GIS analysis. The World Bank, Washington D.C. (Executive summary)

CBS, National Census Report (2011) Kathmandu: central bureau of statistics, Government of Nepal (2012)

Demir I, Raskar R (2018) Addressing the invisible: street address generation for developing countries with deep learning. In: (Conference paper 2018) Research Gate

http://www.researchgate.net (browsed on 06 Aug 2019)

Demir I, Hughes F, Raj A, Dhruv K, Murthi S, Garg M, Sanaym DB, Raskar R (2018) Generative street addressing from satellite imagery. Int J Geo-Inf, ISPRS

https://www.un.org/en/events/citiesday/assets/pdf/the_worlds_cities_in_2018_data_booklet.pdf stics

Jezzini A (2017) Building detection using high resolution satellite imagery (Master Thesis). Computer and communication engineering department, Islamic University of Lebanon. http://www.researchgate.net (browsed on 01 Aug 2019)

KMC and EU Guidelines for Metric Addressing System (2000) Kathmandu valley mapping programme: integrated urban development programme. Kathmandu

LMC home page. http://lalitpurmun.gov.np/en/node/4 (browsed on 11 Aug 2019)

Locke JKSJ (1985) Buddhist monasteries of Nepal, A survey of the Bahas and Bahis of the Kathmandu valley. Sahayogi Press, Kathmandu

Poudel KP (2012) Urban sprawl and socioeconomic change in the Kathmandu Valley, Nepal. In: Dutta AK, Wadhwa B, Thakur B, Costa FJ (eds) Facets of social geography: international and indian perspectives. Foundations, pp 238–251

Thapa RB (2009) Spatial Process of Urbanization in Kathmandu Valley (PhD Thesis) submitted to the graduate school of life and environmental sciences. The university of Tsukuba, Japan

Thapa RB, Murayama Y (2009) Examining spatiotemporal urbanization patterns in Kathmandu Valley, Nepal: remote sensing and spatial metrics approaches. Remote Sens 1:534–556. https://doi.org/10.3390/rs1030534

UN habitat addressing urban and human settlement issues in national adaptation plans—a supplement to the UNFCCC technical guidelines on the national adaptation plan process, (Ms. Maimunah Mohd Sharif Under-Secretary-General and Executive Director, UN-Habitat, Foreword) (2019)

UNESCO home page. https://whc.unesco.org/en/list/121/(browsed on 11 Aug 2019)

World Bank (2005) Street addressing and the management of cities. Washington DC, The World Bank

Migration: An Element of Smart Livable Cities

Braj Raj Kumar Sinha and Priyanka Thakur

Abstract Movement of people from a defined geographical or administrative area to another with an intention to settle down there permanently is called migration. Migration is one of the important elements of smart livable cities because of its citizen-centric behavior. All highly modernized ICTs, institutional, social, economic and physical elements are waste without smart people. In smart or smart livable city migrants as well as smart local citizens play an important role in enhancing the ICTs, inventing new tools and ideas to do better work performance. Migrants are quite open-minded to work according to the plan and create a favorable work environment for themselves and others. Smart livable city being a center of availability of resource opportunities pulls a large number of skilled migrants. Therefore, migrants being a source of creative force provide numerous opportunities to cope with the problems and tackle the obstacles of smart cities in smarter ways. This study aims at understanding the concept, meaning and different dimensions of smart livable city; highlighting the plans, policies and programs of the government in making the cities smart and livable; and demonstrating the role of migration in achieving and maintaining the status of smart livable cities. This paper is mainly conceptual in nature and based on the review of migration and smart city-related literature relevant to the present studies. The study is supported by the chart of different dimensions and components relating to migration and smart livable city.

Keywords Migration · Smart livable city · Dimensions · Information and communication technology

B. R. K. Sinha (✉)
Department of Geography, Institute of Science, Banaras Hindu University, Varanasi 221005, Uttar Pradesh, India
e-mail: sinha_brk@yahoo.co.in

P. Thakur
Department of Geography, Banaras Hindu University, Varanasi 221005, Uttar Pradesh, India
e-mail: priyanka.thakurbhu@gmail.com

L. T. T. Huong and G. M. Pomeroy (eds.), *AUC 2019*, Advances in 21st Century Human Settlements, https://doi.org/10.1007/978-981-15-5608-1_22

1 Introduction

Migration is one of the dynamics of population change. It is an important cultural phenomenon and has a profound relation with socio-cultural, economic, political and environmental aspects in space and time. Migration is generally defined as a movement of person(s) from one administrative unit to another by making preferably a permanent change in the place of abode or residence. Migration is also taken into consideration as an important element of livable and smart city. Because it plays an important role in transforming an ordinary city into a smart one, a city has to pass through several phases to acquire the characteristics of a livable smart city. The words smart livable or livable smart city connote the same meaning. The status of livable smart city or smart livable city depends on several amenities such as water supply, adequate supply of electricity, approaches of good condition wider roads, sewerage and proper sanitary facilities, planned settlements, solid waste management system, parks and open green spaces, safety, security and safe transportation system, adequate and fare administrative vigilance. Apart from these major elements of city habitants as well as migrants contribute a lot in making a city smart. It is humans who regulate several policies for the betterment of their society and to get benefit from it to live a life of peace and comfort. Educated and skilled professionals are generally more attracted to smart cities where such persons continue to do hard and tireless work in enhancing information technologies and communication services. ICTs services facilitate smart and advance solutions for almost all infrastructural problems. For instance, well-planned policies of concerned government and non-governmental bodies, e-learning, pure and clean drinking water for all, smart transportation facilities, smart and safe urban environment, adequate employment-oriented opportunities, inclusive growth, skill and trade development center may be considered as major programs of ICTs. In creating and implementing such programs and policies migrants perform a role of a leader, a manager and a multidisciplinary team creator and are capable enough to make integration and cohesion with other people in their work environment which ultimately opens the path of innovation and transformation of a city into a livable and smart city.

2 Meaning and Concept of Smart City

The word smart city was used for the first time in the 1990s (Albino et al. 2015). It is a relative term which differs from one country to another, for example, ICTs determine the successful status of San Dieago as a smart city. Similarly, California and some other cities are the result of ICTs and smart people involved in that. Southampton is flourished by its multi-application smartcards. Silicon Valley in India is also one of the examples of smart city. All these have social, cultural and economic-based initiatives in making the city livable and smart (Hollands 2008). After studying the work of Shapiro (2006) Winter (2011) defines a city smart that has a huge college degree holder population. Such population is generally found in metropolitan cities.

Sofeska (2017) considered that a smart city is the combination of their citizens having advanced and innovative ideas, good health and education along with the social cohesion and economic integration. All these help increase infrastructure and technologies to be used by the governmental or non-governmental organizations for better development of the city.

In the words of Angelidou (2014) smart cities are those urban settlements that lead to make an effort in creating a new ICTs landscape in a strategic way. Such urban settlements seek to achieve prosperity, effectiveness and competitiveness at multiple socio-economic levels.

According to Komninos (2006) creativities of people (urban dwellers or migrants), their high capacity of learning power and innovative abilities cumulatively turn a non-smart city into a smart one.

In authors opinion, only the availability of ICTs services and modernized technologies is not sufficient to call a city smart unless there is efficient, timely and proper use of required services and technologies.

3 Meaning and Concept of Livable City

The concept of livable city explains those underlying facts which are related to livable city. The term livable is neither static nor measurable phenomenon because of its ambiguous nature. The concept of livable city differs from region to region due to different cultures and people of different perception along with the passes of time.

Daniela (2017) stated that when a city fulfills few criteria such as sufficient income, high productivity, low costs and better economic outcomes, adequate infrastructure and services with hygienic surroundings and proficient government to avail the development strategies then it is considered as livable city.

After having gone through the work of Aristotle, Sofeska (2017) conceptualized that *livability* means living or doing well. Further, she elaborated that a place where people live willingly and can afford easily their living conditions for a stipulated period of time is called a livable city.

Some scholars from economics considered stability of a city, healthcare facilities, suitable cultural activities, sustainable environment, better educational institutions and required infrastructure as indicators of a livable city.

According to Kashef (2016), the concept of livable city mainly deals with the modern and technically advanced version of building and designs of streets and other aesthetically characterized social and physical environment such as better comfortable living, well-beings, level of happiness and tolerance power of habitants.

Yadav and Patel (2015) considered livable city as a function of quality of life, the basic elements of which are water, sewage, solid disposal and wastewater treatment plants, safety, techno-creative society and citizens engagement. Apart from such elements Human Development Index is also taken into consideration as one of the factors of livable city.

Here, author's concept of livable city includes favorable site, situation and proper location of city, good government policies with adequate participation of people living within the city. Proper health and educational facilities, safety and security, mobility services, hard and software facilities and provision of adequate employment should also be made available within the defined city.

4 Different Dimensions of Livable and Smart City

Livable and smart city includes several dimensions such as physical, social, cultural, economic, political and demographic and a variety of components, the details of which are mentioned in the following Chart 1.

5 Role of Migration in Making the Cities Livable and Smart

Most of the smart Cities are meant to stand with high-level human capital which may be from the local city area as well as from its immediate or remote hinterland. In later case, the accumulation of labor force or human resources with critical knowledge, skills and talent can be by the process of movement of people. The local as well as migrant labor force or human capital with critical skills would be a great wealth in fulfilling the needs and smooth running of different sectors of the cities. Therefore, there is a need for an urgency of pulling high-level human resources with critical skills from different areas outside the smart city. Such migrants can contribute their knowledge, skills, talent as management personnel, stakeholders, researchers, city planners, designers as well as practitioners and professionals in leading to make the city livable and smart. For this, government plans, policies and programs also become important. A scholar rightly sited Florida as an example of a smart city in the USA and pointed out that a smart city can better manage and attract large number of skilled labors for diversified and integrated development.

5.1 Migration, ICT Services and Smart City

Migration and ICTs are one of the dimensions of the smart city because the availability of ICT services and its use by the migrants help increase the status and quality of the smart city. Sometimes migrants (especially technical, professionals and skilled persons) also bring with them to the city some specific capabilities for developing

Chart 1 Different dimensions and components of livable and smart city

Different dimensions of smart livable city	Dimensions	Components of livable city	Components of smart city
	Physical	Congenial site, situation and terrain	Pre and Post earthquake and landslide related management, flat plain and terrain with gentle and moderate Slop
	Socio-cultural	Proper health facilities	Availability of adequate number of medical institutions fully equipped with modern technology, infrastructure, services, medicines, technical and professional human resources
		Proper education facilities	Availability of adequate number of formal and non-formal educational institutions fully equipped with modern technology, infrastructures, services, technical and professional human resources for upgradation of quality education
		Safety and security	Perfect law and order situation from the government, proper provision to control crime and social vices, Citizen participation and corporation in maintaining safety and security
		Proper and adequate housing facilities	Provisions of affordable housing for migrants, slum dwellers and local general citizens with required social amenities
		Open public space	Construction and management to maintain green cities
		Hygiene and cleanliness	Availability and management of hygiene and cleanliness related infrastructure to make the city hygienic and cleaned
		Proper planning for land use for different purposes	Proper management and allocation of adequate land for different purposes and human activities

(continued)

Chart 1 (continued)

Different dimensions of smart livable city	Dimensions	Components of livable city	Components of smart city
		Adequate and timely power supply	24 hour continuous supply of electricity, installation and generation of Solar Energy, proper use and monitoring of energy consumption and overall management of different sources of energy
		Affordable transportation and mobility facilities	Provision of smooth transportation facilities and its efficient management, development of a multi-modal transport links of rail, roads, metros, airways etc., availability of adequate space at proper places for parking of vehicles
		Adequate assured supply of potable water	Provision of supply of adequate potable water and monitoring of its quality, prompt and regular observation and monitoring of water supply related problems
		Provision of wastewater management	Proper management to treat wastewater regularly and to supply the treated water to rural side for agricultural purpose
		Provision of solid waste management	Provision of proper and timely collection and smooth transportation of solid waste materials to the proper sites for regular recycling and generation of energy, fuel and making compost to use in further development in different sectors
		Provision to control pollution	Acquisition of proper plans, policies and programs to increase or enhance the awareness among the city people to check and control air, water, dust and noise pollution
	Economic	Economic opportunities and assurance of adequate Employment	Existence of viable economic activities of different types and assured means of employment to generate more income and to produce more output

(continued)

Chart 1 (continued)

Different dimensions of smart livable city	Dimensions	Components of livable city	Components of smart city
	Political	Transparency and good governance	Formulation and implementation of fair policies, plans, programs related to smart cities by the efficient government with good governance
	Demographic	**Flow of population**	**In-flow of highly technical, professional human resources with critical skills under appropriate migration policies to transform, built and create a good center of smart city**
		Optimum population density	Adoption of proper measures towards making a balance between population size-population density and available resources to lead a smart city to easily survive in space and time
		Balanced sex-ratio	Strategies for maintaining balanced sex-ratio from the point of view of social justice in developing knowledge, skills and talent equally among both the males and females for equal participation in producing a use value for satisfying social needs
		Stable population growth rate	Focus on creating situation for making the growth of population stable for better survival of smart city

Source Prepared by authors

specific services relating to information and communication technologies. For a healthy city, there is need of availability of uninterrupted services of mobile and telephone networking system, WhatsApp, viber, snap chat, video-calling, e-mail, websites, etc. from different service providers. For a smart city a continuous service of e-banking, e-commerce and trade should also be available for a smooth functioning of different sectors of smart city. According to Hollands (2008) and Albino et al. (2015) the success rate of any city is highly influenced by the ICT facilities. Similarly, Eremia et al. (2017) also stated that optimum use of infrastructure and achievement of sustainable development is possible only because of e-participation of migrants at local level. If in-migrants are highly skilled and specialized in different areas they contribute in building a smart city in better way. Provision should also be made available to the in-migrants for their required training facilities to develop their functional human resources. In the words of Codagnone and Kluzer (2011) ICTs are prerequisite in raising the level of living standard and to extend social and economic conditions of destination place of the migrants.

5.2 *Migration and Work Environment*

Work environment depends to some extent on the will and participation of in-migrants to the city. It is learnt from Shristi et al. (2017) that migrants are usually cooperative, happy, conscientious, artistically sensitive and more satisfied with their professions. These kinds of efficacious behavior affect their workplace and working environment. They can easily create a favorable working environment and increase the working capability of the technocrats and other workers. Migrants migrating from different areas normally adjust with the people of different ethnic groups in the cities. Migrants with different religion, culture and society are generally the persons of adjusting nature and creative ability to make happier, cooperative and conducive environment. They are generally talented with tolerating capacity and adjust even in odd situation. Migrants are normally the persons of creative attitudes and positive vision and work as a leader or manager to turn the city into smart one. They participate in formulating plans, policies and programs in favor of making the city livable and smart (Bouskela et al. 2016). Amanda et al. (2001) quoted from the work of Saxenian (1994) that people of any smart city try to link with academic, public, private and regional organization that will facilitate group affiliation, cooperation with competing companies and international connections.

5.3 *Migration and Innovation*

Through the process of migration innovation is spread over different parts of the world. Migrants are normally open-minded, extrovert and have adaptive nature and creative ability to create sometimes new and innovative ideas in their surroundings. Shristi et al. (2017) also emphasized that migrants have a tendency to learn everyday something new and participate in public affair happily. Migrants even coming from different cultural backgrounds try to assimilate themselves with the host areas people and become a source of creation of innovative ideas. According to Fassio et al. (2019) highly qualified immigrants act as a key determinant that explores the field of research and discovers innovative ideas such as university-industry linkages, e-governance and create economic and industrial clusters that will support to develop new business which act as the engine for economic growth in the recipient areas. According to Komninos (2006) there is a positive relationship among the skilled immigration, knowledge diffusion, enhancement of complementary ideas and the level of innovation. Kerr and Lincoln (2010) found during their study that, science and engineering workforce migrants constituted 24% and 47% respectively in the United States and secured higher position other than native population. Flow of immigration opens the path for assimilation of creative thinking and enlarges the level of productivity. While quoting the work of Saxexian (1994) Amanda et al. 2001 pointed out that the growth of a city, for instance, Silicon Valley in India is possible because of a variety of factors among them creation of innovative ideas and mobility of workforce are the key elements.

5.4 *Migration and Organized Approach*

Migrants are normally innovative and move in an organized and planned way from their origin to destination places. Innovative migrants can also be considered as a smart people who easily succeed in work and professional activities following organized and planned approach. The growth of a smart city depends on the smart mobility of the people as the commuters. The work of Shristi et al. (2017) also support the stated concept. Sofeska (2017) emphasized that success of the city depends on skill of the citizens which they use in making designs and in creating and improving the technological base for making the city livable and smart. The role of migrants is also noticed in designing, formulating, developing skilled base resources to meet the needs of smart city. Tranos and Gertner (2012) tried to focus on a smart city which "need to be active in attracting talent and precious human capital in order to support policy objectives related to creativity and innovation." Their study also focuses on migration, financial flow, attraction of skills, talents, etc. from other areas to the city. However, there is scarcity of literature in the field of migration and the smart city, because the study of migration and smart city has not yet received due attention of the world's scholars and remains a debatable issues among the scholars of different disciplines, policymakers and politicians.

6 Migration, Smart City and Well-Being of Population

Here, migration, smart city and well-being are chronologically arranged in order of its action. There is a general tendency of the migrants that they move from the areas of resource scarcity to the areas equipped with needed resources. Smart city is the city where necessary, useful and desired resources are appropriately available. The creation and utilization of such resources in smart city act a positive role towards well-being of its local citizens and migrants. Migrants with the help of local people take interest towards the smart solution of physical, social, cultural, economic, political and environmental problems to transform a city into livable and smart in context of social well-being. Picatoste et al. (2017) also have the same opinion regarding smart cities and well-being of migrants and local habitants. They elaborate that smart cities have a major goal to achieve inclusive accessibility for all and to reduce inequalities, exclusion and poverty with a priority of prosperity and well-being of the city people. Migration and employment play a crucial role in creating wealth of prosperity and well-being. The employment of migrants in the smart city depends on the knowledge, skills and talents, the results of educational achievements. That is how employability of migrants is related to education and from this point of view one can say that higher is the level of education higher will be the number of skilled migration and higher is the number of skilled migration higher will be the percentage of employability. In fact, the level of education ultimately determines the employability of migrants. This opportunity is available in the smart city. Tranos and Gertner (2012) also supported that migration of the skilled persons is good for themselves instead of living in areas with less employment opportunities. A city normally adopts a smart development trajectory and formulates such policies that attract large number of urban capital in the form of highly qualified professionals. Thus, creative migrants with high education are considered as a driving force in the development of any smart city because they have more concentration of creative economic outcomes which strengthen the path of growth of regional employment, innovation and high-tech industrial growth. Not only the economic sectors have such advantages, but they also integrate with all forms of creativity to develop social, cultural, technological and environmental aspects. In smart cities, migrants as individuals or communities or a group of people act as a core element because they know technology in a better way. Good management, good organization, different policies, e-governance facilities, economy, various infrastructure as well as environmental sustainability have a great impact on the quality of life at their own (Alawadhi et al. 2012). Eremia et al. (2017) stated that smart development of a city is a future hope for reducing discrimination, inequality, poverty, unemployment as well as for optimum use and efficient management of energy resources. In brief, migration takes place from undeveloped or underdeveloped areas basically rural to the developed areas preferably to the better urban centers for a better quality of life. Therefore, development of the smart city depends to a great extent on in-migration of the skilled, technical, innovative and professional persons. Such persons with the cooperation of the city inhabitants develop ideas to strengthen the social, cultural, economic, political and environmental resources with a view to enhance the level of prosperity and well-being of both the migrants and the local people. In this sense, migrants are an important and valuable human resource, the use of which restructures and develop the society by creating social, cultural, economic and political values in the recipient areas. Migration in most of the smart cities of the world has contributed a lot and made the cities healthy and prosperous, for instance, the Mexico City, Houston in the USA, Singapore and Abu Dhabi in UAE. After having gone through the above and looking into the role of migration it becomes imperative to consider migration as an element of the smart livable city.

References

Alawadhi et al (2012) Building understanding of smart city initiatives. In: 11th International conference on electric government, norway, pp 40–53
Albino et al (2015) Smart cities: definitions, dimensions, performance and initiatives. J Urban Technol 2(1):3–21 (2015)
Amanda et al (2001) E-Governance and smart communities: a social learning challenge. Soc Sci Comput Rev 19(1):80–93
Angelidou M (2014) Smart city policies: a spatial approach. Cities 41:53–511
Bouskela M, et al (2016) The road towards smart cities: migrating from traditional city management to the smart city. IDB, 1–148
Codagnone C, Kluzer S (2011) ICT for the social and economic integration of migrants into Europe. JRC Sci Tech Rep, 1–80
Daneila A (2017) Livable city from an economic perspective, 1–14
Eremia et al (2017) The smart concept in the 21st century. Procedia Eng 181, 12–19
Fassio et al (2019) Skilled migration and innovation in European industries. Res Policy 48, 706–718
Hollands RG (2008) Will the real smart city please stand up? Intelligent, progressive or entrepreneurial? City 12(3):303–320
Kashef M (2016) Urban livability across disciplinary and professional boundaries. Front Arch Res 5:239–253
Kerr WL, Lincoln WR (2010) The supply side of innovation H-1B visa reforms and US ethnic invention. Working paper 15768. National Bureau of Economic Research, Cambridge
Komninos N (2006) Integrating human, collective and artificial intelligence to enhance knowledge and innovative. In: 2nd International proceedings on intelligent environment. The Architecture of intelligent Cities, Athens, pp 13–20
Picatoste et al (2017) Smart cities for wellbeing: young employment and their skills on computers. J Sci Technol Policy Manag
Saxenian A. (1994) Regional advantages: culture and competition in silicon valley and route 128. Harvard University Press, Cambridge
Shapiro JM (2006) Smart cities: quality of life, productivity, and the growth effects of human capital. Rev Econom Stat 88(2):324–335
Shristi G, Mustafa SZ, Kumar H (2017) Advances in smart cities: smarter people, governace and solutions. In: Smart people for smart cities: a behavioral framework for personality and roles, pp 23–30
Sofeska E (2017) Understanding the livability in a city through smart solutions and urban planning towards developing sustainable livable future of the city of Skopje. Procedia Environ Sci 37:442–453
Tranos E, Gertner D (2012) Smart networked cities? Innov: Eur J Soc Sci 25(2), 175–190
Winter JV (2011) Why are smart cities growing? Who moves and who stays. Regional of Regional Science, 1–39
Yadav P, Patel S (2015) Sustainable city, livable city or smart city: what value addition should smart city bring to these paradigms in context of global south? In: 13th International congress of Asian planning schools association. Malaysia

Management of Public Space Towards Livable City: The Case of Hanoi, and Lessons from Singapore

Nguyen Lien Huong

Abstract This paper aims to focus on smart management of urban public space toward livable city. Along with the waves of digital and Information and Communication Technology (ICT) revolutions, smart City model has become a global "hot" trend and is expected to contribute to the livable city. The role of public space is universal, yet its functions and characteristics vary according to the particularities of the area and the community there. Public space sustains the productivity of the city and its social cohesion and inclusion, quality of life and local identity, and ultimately supports urban prosperity. Therefore, they should be planned and managed well and smartly for all kinds of users. Like many fast-developing capital cities, Hanoi has been facing increasing challenges caused by market-based hyper-urbanization, which oftentimes neglects to provide adequate social space for residents. Spontaneous and uncontrolled rapid urbanization results in chaotic urban patterns. This research goes around a central question of how to manage public space in various ways including efficient space usage toward livable city. The research seeks international lessons that show way forward, and lessons learned from Singapore were chosen to introduce. The research adopts qualitative research methodology. The flow of the paper starts off with a review of the discourse on public space and its management and explores the importance of public space in the city and how it can contribute to livable city. Then it introduces urban development in Hanoi with a focus on the Historical metropolitan area as defined in the Primer Minister's Decision 1259/QĐ-TTg, and its current situation regarding public space planning and management. Coming to the demonstration case study, this paper explores experiences and lessons from Singapore, which demonstrate ways to management of urban public space. At last, the paper provides some discussions and concludes that inclusive stakeholder participation with local community involvement under a strategic partnership among public, private and academia can be a key to the success of public space planning and management toward livable city.

Keywords Public space management · Livable city · Hanoi · Singapore

N. L. Huong (✉)
Urban Management Faculty, Hanoi Architecture University (HAU), Nguyen Trai Road, Thanh Xuan District, Hanoi, Vietnam
e-mail: huongnl@hau.edu.vn

L. T. T. Huong and G. M. Pomeroy (eds.), *AUC 2019*, Advances in 21st Century Human Settlements, https://doi.org/10.1007/978-981-15-5608-1_23

1 Introduction

1.1 Background and Literature Review

This paper focuses on the management of public space in the city toward livable city. Public space—such as squares, streets, plazas, markets, parks, playgrounds and waterfront promenades—plays a key role and can become a key drive to enhance urban livability toward sustainablity (To 2017). The role of public space is universal, yet its functions and characteristics may vary according to the particularities of the area and the community there. Public space sustains the productivity of the city and its social cohesion and inclusion, quality of life and local identity, and ultimately supports urban prosperity. Therefore, they should be planned and managed well and smartly for all kinds of users.

In the book "Life between buildings—Using public space" (Gehl 1996), Jan Gehl wrote: "the life between building, one of the most important city attraction." These are therefore vital areas, and planning processes must begin by understanding these spaces between buildings. And the use of urban public spaces is an experience which is not similar for everyone and factors such as age, gender, community groups and ethnic-racial minorities influence the understanding of urban life (Rafieian and Khodaei 2009). Urban public spaces are important elements of modern cities. These spaces, in different parts of human life, play a vital role (Kurniaty 2007).

The role of public spaces in sustainable social development can be important in this sense that today, human societies, due to daily quantitative and qualitative growth and development, has have led to the construction of large and densely populated cities, are facing with new problems. The frequency and density of disparate and heterogeneous cities, despite the physical proximity, has increased the social distance between people day by day and along with it, the solidarity, communication and attention of the people to each other is reduced (Fathi 2012).

According to Mike Douglass, the urban imagination surely needs to be animated with people who are social beings as well as consumers and that we must surely have a greater grasp of city life than to see it only as infrastructure, physical environment and finance. These elements are important, but the implicit posture of the develop mentalist approach appears to be that the living city—its many cultures, people meeting people by chance in public spaces, the unscripted, spontaneous life of the city—is either inconsequential or somehow comes "after" development, whatever that might mean (Douglass 2007).

Public space normally implies public ownership or direct management by the state. What clear is that the quality of the physical environment, and therefore physical public space and space as a social milieu, relates centrally to each of these, yet each is also much broader than a concern for public space management. In this regard, defining public space too widely may result in a nebulous concept that is difficult for those charged with its management to address. Conversely, defining the concept too narrowly may exclude important areas for action which, once omitted from policy, may undermine the overall objective of delivering better managed public space.

Like many fast-developing capital cities, Hanoi City of Vietnam has been facing increasing challenges caused by market-based hyper-urbanization, which oftentimes neglects to provide adequate social space for residents. Spontaneous and uncontrolled rapid urbanization results in chaotic urban patterns. The lack and sometimes decline of public space, lack of attention on public space provision, and poor public space management are alarming. Public space in Vietnam's urban areas has failed to meet demand or create typical characteristics to identify each city. In urban spatial planning, the creating of spaces, particularly public space is important in order to improve the quality of life. However, creating urban public spaces in Vietnam paid little attention to the interests of residents. In localities across the country, planners did not ensure that their planning met national criteria and standards for public space (Phan 2017).

According to Circular 34/2009/TT-BXD of Ministry of construction: "Public spaces are common space, park, square, open space where serve the spiritual life of people". However, in the Vienamese legal document system, there isn't yet an official definition of public space and for whom it caters.

For the researchers, the term public space is a definition which has the relation to philosophy, (urban) geography, visual art, cultural studies and urban design. First of all, public space is a space for everyone to share and that space had to be managed or supported by the state. A public space refers to an area or place that is open and accessible to all citizens regardless of gender, race, ethnicity, age or socio-economic level and no fee of charge (Nguyen 2008).

In the context of Vietnam today, the key task to do is to plan and protect the urban open space system, effectively and reasonably, to create a sustainable premise for development. Planning and protecting the public space system for a reasonable and effective way is the key content to solve.

In addition, with the existing public space in the metropolis, identifying the requirements required to protect the specific value of public space can become an effective guide for urban planning, management and urban development (Nguyen 2011).

1.2 Research Interest and Methodology

My research goes around a central question of how to manage public space in various ways including efficient space usage toward livable city. The research seeks demonstration cases that show way forward. Singapore has been chosen. My qualitative research methodology includes literature review and secondary data collection from diverse sources with a focus on public spaces.

2 Public Space in Hanoi and Its Management

2.1 *Brief Introduction About Hanoi's Urban Development*

Hanoi is the most ancient capital city in Southeast Asia with more than 1000 years history. Over the centuries and through different historical eras, the city has been gradually urbanized and developed. Since the Economic Reform (locally known as *Doi Moi*) in 1986, Vietnam's economy has been taken off, and along with that, urban processes have accelerated most drastically in recent decades. Recent statistical report states that about 37% of Vietnam's total population is now living in dense urban areas (Vietnamese Ministry of Construction, 2017). Together with Ho Chi Minh City, Hanoi is one of Vietnam's two special urban agglomerations with a population of more than 8 millions (General Statistics Office, 2018).

Hanoi's diverse and significant urbanscapes include the following areas: (1) Thang Long Imperial Citadel; (2) Old Quarter; (3) French Colonial Quarter; (4) Neighborhoods built before *Doi Moi*; (5) Private housing built after *Doi Moi*; (6) New urban areas boomed up in the city's outskirt and (7) less urbanized areas (mostly remained urban villages). Except the Citadel, all other areas are mainly residential areas mixed with commercial facilities and offices. In August 2008, Hanoi was officially expanded by merging itself with adjacent municipalities, making it approximately three times larger (Fig. 1).

2.2 *Brief Introduction About Public Space in Hanoi's Historical Inner City*

In the Masterplan toward 2030, Hanoi's Historical Inner City (HIC) is officially specified as from Ring Road 2 to the south bank of the Red River with an area of over 30 km^2, including Ba Dinh, Hoan Kiem, Hai Ba Trung Districts, the north of Dong Da District and a part of the southern part of Tay Ho District. The population is 1.246 million people. This research takes this HIC as the geographical focus.

Over ten centuries, public space in the HIC has been spontaneously evolved as well as properly developed on a planning basis, which carries many architectures, landscape, socio-cultural and historical values. Public spaces in the oldest part of the city—the Old Quater and its adjacent Hoan Kiem Lake—are found in forms of lake (Hoan Kiem Lake), park (Hang Dau Park), traditional market (Dong Xuan Market and Hang Da Market), street-side markets, religious buildings such as the many temples, streets, and so on. In the second oldest part of French Colonial Quarter, public spaces are further developed with new and beautiful tree-lined boulevards, more parks (such as Ly Thai To Park, Dien Hong Park, Hang Trong Park, Co Tan Park, Opera House Park), and new squares (such as Dong Kinh Nghia Thuc Square, State Bank Square), which are nicer and more spacious to cater for the French Colonial administration

Fig. 1 The morphology transformer of Thang long—Hanoi from 1010—1975

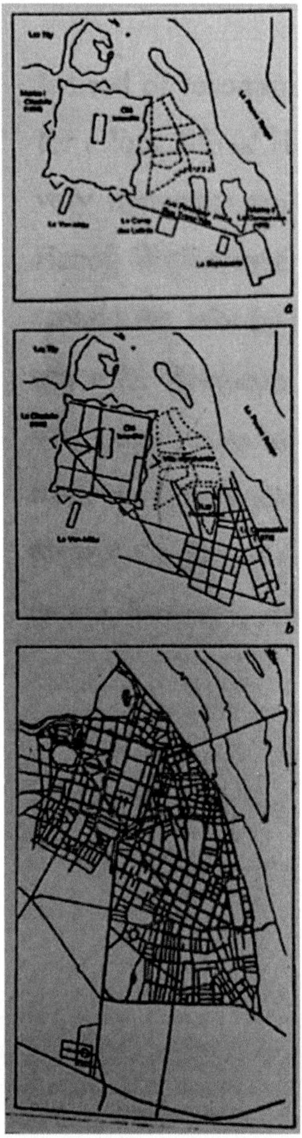

and local elites. To that extent, these new public spaces seemed to be not close to the local people.

After Hanoi is liberated from the French in 1954, following former Soviet Union's planning model, the communist government started to plan and develop "socialist-styled", large-scaled public spaces like Ba Dinh Square, Thong Nhat Park, and so on. These public spaces are spacious and more connected to the local people, and they have actually met people's rising needs for socialization, public life and community life. People's active participation and diverse usage of these public spaces over time help make them become well-used and thus, successful to some extent. From the country's unification in 1975 to the late 1990s, public space has gradually gained attention by urban planners, related governmental agencies and the general public. The period from the year 2000 up to now marks the highly rapid urbanization area in Vietnam. Many new urban areas have been built, and many parts of HIC have been further or re-developed. More new public spaces in forms of parks, squares, plazas, playgrounds, tertiary streets, etc. have been developed, yet the pace of public space development is never catching up with the pace of urbanization. This is partly because in most new residential areas, public spaces are still much less prioritized as compared to high-density residential and commercial spaces (Figs. 2 and 3) (Table 1).

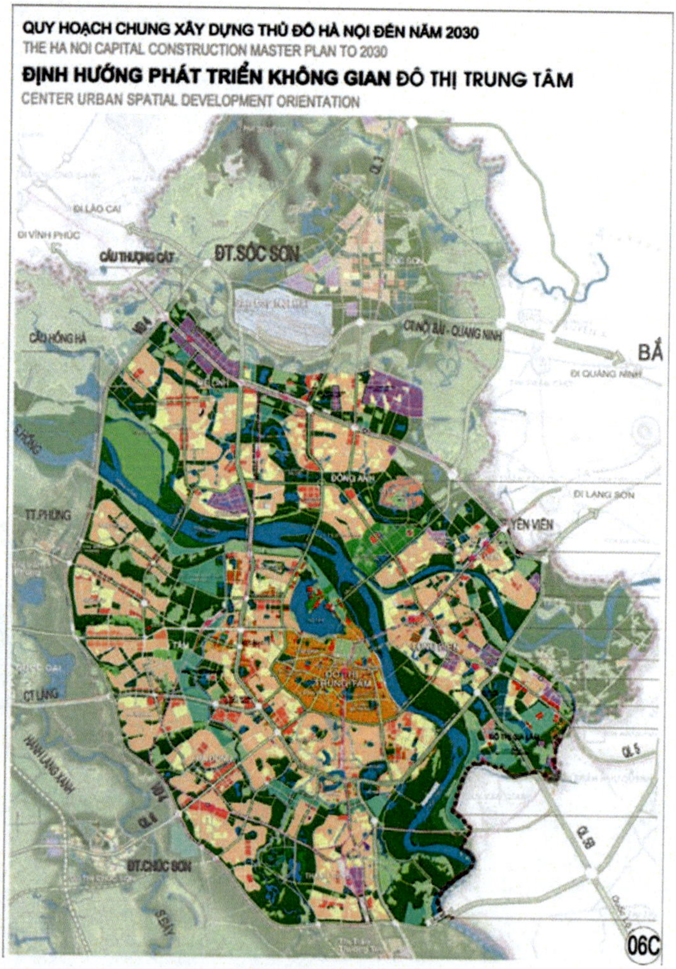

Fig. 2 Masterplan of Hanoi City toward 2030 with a vision to 2050

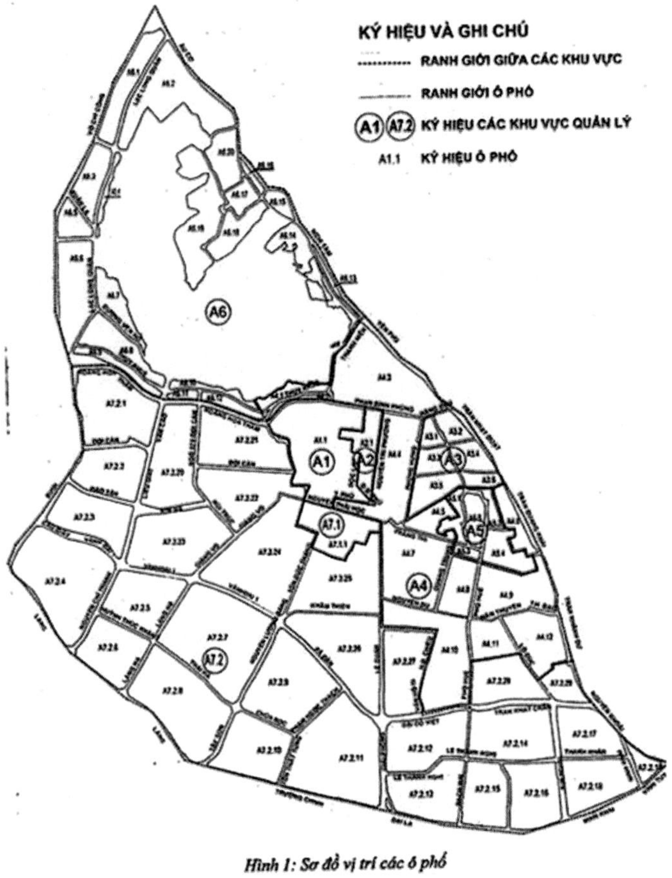

Fig. 3 Plan of HIC

2.3 *Current State of Public Space Development and Management in the HIC*

Today, like many developing cities in Asia who confront with the lack of public funds for infrastructure development (including public space), the lack of a good and long-term development vision, and the weakness of urban governance, public spaces in the HIC are not well managed and maintained. Increasingly intense urbanization stretches the city's border further and further, but not many public spaces are built along its way. Furthermore, as the city government is lacking public funding and calling for the private sector to jump into jointly fund for infrastructure and public space development under a Public–Private Partnership (PPP), a new trend of "privatized public space" has been emerging. Some most typical and popular forms of privatized or "pseudo" public space include shopping malls or mall-front plazas. On

Table 1 Public spaces in HIC

Public space	In the past	Nowaday
Streer side market	 1905	 2018
O Quan Chuong	 1905	 2018
Opera house	 1900	 2018
Hanoi Station square	 1929	 2018
Hoan Kiem lake	 1910	 2018
Thong nhat park	 1980	 2017

one hand, this is economically feasible for many developing countries. But on the other hand, this money-driven development trend potentially causes risks of social segregation, inequality as well as cultural issues.

An inventory of urban public space as part of an urban sociological survey in 2017 shows that in HIC, there are 42 parks, 46 lakes, 5 squares, 3 pedestrian streets, 5-night markets and more than 100 playgrounds. These statistics show that the HIC does have a good number of public spaces, which deserve better management and supporting policies. There are two aspects of public space management to pay attention to including management policymaking and implementation.

In terms of policymaking, after the extension of Hanoi's administrative boundaries in August 2008 and Masterplan of Hanoi City approved by the Prime Minister in Decision No.1259/QD-TTg dated 27/6/2011, zoning planning and detailed planning are being developed to concretize the Masterplan. The zoning plan comprises of 38 zones, 14 of which have been approved and the remained ones are being reviewed. Due to this incompletion of legal basis, the planning management in general and public space management in particular are still limited.

With regard to policy implementation (i.e., legal documents), after the above-mentioned Masterplan was approved, Hanoi City's Committee of the Communist Party (Vietnamese: Thành Ủy Hà Nội) has issued a program called 06-CTr/TU on "Strengthening city planning, development and management of Hanoi City, period 2011–2015" and commanded Hanoi City to implement accross level. Specifically, the detailed plans, urban design, management regulations for the four core districts have been formulated and officially implemented. These processes support public space management. However, the complexity of the areas, the overlapping of mandates and unreasonable, delayed coordination among related governing departments all lead to the widespread of poor urban management situations such as illegal construction, delayed projects, land-use with changed purposes. Poor management coordination among different stakeholders including the local community has caused serious conflicts such as landowners' frustration upon inadequate compensation. Public funding for public space development (with the largest portion being fund for land) is very limited. More pressingly, there is evidence of the overtake of public space land by developers to develop commercial real estate projects.

Nowadays, in Vietnam general and Hanoi particularly, the urban management process is impacted by the subjects of urban management such as economy, construction, land-use environment, population, social facilities, technical facilities. All these subjects made the system and the process of urban management so complicated. Public space is an element of the city. So the public space management process compliance the model of urban management (Fig. 4).

2.4 SWOT Analysis of Public Space Management in the HIC

SWOT (Strengths, Weaknesses, Opportunities, Threats) analysis is a strategic planning technique used to help identify strengths, weaknesses, opportunities and threats

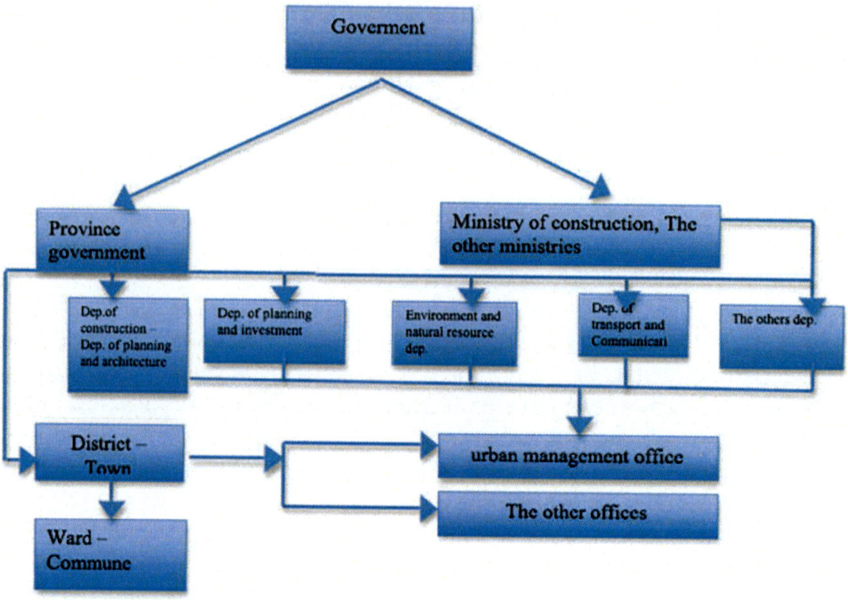

Fig. 4 Model of urban management in Hanoi

related to business competition or project planning (Wikipedia 2018). Today, SWOT analysis has been widely adopted for many other fields including urban development topics. A SWOT analysis of public space management in the HIC has been conducted and briefly summarized in Table 2.

3 Lessons from Singapore

Singapore is a small city-state with a total area of 716 km^2 and a population of 5.6 million (Singstat 2017). Apart from the need to cater for housing, business, social and recreational needs, large-scale industrial, transportation and military uses, which are typically located outside the city, must be housed within this limited landmass with few natural hinterlands. In 2013, the Government has set out a vision to grow its population to 6.9 million in 2030. This scenario requires even more land for residential developments and Singapore needs to be prepared. Singapore has successfully developed large-scale high-rise residential areas (HDB public housing estates) to accommodate its large population.

Table 2 SWOT analysis of public space management in the HIC

Strength	Weakness
• The HIC has a rich history and culture • There is a concentration of numerous public spaces which carry architectural, landscape, historical and socio-cultural values • The HIC has attracted funding from both state and city levels • Vibrant public and street life by people with a rich outdoor lifestyle and diverse group activities • Young population who actively participate in and have a strong need for public spaces • Good human resource in ICT • High popularity of internet usage	• Lack of an official definition of public space and for whom it caters in the legal document system • Lack of public space management regulations • Lack of research on public space management • Fund (especially for land) is poor • Lack of attention on public space provision • Lack of potential space for new public space development because the HIC has been fully built up with high population concentration, while resettlement plans are usually unfeasible • Incomplete detailed planning, top-down urban design only sets out visions but not concretized • Decentralized and overlapping urban management agencies with limited human resources and expertise. • Lack of a mechanism for community participation in urban planning and public space management
Opportunity	**Threat**
• The HIC is a large prime area which has gained extensive attention by the Government, investors, visitors and Hanoi people in general • It can attract more central funding, investment from various sources, and research efforts	• Threat of overtaking public space land for commercial projects by private developers • Threat of privatization of public space • Threat of dangerous traffic situations as well as environmental issues such as pollution that affect the safety and comfortable usage of public space

3.1 Public Space Planning and Provision

Within such land scarcity context, the Government reserves a good amount of public funding for the development and provision of public space in most parts of the island in forms of park, urban park, pocket park, square, plaza, street, playgrounds, waterfront promenades, etc.

As Singapore's built environment grows denser with a higher population, at some certain private properties (such as those located next to a train or bus station), property owners and developers are required to provide dedicated public spaces within their development sites. This applies to both new and redevelopment ones. This kind of privately own public space (POPS) is usually in the form of atrium, courtyard, forecourt, sky garden, pedestrian mall/promenade, etc. Specifically, the most common types of POPS are (1) City room: Covered public spaces located at the first storey of the building. They function as spaces for respite within the dense urban built environment; (2) Plazas: Outdoor public spaces that are typically paved, and defined on two or more sides by buildings and (3) Urban parks: These spaces function as green lungs for the city while enhancing the attractiveness of surrounding buildings. The provision of such well designed, dedicated public spaces within private developments, or POPS, will complement the provision of conventional public space to promote city life and make Singapore a more attractive city to live, work and play in. Therefore, public spaces in Singapore can be categorized into two types: (1) conventional public space (publicly own) and (2) POPS. As part of the Government's efforts to improve the quality of public spaces, including those in private developments, in January 2017, the URA issued a key circular titled "Design Guidelines and Good Practice Guide for Privately Owned Public Spaces (POPS)" (Circular URA/PB/2017/02-PCUDG, see URA, 2017) to provide detailed guidance for the design and management of POPS. These guidelines target Architects, landscape architects, building owners and developers. It sets out to ensure that POPS are Integrated with the overall form of the development; accessible to all users and well connected to the surrounding area; comfortable to use throughout the day and in a variety of weather conditions; well provided with amenities to encourage active and passive recreation and safe and secure for all users (URA 2017).

In another case, the Government has launched a policy to trade Gross Floor Area (GFA) to POPS, namely, to earn GFA exemption for a first-storey covered space if that space meets minimum requirements and is agreed to become public space. Property owners and developers may also voluntarily provide such public open spaces to offer amenity to their development and the surrounding area (URA 2017).

3.2 *Public Space Management*

The management of public space in Singapore, especially type 1, involves a few governmental agencies including the URA (which spearheads urban planning, development control and preservation), the Land Transport Authority (LTA, which spearheads land transport developments), the Singapore Land Authority (SLA, which optimizes land resources for the economic and social development of Singapore), the National Parks Board (NParks, which provides, enhances and manages the greenery of Singapore), Public Utilities Agency (PUB, which is in charge of water supply and management of water surface) and Building Construction Authority (BCA, which shapes safe, high quality, sustainable and friendly built environment). Therefore, certain types of public spaces are managed by one agency (such as lakes by PUB and parks by NParks), but some other types may be managed by multi-agencies (such as squares and plazas by URA, SLA and even LTA), depending on the actual spatial and physical conditions. The management of these public spaces is generally good thanks to a number of factors, including the very strict law and regulations with a heavy penalty system, good role demarcation and smooth coordination among relevant agencies, with the support of extensive utilization of advanced ICT as part of a "Smart Nation" model that Singapore has been promoting.

The management of type 2 involves both public and private sector. For instance, from the public side, the URA's guidelines (URA 2017) requires that the POPS should be accessible to all users and well connected to the adjoining street/pedestrian thoroughfare and neighboring buildings. The access and circulation within the POPS should be barrier-free and comply with BCA's universal design guidelines. It should not be enclosed or fenced in and should always be open to public (Singstat 2017). Furthermore, the Government needs to assure many other factors such as the POPS's location, size, configuration, comfort, safety, landscaping and amenities. Meanwhile, from the private side, the development owner and its management board need to assure other factors including the security and no bad impacts on the retails and smooth operation of business activities. For instance, the WestGate shopping mall in Jurong East area is an award-winning good practice which offers a comfortable, covered public space on its first floor with good landscape designs, urban furniture, a stage for music performances and buskers, which attracts a large number of visitors.

Conventional technology applications such as the extensive deployment of CCTV surveillance cameras in Singapore can indeed support the monitoring and management of public space. Furthermore, the emergent trend of utilization of other advanced ICT in public space management is notable. In the global context, the emergent Smart City model—for which Singapore subscribes—empowers authorities to manage the city's asset intelligently and helps urban planners better design the city to serve the residents. The advancement of Internet-of-Things and big data enables various information to be collected remotely and send to cloud for processing. For example, wireless sensor networks (for detecting noise and motion) have been experimentally deployed in some public housing estates for collecting data and information, which benefits urban planner, developer and relevant public agencies to better understand public space and its usage in those estates, thereby better improve and manage them. The sensors have been deployed at different nodes in typical types of public spaces around the estates such as gateway, urban corridor, park, and playground to detect people's noise and motion, thereby revealing quantitative data of their utilization and activities at those public spaces (Lau et al. 2017).

Another essential angle to look from is the participation of the public space users including residents, on-site workers, visitors and passerby in the making and management of public space. This truly makes sense, given the fact that users are those who are most connected to the public space and know best about them and their utilization state. The Government has launched different initiatives with incentives to encourage local communities to be active in place-making and management, such as the "Our Favorite Place" program by the URA. Moreover, those who initiate or contribute to the making of a public space in their neighborhoods are expected to be the ones who enthusiastically protect the space and help manage them well. With regard to the ICT application, there was an initiative to provide public space users a mobile phone with a pre-installed, user-friendly app and utilize the user data collected from the phone for further analysis. The data generated from the app aimed to answer such questions as "Which spaces are being used, how, and how much?", "How do the people who use these spaces perceive them?", "Are any spaces over- or under-utilized?", "What can be additional/alternative uses for spaces that increase livability?" and "How could spaces be modified (new- or re-design) to improve them in terms of the issues above?" (Tunçer 2018). All these data and information help better understand and manage public space.

4 Discussions and Conclusion

Lessons from Singapore have shown that public space is truly important and gains great attention from public sector, private sector and local communities. Although Singapore is also facing the land scarcity challenge like Hanoi, the country has been successful in (1) reserving an adequate amount of public funding for the development and provision of basic, conventional public space in most parts of the island (in forms of park, urban park, pocket park, square, plaza, street, playgrounds, waterfront promenades, etc.) and (2) involving and enabling the private sector to co-invest in POPS (in other forms of atrium, courtyard, forecourt, sky garden, pedestrian mall/promenade, etc.) to complement and diversify. In the latter aspect, Singapore is a foregoer in view that POPS has recently started to emerge in Vietnam. For example, shopping malls have been developed in Vietnam especially from the year 2000, but because there is no policy, no incentive schemes for trading GFA to POPS and no state-controlled regulations, none of the mall's atriums and plazas are considered as POPS. Most recently, another example has emerged—which can be considered as POPS: The large and central park in the Vinhomes Central Park complex in Ho Chi Minh City. This POPS was a result of a trading of river-side land for that park, and the park shall be permanently and fully open to the public. In principle, this type of emergent POPS in Vietnam shall be promoted, given the context of having limited public funding and the rise of the private sector with rapidly growing developers along with the increasing popularity of public–private partnership (PPP) mechanism. However, in the case of the Central Park, after its opening, there was feedback from local residents that they got trouble accessing the park by the property's security team. This case reveals that emergent POPS in Vietnam truly needs to be paid attention and well monitored and managed.

Singapore also shows that very strict law and regulations with a heavy penalty system, good role demarcation and smooth coordination among relevant agencies are useful for a good management of public space. These are the key to success that Hanoi can learn from. Moreover, Singapore has shown a role model in the region for the utilization of conventional technology (such as CCTV surveillance camera) and foremost, increasing utilization of advanced ICT such as sensor-based, mobile phone-based big data generated from public space users as part of a "Smart Nation" model that the country has been promoting. Although Hanoi has limited financial resources to develop Smart City model, it does have good human resource of young and dynamic ICT experts to leverage from.

The review of the current situation and the SWOT analysis have shown various weaknesses and threats, but also revealed much strength and opportunity. This strength is deeply rooted from the rich and sustainable local culture and lifestyle, whilst opportunity may come from Vietnam's young population, of whom a good human resource in ICT can be utilized as long as good new policies and a good "learn-play-deliver" environment can be made. Only by joining forces to overcome constraints, innovate policymaking and build a better management system leveraging on advanced ICT, public space management in Hanoi City and the HIC in particular will reach a new height toward a better and more sustainable future.

References

Douglass M (2007) Civil society for itself and in the public sphere: comparative research on globalization, cities and civic space in pacific Asia. In: Douglass M, Ho KC, Oii GL (eds) Globalization, the city and the rise of civil society—the social productiob of civic space in pacific Asia. London, Routledge, pp 27–49

Fathi S (2012) The analysis of social relations in sustainable urban space. J Iran Soc Dev Stud 4:47–64

Gehl J (1996) Life between buildings—using public space, 5th edn. Danish Achitectural Press, Copenhagen

Kurniaty R (2007) Local elites and public space sustainability: the local elite roles in the presence and usage of public space in malang raya, Indonesia. Environ Sci 20, 506 –515

Lau BPL, Wijerathne N, Ng BKK, Yuen C (2017) Sensor fusion for public space utilization monitoring in a smart city. IEEE Internet Things J 5(2):473–481

Nguyen M (2008) Chuyen de Khong gian cong cong—Tap chi Kien Truc (Public spaces—Architecture magagine)

Nguyen B (2011) The problems of public space and its control solutions. http://ashui.com/mag/tuongtac/phanbien/5688-khong-gian-cong-cong-nhung-van-de-chinh-va-bien-phap-kiem-soat.html

Phan L (2017) Public spaces in Vietnam's urban areas fail to meet publicneeds. https://english.vietnamnet.vn/fms/society/170812/public-spaces-in-vietnam-s-urban-areas-fail-to-meet-public-needs.html

Rafieian M, Khodaei Z (2009) Evaluation criteria and standards affecting the satisfaction of citizens in urban public spaces. J Rahbord 53, 227–248

Singstat (Singapore Department of Statistics) (2017) Population Trends

To K (2017) Public space as a key drive towards liveable cities for all. In: Proceedings of APPS 2017 international conference. Nagoya, Japan

Tunçer B (2018) Big data for urban design and planning. World Bank Learning Event

URA (Singapore Urban Redevelopment Authority) (2017) Design Guidelines and Good Practice Guide for Privately Owned Public Spaces (POPS). Available at: https://www.ura.gov.sg/Corporate/Guidelines/Circulars/dc17-02

Wikipedia (2018)

Smart Cities: Progress and Challenge in Establishing Liveable Cities

Rini Rachmawati

Abstract Cities are now trying to become smart ones. Various information and communication technology-based applications to support the achievement of a smart city have been implemented. However, its achievement seems to have not been linked to liveable cities. This paper is aimed at describing the progress of the development of smart city, particularly the one to support liveable city. Besides, it is also intended to identify challenges faced in the development of smart cities, especially the ones to support liveable city. This paper is made and based on the analysis of secondary data in the form of smart city masterplan, particularly the one on the strategies, programs, and applications. The location of the study area is Blora Regency, Central Java, one of the cities belonging to the 100-Indonesian Smart City Movement. The data are also obtained by having an in-depth interview about the implementation of the programs and the applications to support the liveable city through the dimensions of smart living and smart environment. The result shows that the progress of the development of smart city is shown through several programs and applications that are planned to support liveable city. In this regard, most programs have been implemented. Some of the applications have been implemented. Challenges in the form of the roles of stakeholders and public awareness to achieve liveable city, human resources in charge of providing services and applications as well as the community's ability to use them.

Keywords Smart cities · Liveable cities · Smart environment · Smart living

1 Introduction

In Indonesia, various efforts to establish smart cities have been done. One of them is through Indonesia's 100—Smart City Movement implemented by the Ministry of Communication and Information of the Republic of Indonesia in the years 2017, 2018 and 2019. Not only cities, but regencies also get the chance to participate in the selection for this movement. There are several requirements for the evaluation in selecting

R. Rachmawati (✉)
Department of Development Geography, Faculty of Geography, Universitas Gadjah Mada, Yogyakarta, Indonesia
e-mail: rinirachma@ugm.ac.id

L. T. T. Huong and G. M. Pomeroy (eds.), *AUC 2019*, Advances in 21st Century Human Settlements, https://doi.org/10.1007/978-981-15-5608-1_24

a city or regency to be included in the Indonesia's 100—Smart City Movement. They are, among others, regional financial ability, green city index, performance index of regional government and sustainable city index (Bappenas RI, Kominfo.go.id 2017). The evaluation is also based on other factors such as the developmental dimension of prominent sectors, developmental dimensions of even distribution and areas, and Nawa Cita as well (Bappenas RI, Kominfo.go.id 2017).

The cities and regencies that are considered a smart city are obliged to make long term, middle term and annual program in the smart city masterplan. Most of the existing programs are activities in the form of ICT-based applications to make the achievement of a smart city come true.

According to the Ministry of Communication and Information of the Republic of Indonesia, the developmental dimension of a smart city includes six aspects of being smart. They are smart governance, smart people, smart economy, smart branding, smart living and smart environment. In Indonesia, the implementation of being smart related to smart governance is the most prominent. Meanwhile, the implementation of smart living and smart environment seems to be still limited. For example, the aspect of being smart of a city is not much related to a city that is properly to live in. Therefore, the discussion on the importance of the relations between the achievement of a smart city and a city that is properly to live in needs to be taken into consideration as well. For this, this paper will discuss how a smart city can be utilized for developing and achieving a liveable city, checking any progress that has been made, and tackling any challenges that are faced. More importantly, how a smart city is certainly successful in making it a proper place to live in will also be discussed in this paper. Besides, progress and challenge in developing a proper city to live in through the application of a smart city will also be discussed. The discussion will begin by describing the concepts related to smart and liveable cities. It will be then followed with several examples related to the programs toward a liveable city in the implementation of a smart city.

This paper is aimed at describing the progress of the development of smart city, particularly the one to support liveable city. Besides, it is also intended to identify challenges faced in the development of smart cities, especially the ones to support liveable city. In describing the progress of the development of smart city and challenges to support liveable city, it is necessary to study the strategies of the supporting programs and applications that have been planned and implemented.

2 Method

This paper is made and based on the analysis of secondary data in the form of smart city masterplan. The location of study area is Blora Regency, Central Java, as one of the cities belonging to the 100-Indonesian Smart City Movement. The data obtained from the smart city masterplan are in the form strategies, programs and applications that are planned to implement, particularly the ones related to smart living and smart environment. Meanwhile, the primary data are obtained by having an in-depth interview for the information on the implementation of the programs found in the smart city masterplan related to smart living and smart environment. Besides, in-depth interview is also related to the applications that have been implemented to support the liveable city through the dimensions of smart living and smart environment.

3 Result and Discussion

3.1 The Concept of a Smart and Liveable City

The dimensions of smart city development, according to the Ministry of Communication and Information of the Republic of Indonesia, include six elements. They are Smart Governance, Smart Economy, Smart Branding, Smart Society Smart Living, and Smart Environment as shown in Fig. 1.

In relation to Fig. 1 above, it can be explained that the description of each related element and activity is broken down as follows (Source: Citiasia Inc. in Direktorat Jenderal APTIKA—2018). The elements of Smart Governance include Public Service, Bureaucracy and Public Policy. The elements of Smart Branding include Tourism, business and city appearance The elements of Smart Economy include industry, welfare and transaction. The elements of Smart Living include harmony,

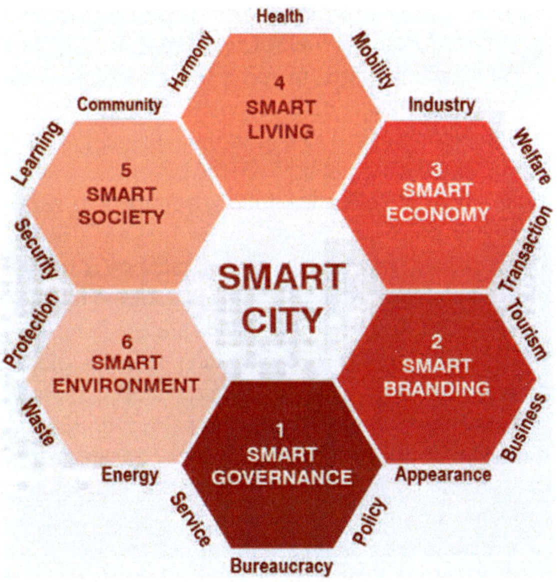

Fig. 1 Developmental elements of smart cities. (*Source* Citiasia Inc. in Direktorat Jenderal APTIKA—2018)

health and mobility. Meanwhile, the elements of Smart Society include community, learning and security. Eventually, the elements of Smart Environment include protection, waste and energy.

One element of a smart city that is closely related to the achievement of livable cities is smart living. Smart living deals with harmonizing a comfortable environment, guaranteeing health facilities and services, as well as transportation and logistics. Meanwhile, Smart Environment is also a smart city dimension that is related to achieving livable city. Smart environment is related to environmental protection, waste and waste management, and developing sustainable energy competitiveness. However, in building a livable city the success of implementing other smart city elements also has an important role.

However, in developing a city as a proper place to live in, the success of implementing other elements of a smart city also plays an important role. The components in a liveable city are multiple and complicated. Not only are they related to environmental development, but they are also related to social, economic and natural aspects (Shouthworth 2016).

3.2 Progress in the Development of Smart City Supporting Liveable City

The progress in the development of smart city supporting liveable city can be seen from several strategies that are planned and the program comprising the dimensions of smart living and smart environment. The data on the strategies and program are summarized from the Smart City Masterplan of Blora Regency.

The strategy related to smart living to support the development of liveable city is spatial harmonization that consists of implementing Regional Spatial Plan in line with the technological advances, implementing Regional Spatial Plan and optimizing the commitment of policy from the government in handling disaster, optimizing the commitment of policy from the government in realizing the spatial harmonization, improving the awareness of society in space use and accelerating the handling for slum areas. Meanwhile, the second strategy is realizing health facilities that consist of optimizing the program health insurance to reduce the occurrence of malnutrition, increasing facilities for health and sports, improving the habit of healthy living, the coverage of clean water service and improving the management for health services. Furthermore, the third strategy is guaranteeing the availability of public transport that consists of improving the quality of public utilities and facilities, managing the traffic routes to meet the need for the means of transportation, optimizing services and easy access to the means of transportation and developing transportation facilities in line with what the society needs.

In relation to smart environment, the strategy is improving environmental protection by optimizing the commitment/policy from the government, institutions and society in charge of handling environment, increasing the need for green-open spaces, increasing the monitoring for the quality status of river water, socializing the culture of health awareness through Blora Green and Clean, increasing the number of water-absorbing wells and biophory holes and increasing the number of green-open spaces, both private and public, to reduce the phenomenon of global heat and climate change. Meanwhile, the second strategy is by improving the garbage and waste management that consist of optimizing the commitment from the government, non-governmental institutions, and society in handling garbage and waste, implementing organic kampongs or city villages as an attempt to reduce the amount of home garbage, increasing the coverage of garbage management and service, developing climatic or decorative kampongs to create the awareness of the habit for healthy living and environmental commitment and socializing the habit of garbage awareness through Blora Green Waste.

Several programs that can be used to support the achievement of a liveable city are the ones related to the management of environment. It covers environmental protection, garbage management and smart public light. Meanwhile, the program that is related to smart living covers harmony, improvement of health services and mobility improvement by providing smart transportation services and other supporting facilities. For the activities, they can be in the form of innovative activities. For example, in garbage management GPS can be used to identify the garbage transport. In several cities in developed countries, the activity of controlling garbage container by applying a sophisticated system has been implemented.

Several programs included in the smart living of Blora Regency are shown in Table 1. Meanwhile, the specific programs related to the applications are shown in Table 2. They comprise the ones divided into short term (1 year), middle term (5 years) and long term (10 years). Besides, there is also Quick Win or a program that needs to be implemented immediately due to its priority since it has something to do with the interest of lots of society and it has direct impact to those lots of society.

From the 21 programs of smart living found in the smart city Masterplan of Blora Regency, there are at least 10 relevant programs supporting liveable city as shown in Table 1. Meanwhile, three of those 21 programs are related to the development of applications as shown in Table 2.

The results of the in-depth interview show that there are two of ten programs that support smart living that have not yet been implemented. Namely, the Program for the Preparation of Information on the Malnutrition from Agency for Health which should have been implemented in 2018 and 2019. Another program that has not been implemented is the Street Light PJU Smart System Installation from the Agency for Housing, Settlement and Transportation. This program should be started in 2019.

Table 2 shows three applications that are developed in Blora Regency to support smart living. Those three are related to health. One of the applications planned and implemented in 2018 is Sedot A Mas Application (Online Admission System, Ward Bed Availability, Public Complaint). Meanwhile, the other two applications are planned to be implemented in 2022.

> Information from the Regional Public Hospital of Blora Regency says that the Sedot A Mas Application can be easily, freely and fast downloaded and available on playstore and google play on android smartphone. This application is aimed at increasing the quality of health

Table 1 The programs of smart living in blora regency

No	Activities/Project	Implementer/Institution	Year 2018	Year 2019	Year 2020	Year 2021	Year 2022	10th Year 2023–2028
1	Composing spatial information system	Agency for public works and spatial planning				v	v	v
2	Communal sanitation	Agency for housing, settlement and transportation				v	v	v
3	Increasing the role of society in spatial planning	Agency for public works and spatial planning	v	v	v	v	v	v
4	Slum area handling	Agency for housing, settlement and transportation	v	v	v	v		
5	Composing information map of people with malnutrition	Agency for health	v	v	v	v	v	v
6	Developing promotion and information media for the awareness of healthy living	Agency for health	v	v	v	v	v	
7	Clean water project	Agency for housing, settlement and transportation	v	v	v	v		
8	Providing drinking water and public-based sanitation	Agency for housing, settlement and transportation	v	v	v	v		
9	Developing and maintaining road	Agency for public works and spatial planning	v	v	v	v	v	v
10	Installing street lights with smart system	Agency for housing, settlement and transportation		v	v	v	v	

Source Summarized from the smart city masterplan of blora regency, Year 2018 (Regional Government of Blora Regency 2018)

services for public satisfaction (mainly patients). Meanwhile, one of the benefits of this application according the informas is that it can increase the quality of health services since there is no need for patients to be in line at the admission (Muiz 2018).

By having the program of Sedot A Mas, the problem of long line for the patients at the admission can be solved. Besides, this application can also provide information

Table 2 Programs related to the applications supporting smart living in blora regency

No	Activities/Project	Implementer/Institution	Year 2018	Year 2019	Year 2020	Year 2021	Year 2022	10th Year 2023–2028
1	Sedot A Mas application (Online admission system, ward bed availability, public complaint)	Regional public hospital of blora regency	v	v	v	v	v	v
2	SPGDT application (Development of integrated emergency handling system)	Agency for health					v	v
3	Simpus application (Information system of public health center management) linked to BPJS	Agency for health					v	v

Source Summarized from the smart city masterplan of blora regency, Year 2018 (Regional Government of Blora Regency 2018)

on the availability of ward bed for public to access. Furthermore, this application can also be a medium to accommodate complaint from society so that they can make complaint, if any, directly to the hospital.

> The Sedot A mas Application is the one based on android. However, for those who do not have an android-based smartphone, they still can use the queue machine manually at the admission and SMS getway. Displayed in the waiting room of the admission is also the information on the availability of ward bed. (Smart City Masterplan of Blora Regency 2018).

From the 13 programs of smart environment found on the Smart City Masterplan of Blora Regency, there are 11 programs relevant to support liveable city as shown in Table 3. For the dimension of smart environment of the programs planned and implemented, there is no application that will be developed. This is due to the difficulty to relate between environment and application. However, the program stresses more on how to educate the society in environmental management, such as the establishment of organic kampong, the activation of garbage bank, and the making of biophory absorbing wells. One of the programs related to the use of ICT is the information system of online-based living environment. Yet this program will go into effect in 2020.

The results of the in-depth interview show that there are two of eleven programs that support smart environment that have not yet been implemented. Namely, the

Table 3 Programs of smart environment in blora regency

No	Activities/Project	Implementer/Institution	Year 2018	Year 2019	Year 2020	Year 2021	Year 2022	10th year 2023–2028
1	Blora green and clean	Agency for living environment			v	v	v	v
2	Kampung berhias (Decorating kampong)	Agency for living environment	v	v	v	v	v	v
3	Kampung Iklim (Climatic kampong)	Agency for living environment	v	v	v	v	v	v
4	Information system of online-based living environment	Agency for living environment			v	v	v	v
5	Air quality monitoring system	Agency for living environment						v
6	Developing city parks	Agency for housing, settlement, and transportation	v	v	v	v	v	v
7	Developing green building	Agency for housing, settlement, and transportation			v	v	v	v
8	One million biophories and absorbing wells	Agency for living environment			v	v	v	v
9	Blora green waste	Agency for living environment	v	v	v	v	v	v
10	One city village, one garbage bank	Agency for living environment		v	v	v	v	v
11	Organic kampong	Agency for living environment	v	v	v	v	v	v

Source Summarized from the smart city masterplan of blora regency, Year 2018 (Regional Government of Blora Regency 2018)

Information System of Online-based Living Environment program. However, this program will be started in 2022. Another program that has not been implemented is Air Quality Monitoring System. This program will be started in 2023. Both of these programs are under the responsibility of Living Environment Agency.

The following is an explanation from the informant regarding the non-operation of several programs in the smart environment dimension:

> In relation with smart environment, there are some programs that do not work, constraints require large budgets such as air quality control systems, it is planned to wait for special allocation funds. (Source: Informan 2020)

In relation to the Quick Win program, Blora Regency has three Quick Win programs, two of which are relevant to support liveable city. One of the two is Sedot A Mas program as stated earlier. It is a kind of service that provides online admission and information on the availability of ward bed and complaint from society as well. In the document of Smart City Masterplan of Blora Regency, besides being a Quick Win, Sedot A Mas is a program that will be implemented sustainably for short term (year 2018), middle term (five years, 2018–2022) and long term (2nd five years, 2023–2028). Besides covering the dimension of smart living and smart environment, Sedot A Mas also covers the dimension of smart society.

The second relevant quick win program to support liveable city is Smart Eco Village. Nevertheless, in the document of Smart City Masterplan of Blora Regency, this program belongs to the dimension of smart economy and smart branding. Irrespective of this all, it is more appropriate to include the dimension of smart environment. This is due to the initial program called Climatic Kampong implemented by the Agency for Living Environment as shown in Table 3. In the context of controlling the impact of climate change, it is necessary to establish a program to support the improvement of adaptation capacity and mitigation of climate change at the local level. Climatic Kampong Program is a national program intended to encourage society to undergo the improvement of adaptation capacity toward the impact of climate change and a decrease of gas emission. Besides, it is also an adaptation attempt and mitigation of climate change that has been implemented at the local level in line with the regional condition. Climatic Kampong is a location where the people have done an attempt for adaptation and mitigation of climate change in a sutaibnable way.

4 Challenges in Developing Smart City Supporting Liveable City

Several opportunities are open to be available by creating applications that can support the achievement of smart cities. The one that is used to support services in hospitals is, for example, online admission, online information on ward availability, and online health services for emergency. This kind of application can make health services efficient. However, the challenge lies on the aspects of people in making use of those services. Therefore, the achievement of digital literacy for people needs to be improved. Another challenge lies on the ability of human resources in operating ICT-based services. Besides, one thing related to the achievement of digital literacy is also a challenge to prepare smart people in accessing public services. Related to the use of ICT, in fact, Google Street View gives the impression that places which are proper to live in are larger than those taken from direct audit, showing a carefulness in relying on virtual experience as a proxy in fieldwork (Ruggeri et al. 2018).

The results of in-depth interview show that one of the limitations that will be a challenge is as shown below:

> Overall the biggest obstacles are in the policy sector, the lack of competent human resources and participation from the community. This is because the smart city program is a new program, so there are still many people who do not understand it. (Source: Informan 2020)

To prepare a well-managed infrastructure is also another challenge. The challenge includes the qualities of roads, public transportation, housing availability, telecommunication network, drinking water, and the sustainability in providing electricity and water.

5 Conclusion

Through a smart city, it is easier to establish a liveable city because to achieve a smart city needs to implement smart living and smart environment first. Both constitute an aspect in establishing a liveable city. The progress that has been made is in the form of program supporting liveable city. It has been planned through the smart city masterplan and many of them have been implemented. It has so far also been running well due to the supporting application that has been in existence. However, the applications to support the achievement of liveable city based on the dimensions of smart living and smart environment are still limited. This becomes a future challenge. Meanwhile, the challenges that are not less important are the role of stakeholders and public awareness to achieve an intelligent environment and the fulfillment of limited city infrastructure in achieving smart living.

References

Muiz A, Sedot A Mas, Pasien Tak Perlu Antre Mendaftar (2018). https://www.suaramerdeka.com/news/baca/116208/sedot-a-mas-pasien-tak-perlu-antre-mendaftar, 23 Feb 2020

Bappenas RI, Kominfo.go.id, Indikator pemilihan smart city (2017). http://indonesiabaik.id/infografis/indikator-pemilihan-smart-city-1

Direktorat jenderal aplikasi informatika kementerian komunikasi dan Informatika, Persiapan bimbingan teknis gerakan menuju 100 smart city (2018)

Regional Government of Blora Regency (Pemerintah Kabupaten Blora), Masterplan Smart City Kabupaten Blora, (2018)

Ruggeri D, Harvey C, Bosselmann P (2018) Perceiving the *Livable City* cross-cultural lessons on virtual and field experiences of urban environments. J Am Plan Assoc 884(3–4):250-262

Shouthworth M (2016) Learning to make Livable Cities. J Urban Des 21(5)

Studying Urban Expansion and Landscape Surrounding Monuments for Conservation the World Cultural Heritage in Hue City—View from GIS and Remote Sensing

Thi Dieu Dinh and Van Manh Pham

Abstract The socio-economic development and the growth of urban population are direct causes putting pressure for air quality, noise, and quality of life. The cultural heritage in the world is under the influence of urbanization and tourism activities due to the growing of hotels, restaurants, and associated services. This has a direct impact on the landscape of Heritage with many cultural values. Hue City is recognized as a world cultural heritage since 1993 by UNESCO. This is the only heritage urban in Vietnam composed of three elements: (i) heritage site, (ii) urban, and (iii) landscape (including Green Space), and not out of the impact of urbanization and tourism development. This paper focuses on quantifying the impact of urbanization and tourism development to the changing landscape of heritage sites in Hue by using all the data of remote sensing, statistical and spatial analysis tools in GIS. The results showed a positive relationship between the increase in population, developing tourism infrastructure with the expansion of urban land and thus causes deterioration of Green Space—landscape space of the world cultural heritage. From that we can see the trends of spatial changing can affect the quality of life and the tourism industry in the future of Hue City.

Keywords Spatial analysis · Urbanization · Tourism development · Heritage conservation · Changing landscape

1 Introduction

The urbanization process is the expanded space construction area (residential, industrial, infrastructure, and so on) in direct connection with urban areas. City boundaries

T. D. Dinh (✉)
Vietnam National Space Center, Vietnam Academy of Science and Technology (VAST), Hanoi, Vietnam
e-mail: dtdieu@vnsc.org.vn

V. M. Pham
Faculty of Geography, VNU University of Science, Vietnam National University, Hanoi, 334 Nguyen Trai, Thanh Xuan, Hanoi, Vietnam

L. T. T. Huong and G. M. Pomeroy (eds.), *AUC 2019*, Advances in 21st Century Human Settlements, https://doi.org/10.1007/978-981-15-5608-1_25

continue to expand, consuming more and more rural areas, forests, heritage places, and other important non-urban areas. Urbanization processes as a result of population growth, migration, and infrastructure initiatives (Agapiou et al. 2015).

World Cultural Heritages are under pressure from urbanization, including urban expansion, the pressure of population growth, migration, infrastructure construction, and tourism activities, which are due to increasing of hotels, restaurants, and associated services (Agapiou et al. 2015; McKercher et al. 2005). With many cultural heritages, visitors overload in the peak months and the festive season partly affect the life of the relic and pollute the relic environment. The heavy traffic of large vehicles carrying tourists near the relics also has a great impact on the safety of the relics (Qian et al. 2012). All these activities directly affect the landscape of the heritages (Agapiou et al. 2015; McKercher et al. 2005; Qian et al. 2012; Taylor 2013; Koers et al. 2012). The preservation of the values and landscape of the heritages is governed by the Convention on the Protection of the World Cultural and Natural Heritage in 1972, in Vietnam by the National Cultural Heritage Law (2001) that adjusted in 2009 and many other regulations and decisions at the local level. They emphasize that the protection of the heritage integrity is closely related to the intact preservation of the elements of the feng shui landscape of the heritage. Especially, surrounding Green Space are attached to the architectural clusters of the heritage (Marafa 2003; Van Huynh 2013; UNESCO 2011; Gillieson et al. 2006).

Moreover, there is an integration of strategies into local development processes and urban planning regarding cultural heritage management and planning. Therefore, an accurate assessment of urban expansion and its relationship with the landscape of heritage is very important for urban planners and cultural heritage managers. For zoning, protect monuments and design, plan urban development for these sensitive areas.

The applications of remote sensing (RS) and GIS in research for conservation and sustainable exploitation of heritage in the world is very diverse both in terms of an object with both natural heritage (Palomo 2012; Gross 2006; Melis 2008; Turner 2003; Luo et al. 2019; Luo 2018) and cultural heritage (Luo 2018). Data types (different resolutions) include photography (Agapiou et al. 2016; Aqdus et al. 2012), multispectral and hyper-spectral imaging (De Laet et al. 2015; Chen et al. 2017), synthetic aperture radar (SAR) (Cigna 2014; Chase et al. 2011), and light detection and ranging (LiDAR) (Goncalves 2009). The spatial resolution of RS data has been improved to allow determining more detail levels of the objects, describing more features so that the RS image is increasingly being applied widely, especially in the field of heritage conservation (Luo 2018; Fischer and Ioanna 2006). There is the application for long-term monitoring, for large scale conservation areas, some projects using medium resolution data such as Landsat, MODIS (Luo et al. 2019), (Fischer and Ioanna 2006; Pakimbo and Powlesland 1996) or using SPOT to provide general environmental such as the study of temperate forest in Europe and the arid desert in New Mexico (Mathieu et al. 2007). In order to study the variability of objects in heritage, other authors used high-resolution satellite images such as QuickBird and IKONOS (Qian et al. 2006; Ahmadzadeh et al. 2013). Thus, RS technology provides an effective tool for identifying heritage areas and setting up databases to support efficient management and monitoring of protected areas (Government web portal 2014).

The Complex of Hue Monuments has been recognized by UNESCO as a World Cultural Heritage since 1993 due to the Outstanding Universal Value of architectural landscapes planned under the Nguyễn Dynasty. The area with the monuments and feng shui landscape is also a densely populated place and facilities for people's livelihood. Report on adjusting the planning of Hue City to 2030 with a vision to 2050 (Government web portal 2010), and the scheme of adjusting the conservation plan and promoting the value of the ancient capital of Hue in the period 2010–2020 (Statistical Yearbook of Thua Thien Hue Province 2013) emphasized the planning of Green Space and green belt construction and functional sub-areas will be built separately. In addition, to preserving this space of cultural and architectural art, preserving the Heritage against influences of development pressures, environment, natural disasters, tourism activities, and people living in the heritage areas, also it is necessary to conserve the natural space from the urban compositions such as Huong River, garden house space, and urban–rural morphology. Assessing the change of those spaces over time and the factors affecting is important in determining the zoning of the protected area, which is a reference for planners, policy, and urban planning.

This paper focuses on quantifying the change of Green Space and urban space in the process of urbanization and tourism development around the heritage sites in the Complex Hue Monuments by using all the data of remote sensing, statistical and spatial analysis tools in GIS. The research question is (1) How is the change of Green Space and urban space around the monument area; (2) How is the relationship between these spatial changes related to natural conditions, population development, and tourism infrastructure; and (3) Can we predict the trend of changing Green Space and urban space around the monument area in the future to orient the protected zoning to preserve landscape and cultural heritage of Hue City?

2 Materials and Methods

2.1 Description of the Study Area

Hue City is the unique city in Vietnam that is a heritage city, incorporating heritage components: Nguyen Dynasty architecture; town architecture; traditional villages; colonial architecture; cultural and natural landscape; and urbanization. These historical–cultural areas were built by Nguyen Dynasty from the early nineteenth to the middle of the twentieth century, locating along both sides of Huong River and near some neighbor districts (Fig. 1). Huong River has been considered as a main route of the Capital, dividing the architecture into two large areas: on the northside of the Huong River are the citadel, palaces, and royal offices, to the far southside of the

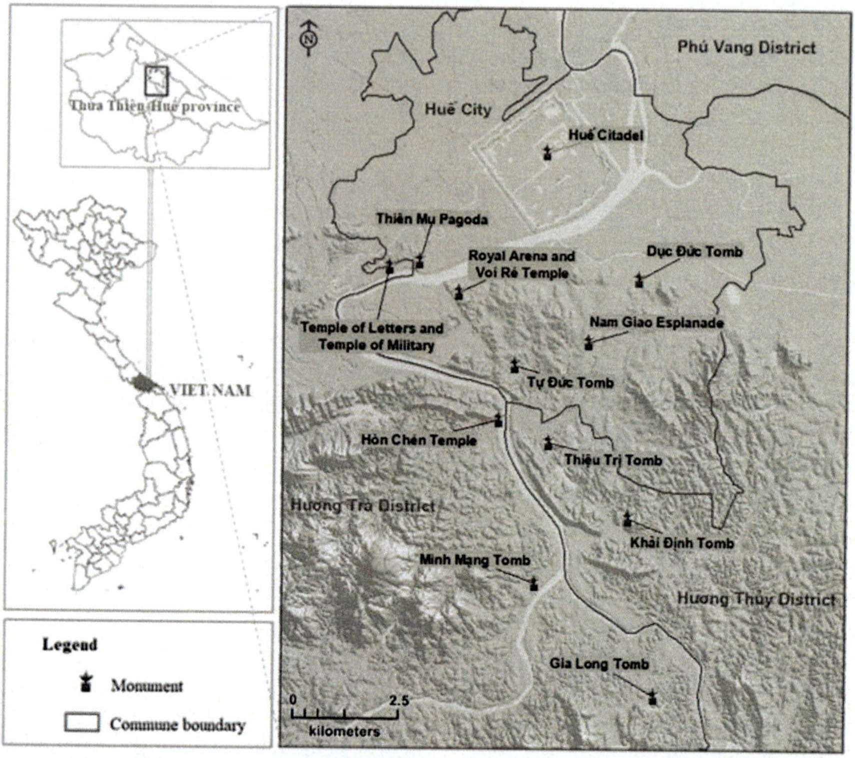

Fig. 1 Overall map of the study area in Hue City, Huong Tra, and Huong Thuy town of Thua Thien—Hue province; and locations of 12 historical sites in Complex of Hue Monuments

Huong River is a quiet mountainous area where temples and shrines were constructed, also with kings' tombs.

However, like another world cultural heritage, the Complex of Hue Monuments has been under pressure of urbanization and constructions such as hotels, restaurants, and many other infrastructures, which served developing tourism. From 1999 to 2015, the population of Hue has increased from 297,626 to 354,124 (Schneider and Woodcock 2008). Tourists also increased rapidly since the Hue Festival held in 2000 with the number of visitors from 920 thousand to 3,800 thousand by 2017.

2.2 *Data and Methodology*

2.2.1 Mapping Landcover Change from Remote Sensing Images

Remote sensing image classification method toward (i) generalizing a large set of data to distinguish objects; (ii) boundaries or land cover change identification according to the spatial uniformity of spectral property; and (iii) transforming that information into data by GIS. Object-based classification technique overcomes the limitation of pixel-based classification technique in the information integration of textures, shapes, or land-use object relationships; Object-based classification method allows accurately identifying objects on digital images.

The study focused on evaluating changes in historic landscapes in the period from 1995 to 2016. Specifically, SPOT remote sensing image data were collected (SPOT3—dated 1995/3/17 with the spatial resolution is 10–20 m, and SPOT7—dated 2016/9/20 with the spatial resolution is 1,5–6 m). All the images data were pre-processed and classified according to the steps.

(i) Satellite image preprocessing: In the study, the multi-time satellite image data were calibrated for radiation/atmospheric influence on reflection images by the COST method (Cosine of the Solar Zenith Angle). The satellite image data is geometrically calibrated based on the relationship between the coordinates of the points in the image and on WGS84-Zone 48 N coordinate with a precision error smaller than ±0.5 pixels. The data is then conducted to balance the spectrum at the same time to eliminate environmental differences to the spectral radiation values for the purpose of assessing urban fluctuations.

(ii) Object classification: Object segmentation process is conducted based on customizing the shape, compactness, scale parameters—which directly affects the size of objects. After the number of tests, the segmentation result identifies the optimized value of parameters Scale (15), Shape (0.8), and Compactness (0.5) to minimize the mistaking between objects in the image. The classification was established into eight types of land-use: Agriculture (AGR), Built-Up High-Density (HDB), Built-Up Low-Density (LDB), Forest and Shrubs (FR), Water (WT), Urban Green Space (UGS), Cemetery (CEM), and Other Land (BS).

(iii) Modification and accuracy evaluation of classification results: This process is conducted based on statistics on the number of "mistaking" individual land cover areas. Accuracy evaluation is a crucial step to evaluate the results of the classification process and is determined by assuming that the objects are distributed evenly throughout the study area and calculated using the overall accuracy, which was established through the Kappa coefficient.

We extracted two trends of land-use change from 1995 to 2016: Urban expansion and Green Space reduction. Urban expansion is the change of any type of land-use to urban areas. Green Space reduction is the change from Green Space (Urban Green Space, Forest and Shrubs, Agriculture) to another type.

2.2.2 Analysis of the Relationship Between These Spatial Changes Related to Natural Conditions, Population Growth and Tourism Infrastructure

In order to find the relationship between Dependent variable (urban expansion, Green Space) reduction-related to Independent variables (Tourism infrastructure, Natural Condition, Distance to monuments, and Population growth) described in Table 1.

Table 1 Variables used for the logistic analysis

Variable	Description	Type	Unit	Source
Dependent	Urban expansion	Binary	0–1	SPOT images 1995, 2016
	Green Space reduction	Binary	0-1	SPOT images 1995, 2016
Independent				
Tourism infrastructure	Average of the distance to restaurant calculated at pixel-level	Continuity	meters	Information system of Thua Thien-Hue province
	Average of the distance to hotel calculated at pixel-level	Continuity	meters	Information system of Thua Thien-Hue province
	Population density 2016	Continuity	Person/km^2	Yearbook in 2016-GSO
	Average of the distance to main roads calculated at pixel-level	Continuity	meters	
Natural Condition	Land-surface temperature 2016	Continuity	T^0	Landsat 8 OLI image 2016
	Average slope calculated at pixel-level	Continuity	Degree	
	Average of the distance to river calculated at pixel-level	Continuity	meters	
Distance to monuments	Average of the distance to relics calculated at pixel-level	Continuity	meters	
Population growth	The population growth rate in the period 1999 - 2016	Continuity	%/year	Yearbook in 1999 and 2016-GSO

Table 1 shows all the variables, and their sources collected and accounted for in a radius of 2 km from the boundaries of Buffer zone protection of all relics. Core zone and Buffer zone protection boundaries were collected from Hue Monuments Conservation Center.

Population statistical data were collected from Yearbook in 1999 and 2016, General Statistics Office, The digital elevation model (DEM 2.5 m) was used to extract elevation and slope. Tourism infrastructure data and natural conditions were taken into account into distance variables to evaluate the potential effect of tourism activities and natural around the monument areas (see Fig. 2).

Expand buffer zone around monuments: This method is developed based on the methods that have been used a lot before for evaluating urban expansion (Kontgis et al. 2014; Stewart et al. 2004; Nong 2014) in which analyzing the extension is outward from the Core to Buffer zone in large metropolitan areas. In Vietnam, there are only

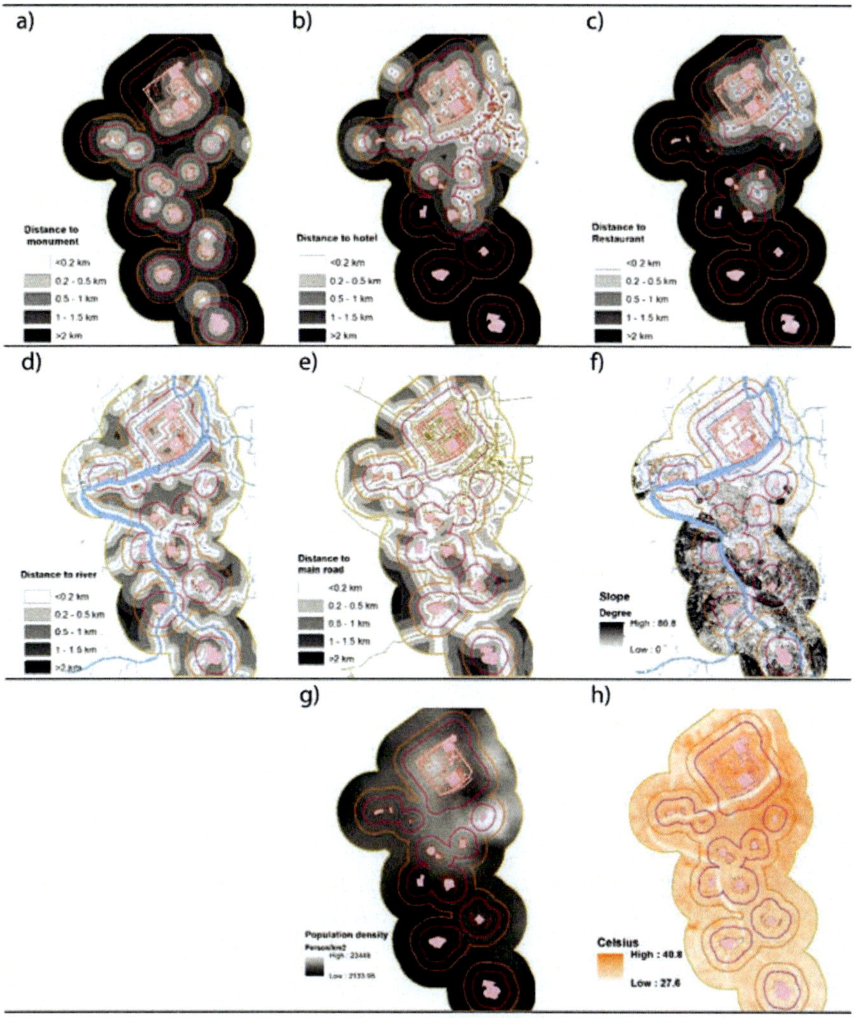

Fig. 2 Spatial patterns of Core zone, Buffer zone protection, Buffer-500 m, Buffer-1,000 m, and Buffer-2,000 m by independent variable: **a** *distance to relics calculated at pixel-level;* **b** *distance to hotel calculated at pixel-level;* **c** *distance to restaurant calculated at pixel-level;* **d** *distance to river calculated at pixel-level;* **e** *distance to main roads calculated at pixel-level;* **f** *slope;* **g** *population density 2016;* **h** *land-surface temperature 2016*

buffer zone studies that have been conducted in Hanoi (Hair 2010) and Ho Chi Minh (Stewart et al. 2004). According to Article 32 of the 2001 Cultural Heritage Law, Protected Area I (Core zone) consisted of monuments and areas defined as original constituent elements of monument, which must be protected as original status; and protected area II (Buffer zone protection) is the area surrounding the Core zone of the monuments. So, we used the Core zone and Buffer zone protection of the monument to expand around the relics by a buffer-500 m; buffer-1,000 m; and buffer-2,000 m.

Spatial statistical analysis: Logistic regression analysis (LR) was performed to assess the relationship between natural conditions, population development, tourism infrastructure, and urban expansion, Green Space reduction around the monument areas. Logistic regression is a specialized form of regression that is formulated to predict and explain a binary (two-group) categorical variable rather than a metric dependent measure (Thu 2015). These coefficients can be used as parameters to calculate transition probability maps. If X_k is a collection of independent variables and y is a binomial outcome variable with the probability of success $p = Pr(y = 1|X)$ with $y = 0$ meaning the absence of change, $y = 1$ meaning the presence of change, a linear logistic model has the following form:

$$\text{Logit}(p) = \text{Ln}(p/(1-p)) = a + b_1X_1 + \cdots + b_kX_k \quad (1)$$

where a is the intercept, $b_1, \ldots, b_k$ are parameters.

This has been used extensively in studies of land-use changes about socio-economic factors, population, ethnicity, ecology, and urbanization (Vu 2007; Serneels and Lambin 2001; Huong 2014; Domencich and MaFadden 1975). In this study, logistic regression is used to determine which factors motivate and which factor restricts the transformation of urban and Green Space. Urban expansion and Greenspace reduction as the dependent variable Y with the reference state is 0–1 (0: no change, 1: change) Three group of independent variables: population growth, tourism infrastructure, natural conditions as numerical variables. The model was calculated with 10,000 randomly sampled points of urban expansion and Green Space reduction within a radius of 2 km from the boundaries of Buffer zone protection.

The suitability of the model was determined by the coefficient of determination of pseudo R^2, which is in the range of 0–1. Pseudo-R^2, in the range of 0.5–0.1 shows the high degree of data fit with the model (Domencich and MaFadden 1975). Suitability of the model also requires considering the significant regression coefficients when p-value ($>Chi^2$)< 0.05 with 95% confidence level, p-value <0.1 with 90% confidence level, and the line ROC curve are used to explain the relationship between true and false prediction for the default value of 0.5 as the area under the curve. The area under the curve varies from 0.5 to 1 indicating the degree of fit between the model's ability to transform and the actuality.

Finally, the landcover change process can be predicted by calculating the probability of the analysis. Based on the LR results, we can calculate the probability of whether a pixel might change to urban or Green Space change to other pixel using the following equation:

$$P(Y = 1) = \frac{e^{(a+\beta_1 x_1+\beta_2 x_2+\ldots+\beta_k x_k)}}{1 + e^{(a+\beta_1 x_1+\beta_2 x_2+\ldots+\beta_k x_k)}} \quad (2)$$

Then, a forecast map of the possibility of expanding urban land and the possibility of declining Green Space within a radius of 2 km according to monument areas were built based on Interpolates a raster surface from points using an inverse distance weighted (IDW) technique in ArcGIS 10.3.

3 Results and Discussion

3.1 The Change of Landscape and Urban Factors Surrounding Monuments

Based on SPOT satellite image data of the Complex Hue Monuments in 21 years (1995–2016), accuracy of LCLU classification results (Table 2) WT, HDB, AGR, UGS, and FR types have an average post-classification accuracy above 80%. Mainly LDB type seems to have problems in classification, and the classification results show that the average accuracy is only 79.2%. Meanwhile, the Kappa coefficient achieved an accuracy of 0.76 in 1995 and 0.80 in 2016, respectively. The classification results created LCLU types with relatively high accuracy for the two periods used to analyze fluctuations in terms of space and time.

The results of land-use changes from two typical land-use groups were extracted from SPOT satellite image data (1995 and 2016) (see Fig. 3). There was a dramatic change between (i) urban land and (ii) Green Space land (UGS, AGR, and FR). Agricultural land areas had a tendency to decrease mainly and mostly converted to urban land and some other types of land-use

New urban areas have been formed around the Hue Monuments. These developments include the expansion of boundaries, urban land area, and increasing population density which have narrowed Green Space and water surface. The urban land area accounts for more than 4,500 ha and is mainly distributed in Hue City, and a small part is located at Huong Thuy and Phu Van districts. Because the urban area does not have clear boundaries between inner and suburban areas, urban land not

Table 2 Aggregated class accuracy, OA, and Kappa coefficient

Unit: %

Year	LCLU									
	HDB	LDB	UGS	AGR	WT	FR	CEM	BS	OA	Kappa
1995	79.8	77.7	80.6	81.1	84.5	76.5	73.7	76.2	78.7	0.76
2016	81.5	80.7	83.7	83.4	88.3	87.2	77.6	80.1	82.8	0.80
Average	80.65	79.2	82.15	82.25	86.4	81.85	75.65	78.15	80.75	0.78

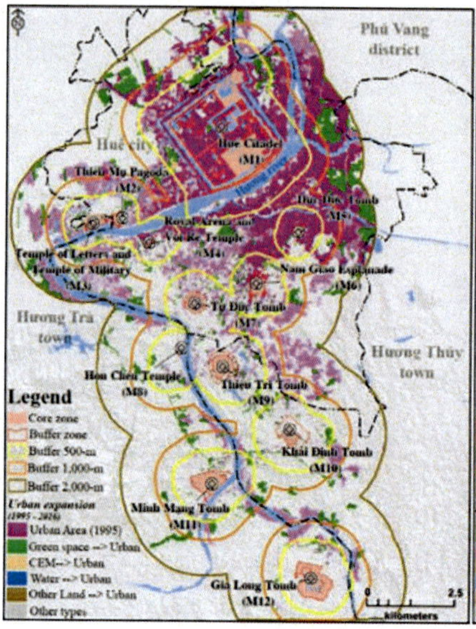

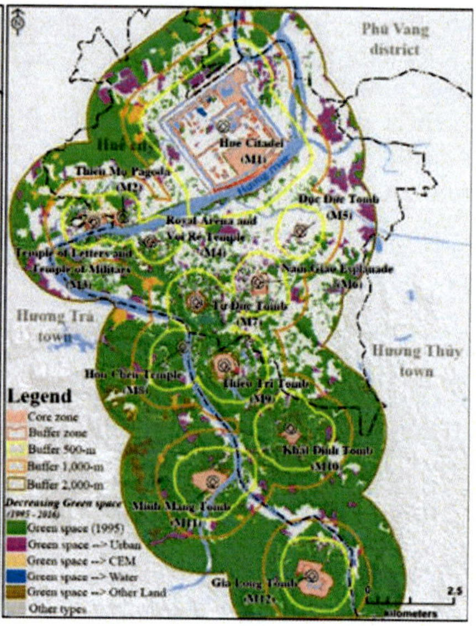

Fig. 3 Spatial distribution of Landscape surrounding the complex of hue monuments in the period of 1995–2016

only extends to the East and the South of the area but also appears in the connection between Hue City and surrounding development zone. The urban area expanded to over 6,000 ha in 2016. The rapid increase of urban area is partly due to the upgrades of Huong Thuy and Huong Tra districts to towns. The urban land area of the Complex of Hue Monuments has increased by nearly 1,500 ha with an annual expansion of roughly 72 ha.

Furthermore, there was also a significant reduction of Green Space and an increase of urban land in each Buffer zone of the monuments (see Fig. 4). With Buffer zones, there was no sight of Green Space reduction, but there are still minor increases of urban area. With 500 m and 1,000 m buffer zones, urban area expanded more 4,000 ha, and 7,200 ha and Green Space area reduced almost 4,300 and 8,700 ha, respectively. With 2,000 m buffer, there was an increase of over 13,500 ha in an urban area, and Green Space area witnessed a loss up to 18,000 ha.

3.2 *Relationship Between Spatial Changes Related to Natural Conditions, Population Development, and Tourism Infrastructure*

The result from the Logistics Regression Models to assess the relationship between natural conditions, population development, tourism infrastructure, and urban expansion, Green Space reduction around the monument areas are presented in Table 3.

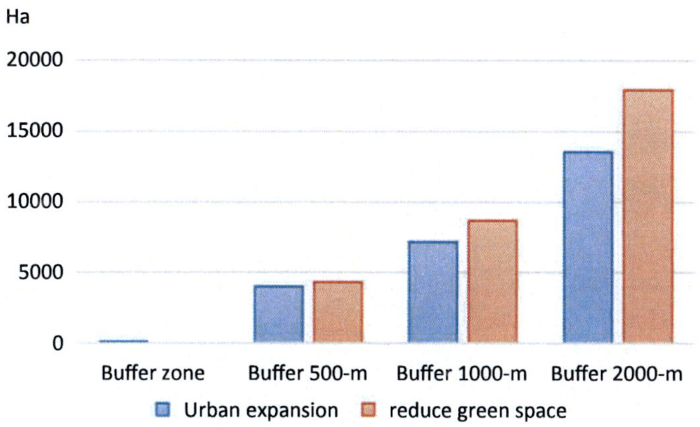

Fig. 4 Urban area and green space according to buffer

Table 3 The result of Logistics Regression Models

Variable	Impact factor	Estimated coefficient	Standard deviation	Pr > Chi^2	Estimated odds ratio Exp (B)	95% C.I for exp (B)	
						Lower	Upper
Urban expansion (URX) R^2 : 0.575	Intercept	−3.463	0.094	0.000	0.031		
	Tourism infrastructure (TI)	−1.011	0.05	0.000	0.364	0.330	0.401
	Natural condition (NC)	−0.518	0.049	0,000	0.596	0.542	0.656
	Distance to monuments (DM)	0.001	0.000	0,000	1.001	1.000	1.001
	Population growth (PG)	0.231	0.032	0.000	1.260	1.183	1.341
Green space reduction (GSR) R^2 : 0.42	Intercept	−3.040	0.086	0.000	0.048		
	Tourism infrastructure	−0.458	0.045	0.000	0.632	0.579	0.691
	Natural condition	−0.043	0,037	0.239	0.958	0.891	1.029
	Distance to monuments	0.001	0.000	0.000	1.001	1.000	1.001
	Population growth	−0.012	0.036	0.730	0.988	0.920	1.060

If (Pr > Chi^2) < 0.05 then the variable is a significant variable in the model. If (Pr > Chi^2) > 0.05, the variable is not significant and can be ignored. Therefore, natural conditions and population growth were left out of Green Space reduction model.

The results show that the ability of urban expansion is dependent on all of the input impact factors: tourism infrastructure factors, natural condition, and population growth with the dependencies of -1.011 and -0.518 and 0.231. The shorter the distance to areas with tourism infrastructure, the higher the population growth, the higher the possibility of expanding urban land. The equation for the regression model of the relationship between the possibility of urban land expansion and the influencing factors can be written as

$$P(Y = URX) = \frac{e^{(-3.463-1,011*TI-0518*NC+0,001*DM+0,231*PG)}}{1 + e^{(-3.463-1,011*TI-0518*NC+0,001*DM+0,231*PG)}} \quad (3)$$

As for the possibility of Green Space reduction, the factors that influence include Tourism infrastructure and Distance to monuments. This means that in areas with a higher population density and the shorter the distance to hotels or restaurants, the higher the possibility of Green Space reduction. So, the equation for the regression model of the relationship between the possibility of Green Space reduction and the influencing factors can be written as

$$P(Y = GSR) = \frac{e^{(-3.040-0,458*TI+0,001*DM)}}{1 + e^{(-3.040-0,458*TI+0,001*DM)}} \quad (4)$$

Then, a forecast map of the possibility of expanding urban land and the possibility of declining Green Space within a radius of 2 km according to monument areas were built and presented in Fig. 5.

The ability to reduce Green Space reduction and urban expansion has the difference between Buffer zones and different monuments. In general, the probability of Green Space reduction in all areas has a low probability. The ability to expand urban land is highest in An Cuu ward—the area near Duc Duc Tomb and Huong So and Tay Loc wards close to Hue Citadel. The farther away from the city center, closer to Gia Long Tomb, Minh Mang Tomb, and Khai Dinh Tomb, the smaller the potential for urban expansion and the reduction of Green Space.

4 Conclusions

Remote sensing and statistical analysis methods, the paper has demonstrated the changes, the ability to change of Green Space and urban space in the process of urbanization and tourism development around the heritage sites in Hue within 21 years from 1995 to 2016. We can see that the ability of urban land expansion is to take place more strongly than the decline of Green Space. The expansion of urban land is concentrated in the inner-city wards near Hue Citadel, located in buffer-500 m, with a rapidly

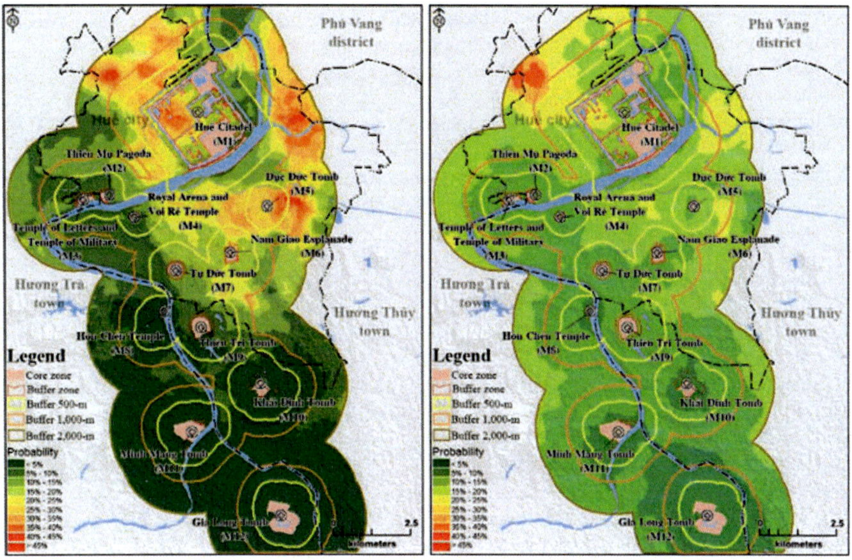

a. Probability of urban expansion b. Probability of Green Space reduction

Fig. 5 Map of the ability to landscape change around heritages of Hue period 1995–2016

increasing population and infrastructure for tourism activities. These changes and trends indicate that urbanization pressures are having a substantial impact on heritage areas, even in areas where conservation is required. This result will be suggestions for heritage conservation to planners, policymakers, and Hue Monuments Conservation Centers. The limitation of this study is not yet provided with a full of factors affecting landscape changes around the monument. Therefore, the next research orientation will be to continue indexing the impact factors and forecasting the changing heritage landscape in the future.

References

Agapiou A et al (2015) Cultural heritage management and monitoring using remote sensing data and GIS: the case study of Paphos area, Cyprus. Comput Environ Urban Syst 54:230–239

Agapiou A et al (2016) Colour to greyscale pixels: re-seeing greyscale archived aerial photographs and declassified satellite CORONA images based on image fusion techniques. Archaeol Prospect 23(4):231–241

Ahmadzadeh R et al (2013) Using satellite images to measure ecological metrics of landscape of the conserved area of North Iran. J Biodivers Environ Sci 3:69–76

Aqdus SA, Hanson WS, Drummond J (2012) The potential of hyperspectral and multi-spectral imagery to enhance archaeological cropmark detection: a comparative study. J Archaeol Sci 39(7):1915–1924

Benz UC et al (2004) Multi-resolution, object-oriented fuzzy analysis of remote sensing data for GIS-ready information. ISPRS J Photogramm Remote Sens 58(3–4):239–258
Chase AF et al (2011) Airborne LiDAR, archaeology, and the ancient Maya landscape at Caracol. Belize J Archaeol Sci 38(2):387–398
Chen F et al (2017) Radar interferometry offers new insights into threats to the Angkor site. Sci Adv 3(3):e1601284
Cigna F et al (2014) Persistent scatterer interferometry processing of COSMO-SkyMed StripMap HIMAGE time series to depict deformation of the historic centre of Rome, Italy. Remote Sens 6(12)
De Laet V et al (2015) Integrated remote sensing investigations of ancient quarries and road systems in the Greater Dayr al-Barshā Region, Middle Egypt: a study of logistics. J Archaeol Sci 55:286–300
Domencich TA, MaFadden D (1975) Urban travel demand: behavioural analysis. North-Holland, Amsterdam
Fischer C, Ioanna K (2006) Multispectral and hyperspectral imaging technologies in conservation: current research and potential applications. Int Inst Conserv Hist Artist Work 7(7):3–16
Gillieson DS, Lawson TJ, Searle LE (2006) Applications of high resolution remote sensing in rainforest ecology and management. Cooperative Research Centre for Tropical Rainforest Ecology and Management, Cairns. p. 54
Goncalves LMS et al (2009) Application of remote sensing to heritage conservation: a case study in central Portugal. Remote Sens Chang Eur 239–246
Government web portal (2010) Decision No. 818/QD-TTg, Approving the scheme on adjusting the conservation planning and promoting the value of Hue ancient ruins in the 2010–2020 period
Government web portal (2014) Decision No. 649/QD-TTg. Adjusting the general planning of Hue City to 2030 with a vision to 2050
Gross JE et al (2006) Remote sensing for inventory and monitoring of the U.S. National Parks. In: Wang YQ (ed) Remote sensing of protected lands. Taylor & Francis, Boca Raton, FL
Hair JF et al (2010) Multivariate data analysis. Pearson Prentice Hall
Huong HTT et al (2014) Changing human-landscape interactions after development of tourism in the Northern Vietnamest Highlands. Anthropocene 50
Koers W et al (2012) Future of cultural heritage—impact of external developments, in European policy workshop. European Foresight Platform, Brussels
Kontgis C et al (2014) Monitoring peri-urbanization in the greater Ho Chi Minh City metropolitan area. Appl Geogr 53:377–388
Luo L et al (2018) Auto-extraction of linear archaeological traces of tuntian irrigation canals in miran site (China) from Gaofen-1 satellite imagery. Remote Sens 10(5)
Luo L et al (2019) Airborne and spaceborne remote sensing for archaeological and cultural heritage applications: a review of the century (1907–2017). Remote Sens Environ 232:111280
Marafa L (2003) Integrating natural and cultural heritage: the advantage of feng shui landscape resources. Int J Herit Stud 9(4):307–323
Mathieu R, Aryal J, Chong AK (2007) Object-based classification of ikonos imagery for mapping large-scale vegetation communities in urban areas. Sensors 7:2860–2880
McKercher B, Ho PSY, Du Cros H. (2005) Relationship between tourism and cultural heritage management: evidence from Hong Kong. Tour Manag 26:539–548
Melis MT et al (2008) Remote sensing application to the knowledge of environmental dynamics for a project of management planning in central karakoram national park, Pakistan. In: The international archives of the photogrammetry, remote sensing and spatial information sciences. Beijing
Nong D et al (2014) Built-up area change analysis in Hanoi using support vector machine classification of Landsat multi-temporal image stacks. In: Environment, population, and health series No 1. East-West Center working paper

Pakimbo G, Powlesland D (1996) Remote sensing and geographic database management systems applications for the protection and conservation of cultural heritage. Remote Sens Geogr, Geol, Land Plan, Cult Herit 2960:124–128

Palomo I et al (2012) National parks, buffer zones and surrounding lands: mapping ecosystem service flows. Ecosyst Serv

Qian Yu et al (2006) Object-based detailed vegetation classification with airborne high spatial resolution remote sensing imagery. Photogrametric Eng Remote Sens 72(7):799–811

Qian J, Feng D, Zhu H (2012) Tourism-driven urbanization in China's small town development: a case study of Zhapo Town, 1986–2003. Habitat Int 36:152–160

Schneider A, Woodcock CE (2008) Compact, dispersed, fragmented, extensive? a comparison of urban growth in twenty-five global cities using remotely sensed data, pattern metrics and cencus information. Urban Stud 45(3):659–692

Serneels S, Lambin EF (2001) Proximate causes of land-use change in Narok District, Kenya: a spatial statistical model. Agric, Ecosyst Environ 17

Statistical Yearbook of Thua Thien Hue Province (2013) Statistical Publishing House of Thua Thien Hue

Stewart DJ et al (2004) Assessing the spatial structure of urban and population growth in the Greater Cairo area, Egypt: a GIS and imagery analysis approach. Urban Stud 41(1):95–116

Taylor KAM (2013) Cultural heritage and urbanisation in China. China Story

Hoang TTH (2015) Researching the impact of urbanization on the structure of agricultural land use in Dong Anh, Hanoi. The doctoral thesis on the map survey. University of Mining and Geology. Hanoi

Turner W et al (2003) Remote sensing for biodiversity science and conservation. TRENDS Ecol Evol 18(6)

UNESCO (2011) A new international instrument: the proposed UNESCO recommendation on the Historic Urban Landscape (HUL)

Huynh TAV (2013) Hue's striking global values are viewed from the integrity of its landscape. Hue Monuments Conservation Center

Vu KC (2007) Land use change in the Suoi Muoi catchment, Vietnam: disentangling the role of natural and cultural factors. In: Geography—geology. Katholieke Universiteit Leuven, p 186

Challenges in Developing and Implementing Smart City in Palangka Raya

Fikri Rafif Suprayitna, Latifah Asri Munawaroh, Mustafa Al Azmi, Aidha Imtinan Besari, and Rini Rachmawati

Abstract The implementation of smart city in Palangka Raya is interesting to be studied because Palangka Raya has both urban and rural landscape. The most extensive of land use is forest with percentage around 72%. There are many opportunities and challenges to implement the concept of smart city in Palangka Raya. Despite this uniqueness, the city government strives to implement the smart city concept in planning the city of Palangka Raya especially to face the challenges of urban development. The city government has a mission focused on Smart Environment, Smart Society, and Smart Economy as the concrete manifestation in supporting the realization of smart city. The purpose of this research is to describe the implementation of smart city and the challenges in implementing and developing smart city in Palangka Raya. This research uses a mixed qualitative and quantitative approach by using in-depth interviews, observation, and literature to collect data. Implementing smart city has become one of the development priorities in the city of Palangka Raya. However, the implementation has not been optimal since it is only focused on developing Internet networks in public spaces. The local government is also trying to develop applications and websites for public services. Various challenges are faced by the local government to develop smart city concept in Palangka Raya, especially to provide infrastructure which is constrained by the geographical conditions, limited budget, and the low quality of human resources.

F. R. Suprayitna (✉) · L. A. Munawaroh · M. A. Azmi · A. I. Besari · R. Rachmawati
Department of Development Geography, Faculty of Geography, Universitas Gadjah Mada, Yogyakarta, Indonesia
e-mail: f.rafif@mail.ugm.ac.id

L. A. Munawaroh
e-mail: latifah.a@mail.ugm.ac.id

M. A. Azmi
e-mail: mustafa.a@mail.ugm.ac.id

A. I. Besari
e-mail: aidha.imtinan.besari@mail.ugm.ac.id

R. Rachmawati
e-mail: rinirachma@ugm.ac.id

L. T. T. Huong and G. M. Pomeroy (eds.), *AUC 2019*, Advances in 21st Century Human Settlements, https://doi.org/10.1007/978-981-15-5608-1_26

Keywords Smart city · Implementation · Challenge

1 Introduction

The city has experienced rapid development on various sides of city life in the recent decades. This condition is in line with population growth. According to the United Nation (2018), about 55% of the entire world population lives in urban areas. It is projected that the number will increase due to the increasing rate of urbanization. This also happens in developing countries like Indonesia. A total of 53.3% of Indonesia's population lives in urban areas (BPS 2014). The city is still a magnet of attraction that is able to offer a variety of important opportunities to improve the economic and social level (Cohen 2006).

Palangka Raya is the capital of the Province of Central Kalimantan with about 36.6% of the city population lives in urban areas (BPS 2014). BPS predicts that in 2020 the number of population will increase by 40.2%. The growth number of residents has led to the emergence of various new problems that require well-planned city management. Smart city is one solution to create a city that is suitable for living.

The city of Palangka Raya has a unique city characteristics consisting of urban and rural area. The most extensive of land use is forest with a percentage around 72%. This condition is a challenge in the development of the city so it requires careful and specific planning. The development of network infrastructure related to technology in applying the smart city concept has become the main focus of the government. At present, there are certain areas that have not been connected to the telephone and Internet networks, especially in the areas far from the city center. Jekan Raya and Pahandut Subdistricts are urban areas in Palangka Raya City which have a function as a center of government, services, and economy. Rakumpit, Sebangau, and Bukit Batu subdistricts are functionally classified as rural areas, but administratively classified under urban administration. The dominant land use in the three districts is dominated by conserved area and croplands. About 72% of land use in Palangka Raya City is covered by forest, while the remaining 18% is used for other land uses. This condition is a limitation in infrastructure and network development, especially to support the implementation of the smart city concept.

Various regions in Indonesia are trying to implement the concept of smart city, including Palangka Raya. However, the development of the concept is still focused on infrastructure development. There are various challenges that must be faced in realizing the concept. Therefore, this research was conducted to describe the implementation of smart city in Palangka Raya and to describe the challenges faced in developing the smart city concept. This research was conducted in Palangka Raya as an "over bounded city" with a unique city characteristics consisting of urban areas, rural areas, and forests.

2 Literature Review

2.1 *Smart City*

The concept of smart city arises as a discussion by various cities in the world because it is considered as a solution to solve various complex city problems (Chourabi et al. 2012; Pourahmad et al. 2018). Complex city problems cause the need for effective and efficient urban planning and management. This is required by any city in the world to ensure the comfort for the citizen. The smart city concept relies on the effort to create a city that is livable by meeting the needs, security, and comfort of each individual who lives inside (Caragliu et al. 2011; Nam and Pardo 2011). Smart city has six characteristics consisting of smart economy, smart mobility, smart environment, smart people, smart living, and smart governance (Albino et al. 2015).

2.2 *Smart City Challenges*

There are various challenges faced by cities in developing smart cities, so cities must have the courage to take the main risks from the use of technology (Hollands 2008). Technology can indeed accelerate the development of cities, especially to help public services to run more effectively and efficiently. However, to realize the convenience of public services, the government needs to consider the security and privacy aspects of the community. The government needs to establish a strict safeguard system to ensure the security of community data (Robello et al. 2014).

The challenges that many developing countries face in implementing smart cities are financial and partnership problems (Pierce and Anderson 2017). Enough financial support plays an important role in the successful implementation of the concept, because infrastructure development, human capacity development, and integrating it with technology require substantial funds. In addition, partnerships also play an important role. Building good partnerships with various parties becomes a challenge for the government in a city to assist the running of programs related to smart city implementation; therefore, the government also needs to make a pro-investment and pro-people policy.

2.3 *Smart City Implementation*

Policy plays an important role for the smart city implementation (Mursalim 2017). The central role of the government as a policyholder influences the application of smart city to the development of a city that can be supported by the availability of ICT so that the city works efficiently and effectively (Rohmah and Rachmawati 2019). Technology support including IoT, Big Data, and Cloud Computing will be main elements so that smart city policy can be run well (Anindra et al. 2018). The smart city implementation is carried out in a structured and inclusive manner involving the roles of government, academics, the private sector, and the community in order to achieve the development goals achieved together. It is necessary to involve various sectors on implementing smart city to ensure its sustainability (Ni Loideain 2017). During the smart city implementation, it takes a lot of cost, energy, and time, so careful planning is needed. The availability of skilled workers technically determines the competition locally and globally of a city that is said to be ready for smart city. Nowadays, some cities are starting to be pioneer of implementing smart city concept as a part of response to the growing challenges of the city.

3 Research Method

The research was conducted in April 2019 in Palangkaraya, Central Kalimantan, Indonesia. This research uses a qualitative and quantitative approach by using in-depth interviews, observation, and literature to collect data. Primary data were obtained from observations in the field for one week. This research used purposive sampling method to collect data in the society. Total sample of 58 were taken during the observation. In addition, primary data were collected through in-depth interviews with government agencies as key informants. The agencies were Regional Planning Agency of Palangkaraya and Department of Communication and Informatics of Palangkaraya.

4 Result and Discussion

Palangka Raya covers 2,853.52 km^2 land area. Its population grow with 2.22% rate in 2017. It is higher than the national population growth rate which is only 1.36 percent, and the growth rate of Central Kalimantan Province which is 2.16% (BPS 2014). The increase of population in the City of Palangka Raya is also accompanied by an increase in community activities so as to spur development in various aspects both economic, social, and cultural.

Population growth can contribute to the development of the city as long as it can be anticipated and controlled properly. Unplanned population growth will cause various complex problems. Palangka Raya City Government since 2018 began to realize the importance of city management to anticipate the emergence of various city problems caused by population growth. Therefore, the government is trying to implement the smart city Concept by using ICT in the city development (Rachmawati et al. 2018).

The serious effort shown by the government stated in the vision of the Palangka Raya City development plan is "The realization of the City of Palangka Raya Becoming a City that is Progressive, Peaceful, and Prosperous for All" (Mayor's

Mission, Vision and Goals of Palangkaraya) (BAPPEDA 2018). The implementation of the smart city concept is realized by prioritizing three main indicators, namely smart environment, smart people, smart economy.

> "… The vision and mission of the mayor of Palangkaraya now has three focus, smart city, smart economy, and smart environment …" Informant (Department of Communication and Informatics)

4.1 Smart Government

Smart Government is a city development effort by the government toward sustainable development. The government has a vital role in this regard to carry out their vision (Govada et al. 2019). In Palangkaraya, the visions are mainly focused on smart environment, smart people, and smart economy (Albino et al. 2015). The government in this case functions to integrate the use of technology in the internal development of government, as well as for community development and seeks to partner with other institutions (Table 1).

Table 1 Application/website managed by the local government of the city of Palangka Raya

Website/application	Function	Executant
SLIKOP	Propose road improvements, drainage development and operations and maintenance	DPRKP (Department of Housing and Residential Area of Palangka Raya)
Siromi	Development and promotion of small and medium enterprises	Diskominfo (Department of Communication and Informatics of Palangka Raya)
'Detektif Sampah	Monitor the garbage truck lane and landfill	DPRKP (Department of Housing and Residential Area of Palangka Raya)
SIMPEG	Employee management	BKPP SIM Kepegawaian (Staffing Agency of Palangka Raya)
simerin.Palangka Raya.go.id	Archive of death and burial location	SETDA Bagian Kesra (Community Empowerment Division of Local Government Secretariat of Palangka Raya)
Sipander	Facilitate administrative services at the subdistrict level to the community	Diskominfo (Department of Communication and Informatics of Palangka Raya)
Siperjaka.Palangka Raya.go.id	Position and rank of the officer consideration system	BKPP (Staffing Agency of Palangka Raya)
Sipproplus-k2.Palangka Raya.go.id	Service request for procurement of agency services	SETDA (Local Government Secretariat of Palangka Raya)
Simentel	Telecommunication tower management	Diskominfo (Department of Communication and Informatics of Palangka Raya)
mediacenter.Palangka Raya.go.id	Presenting the latest news and publications	Diskominfo (Department of Communication and Informatics of Palangka Raya)

Source Primary Data Analysis

"... we are targeting in 2020 Palangkaraya already has an ICT master plan ... if we are smart city, it is not just digitalization, therefore it is also called smart city, meaning how government services can serve the community effectively and efficiently. In this case the government is trying, although limping to make the city of Palangkaraya a smart city, the vision and mission of the mayor of Palangkaraya also supports achieve the label of smart city" Informant (Department of Communication and Informatics)

The total applications that have been successfully developed and actively used by the City Government are 24 applications. These applications are actively used for internal office management and employee management in existing agencies. Some applications are also used to facilitate public services.

"there were 58 applications, only from the data collection. But, have to classify again. It's not necessarily all of them use it, only a few are actually use it, maybe 20s, so it needs re-data collection" Informant (Department of Communication and Informatics)

The role of government is needed in the management and development of the concept. The challenge that must be faced by the government is the limited human resources in the government itself that can manage and develop smart cities so that recruitment of experts and employee training are needed. The government has a big challenge and task to integrate between government agencies into the same server, even though the distance between these agencies is far apart.

"In addition, the use of applications made by the government is not optimal because there are only a few people who use it, so the government needs to conduct socialization related to the use of these applications. Another challenge that must be faced by the government is related to complicated rules that makes it difficult to access various population data needed to develop applications and other public services" Informant (Department of Communication and Informatics of Palangka Raya)

The government is also actively building a telecommunications network because so far it is still concentrated in two main districts, namely Pahandut and Jekan Raya. Therefore, in order to support the smart city concept, the government is expanding telecommunications networks to other subdistrict centers. Palangka Raya's Government is also trying to improve the digital literacy of the community through providing Wi-Fi networks in public places such as city parks and other city centers. However, there are challenges in its development because the installation of a Wi-Fi transmitter requires high security by considering the damage and loss of these devices. When the government wants to build Wi-Fi transmitters and other telecommunications networks, they need to be brave to take risks because the more systems that are connected, the more complicated the management will be.

4.2 Smart Environment

Smart Environment is an effort to sustainable city development by implementing a commitment to the wise use of environmental resources. This can be realized by a city through improving the esthetics of the city by not damaging the environment. Cities can also make certain innovations in various urban development plans using technology. Apart from that, smart environment also considers how a city can maintain its environment either through the use of energy wisely or reduce pollution (Colldhal et al. 2013).

The development of smart city in Palangka Raya that related to the smart environment indicator has not yet led to a focus on the use of energy wisely. However, efforts have begun to reduce the air pollution in urban centers by building city parks. City parks and road borders in cities are well designed to be able to increase the esthetic value of the city by paying attention to the principle of sustainability. The city parks include the Pasuk Kameluh Park and Yos Sudarso Park, as well as many other small parks. Palangka Raya government in designing the development of the city is committed to save the existing ecosystem. Although classified as a city, 72% of the land is forest so that the preservation of the forest takes precedence, and natural parks such as Sebangau National Park are optimized for its function while maintaining the ecosystems.

Waste management is a challenge for the government. There is a waste detective application that is used for taking garbage from the settlements area managed by the government, but the obstacle is that it still does not cover all areas of Palangka Raya. This application is used to transport garbage to the landfill as well as to monitor the truck lane because the truck is often being misused.

Palangka Raya has a relatively wide area so that it becomes a challenge in development; besides that not all of its areas can be used for land use because it is still dominated by land use in the form of forests. Palangka Raya Government has a mission in the Regional Medium-Term Development Plan namely "Realizing the Progress of Palangka Raya with Smart Environment covering the development of infrastructure, information technology, water management, land management, waste management, building management and spatial planning and transportation" (BAPPEDA 2018). The challenges in implementing a smart environment are constrained to physical and geographical conditions, so that development is uneven and concentrated in the city center.

> "… it is still premature, there are still many constraints on infrastructure, superstructure, human resources, budget and others that are limited so that it affects the achievement of smart city going forward …" Informant (Department of Communication and Informatics of Palangka Raya)

Geographical conditions are a challenge for the infrastructure development because most of the land uses are protected forests and swamps with peat soils. The main challenge for the government is connecting the two subdistrict centers with other subdistricts so that people in the other subdistricts can get the equity to access the same public services. The challenge of protecting ecosystems and carrying out equitable development throughout the region is a big task for the Government.

> "Palangka Raya is unique, our province has three faces, the face of the city, the face of the countryside, the face of the forest. So do not think our province like the other provinces in Java, maybe five and seven kilos have already entered other regions in Java, but here it is not something like that, even there are still areas that have not been connected by road

because we need to cross the river and it took more than one hour. So when it comes to blank spots automatically there are areas that are indeed blank spots. Especially for areas that have not yet been reached by the telecommunications network from vendor providers, we also continue to be constrained by the electricity side, there are several regions that do not yet have electricity, but for 2017, all of Palangka Raya has been electrified, for blank spot areas, especially in Sebangau Subdistrict, Rakumpit Subdistrict there are several, Bukit Batu Subdistrict there are some areas that are still unreached" Informant (Department of Communication and Informatics of Palangka Raya)

4.3 *Smart Economy*

Smart Economy is an effort to sustainable economic development with new innovations in various economic activities so that an area will have a good economic performance. Smart Economy is very closely related to the third mission of Palangka Raya, namely Achieving Urban Community Welfare and Regional Communities with Smart Economy which includes industrial development, small and medium-sized businesses, tourism, and banking.

The implementation of smart economy in Palangka Raya is mainly focused on developing micro and medium industries. The Government built an application system called SIROMI which is used to integrate all small and medium-sized industrial businesses in Palangka Raya. The application was created to help marketing products from small and medium industries and to integrate between small and medium industries. However, the people themselves do not even know to use the application. Incessant product promotion should be in line with the dissemination of information about the application and the contents contained therein so that the content created is factual.

"Actually, many applications developed by the city government to improve the economy, such as SIROMI are now widely used. We also cooperated with Gojek and Grab where SIROMI is also used for tourism, culinary tourism and others. We have many applications, but we can't manage it well" Informant (Department of Communication and Informatics of Palangka Raya)

Palangka Raya government focuses on increasing community productivity through coaching through funds and various counseling for small and medium industries. The funding assistance is given as a form of government investment so that small and medium industries in Palangka Raya can develop faster and be able to increase their productivity, so that their hopes in the future will come to the marketing of small and medium industrial products that will be able to reach large areas, even can be exported. The economic chain of products that are still short needs to get the attention of the government along with the rapid globalization so that products must be able to compete. By creating product competitiveness in a wider scope, it is certain that the regional economy, especially from the informal sector will increase while opening new job vacancies for unemployed people. Palangka Raya has also improved to present a convenient shopping place by fixing the traditional markets so that it can become a convenient center of economic activity for the community.

Economic development is also carried out from the tourism sector. Because the tourism sector is one of the potential sectors in Palangka Raya. The most attractive tourist attractions are the Kahayan River and Sebangau National Park. Both of these attractions can be used as a branding area of the City of Palangka Raya through the collaboration of the Palangka Raya Government with the National Government. Promotion of tourism through online media by the government also includes a component of the smart economy that utilizes current technology and information. The high intensity of promotion can have a good impact with the increasing number of tourists in Palangka Raya.

The challenge in developing tourism is the lack of access due to the geographical condition of the city which is right in the middle of the island and far from other big cities. In addition, access to tourist sites is also still minimal because there is no mass public transportation. The existing tourism objects have not been able to optimize their management and development. This causes a lack of tourist attraction to visit these places.

4.4 *Smart People*

Smart people are the social capital needed by Palangka Raya to run the dream to be a smart city. Social capital in the form of quality human resources has a vital role for the development of cities, not only about their interaction to technology but also the quality of social interaction (Giffinger et al. 2007). Palangka Raya has human resources that are starting to develop, although it is not included in one of the major cities in Indonesia. Palangka Raya is the growth center of Central Kalimantan Province. Therefore, Palangka Raya is the center of various economic and educational services. Palangka Raya welcomes the concept of smart city by improving educational services such as schools and colleges. Efforts to support the smart city, the people also began to be directed to work in the creative industries, especially in the culinary and handicraft sub-sector. They also began to experience modernization and began to be literate in technology.

The Government has installed several Wi-Fi facilities in public spaces around Yos Soedarso Park and Pasuk Kameloh Park. The existence of Wi-Fi that can be accessed is open and free with a quota of up to 50 GB. However, it has not been optimally utilized by the community because some people are not aware yet of the existence of Wi-Fi facilities, from 58 respondents only 15 who know of the Wi-Fi facility. There are seven cellular network operators in Palangka Raya and one stationary telephone network operator. Only a portion of the area can be reached by the operator. Not all cellular networks can be utilized, only Indosat and Telkomsel networks can be utilized. Indosat and Telkomsel networks can only be utilized in four districts, namely Pahandut, Jekan Raya, Sebangau, Bukit Batu, and Rakumpit Districts. While the cellular network with Telkom Indonesia operators can be utilized only in two districts, namely Pahandut and Jekan Raya Districts (BAPPEDA 2018).

People in urban areas of Palangka Raya began to use smartphones to support their activities. However, the use of smartphones is still limited for social media only. Meanwhile, applications and public service websites made by the government have not been utilized by the public. Among 58 respondents interviewed, only 5 of them had used the application/website to support their activities. Even for commercial private-owned applications, their use is still minimal, some respondents stated that they do not need these commercial applications.

> "I know one of the online application that is provided by the government, Sipander, I know from my friends, but it still can't be used like it seems, or indeed I don't understand how to use it, there's no socialization from the government either" Informant (citizen of Palangka Raya)

> "I don't need any online transportation application, I have my own vehicle, and Palangkaraya doesn't have any traffic jam also so I don't think it is necessary" Informant (citizen of Palangka Raya)

> "I know SIROMI application already, but it seems that it has not been effective yet, I prefer to use online transportation application to deliver my products to my costumer" Informant (citizen of Palangka Raya)

5 Conclusion

Smart city implementation is one of the city development solutions to create a city that is comfortable and able to support city activities efficiently. Implementation of smart city in Palangka Raya has become one of the development priorities and has become part of the vision and mission of the city development. The implementation of smart city concept in Palangka Raya is not optimal because it is still focused on developing and structuring public spaces to connect to the Internet and by building various websites and applications. The government of Palangka Raya faces various challenges to develop smart city in Palangka Raya especially in the development of infrastructure which is constrained by the geographical conditions, limited budget, and human resources in both the local government and the community.

Acknowledgements The author gratefully acknowledges to Department of Geography Development, Faculty of Geography, Universitas Gadjah for the field study opportunity.

References

Albino V, Berardi U, Dangelico R (2015) Smart cities: definitions, dimensions, performance, and initiatives. J Urban Technol 4–21

Anindra F, Warnars HLHS, Min DM (2018) Smart city implementation modelling in Indonesia with integration platform approach. In: International conference on information management and technology, pp 2–7, Jakarta (2018)

BAPPEDA (2018) Medium-term regional development plan of Palangkaraya 2018–2023. Badan Perencanaan Pembangunan Daerah, Palangkaraya

BPS (2014) Presentasi Penduduk Perkotaan 2010–2035. Badan Pusat Statistik, Jakarta

Caragliu A, Nijkamp P, Del Bo C (2011) Smart cities in Europe. J Urban Technol 65–82

Chourabi H, Nam T, Walker S, Gill-Garcia J, Mellouli S, Nahon K (2012) Understanding smart cities: an integrative framework. In: Proceedings of the 45th annual Hawaii international conference on system sciences, HICSS-45, pp 2289–22970. IEEE Computer Society

Cohen B (2006) Urbanization in developing countries: current trends, future projections, and key challenges for sustainability. J Technol Soc 28:63–80

Colldhal C, Frey S, Kelemen JE (2013) Smart cities: strategic sustainable development fo urban world. Blekinge Institute of Technology, Sweden

Giffinger R, Fertner C, Kramar H, Kalasek R, Pichler-Milanovic N, Meijers E (2007) Smart cities. Ranking of European medium-size cities, Final Report. Centre of Regional Science, Vienna UT

Govada S, Rodgers T, Cheng L, Chung H (2019) Smart environment for smart and sustainable Hong Kong. In: Advances in 21st century human settlements, Scopus, pp 57–90

Hollands R (2008) Will the real smart city stand up creative, progressive or just entrepreneurial? 12(3):302–320

Mursalim SW (2017) Smart city policy implementation in the city of Bandung. Jurnal Ilmu Administrasi 14(1):126–138

Nam T, Pardo T (2011) Conceptualizing smart city with dimensions of technology, people, and institutions. In: Proceedings of the 12th annual international conference on digital government research, pp 282–288

Ni Loideain N (2017) Cape Town as a smart and safe city: implications for governance and data privacy. Int Data Privacy Law 7(4):314–334 (2017)

Pierce P, Anderson B (2017) Challenges with smart cities initiatives——a municipal decision makers' perspective. In: Proceedings on 50th Hawaii international conference on system sciences, pp 2804–2811, Hawaii USA

Pourahmad A, Ziari K, Hataminejad H, Pashabadi SP (2018) Explanation of concept and features of a smart city. Sci J NAZAR Res Center (Nrc) Art, Architect Urban 15(58):5–26

Rachmawati R, Hapsari S, Cita A (2018) Virtual space utilization in the Digital SMEs Kampongs: implementation of smart city and region. Hum Geogr—J Stud Res Hum Geogr 12(1):42–52 (2018)

Robello M, Bartoli D, Serrano H, Forne J, Soriano M (2014) Reconciling privacy and efficient utility management in smart cities. Trans Emerg Tellecomun Technol 25(1):94–108

Rohmah A, Rachmawati R (2019) Utilization and quality of information system for administration services based on ICT in Patehan, Kraton, Yogyakarta. Indones J Sci Technol 4(1):55–63 (2019)

United Nation (2018) 68% of world population projected to live in urban areas by 2050, says UN. https://www.un.org/development. Last accessed 05 September 2019

Online Real Time (ORT) Waste Management Through "Si Detektif Sampah" Application in Implementing Smart City in Palangka Raya City

Haddad Al Rasyid Sukawan, Bias Osean Ali, I Made Arya Widhyastana, and Rini Rachmawati

Abstract Waste management is the main concern in handling environmental problems in urban areas. This management includes transportation and processing of domestically generated waste. Palangka Raya City as the capital of Central Kalimantan Province is increasingly finding it difficult to manage waste with the safety of misuse of trucks which causes an increase in landfill that cannot be transported and processed properly. Palangka Raya City Government through the Office of Public Housing and Settlement Area then developed "Si Detektif Sampah" (an Information, Detection, and Education System) application that displays route visualization for the most effective route transport transportation and route planning that uses online real-time fleet control. This application is GPS based which could be accessed anytime and anywhere and thus monitors waste transportation to be carried out sustainably. The method used in this research is descriptive qualitative through in-depth interviews. The results of the study show that this application could cut the misuse of trucks to reduce landfill waste while implementing technology-based smart city in Palangka Raya City.

Keywords Waste management · Technology · Smart city

H. A. R. Sukawan (✉) · B. O. Ali · I. M. A. Widhyastana · R. Rachmawati
Department of Development Geography, Faculty of Geography, Universitas Gadjah Mada, Yogyakarta, Indonesia
e-mail: haddad.al.rasyid.sukawan@mail.ugm.ac.id

B. O. Ali
e-mail: bias.osean.ali@mail.ugm.ac.id

I. M. A. Widhyastana
e-mail: i.made.arya.widhyastana@mail.ugm.ac.id

R. Rachmawati
e-mail: rinirachma@ugm.ac.id

L. T. T. Huong and G. M. Pomeroy (eds.), *AUC 2019*, Advances in 21st Century Human Settlements, https://doi.org/10.1007/978-981-15-5608-1_27

1 Introduction

1.1 Research Background

Waste becomes an environmental problem that has not been being resolved. Moreover, the problem of waste is also exacerbated by the widespread paradigm that is waste is only waste that is not useful (KNLH 2008). Based on Indonesia Law No. 18/2008, the Paradigm must view waste as a resource that has economic value and can be utilized. Waste problems that cannot be handled properly can cause other problems such as human health problems. The existence of waste is often found especially in urban areas with dense population so that waste management in urban areas is a must. In line with the complexity of the urban area, the waste appears as a problem that requires special attention and treatment. But, waste management system is still weak mainly because of policies or management programs that have not been integrated and also the lack of support and participation from the community, both the business world and society in general.

Palangka Raya City as capital of Central Kalimantan Province has a very wide administrative area of 2,853.52 km^2. Palangka Raya City is also one of the cities in Indonesia which is currently developing and is not spared from waste problems both from the management system and facilities and infrastructure. Based on data from the Central Statistics Agency of Palangka Raya City, the volume of waste transported in Palangka Raya City in 2017 was 159,774.92 m^3 where 158,961 m^3 was collected by the sanitation service, 143 m^3 of garbage was collected from the private sector, and 670.92 m^3 was obtained outside collected by the cleaning service and private parties. In general, the rate of waste production is faster than the response effort. This situation makes waste management and cleanliness program become one of the top priorities. This also caused Palangkaraya City not to pass the first stage of evaluation (P1) to get the 2015 Adipura award.

According to the Head of Office of Public Housing and Settlement Area Cleanliness Division, the City of Palangka Raya has many shortcomings in improving the quality of processing in the Final Disposal Site (TPA), especially the problem of roads that are still muddy and inadequate. Then there is still a lack of heavy equipment used to decompose piles of rubbish because only one heavy equipment is used while ideally to decompose rubbish there is a need of two heavy equipments. In addition, the main problem of garbage in the City of Palangka Raya is the lack of a fleet used to transport waste to the Final Disposal Site (TPA). Therefore they need a good management of waste management so that waste services can be carried out optimally. The waste management system includes storage, collection, transfer of depo transfers, and transportation to the landfill (KLH 2009).

The concept of smart city is one alternative solution that can be tried to be implemented in pioneering and realizing the quality of urban life that is livable, comfortable, and healthy. Smart city is expected to be able to solve the problems faced by several big cities in Indonesia, including the City of Palangka Raya, especially in environmental issues, especially waste problems. Because with the growing era, especially in the digital era like today, the use of information technology is something that can not be avoided. So that managing urban waste management also needs to take advantage of modern technology. One of the uses of technology in waste management is in terms of supervision and management of waste transportation management. Utilization of technology in transportation management is believed to be able to improve the efficiency and effectiveness of the fleet management. Based on this background, the formulation of the problem in this research is "How is the waste management system carried out by the Government of Palangka Raya City in an effort to realize a smart city in Palangka Raya City?".

1.2 Research's Purposes

Based on the background and formulation of the problem, the goal to be achieved is to know the waste management system that is carried out by the City Government of Palangka Raya in an effort to realize a smart city in the City of Palangka Raya.

2 Literature Review

2.1 Online Real Time

Microsoft Press Computer Dictionary 3rd Edition [DICT97] defines real time as a machine activity could match human perception in terms of time or real circumstances. It could be said that real time is a state that is happening at that moment. As for if it is described that as in filling out an online registration form on a computer, the computer will immediately display the form that has been filled in and all data will be stored at that time (Setiawan 2016). Online Real Time (ORT) system allows to simplify the work of monitoring which is supported by the Internet network. As for the ORT, all the desired activities on other computers that have been connected to the system will be stored and recorded as data that can display all activities at that time. One application of ORT is implemented by the government as a monitoring system to avoid misuse, especially in activities related to mobilization effectively.

2.2 Digital Based Management Information System

Setiawan (2016) defines a data-based management information system functions to facilitate the organization in obtaining accurate, relevant, and fast information in making decision/policy in operations and managerial activities, including planning, organizing, directing, and controlling to achieve organizational goals effective and efficient. In implementing government policies, management information systems with computer commands can help the government solve problems or constraints quickly. In addition, management information systems make it possible to take instructions and procedures more wisely, that is, so that problems can be resolved but do not open gaps for new problems or so-called digging holes.

2.3 Implementation of Waste Management

The implementation of waste management is carried out with waste reduction and handling activities. Waste reduction includes restriction, reuse, and recycling activities, while waste management activities include sorting, collecting, transporting, processing, and final processing (Law No. 18 of 2018). The implementation of waste management is not only carried out by the government but must also be followed by the awareness of the community. Which is in line with the objectives of waste management, which is to improve public health and environmental quality and make waste a resource. The role and function of the government in managing waste management is stipulated in Law No. 18 of 2018, namely the provincial government has the authority: (a) determine policies and strategies in waste management in accordance with Government policies; (b) facilitate cooperation between regions within one province, partnerships, and networks in waste management; (c) organize coordination, guidance, and supervision of district/city performance in waste management; and (d) facilitate the settlement of disputes on inter-regency/inter-city waste management in 1 (one) province.

2.4 Ideal Urban Waste Management

Indonesia Law Regulation with no. 18 of 2018 states that waste management is carried out based on the principle of responsibility, the principle of sustainability, the principle of benefits, the principle of justice, the principle of awareness, the principle of togetherness, the principle of safety, the principle of security, and the principle of economic value. Aboejoewono (1999) mentions things that need to be considered in the policy of urban waste management, namely (1) Applying appropriate technology related to waste management, (2) Community participation in waste management, (3) Profit mechanism in waste management, (4) Optimization of garbage landfill, (5) The existence of an integrated system of waste management.

3 Research's Method

The method used in this research is descriptive qualitative. The qualitative descriptive method is chosen to explore waste management in the City of Palangka Raya Indonesia deeply. Data collection method used in this study is in-depth interview.

4 Results

Various efforts have been made by the City Government in overcoming the problem of waste in Palangka Raya in achieving sustainable development. Waste management is not only for selecting and collecting but also for transporting and eventual disposing of the waste. In an effort to develop the concept of smart city in Palangka Raya, the Department of Public Housing and Settlement Areas developed an application called the "Si Detektif Sampah." The Waste Detective is an abbreviation of the Information, Detection, and Education System for Waste Management online and in real time in the City of Palangka Raya. This system is able to display route visualization for planning trips of garbage transport vehicles based on the most effective route and periodically controlling the fleet. So that diversion in performing duties and obligations can be minimized.

This "Si Detektif Sampah" arises based on problems created, especially in the field of waste management in the Palangka Raya. This application was developed in 2016 as a pioneer in the development of the first garbage monitoring application in Kalimantan then received the best award of 1 Innovation in Borneo in 2017. This application only applies to internal areas of the Cleanliness of the Public Housing and Settlement Areas of the City of Palangka Raya. Based on data from the sanitation sector of the Department of Housing and Settlement, it can be seen that the garbage heap of Palangkaraya City currently reaches an average of 133,879 kg/day or around 133.88 tons/day. All of the garbage piles are not fully served by the garbage transportation fleet owned by the Palangkaraya City Government because the fleet is very limited, namely around 21 garbage trucks where the fleet management for garbage collection services has not been running optimally. There are several problems or obstacles in the process of monitoring waste transportation services, including difficulties in monitoring the actual (actual) amount of transportation from each TPS and being dumped to the TPA, difficulties in tracking compliance with work schedules and plans, difficulties in monitoring TPS status including effectiveness placement, there is no estimation of the amount of waste that is in the collection and distribution, the difficulty in preparing a report on the status of the transport location because it uses a manual process and takes a lot of time and resources. As a result of the above, it is necessary to create a system that is able to display the route visualization for the planning of garbage transportation vehicles based on the most effective route, so this application of the "Si Detektif Sampah" was created.

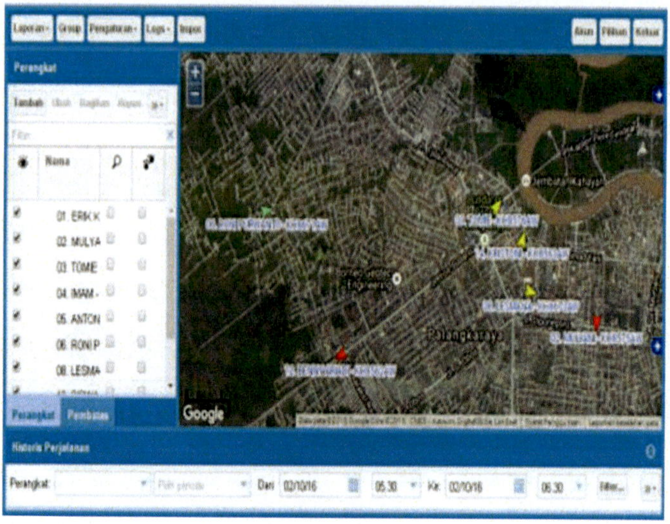

Fig. 1 Online main display in monitoring the movement. *Source* Alfath (2018)

This application was developed using GPS, where every truck is installed with GPS to facilitate monitoring and checking the activities or movement of the garbage truck every day. The first function of this application is to be able to supervise these trucks online and in real time. In the previous years, the Cleanliness Division of the Palangka Raya City Public Housing and Settlement Areas often experienced difficulties in monitoring the movement of trucks, so with this system garbage trucks could be monitored more effectively. Monitoring can be done via PC or via mobile phones, so that wherever and whenever each supervisor can monitor the movement of the garbage truck during transportation (Figs. 1 and 2).

Prior to this application, garbage trucks were often misused to run the business of every truck driver starting from transporting sand and carrying out their personal activities. When there is this system, the Sanitation Department of the Public Housing and Settlement Area of Palangka Raya City can delineate, so that it can know which paths the truck is traveling, whether the truck is moving anywhere, out of location or not, its final position, latitude of its position, where the distribution or how many kilometers the distance traveled on that day, how many times the truck stopped, which locations did the trucks stopped, the maximum speed used by the truck because each truck was given a maximum speed limit of 60 km/hour. If the truck exceeds the set speed, there will be an alarm and the violation can be monitored or known by the supervisor.

Punishment will be given on an offense by the driver based on violations done. If included in grave violations, the driver will be dismissed or replaced; if the violation is moderate, a penalty such as suspension can be given. Examples of moderate violations, for example, those that occur due to accidents due to the use of speed beyond the maximum limit, and garbage disposal is not done where it should be.

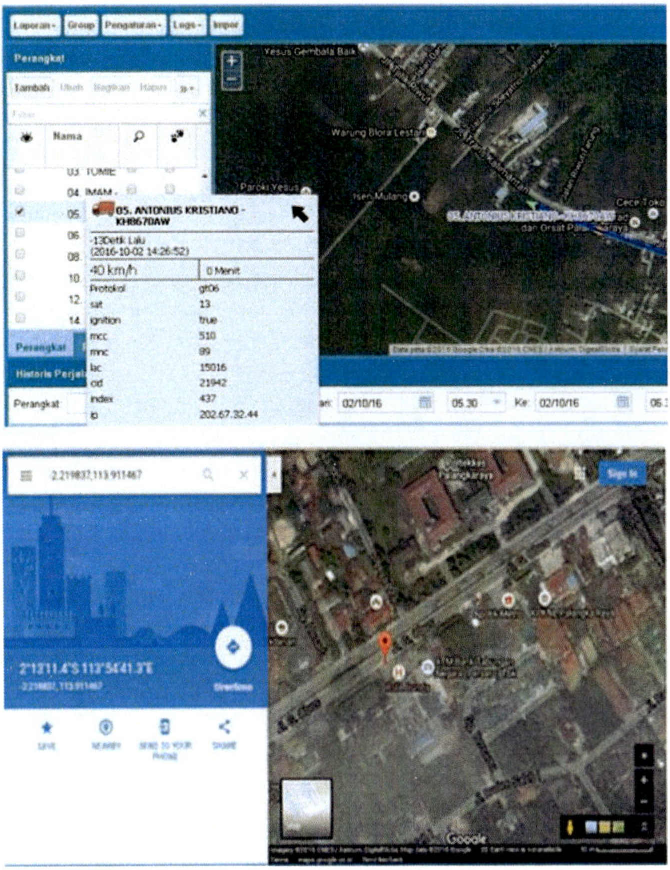

Fig. 2 Multiple menu display in tracking truck's route. *Source* Alfath (2018)

Minor violations would be given a Warning Letter, with a maximum of 3 times of the Warning Letter. If there are more than 3 times violations, the driver will be fired by the government directly. For the number of trucks installed with GPS so far, only 12 of the total 21 trucks are available because of the limited cost. The main reason for the number of GPS installed only 12 trucks because the truck is considered to have the most crowded lane and the path that is often problematic. Whereas 9 other trucks have no problems or have a good track record. The Sanitation Department of the Public Housing and Settlement Areas will try to install GPS in all trucks in 2019 and will fully operate all of them.

The second function of the "Si Detektif Sampah" application is to be able to analyze the fuel requirements of each vehicle based on the traveled distance. With the mileage data record of each truck, it can be calculated the needs of fuel (fuel oil) of each truck. The longer the transportation route, the higher the fuel demand will be, and vice versa. With the analysis of the data, the waste management section can determine the amount of fuel consumption that must be provided in each truck on each transport line. Prior to this system, all fuel consumption trucks were beaten flat. This has drawn protests from several drivers because the movement and distance traveled by each truck are different, far, medium, and close. Therefore, with the application of the "Si Detektif Sampah," the zoning system and fuel consumption can be adjusted to the distance traveled on each vehicle. In addition, the determination of truck vehicle mileage can also be used for maintenance or scheduling of oil changes or other special services for each transport.

In the development of this application there are still some shortcomings, namely Internet connection that is still unstable (still dependent on Telkom's network), causing the data received by the server to be incomplete. Secondly, the Clock on the GPS system is still GMT 0, which should be GMT +7 so that it has an impact on the record time not in accordance with the time in the Palangka Raya City area, so it needs to be adjusted. Third, the TPS location still does not appear in the monitoring system so the TPS location is not visible on the main map. Fourth, the haul road cannot be included so the haul road that has been determined by the Dinas does not appear on the main system map. Shortcomings of the system that has been built will continuously be tried to be refined by the team. Any deficiencies that arise during this system will be followed up and become a record for IT team members. Evaluation and improvement of the system absolutely must continue to be done so that the purpose of the construction of this system can be achieved properly. In addition, the main obstacle in developing this application is limited human resources. The Sanitation Department of Public Housing and Settlement Areas Palangka Raya City has only one technician or expert in managing the system.

The system developed by the Department of Public Housing and Settlement Areas in addition to the "Si Detektif Sampah" application contained a public complaint service called the SMS center. The Sms Center is a public complaints service especially related to the presence of illegal waste which is generally very disturbing around the area of residents which was developed in 2017. Through this SMS center, citizens of the city can report the presence of illegal garbage in their environment so that it can be immediately followed up by the garbage transport team on duty.

Over time, more or less in 2018 the SMS center was closed because the user of the SMS center is increasingly down because it is considered less effective and less real time. In addition, another reason for the closure of the SMS center is the development of a centralized complaints service application managed by the Office of Communication and Information (DISKOMINFO) of Palangka Raya, namely E-LAPOR (Fig. 3).

According to Head of Cleanliness of the Public Housing and Settlement Areas, who said that each department did not need to issue a complaints service, it was enough to use one centralized application or rely on big data to be more effective and efficient, for example, the E-REPORT application. If there are public complaints relating to waste and sanitation complaints, the complaint will be forwarded to the head of the office. Then in less than 6 h it is certain that the report will go to the system owned by the Cleanliness Section of the Public Housing Agency and the Permukiman Area because the system has been integrated. If every service has its own complaints

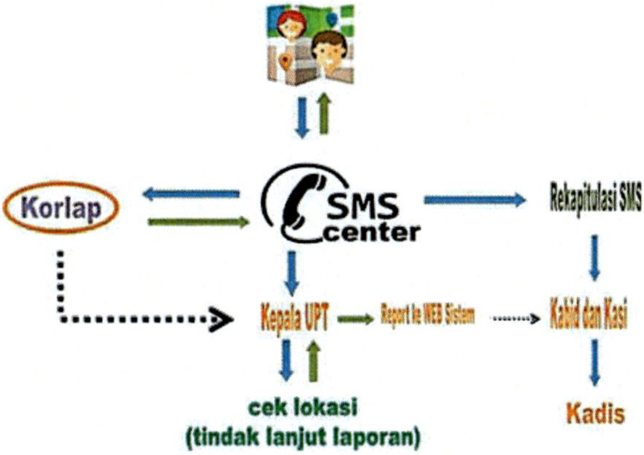

Fig. 3 SMS center's mechanism. *Source* Alfath (2018)

service, besides making it difficult, it will also confuse the community. In addition, the more systems or applications, the greater the operational budget spent, as will the maintenance or repair budget.

5 Conclusion

From the research result related to waste management in Palangka Raya City, conclusions can be drawn. The management of garbage in Palangka Raya City can run optimally by developing the application of the Trash Detective. This system can solve problems or obstacles in the process of monitoring waste transportation services, difficulties in tracking compliance with work schedules and plans, difficulties in monitoring the status of the TPS including the effectiveness of its placement. There is no estimation of the amount of waste present at the collection and distribution site, difficulties in preparing transport location status reports. The development of the SMS center application cannot run optimally in the management of waste in the City of Palangka Raya due to a trend that is increasingly decreasing and less real time in reporting. The presence of the E-Report service application replaces the SMS center system because this service application is more centralized, integrated, and easily monitored.

References

Aboejoewono (1999) Pengelolaan Situ-Situ di Wilayah DKI Jakarta. Bogor: PPLH-LP. Institut Pertanian Bogor

Alfath M (2018) Pemanfaatan Teknologi Informasi dan Pemberdayaan Masyarakat dalam Meningkatkan Pengelolaan Persampahan di Kota Palangka Raya. Jurnal Boreneo Administrator Vol 14 Tahun 2018 halam 1–16

BPS Kota Palangka Raya (2017) Kota Palangka Raya Dalam Angka 2017. BPS Kota Palangka Raya, Kota Palangka Raya

KLH (2009) Status Lingkungan Hidup Indonesia. KNLH, Jakarta

KNLH (2008) Status Lingkungan Hidup Indonesia. KNLH, Jakarta

Microsoft Press Computer Dictionary Third 3rd Edition Paperback—1997

Pokja Sanitasi Kota Palangka Raya (2014) Buku Putih Sanitasi Kota Palangka Raya Provinsi Kalimantan Tengah. Pemerintah Kota Palangka Raya

Setiawan D (2016) Implementasi Penerimaan Peserta Didik Baru Sekolah Menegah Atas Sistem Real Time Online (RTO) Di Kabupaten Bantul Tahun Pelajaran 2015/2016. Skripsi. Universitas Negeri Yogyakarta, Kota Yogyakarta, 37 pp.

Undang-Undang Republik Indonesia Nomor 18 Tahun (2008) Tentang Pengelolaan Sampah

Utilization of Human Resources Management Information System (SIMPEG) Application to Support E-Government in the BKPP at Palangka Raya Municipality

Puja Dania Almira, Bergita Gusti Lipu, Aditya Widya Pradipta, and Rini Rachmawati

Abstract Dynamic city development produces opportunities for innovation in managing the government sector. Technological innovation in the governmental sector is implemented in the form of e-government. In the process of e-government, the government of Palangka Raya Municipality through Personnel, Education, and Training Agency (BKPP) creates and implements Human Resources Management Information System (SIMPEG). The purposes of this research are to find out the application of SIMPEG in the local government and to identify the driving factors and inhibitors of its application. This research uses a qualitative descriptive method, where data is obtained from the results of in-depth interviews with employees responsible for implementing SIMPEG on BKPP. The research results show that the application of SIMPEG in the e-government of Palangka Raya Municipality could help in assessing the performance of the State Civil Apparatus (ASN), and also able to present information on each ASN staff in detail.

Keywords E-Government · Human Resources Management Information System (SIMPEG) · Personnel, Education, and Training Agency (BKPP)

P. D. Almira · B. G. Lipu (✉) · A. W. Pradipta · R. Rachmawati
Department of Development Geography, Faculty of Geography, Universitas Gadjah Mada, Yogyakarta, Indonesia
e-mail: bergita.gusti.lipu@mail.ugm.ac.id

P. D. Almira
e-mail: p.dania@mail.ugm.ac.id

A. W. Pradipta
e-mail: aditya.widya.pradipta@mail.ugm.ac.id

R. Rachmawati
e-mail: rinirachma@ugm.ac.id

L. T. T. Huong and G. M. Pomeroy (eds.), *AUC 2019*, Advances in 21st Century Human Settlements, https://doi.org/10.1007/978-981-15-5608-1_28

1 Introduction

Utilization of applications based on information and communication systems connected to the Internet has begun to be applied in the environment or institutions of the City of Palangka Raya, one of which is Personnel, Education, and Training Agency (BKPP). By the Regional Regulation of Palangka Raya City Number 11 of 2016, BKPP carries out strategic tasks and functions in government affairs, namely the Administration and Management of Regional Civil Service based on the principle of autonomy and assistance tasks. BKPP has to recap all types of personnel data in dozens of agencies or institutions totalling more than 4,964 employees in the Palangkaraya City Government.

Previously, the data collection of employees in the Palangkaraya City Government was carried out in a manly manner through paper media, but this was felt to be inefficient in terms of manpower and time if suddenly there was a need to process the personnel data. Therefore, to support the tasks of BKPP and overcome these problems, the BKPP needs a Human Resources Management Information System (SIMPEG) as a platform for managing all personnel data. (SIMPEG) is an information system that plays an important role in providing services to State Civil Apparatus (ASN) with the main function of recapitulating ASN detailed data from basic data to personnel history data such as information on work mutations, promotions, salary increases, pensions, awards, penalties, penalties discipline, and so on. The SIMPEG application began to be implemented in the Palangkaraya City Government environment in mid-2017 to realize better personnel data services by the Palangkaraya City BKPP. This application uses web-based which can be accessed via a computer, tablet, or smartphone while connected to the Internet.

2 Literature Review

2.1 *Understanding E-Government, Objectives, and Benefits*

E-government is an effort to transfer existing information and service in government to the public by the usage of electronic (Mediaswati and Sidik 2013). E-government provides benefits for both the institutions and the public. Institutions through E-government can optimize their role in sending information to the public easily and effectively. Meanwhile, the benefit that will be got by the public is information that can be accessed quickly according to the needs they are looking for (Amandita and Rachmawati 2019). E-government is not just about automating various ineffective processes but creating a new relationship between government and society (Bovaird 2005). E-government is a new solution that is revolutionizing the governance process by improving the relationship with citizens (Garcia 2016). The type of services provided by E-government are divided into three types; they are to publish, interact, and transact services (Indrajit and Richardus 2004). The utilization of SIMPEG in this journal is mainly part of interact services because SIMPEG provides two-direction communications between government and employee.

2.2 *Human Resources Management Information System (SIMPEG)*

Babaei and Beikzad defined management information systems as an integrated, computerized, and machine user system to help the institution in operation and decision-making by providing information (Babaei and Jafar 2013). Management Information System consists of three parts: management, information, and systems. Management information systems are used to facilitate the work procedures and improve efficiency, productivity, and performance in an organization such as human resources affairs (Al-Mamary et al. 2014).

Human Resources Management Information System (SIMPEG) is a system capable of containing information on personnel data in a company or agency. SIMPEG is composed of processing devices, storage devices, and communication devices (Mediaswati and Sidik 2013). SIMPEG based on the Minister of Home Affairs Decree No. 17 of 2000 is defined as an integrated system, which includes employee data, data management, procedures, work procedures, human resources, and information technology to produce fast, complete, and accurate information in supporting internal personnel administration (The Regulation of the Minister of Home Affairs No.17/2000).

2.3 *Indicator of Utilization*

SIMPEG application is used as one of the products for smart city, rated by indicators that relate to E-government and E-government successfully. There are 4 indicators of e-governance successfully (Rianto and Lestari 2012):

a. Availability of the data and information in the data center
b. Availability of the data and information for regional promotion
c. Availability of the government application to support office work and public service
d. Availability of the public dialog application.

Meanwhile, the indicators relating to the implementation of E-government according to Indrajit (2002) are six (Indrajit 2002). The six indicators include data, infrastructure, institutional, human, technology, and thought leaders.

3 Research Method

This research is a combination of qualitative and quantitative methods. The primary data collection method in this study was carried out by in-depth interviews while the secondary data was obtained in the form of statistical data about the use of SIMPEG by State Civil Apparatus (ASN) in the scope of the government of Palangka Raya City. The in-depth interview activity was carried out with one of the employees at the Personnel, Education, and Training Agency (BKPP) Office of Palangka Raya City. The data analysis technique is descriptive qualitative. The results of the in-depth interview in the transcript are then synthesized until the core answers are obtained from each question asked.

4 Discussion Results

4.1 SIMPEG Utilization

The Palangkaraya City Personnel, Education, and Training Agency (BKPP) uses the Human Resources Management Information System (SIMPEG) to deal with complex issues related to the availability of employee data. The data of the State Civil Apparatus (ASN) in the City Government of Palangka Raya amounted to 4,964 employees. The manual data recording process as before is not very efficient if it is still maintained. The number of ASN is 4,964 employees registered in SIMPEG, coming from various agencies or institutions such as 1 inspectorate, 1 class D hospital, 1 civil service police unit, 3 secretariats, 20 offices, 7 agencies, 7 agencies, 3 levels of schools up to 5 districts which is in the City of Palangka Raya.

There are 3 main actors who participated in the utilization of SIMPEG. They are employees, operators, and administrators as seen in Fig. 1. The SIMPEG utilization mechanism starts with collecting employee data manually. The most important role is the operator. Operators play an important role such as collecting, data scanning, and updating data. The result of the operator's work will be monitored and evaluated by the admin to guarantee the data updates. Those data then can be processed as material for decision-making.

The pie chart below shows the percentage of the number of employees recorded by SIMPEG in various agencies or institutions with the highest number of registered employees at the School by 47%. Data on the number of employee details in the Palangkaraya City Government as of December 31, 2018, was obtained directly from the BKPP Office in the form of hard files. A total of 2,344 out of 4,964 employees were in the school environment from kindergarten, elementary to the junior high school level. Then the offices took second place with 34% or 1,699 employees. Followed by the number of employees in 5 districts in the City of Palangka Raya by 7% or the number of 324 employees while the number of employees in 7 Government agencies of the City of Palangka Raya is as many as 273 employees, with a percentage of 6%.

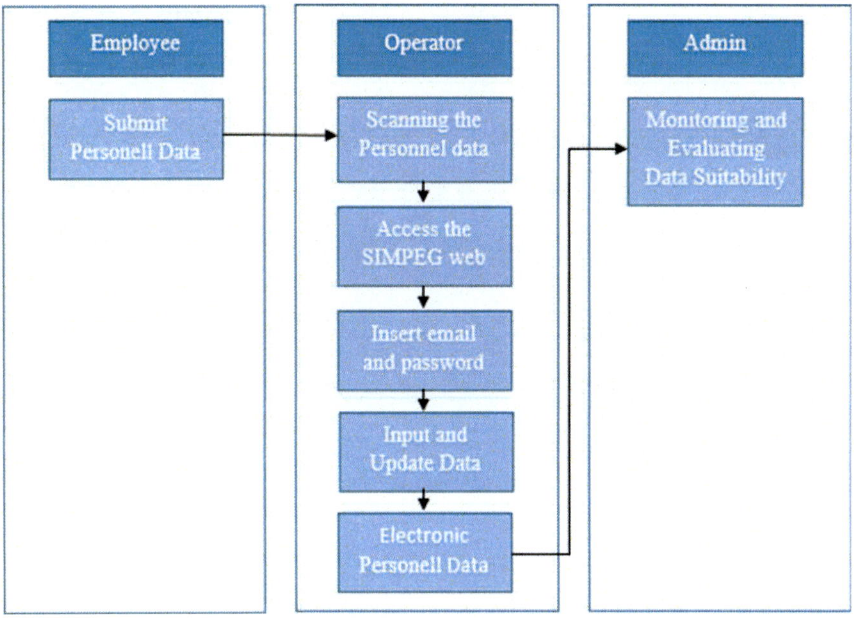

Fig. 1 The mechanism and flow of SIMPEG utilization task

Others have a percentage of less than 3%, namely employees in the Civil Service Police Unit, class D hospitals, and the inspectorate, as seen in Fig. 2).

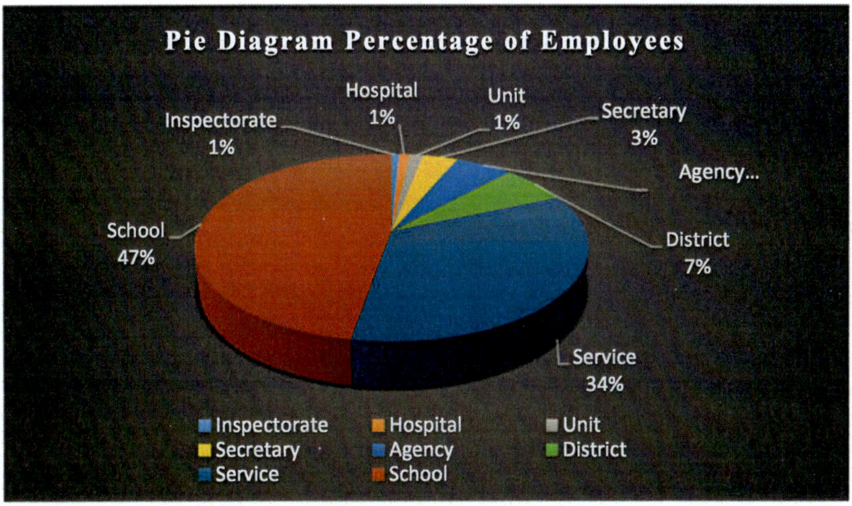

Fig. 2 A pie diagram percentage of employees recorded by SIMPEG in the Palangka Raya City

If agencies still use manual systems, in the future agencies must face various obstacles such as limited services and more complicated administrative procedures so that they are less efficient (Amandita and Rachmawati 2019). These constraints can be overcome by the use and application of information technology, namely the Human Resources Management Information System (SIMPEG). The SIMPEG application is very helpful in providing good service to every ASN in the Palangka Raya City Government. The use of SIMPEG is carried out to take care of personnel administration services from basic data, personnel history to personnel reports as follows:

- Promotion
- Position mutation
- Study permit/assignment
- Leave
- Training on classification
- Employee card
- Husband or wife card
- Health insurance data/BPJS
- Taspen
- Marriage
- Formal Education
- Informal Education
- Periodic salary increases
- Retirement
- Family history data
- Awards
- Disciplinary punishment
- Organization
- Family data
- Child data.

All features in this SIMPEG application function to facilitate the Palangka Raya City Government in the process of managing employee data. The list of basic historical data to the list of personnel history must always be updated by each ASN through the operator in each agency/office/SOPD/district and will be a basic reference in the consideration of granting policies to each ASN, when it will receive promotions, salaries, study permits, leave or other things. This can avoid ASN data manipulation as it is normally encountered. The use of SIMPEG in the implementation of E-government is assessed based on established indicators. Indicators that have been selected to assess the extent to which SIMPEG is based on the successful implementation of E-government (Indrajit 2002), include:

Data Availability Indicator. It is the legal basis for the application of the Personnel Management Information. One indicator in assessing the success of E-governance in an urban area is the availability of data and data center information. This indicator has been achieved for the SIMPEG application because the personnel data owned by the City Government of Palangka Raya have been held or managed in an integrated manner by the Personnel, Education, and Training Agency (BKPP). This personnel data integration has an important role in accelerating the data management. Data servers that have been held by BKPP are deemed appropriate because they are the main tasks and functions of the agency.

Access of personnel data in SIMPEG is limited; it can only be done by certain parties, namely by the State Civil Apparatus (ASN). BKPP also has a platform in displaying and delivering personnel information to the general public, namely through TV Personnel Info. This platform also displays personnel data that are part of SIMPEG. However, this platform has not displayed detailed personnel data in SIMPEG, the display is only limited to statistical data on the number of promotions, the number of periodic salary increases, and the number of pensions. Statistical data that is displayed cannot be downloaded by the general public. Besides, another appearance in this TV Personnel Information is the Rank Order Departure (DUK), the announcement of CPNS acceptance information within the Palangka Raya City Government, and there is also a photo gallery containing photos of activities carried out by the BKPP Office, as seen in Fig. 3.

Legal Infrastructure. The legal basis for the application of the Human Resources Management Information System in the Palangkaraya City Government environment has been regulated in the Minister of Home Affairs Decree No. 17 of 2000 concerning the Human Resources Management Information System of the Ministry of Home Affairs and Regional Governments which was later revised through Minister of Home Affairs Regulation of the Republic of Indonesia No. 125 of 2017 concerning the Ministry of Home Affairs Civil Service Management Information System (Minister of Home Affairs Regulation of the Republic of Indonesia No. 125 of 2017). This SIMPEG application has been utilized at the national level to the district/city level in Indonesia. Therefore, most district/city governments throughout Indonesia have used SIMPEG as an application that supports the implementation of E-government.

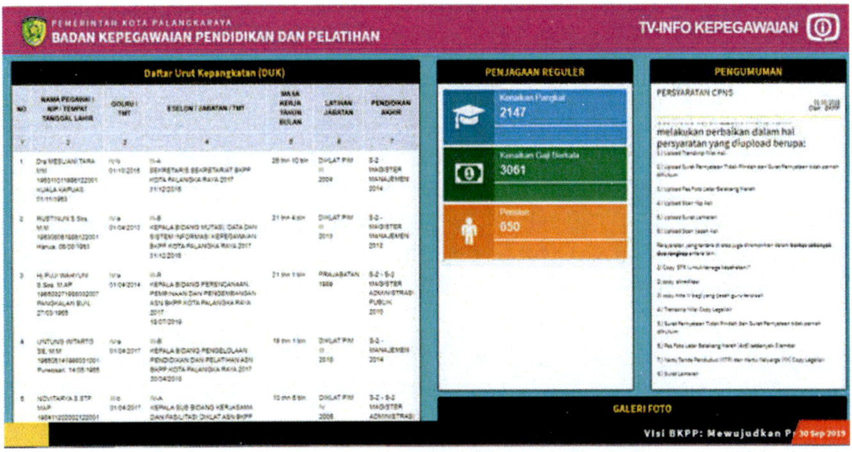

Fig. 3 TV display of personnel data info at BKPP Palangka Raya City

Functionally, the management of SIMPEG is directly carried out by Personnel bureaus at the central and regional levels so that there is a great possibility of SIMPEG interaction between these levels of authority based on their needs. The substance of the Regulation of the Minister of Home Affairs of the Republic of Indonesia No. 125 of 2017 in detail explains the management and development of SIMPEG. The application of SIMPEG as an application that supports E-government in the City of Palangka Raya has met the criteria on legal infrastructure indicators because it officially has a legal policy covering it. So far, the application of SIMPEG in Palangka Raya City Government agencies or institutions has been running on the right track although sometimes there is a Certificate issued by the Mayor to regulate the implementation of SIMPEG to keep it going well.

Institutional Infrastructure. Institutional infrastructure in implementing E-government is related to the role of institutions that is realized by the existence of coordination and communication between agencies. SIMPEG is generally managed by order by Personnel, Education, and Training Agency (BKPP). The main task of BKPP is to carry out functions or management functions in the field of administration and regional personnel, such as personnel development and personnel administrative services

The existence of institutional infrastructure has supported the implementation of E-government well because it has established cooperation between agencies. The form of cooperation between agencies is the reconciliation of SIMPEG data between SIMPEG operators in all agencies/offices/SOPD/districts in the Palangkaraya City Government which was carried out in 2019. This data reconciliation aims to harmonize State Civil Apparatus (ASN) data in every SIMPEG application in each agency/office/SOPD/district with data in the SIMPEG application belonging to BKPP Palangkaraya City. In addition to data reconciliation, BKPP also conducts socialization on the use of SIMPEG for operators in other agencies and there is also a SIMPEG technical guidance activity organized and participated by several people from offices, agencies, offices, and districts in Palangka Raya City.

Human Infrastructure. Human infrastructure is an important capital in managing E-government well. Human resources have the role of operator and managerial of this SIMPEG application. Human resources owned by the City of Palangka Raya in managing SIMPEG applications already exist, with the placement of two experts in the field of information technology for operational SIMPEG applications in the City of Palangka Raya. The two experts in the field of ICT are placed in every government institution in the City of Palangka Raya that has the task of updating the personnel data in each agency/office/SOPD/district. The informant determines the mechanism of the task of human infrastructure in SIMPEG as seen down below:

> *…The task mechanism of each institution has two operators who input and update data. In addition there is one admin that is only available in BKPP whose task is to monitoring and evaluating all the data in each institutions*" informant (Mrs. Desi Dotriana Novita, Head of BKPP Civil Service Subdivision, Palangkar Raya City).

SIMPEG application operators are placed in every agency in the City of Palangka Raya. This is because personnel data requires faster and integrated data input through horizontal relationships between agencies/offices/SOPD/districts. Besides, there is also the staff who act as SIMPEG admins. Admin is placed only in the Personnel, Education, and Training Agency (BKPP). The task of the admin who is in BKPP is monitoring and evaluating all SIMPEG operators in each agency/office/SOPD/district.

Technology Infrastructure. Technological indicators are related to the availability of technological tools that support the implementation of E-government. The technology available at Personnel, Education, and Training Agency (BKPP) is considered to be good enough because each agency has adequate technological devices. Each service is connected by the same Internet network so that SIMPEG can be easily accessed by all agencies. Based on the results of interviews conducted with guest speaker Desi Dostriana Novita as Head of the Civil Service Subdivisions BKPP Palangka Raya City, she said that the use of SIMPEG in the Internal government of Palangka Raya City had never experienced problems such as Internet network constraints for the uploading of employee files or other things

> "*…We do not use indihome (internet network provider), the office has its own internet network facilities that are directly connected to Telkom Balikpapan. The name of the facility is Astinet so that if indihome experiences a disturbance then we are not affected because we are directly from the central*" informant (Mrs. Desi Dotriana Novita, Head of BKPP Civil Service Subdivision, Palangkar Raya City)

The BKPP operator has optimized well the existence of Internet network. All of the data processes mainly require Internet to have faster data integration. The main challenge of the technology aspect is the fastest revolution of technological infrastructure, lack of knowledge, resources, and competencies of operators available in the public sector.

Leader's Thought Infrastructure. The next indicator used is a leader though who plays a role in directing the efforts of the Regional Organization Unit (SOPD), so the desired goal of E-government implementation can be achieved properly in the SIMPEG application. The leather's thought for SIMPEG utilization is attached in the vision and mission of the Mayor of Palangka Raya City in the 2018–2023 Medium-Term Development Plan (RPJM). Palangka Raya Government is now trying to do the realization of a smart city in Palangka Raya City with E-government as one of its applications (Palangka Raya Medium Term of Development Plant 2018). Indrajit (2002) explains the related types of e-governance which include SIMPEG applications with the type of relationship—Government to Government. This is because the SIMPEG application is only used for employee personnel activities and data in the city of Palangka Raya. It also allows for the exchange of various information between SOPDs in the City of Palangka Raya in terms of personnel data. The personnel data provides information related to the number of employees in Palangka Raya City of more than 4,964 employees. The number of employees must be balanced with an integrated employee data collection system and can be obtained more easily.

4.2 Supporting and Inhabiting Factors

All types of financing used for the development and management of SIMPEG are supported by the allocation of funds in the City budget of Palangka Raya. The budget can change according to government funding priorities. An adequate budget is needed to provide technology infrastructure and operator socialization as the most important support in the management of SIMPEG. The existence of SIMPEG is supported by the highest government institution in the City of Palangka Raya. The City Government of Palangka Raya issues a Mayor Decree regarding SIMPEG application operators in every agency or office and district. The existence of this Mayor Decree should make the implementation of SIMPEG run optimally, but since this application is implemented or run from mid-2017, it has not run optimally. This is evident from the lack of updating of personnel data at the agency or office and district according to the State Civil Apparatus (ASN) who has moved or transferred, retired, on leave, received awards to disciplinary penalties.

Barriers to data background are motivated by human resource problems. The availability of skilled and expert human resources on the application of SIMPEG is still limited. Human resources owned by the government tend not to master information technology or even have a proper understanding of knowledge about E-government. The application operators in several agencies or offices or districts do not yet understand how to operate SIMPEG properly or the lack of human resource competence in information technology. The unequally distributed distribution of ASN competencies within the Palangka Raya City Government is one of the reasons for the lack of understanding of SIMPEG.

Another obstacle that is owned is access to information about personnel made by the Personnel, Education, and Training Agency (BKPP) of Palangka Raya City, which is very limited to the wider community. Detailed access to information on personnel can only be accessed by ASN, so if there is a need for that data, it is necessary to ask directly to the relevant office. Therefore, to overcome this obstacle, the existing TV Info media that can be accessed widely by the public through the website should increase the list of staff data displayed and it would be better if the data can also be downloaded in the form of a soft file as a form of data transparency and make it easy for people who want to use it.

5 Conclusion

Human Resources Management Information System (SIMPEG) is a form of E-government application in the City Government of Palangka Raya. The application that has been used in the Palangka Raya City Government since mid-2017 is a concrete manifestation of the application of one of the smart city achievement indicators and is in line with the vision and mission of the Mayor of Palangka Raya in 2018–2023 to realize the smart city in Palangka Raya City. There are six indicators of successful implementation of E-government according to Indrajit (2002) namely infrastructure of data availability, legal infrastructure, human infrastructure, the thought of leader infrastructure, technology infrastructure, and institutional infrastructure.

The result analysis shows that the application of SIMPEG has succeed to support the e-government implementation criteria in each indicator. The application of web-based SIMPEG is very helpful in recording all employee data from several agencies or institutions within the Palangkaraya City Government, especially for Personnel, Education, and Training Agency (BKPP), which is the body in charge of recording employee data as a whole. Even so, there are several driving factors and inhibiting factors during the application of this application. The driving factor is the guarantee of funding that comes from the budget allocation of the Palangka Raya City as a priority while the main inhibiting factor comes from human infrastructure indicators where operators in several agencies/offices/SOPD/districts do not yet understand in detail the operation of SIMPEG so that employee data in the relevant agencies or institutions do not update.

Acknowledgements This journal was made possible through the help and support from the smart city field trip team 2019 at Palangka Raya City and all of the Regional Development 2016 students, who had an important role in the successful realization of this journal. The greatest appreciation also goes to Mrs. Desi Dotriana Novita, Head of BKPP Civil Service Subdivision as the informant.

References

Al-Mamary YH, Shamsuddin A, Aziati N (2014) The meaning of management information systems and its role in telecommunication companies in Yemen. Am J Softw Eng 2(2):22–25

Amandita AR, Rachmawati R (2019) Utilization and quality of information system for administration services based on ICT In Patehan, Kraton, Yogyakarta. Indones J Sci Technol 4(1):55–63

Babaei M, Jafar B (2013) Management information system, challenges and solutions. Eur Online J Nat Soc Sci 2(3)(s):374–381. www.european-science.com

Bovaird T (2005) Performance measurement and evaluation of e-government and e-governance programmes and initiatives. In: Khosrow-pour M (ed), Practicing e-government: a global prespective. Idea Group Publishing, United States, pp 16–42

Garcia, LM (2016) Ucer Centric e-government: The modernization of national migration of National Migration Institute in the Southern Mexican Border. In: Scholl HJ (ed) Electronic government and electronic participation. IOS, Amsterdam, pp 328–335

Indrajit RE (2002) Linux smart book: building e-government application. PT Elex Media Komputindo, Jakarta

Indrajit RE, Richardus E et al (2004) Electronic government: development strategy and development of digital technology based service system. Andi, Yogyakarta

Mediaswati R, Sidik F (2013) Analyzation the implementation of the web based employees reporting application in the regional employees board in Cilacap District. J Public Policy Adm (JKAP) 4–14

Minister of Home Affairs Regulation of the Republic of Indonesia No. 125 of 2017

Palangka Raya Medium Term of Development Plant 2018–2023 RPJM

Rianto B, Lestari T (2012) Polri & E-government application on public services. CV. Putra Media Nusantara (PMN), Surabaya
The Regulation of the Minister of Home Affairs No. 17/2000 about the Human Resources Management Information System of the Ministry of Home Affairs and Regional Governments

Evaluation of Urbanites' Perception About Livable City Using Analytic Hierarchy Process (AHP): A Case Study of Dhaka City

Raisa Sultana and Afrida Asad

Abstract Urban livability, an implicitly and intensely debated concept, is extensively being used by the urban planners and policymakers for incorporating sustainable, inclusive city for all. But the conception of the term should not be rigidly limited under the boundaries of academics rather the perception of general people is highly necessary to diagnose the prime constraints for ensuring a livable city. Due to rapid and unplanned urbanization, Dhaka city has been experiencing extreme problems creating serious nuisance of urban livability. The objective of this study is to statistically evaluate the conception of dwellers of Dhaka city regarding the urban issues they face on regular basis as an impediment for ensuring livability. An individual resident's perspective has been studied qualitatively and purposively using Analytical Hierarchy Method (AHP) from 100 students of age group 19–20 years. The factors influencing the livability of urban environment are studied under these parameters—Healthy Physical Environment, Economic Stability, Availability of Fresh Food, Transportation and Mobility, Utilities and Services Functionality, Social Equity and Justice, Accommodation Convenience and Quality Health Care. After analyzing, different factors under each issue were ranked. Major complications have been explored, e.g., noise pollution, corruption, unhealthy food consumption, sexual harassment, waterlogging, unsafe and insecure environment, lack of standard medical service, high house rent, etc. which are creating the nuisance for Dhaka city's livability and making Dhaka as the second least livable city in the world.

Keywords Livability · Analytical Hierarchy Process (AHP) · Livable city

1 Introduction

Livability refers to an assemblage of qualities considered desirable by inhabitants of a certain area. It is concerned mainly with the experience of an individual resident's perspective (Draft Dhaka Structure Plan 2015). The word "Urban livability" is widely

R. Sultana (✉) · A. Asad
Department of Geography and Environment, University of Dhaka, Dhaka, Bangladesh
e-mail: raisasultana@du.ac.bd

L. T. T. Huong and G. M. Pomeroy (eds.), *AUC 2019*, Advances in 21st Century Human Settlements, https://doi.org/10.1007/978-981-15-5608-1_29

now recognized as an interdisciplinary concept which integrates many factors of interests.

Burton (2014) defined livability as quality of place and extended it as the "*the physical characteristics of community, the way it is planned, designed, developed, and maintained* (Burton 2014)." Often it has some synonym between the notion of "quality of life," "urban quality or environment quality," "standard of living," "satisfaction," "happiness," "well-being" of a population (Burton 2014; Ahmed et al. 2019). Hence, quality of life can be described as a standard of living state where a person can live in a clean and secured environment, ease of access to the basic amenities and services, have social and economic stability, move comfortably, and acquire quality food and health care with satisfaction in every sector of a particular area or region. Public policies, plans are aimed to enhance the livability of the cities. Journalists, politicians, planners, and commentators foment an ongoing public discussion about the meaning of livability, who possess it and who does not.

Therefore, it is important to validate residents' assessment of attaining a healthy and sound life where people can enjoy and access all the basic amenities of life. Residents' perception within a community makes identifying the problems and directs the shape for enhancing the quality of life in a subjective way. The urban ecosystem is based on two pillars—sociocultural fabric and physical environment which are highly interdependent on each other's sustainability. Livability is presently an equivocal and multi-dimensional concept comprising multiple aspects related to the economic, social, physical, and general well-being of the citizens. It includes perceptions and expectations of people about their neighborhood, safety, accessibility, etc. (https://www.wur.nl/en/article/Msc-thesis-subject-Modelling-Urban-Livability.html). A common set of guiding principles that give substance to the term "livability" are equity, dignity, accessibility, conviviality, participation, empowerment (CitiesPLUS 2003), safety, comfort, transit, walkability, etc. (Ahmed et al. 2019; Balsas 2004; Knox and Heike 2013).

Some big international institutions researched about livable cities' like Mercer Quality of Living Survey, Economist Intelligent Units (EIU), Monocle's Magazine "Most Livable Cities Index," and Philips reaps. Though the assessment methods and system do overgeneralization on examining the conditions of the cities, however, these are considered as widely accepted and approached (Ahmed et al. 2019; Setijanti et al. 2015). Recently, The Economist Intelligence Unit has surveyed livability index and ranked 140 cities on which Dhaka positioned as 137, which makes Dhaka as one of the least livable cities in the world (The Economist Intelligence Unit 2017). Though the livability varies from area to area and in different levels of region (neighborhood, city, and national), the worst sufferers are the slum people than those in rural areas, stated by World Bank (2018).

This paper aims to explore the perception about the urban livability of urbanites under the following questions: What are the factors that hinder the urban resident's to achieve a quality of life in Dhaka city? Out of various problems, what are the dominant factors that make city life more miserable?

To address these questions, it is needed to understand that developing city like Dhaka, is fabricated with a wide range of problems that are making the city life worse day by day. Therefore, it is very difficult for an urban dweller to pick one specific problem which they face in their life. People come to the city for different reasons, e.g., in search of jobs, education, better life facility but ended up struggling and entangled with dealing the problems in every sector. Hence, it is quite impossible to choose one dimension and understand their perception.

2 Literature Review

2.1 Urban Environment, Livability, and Quality of Life

In the field of urban planning, two terms "quality of life" and "livability" are used reciprocally. Where livability is referred as the presence and qualities of amenities of built and natural environment, "quality of life" denotes the individual's experience of those services and facilities (Ahmed et al. 2019). Lyndhurst defines "livability" as the quality of life where clean, secure, and green environmental condition prevails (Lyndhurst 2004). Synthesis of existing literature brought in forward the two approaches of quality of life and its determinants: "objective," "subjective," (Ballas and Dorling 2013; Marans and Stimson 2011) and normative (Ahmed et al. 2019). The first one is the theoretical approach of evaluation of social and physical environmental factors (Apparicio et al. 2008). In contrast, the second one focuses on individuals' perception about perceiving life combining with one-dimensional facets of life or qualitative measurements, e.g., happiness or satisfaction (Douglas et al. 2018; Lesli et al. 2010), whereas "normative" appraises about the philosophical view of a good life.

Din et al. (2013) studied theoretically and classified urban quality of life into seven dimensions including environmental, physical, mobility, social, psychological, economical, and political. All these main dimensions are divided into thirty basic principles that can be applied in various combinations to achieve quality of life for communities.

Veenhoven (2004) comprised the quality of life with four criteria in a matrix, in which opportunity and achievement of life set in vertical position and horizontal position is represented by internal and external qualities. So Livability is defined as environmental or living conditions which produce a combination of external opportunities and quality of life (Veenhoven 2004).

Aulia (2014) opined that various components of a livable community: land use, social and cultural opportunities, economic resources, access and mobility, safety and aesthetics, environmental concerns, cumulative aspects, etc. can affect a variety of human and ecological resources (Aulia 2014).

2.2 *Livability and Dhaka City*

A livable community is an area in which the dwellers have "*affordable and appropriate housing, supportive community features and services, and adequate mobility options, which together facilitate personal independence and the engagement of residents in civic and social life.*" (Kihl et al. 2005). Dhaka city has undergone a rapid change of its demographic, economic, social, environmental, and land use after the independence of the country. Due to rapid urbanization, developing cities like Dhaka is experiencing tremendous transformation which effects on consumption, behavior, and mobility pattern of urbanites. Every year, about 6000 people migrate to this city without any job or settlement plan (The Daily ProthomAlo 2015). Dreadful traffic congestion is endemic, inadequate infrastructure, public and open spaces, and public transport relative results as a consequence of the high population and economic density. Piped water covers only two-thirds of the city, only a mere 3–4% of all wastewater generated is treated, and only 60% of municipal solid waste is collected (World Bank 2018). Air pollution levels eight times higher than WHO guidelines, which makes the city as one of the world's most polluted cities (World Bank 2018). The absence of cycling paths, accessible walkway, excessive rickshaw, well-connected transportation network, etc. hinders the mobility of the urban dwellers tremendously. The city has become like a cage of concrete infrastructure due to a rigorous work schedule, lack of green open space and recreational activities.

Although Dhaka city possesses picturesque sites, establishments, and some planned areas, the inconveniences of city obstruct to function as an efficient modern city delivering desired services. The dwellers experience difficulties in most anterior of life as it lacks an appropriate city plan to offer the desired quality of life. The city needs to improve its basic infrastructure including transportation network, drainage system, environmental conditions, health services, and community facilities. Also, the city has to ensure a society with no discrimination regardless of power, position, age, and gender. A city plan should offer economic opportunities to strengthen a city in all dimensions. Economically, a city should be able to compete for investment in industrialization, trade, and commerce (Kalam 2009).

3 Conceptual Framework/Selection of Components and Variables for Livability Evaluation

The quality of life of an individual has many dimensions, including family, job, finance, health, societal values, movability, etc. Every urban environment that people live in different places has numerous environmental attributes which contribute significantly to the livability index. The places ranging in size or scale from the individual dwelling to the local area or neighborhood, to the city, to the broader region, or even to the state or nation—and it has been documented that where people live do influence and ultimately shape their lives and, therefore, their overall livability

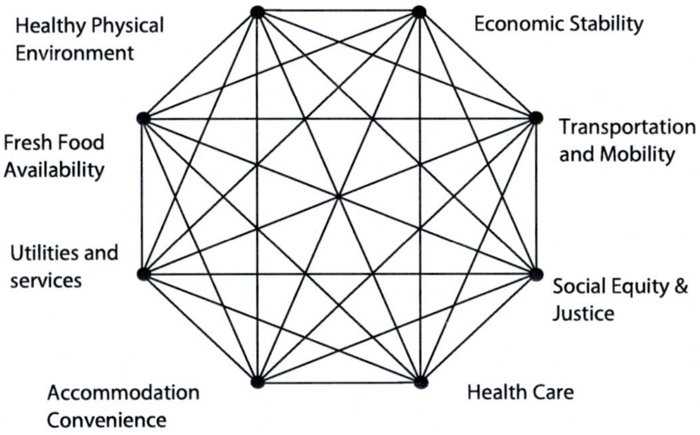

Fig. 1 Multi-aspects of livability (the researcher, 2019)

(Marans 2012). Hence, in this paper, the perception of livability is evaluated in the areas of intracity context. Even if a sample of urbanites were asked to rank in order of importance a list of items relating to livability, the information thus integrated does not necessarily allow one to evaluate the components explained by any one factor. The octagon theoretical framework enables to guide the process of the complex relationships between different factors of urban environments at different scales. The primary factors are then again categorized into 29 sub-factors to understand the perception of urban dwellers (Fig. 1).

The first factor, generically titled Healthy Physical Environment, refers to such condition where urbanites have a healthy environment which enables them to breathe pure and air, urban infrastructure with ample amount of greeneries, a pollution-free habitat. A clean landscape is most desirable in big cities; they impede pollution, relief us with shade, and present a pleasant visual picturesque environment.

The second factor, Economic Stability, denotes corruption, job security and employment issues, and affordability problems

The third factor, Availability of Fresh Food, addresses the quality, hygiene, and price issues.

The fourth factor, Transportation and Mobility, refers to the overall traffic condition along with the eve-teasing matter which is serious among the female residents. Streets should be secured from traffic congestion and speed, pedestrian should be able to walk safely with suitable sidewalks through the major streets as well as allies. All the roads should be well connected in which the dwellers can move flawlessly with affordable transport mode.

The fifth factor, Utilities and Services Functionality, discusses the efficiency of basic services like gas, water, electricity, and drainage facilities.

The sixth factor, Social Equity and Justice, comprises the issues concerning the feeling of toward their city regarding equity and justice like protection from crime and violence, security, nepotism, and access to the recreational activity.

Table 1 Domain and sub-factors of urban livability to evaluate perception of urbanites

Domain factors	Sub-factors	Code
Physical environment A	Dirty environment	A1
	Dusty air	A2
	Lack of green open space	A3
	Noise pollution	A4
Economic stability B	Corruption	B1
	Unemployment	B2
	High living cost	B3
Availability of fresh food C	Poor food quality	C1
	Unhygienic food	C2
	High food price	C3
Transportation and mobility D	Poor road network	D1
	Unavailability of cheap transport mode	D2
	Poor road quality	D3
	Inaccessible walkway	D4
	Eve teasing/harassment	D5
Utilities and services functionality E	Gas inaccessibility	E1
	Pure water unavailability	E2
	Poor Drainage Facilities	E3
	Electricity inconvenience	E4
Social equity and justice F	Nepotism	F1
	Lack of security/Safety	F2
	Crime and Violence	F3
	Lack of Recreational activity	F4
Accommodation convenience G	Quality housing space	G1
	High house rent	G2
	Housing unavailability	G3
Quality health care H	High medical Price	H1
	Poor health care quality	H2
	Medical service unavailability	H3

The seventh factor, Accommodation and Convenience, refers to the quality, rent, and availability of accommodation and their respective space.

The last factor, Quality Health care is comprised of availability, quality, and price in medical sector. All these factors are interconnected and make the term "livability" as integrated and inclusive approach in urban context.

Table 1 explores the factors influencing the livability on urban environment that are studied under these parameters.

4 Methodology

Students from a renowned institution in Dhaka have been chosen as sample where they have been categorized into two types representing two different classes of people living in the city.

- Who have recently migrated to city for education or livelihood purpose and living proximity to their respective institutions and have less exposure to the broader spectrum of city.
- Who have been living since their birth or longer period in the city and have greater mobility across the city than to first group.

There were mainly two reasons for choosing this category of urbanites. First, urban dwellers by birth have witnessed the gradual change of Dhaka city for the past several years, and they have some better and valid opinions. Second, the people who are not used to live in a megacity like Dhaka might have undergone some serious changes in their standard of living. The study has been developed in the following steps:

- Establishing the dynamic and operable main factors which are integral part in evaluating urban livability was the first step. The expert judgments from urban planners have been taken to set equally important factors under which the sub-factors are chosen for further ranking or scoring.
- Analytical Hierarchy process (AHP) is a type of multi-criteria decision analysis which has been developed by Thomas L Saaty in 1980 where individual decision within a framework will be evaluated through pairwise comparison for scoring or ranking multiple decisions (Thomas 1980). Respondents have ranked all the sub-factors with each other and put their decisions on ranking urban issues as sub-factors under main factors (Fig. 2).

4.1 AHP Modeling and Data Collection

The questionnaire requires each respondent to make pairwise comparisons of the entire criteria (Thomas 1980) according to the AHP model (see example question style in Table 2). In this work, we collected 100 questionnaires in total according to Taro Yamane sample size for infinite population for 10% precision level where confidence level is 95% (Taro 1967). Compiling the data using AHP method, comparisons have been made and different factors have been ranked.

The AHP method starts by calculating the Consistency Index (C.I.) and Consistency Ratio (C.R.) for each respondent using the formula stated in (Thomas 1980).

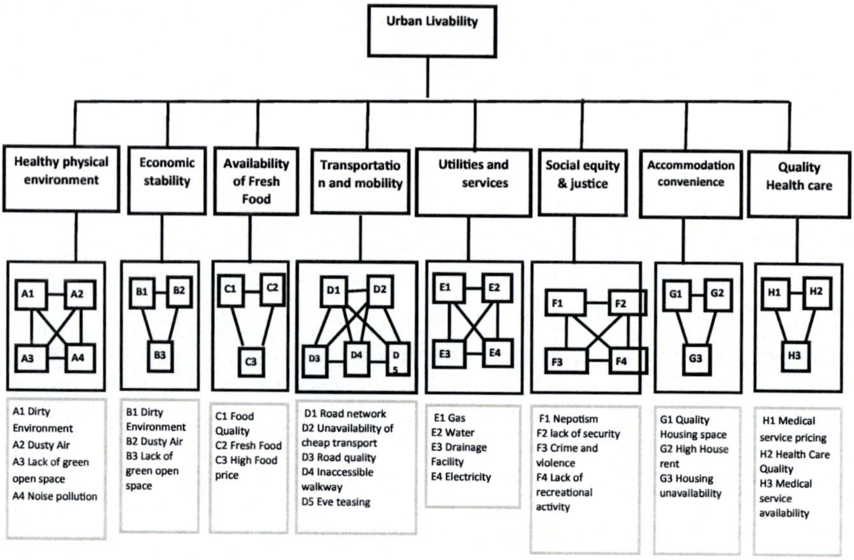

Fig. 2 Structured AHP model of urban livability

For each respondent of questionnaire, if the value of C.R. is less than 10% (or 0.1), the response is considered to be acceptable. In this experiment, there are 82 responses having C.R. less than 0.1. The hierarchical ranking calculation using respondents' choice is then performed according to (Irfan and Junseok 2009) for all alternatives in each level of the AHP model.

5 Results and Discussion

All the main factors have equal weightage value putting the same importance to make the city livable. However, the perspective of people who have been living in the city for almost entire life (non-resident) differs greatly from those who have exposed to the city life (resident) recently and the pairwise comparison among sub-factors has elucidated the matter. Moreover, the people who are resident are confined in the university campus area having limited exposure to the whole city area.

Problems and complexity arise when the pace of development in areas of housing, medical, utility and services, mobility, infrastructure, connectivity, etc. cannot match with the momentum of rapid growth of urbanization. Though the factors have been analyzed and ranked based on the responses (resulted in percentage), however, it is mentionable that very small variations have been found between those factors. This depicts that almost all the urbanites experience these problems more or less in Dhaka city. The analysis result is displayed in Table 3. For graphical representation, radar

Table 2 Pairwise comparison

With respect to URBAN PHYSICAL ENVIRONMENT, Using the scale from 1 to 9 (where 9 is extremely and 1 is equally important), please indicate (X) the relative importance of options A (left column) to options B (right column)

A options	Extremely	Very Strongly	Strongly	Moderately	Equally	Moderately	Strongly	Very Strongly	Extremely	B options
Dusty air	9	7	5	3	1	3	5	7	9	Noise pollution
Dusty air	9	7	5	3	1	3	5	7	9	Lack of green open space
Dusty air	9	7	5	3	1	3	5	7	9	Dirty environment
Noise pollution	9	7	5	3	1	3	5	7	9	Lack of green open space
Noise pollution	9	7	5	3	1	3	5	7	9	Dirty environment
Lack of green open space	9	7	5	3	1	3	5	7	9	Dirty environment

Table 3 The hierarchical ranking calculation by respondents using AHP method

Primary factors/domain Factors	Sub factors with code	Urbanites migrated (%)	Rank	Non migrated urbanites (%)	Rank
Healthy physical environment	Dirty environment (A1)	38.1	1	21.1	4
	Dusty air (A2)	26.9	2	30.3	1
	Lack of green open space (A3)	20.3	3	25.5	2
	Noise pollution (A4)	14.6	4	23.1	3
Economic stability	Corruption (B1)	49.2	1	45.9	1
	Unemployment (B2)	32.1	2	27.5	2
	High living cost (B3)	18.64	3	26.6	3
Fresh food availability	Unhygienic food (C1)	40.9	1	35.6	2
	High food price (C2)	35.6	2	36.5	1
	Poor food quality (C3)	23.5	3	27.9	3
Transport and mobility	Poor road network (D1)	30.0	1	13.4	5
	Poor road quality (D2)	21.2	2	23.4	2
	Unavailability of cheap transport mode (D3)	20.6	3	15.5	4
	Inaccessible walkway (D4)	16.1	4	21.3	3
	Eve teasing/harassment (D5)	12.0	5	26.7	1
Utilities and services	Pure water unavailability (E1)	41.8	1	32.9	1
	Gas inaccessibility (E2)	30.7	2	21.7	3
	Poor drainage facilities (E3)	14.9	3	31.8	2
	Electricity inconvenience (E4)	12.6	4	13.6	4
Social equity and justice	Nepotism (F1)	31.4	1	22.1	3
	Lack of security/safety (F2)	29.0	2	32.1	1
	Crime and Violence (F3)	24.2	3	25.6	2

(continued)

Table 3 (continued)

	Lack of recreational activity (F4)	15.4	4	20.5	4
Accommodation and convenience	Quality housing space (G1)	40.9	1	41.8	1
	High house rent (G2)	37.1	2	24.6	2
	Housing unavailability (G3)	21.9	3	33.6	3
Quality health care	High medical price (H1)	43.9	1	37.1	2
	Poor health care quality (H2)	29.1	2	37.5	1
	Medical service unavailability (H3)	27.0	3	25.4	3

Source Field Survey, 2019

chart (Fig. 3) has been prepared for displaying multivariate data of three or more quantitative variables represented on axes starting from the same point.

5.1 *Environmental Attributes*

In case of judging physical environment, resident people have given the utmost concern in dirty environment around the city which is mainly the open garbage, sewerage system hampering daily life. On the other hand, Resident people have given dusty air and air pollution as their main problem to be faced and secondly the lack of green space. Surprisingly the dirty environment is their least concern to be solved. Exposure to dirty, unhygienic city environment for a longer period of time may have made them resilient to such issue, but non-resident people's maximum mobility coverage around the city can be the reason for dusty air and pollution being in the top rank (Fig. 3, Radar Chart 5.1).

5.2 *Economical Attributes*

Understanding the economic stability in city institutions and economic dimension of city life has been studied in three sub-factors but surprisingly corruption has been nominated as the first problem both in resident and non-resident perspectives (Fig. 3, Radar Chart 5.2).

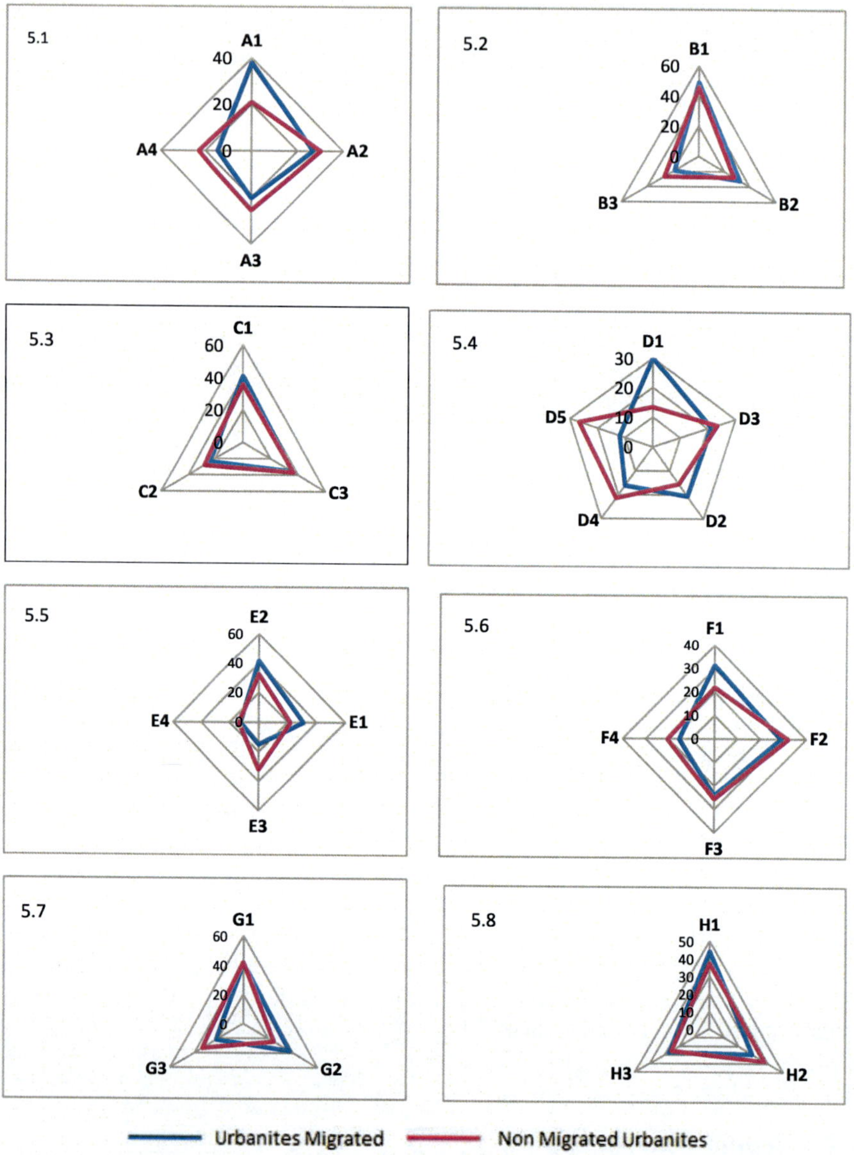

Fig. 3 Graphical presentation of urbanites perception about livability using radar chart

5.3 *Healthy Food Consumption*

Food security is a very important aspect in a livable city where ensuring food quality, availability, and pricing all should equally be prioritized. Due to comparatively low pricing in campus area where resident people live, the availability of food has become first in concern, whereas the food pricing is main issue for people living in the city for longer period (Fig. 3, Radar Chart 5.3).

5.4 *Transport and Mobility*

Transportation and mobility is one of the major indicators of urban livability. However, the urbanites recognized it as the top problem that affects significantly in their life. According to the resident people, the improper road networks are the main problem in the city, and harassment is the least but for the non-resident people who travel more than resident people have given a reverse scenario in outlining city's problem. From a daily basis commuter's perspective, eve-teasing, harassment is the main problem leading to road quality and lack of accessible walkway (Fig. 3, Radar Chart 5.4).

5.5 *Utility and Services*

The perspective of resident and non-resident people is quite similar in the utility and service scenario of the city. Water problem has been outlined by both major issues around the city. As drainage and waterlogging issue is not serious in the limited campus area, resident people have given it low rank than non-resident people who face these drainage issues more (Fig. 3, Radar Chart 5.5).

5.6 *Social Equity and Justice*

According to the resident people, nepotism has been outlined as a major concern in city's social cementing factor but non-resident people have given lack of security and safety issue as their major problem. The existence of institutional politics and nepotism in small institutional sphere may be the reason behind such scenario which is contrast to the broader city sphere (Fig. 3, Radar Chart 5.6).

5.7 *Accommodation Convenience*

Accommodation Convenience around the city and inside the small institutional campus is completely similar for both newcomers and people living around the city for longer time. Lack of quality housing space has considered as major concern in accommodation choices around the city, leading to high price of house rent (Fig. 3, Radar Chart 5.7).

5.8 Quality Health Care

Finally, the health care services around the city and proximity to campus area have been analyzed. The people living in the campus area having available, quality medical service have given pricing as the main reason but people coming from all over the city have outlined lack of quality medical service as main concern. The quality of medical service outside the capital is far worse which may create the biased perspective in people newly coming to the city (Fig. 3, Radar Chart 5.8).

6 Conclusion

Livability in cities through achieving a standard quality of life and sustainable environment has become one of the imperatives of modern planning and of increasing concern to the public; this term nowadays is linked to the rapid urbanization process in cities especially among the developing and less developing nations. Urbanization in Bangladesh is a growing phenomenon, which is steady in nature but fretfully affects the urban dwellers' life. This paper provides a general understanding of urban livability and urbanites' evaluation and tries to embrace related factors which shape the quality of life and its challenges hindrance to experience a sustainable and standard living condition in Dhaka city. The study classified urban quality of life into eight dimensions: physical environment, economic stability, fresh food availability, transportation and mobility, utilities and services, social equity and justice, accommodation and services, and quality health care. These main dimensions are divided into twenty-nine sub-factors that are applied in various combinations to achieve the quality of life for communities. These seven dimensions are studied theoretically and can be subjected to an applied study.

References

Ahmed NO, Halafawy AME, Amin AM (2019) A critical review of urban livability. Eur J Sustain Dev 8, 1:165–182

Apparicio P, Séguin A-M, Naud D (2008) The quality of the urban environment around public housing buildings in Montréal: an objective approach based on GIS and multivariate statistical analysis. Soc Indicat Res 86:355–380

Aulia DN (2014) Planning livable community with social systems approaches: Medan, Indonesia. In: Presented at 2nd international conference on sustainable innovation (ICoSI 2014). Universitas Muhammadiyah Yogyakarta 3rd–5th June 2014

Ballas D, Dorling D (2013) The geography of happiness. In: The oxford handbook of happiness, pp 465–481

Balsas CJ (2004) Measuring the livability of an urban centre: an explanatory study of key performance indicators. Plann Pract Res 1(19):101–110

Burton M (2014) Quality of place. In: Encyclopedia of quality of life and well-being research. Springer, pp 5312– 5314

CitiesPLUS (2003) A sustainable urban system: the long-term plan for greater Vancouver. citiesPLUS, Vancouver, Canada

Din HSE, Shalaby A, Farouh HE, Elariane SA (2013) Principles of Urban Equality of life for a neighborhood. HBRC J 9:86–92

Douglas O, Russell P, Scott M (2018) Positive perceptions of green and open space as predictors of neighbourhood quality of life: implications for urban planning across the city region. J Environ Plan Manage 1–21

Draft Dhaka Structure Plan (2015–2035) Rajdhani Unnayan Kartripakkhya (RAJUK), Dhaka Improvement Trust (DIT), Dhaka

Irfan S, Junseok H (2009) The application of AHP model to guide decision makers: a case study of E-banking security. In: 2009 fourth international conference on computer sciences and convergence information technology

Kalam AKAM (2009) Planning Dhaka as a global city: a critical discourse. J Bangladesh Inst Plan 2:1–12. Bangladesh Institute of Planners

Kihl M, Brennan D, Gabhawala N, List J, Mittal P (2005) Livable communities: an evaluation guide. AARP Public Policy Institute Washington DC

Knox P, Heike M (2013) Small town sustainability: economic, social, and environmental innovation. Walter de Gruyter

Lesli E, Sugiyama T, Ierodiaconou D, Kremer P (2010) Perceived and objectively measured greenness of neighbourhoods: are they measuring the same thing? Landsc Urban Plann 95:28–33

Lyndhurst B (2004) Livability and sustainable development: bad habits and hard choices. Final report for the ODPM, UK

Marans RW (2012) Quality of urban life studies: an overview and implications for environment-behaviour research, Elsevier, Procedia—Soc Behav Sci 35:9–22

Marans RW, Stimson R (2011) An overview of quality of urban life. In: Investigating quality of urban life. Springer, pp 1–29

Setijanti P, Defiana I, Setyawan W, Silas J, Firmaningtyas S, Ernawati R (2015) Traditional settlement livability in creating sustainable living. Procedia—Soc Behav Sci 179:204–211, Elsevier

Taro Y (1967) Statistics an introductory analysis, 2nd edn, New York Harper and Row

The Daily ProthomAlo (2015) Dhaka city and miserable life Published on 23 September 2015

The Economist Intelligence Unit (2017) The global liveability report 2017: a free overview. https://pages.eiu.com/rs/753-RIQ-438/images/Liveability_Free_Summary_2017.pdf

Thomas LS (1980) The analytic hierarchy process. McGraw-Hill International

Veenhoven R (2004) Happiness as an aim in public policy: the greates happiness principal, positive psychology in practice. Chapter 39. John Wiley and Sons, Inc, USA

World Bank (2018) https://blogs.worldbank.org/endpovertyinsouthasia/toward-livable-dhaka

https://www.wur.nl/en/article/Msc-thesis-subject-Modelling-Urban-Livability.html

Integrated Planning and Development

Urban Migrant Labor: Public Spaces and Social Integration (Review on Studies in Vietnam)

Pham Quynh Huong, Hoang Vu Linh Chi, Nguyen Tuan Minh, Luong Thuy Duong, Do Thi Ngan, and Phan Thi Song Thuong

Abstract Urban migration is an indispensable tendency of urbanization. Migrant workers contribute much to the development of cities as well as the welfare of communities and families. Based on analyzing research on four migrant workers including street vendors, industrial workers, domestic helpers, small-service sector workers, the article shows up a picture of living conditions, jobs, and use of public spaces of migrant workers. The approaches and use of public spaces for their earnings for living, relaxing, and widening their social relations play a significant role toward social integration in their new destination. Research shows that unskilled migrant workers are a group suffering much discrimination in social relations and using public services. The differences among culture, lifestyles, and social prejudice as well as shortcomings in urban management policies are huge obstacles in social integration of migrant workers. Moreover, the article initially identifies factors motivating and hindering migrants' access and use of public spaces.

Keywords Migration · Rural-urban · Unskilled labor · Social integration · Public spaces

1 Introduction

Important rural-urban migration flows are a defining feature of contemporary Vietnam's urbanization and modernization process (Nguyen 2002; UNFPA and GSO 2015). This phenomenon shapes the country's socioeconomic development by reallocating the national population in space, restructuring the labor economy, creating new jobs, and reducing poverty. Studies of rural-urban migration evolved significantly in the last decades. Throughout most of the 1990s and 2000s studies tended to

P. Q. Huong (✉) · H. V. L. Chi · N. T. Minh
Institute of Sociology, Vietnam Academy of Social Sciences, Hanoi, Vietnam
e-mail: phamquynhhuong2@gmail.com

L. T. Duong · D. T. Ngan · P. T. S. Thuong
Institute of Regional Sustainable Development, Vietnam Academy of Social Sciences, Hanoi, Vietnam

L. T. T. Huong and G. M. Pomeroy (eds.), *AUC 2019*, Advances in 21st Century Human Settlements, https://doi.org/10.1007/978-981-15-5608-1_30

focus mainly on documenting and characterizing migration flows, understanding the roles that migration in the economic development process, and identifying the challenges migrants face in accessing social services (Dang 1998; Dinh 2013; Le 2014). For example, rural-urban migration flow is the largest one among all of the migration flows. Over 80% of migrants come from rural areas moving to big cities and industrial zones from the North to the South to find job opportunities and improve their income as well as their family's living standard (UNFPA and GSO 2015). These works concentrated primarily on describing migration's general patterns and trends and on the situation and hardship experienced migrant laborers (Dang 2012; Dinh 2013; UNFPA and GSO 2015). They, however, paid less attention to the prejudice and social inclusion, which is significantly important in supporting an equal life for every individual in society. Only recently did scholar begin to look beyond issues related to migrants' jobs and material lives and consider questions of prejudice and social inclusion (UN Vietnam 2010; Nguyen 2012; Nguyen et al. 2016; Tran 2014).

Many studies defined rural-urban migration as the movement of a person from rural-to-urban areas in searching for employment or seeking opportunities to improve their income (Cu 2005; Tran 2014; Vu 2013). These studies often classify migrant workers as either long- or short-term migrants. Long-term migrants are generally identified as rural people who are registered as permanent or temporary residents in their urban destination (Oxfarm 2015; UNFPA and GSO 2015). Short-term migrants have a temporary residence registration in cities or are not registered at all (in the case of pendulum or seasonal migrants) (Bui et al. 2013; Dang 2012; UNFPA and GSO 2015).[1] Large surveys, including national surveys conducted by Vietnam's General Statistic Office with the assistance of foreign agencies, only account for three kinds of migrants: permanent, long-term, and registered migrants. As such, they overlook large number of short-term, pendulum, and seasonal migrants (Dang 2012; UN Vietnam 2010).

Many studies indicate that migrants play an indispensable role for economic development, contribute to urbanization, transformation of labor structure as well as improve qualifications of labor force in place of destination (UNFPA and GSO 2015). However, migrants are not in good positions in their destination with an aim to fight for a better life. Therefore, facilitating equality and eliminating institutional, economic, social discrimination are important missions for win-win development not only for migrant individuals, households but also for communities. This article reviews studies concerned with rural-urban migration in Vietnam and other countries in the region conducted over the past 20 years. The discussion begins with a review of the methods used by this scholarship. It then turns to a discussion of this literature's treatment of the sociodemographic characteristics of rural migrants, their access and utilization of urban public spaces, and their social inclusion in cities. The aim of this article is to draw a general painting about rural-urban migrants' life, to prove some

[1] In Vietnam, the household registration system (ho khau), has been issued since 1960s to control public security, to ration food and allocate jobs during the subsidy period and a way to limit urban migration after economic reforms in the Doi Moi period. The registration book is very important in many basic document procedures, including birth certificates, enrollments in public schools, health care, insurance.

reasons which make them feel difficult to access public spaces and to integrate the life in big cities.

2 Methodology

A large number of studies on rural-urban migration in Vietnam are based on national survey data about migrant workers produced by Vietnam's General Statistics Office or by international organizations such as UNFPA, UN Vietnam or Oxfam (Cirera and Lakshman 2017; UN Vietnam 2010; Pasquier-Doumer et al. 2017; Truong 2016). These studies tend to adopt quantitative analytical approaches. But the scholarship on rural migration in Vietnam also includes studies using qualitative, and mixed methods—the latter being increasingly common. A small number of studies rely on critical reviews (Bui 2013, 2015; Luong 2014). Studies using purely qualitative information collection methods are relatively few. Qualitative observations and in-depth interviews have been used in studies conducted in the early 2000s that explored sensitive topics and psychological issues experienced by migrant workers. This includes Nghiêm (2003) research on garment workers, Dao (2009) investigation on the employment status and life of female migrants working as domestic workers in Ho Chi Minh City, and Nguyen et al. (2010) study of stresses that migrant workers face in urban Vietnam in terms of livelihoods, legal challenges at destination, health and of their responses and coping strategies.

As mentioned above, studies mixing quantitative and qualitative data collection methods are the most common. These studies often combine quantitative questionnaire surveys with in-depth interviews and group discussions These studies often look at migration in general, at the economic and social impacts of migration flows on migrants' places of destination and departure, and at migrants' family dynamics in general (for example, Nguyen et al. 2012; Vu 2013). Mixed methods have also been used in studies focused on migrant factory workers (Le 2017; Nguyen 2016; Phan 2017) and on street vendors and informal sector workers (Nguyen 2009, 2016, 2017).

3 Social and Employment Characteristics of Migrant Workers

The results of 1999, 2009 Census and mid-term Census conducted by the General Statistics Office and the United Nations Population Fund (2016) show that domestic migrants in Vietnam are young. The average and median age of migrants has consistently remained lower than those of non-migrants. The average age of migrants who migrate in the same district is 29.1 years; 29.6 years old for migrants who migrate among districts in the same province; 27.8 years old for migrants who migrate among different provinces or areas. Women migrants tend to be younger than men migrants.

Some of the smaller studies indicate that migrant workers in industrial sectors tend to be younger than the general migrant population (Nguyen et al. 2013). According to research by the (Ministry of Labour, Invalides and Social Affairs (MOLISA) (2013) the age of migrant factory workers is about 23 years old. Research by Tuan (2016) shows that the age of this migrant factory worker group is even lower, from 15 to 29 years old.

Regarding group of freelance and service workers, the average age is from 20 to 24 years old. They are often unmarried. In terms of domestic rural-to-urban migrant workers, they often start working at a younger age, 15–16 years old, or even earlier (Chu 2005). The age of female migrants who work in vendor sectors, however, often middle age. Most of them are farmers who switch from agricultural sector to street vendor sector or still maintain agricultural activities to ensure their family's income while agriculture sole cannot bring much economic efficiency (Nguyen 2017). This difference indicates the difference among the age groups in the whole country, compared to the age of each specific occupational group.

Some studies suggest trends of feminization and rejuvenation among migrant workers (Dang 2005; Dao 2016; Nghiem 2003; Vu 2013). The feminization trend is most evident among rural-to-urban migrants who work as street vendors in cities. Studies show that the majority of these workers are female who work as seasonal or short-term migrants (daily, monthly) (Rolf 2011; Rolf et al. 2009; Nguyen 2017). Another study, conducted in Hanoi, showed that, although the proportion of female migrants tends to increase among migrants workers employed in the formal sector, men still dominate. These studies point to clear patterns linking types of employment in urban areas to age groups and genders.

Migration studies have repeatedly shown that the member of rural households who migrate to cities have a higher educational level than those who don't. Rural migrants are often the members of these households that display the best employability characteristics: young, single, healthy, and with the highest education. This underpins households' decision to arrange for these members to go to the city to work as the evaluation that they will bring the highest economic benefits to their family (Le and Nguyen 2011). Some authors also suggest that the education of permanent migrants is often better. The majority of short-term migrants have a high school degree or a lower educational level. Those migrants who come to the city with a low education and no training tend to work as street vendors or in other informal, low-income jobs such as domestic workers, carpenters, tailors, porters, or motorbike taxi that provide them with no social insurance. The education level of female rural-to-urban migrant workers is slightly lower than that of men, and women tend to do more business activities than men (Dinh 2010; Luu and Nguyen 2010; Nguyen 2016).

There are several possible reasons why a person or a household living in rural areas takes the decision to migrate to an urban area. The main and often the first reason relates to individuals' or families' difficult economic situation. Migrant workers mainly move from rural-to-urban areas to seek employment in order to increase their families' income. The difficult economic situations that households face, and which motivate the decision to send one or more member to work in the city, can be due to a rural areas' lack of jobs, lack of arable land, or low income (Chu 2005; Luu and Nguyen 2010; UN Vietnam 2010; Nguyen 2017; Pham and Tran 2015a, b, 2016). Rural individuals and families can also decide to move to an urban area to improve the quality of life, the quality of education and health care for the rest of the family who are still in the countryside, by sending them remittance (Elias 2010; Luu and Nguyen et al. 2010; Vu 2013). In all these cases, migration is generally considered a livelihood strategy of the whole family and is generally the outcome of a collective decision and agreement among family members (Luu and Nguyen et al. 2010; Vu 2013).

Migrants often start looking for work immediately after arriving in their urban destination or have applied for a job before arriving. They tend to concentrate in certain occupations which are not guaranteed jobs with low salary. Those who do not sign labor contracts are not entitled to social benefits, health insurance, or unemployment insurance (Dao 2014).

Compared to local residents, migrant workers' living conditions are often much worse. They have to live in small, cramped, crowded housing with poor sanitation facilities (Dao 2009; La and Nguyen 2011; UN Vietnam 2010; Phan 2017). In addition, supplying electricity and water for migrant workers who rent house/apartment is not a guarantee. They have to pay a higher price compared to local residents for accessing clean water and electricity. The apartments/houses where they rent are very simple with metal roofs and as less home equipment as possible and polluted environment because they consider this place as their temporary shelter (Bui 2015; Nguyen 2012). However, they tend to accept living in such conditions because they need to live close to their workplace or because the rental price is cheap. For workers in the industrial parks, there is often no dormitory. And when companies offer them such accommodation, some migrants feel inconvenient with the time regulations of the dormitory as their working time is in shift (Le 2017).

4 Public Spaces and Social Integration

4.1 Social Prejudices Against Migrant Workers

The International Labor Organization (ILO) defines the concept of discrimination as being based on fair standards and is protected by law (ILO 2012). Follow that, discrimination is any distinction, exclusion or preference made on the base of race, color, sex, religion, political opinion, national extraction, or social origin (among other characteristics), which has the effect of nullifying or impairing equality of opportunity and treatment in employment or occupation.[2] Generally speaking,

[2]See Discrimination (Employment and Occupation) Convention, 1958 (No. 111) and its accompanying Recommendation (No. 111), Article 1(1a)). As of May 2003, ratified by 158 of the ILO's 176 member States.

compared to foreign scholars Vietnamese social scientists are paying scant attention to local people's prejudices migrants. And yet, according to a study conducted in China by Mou et al (2011), besides lack of access to social security services such as education, health, social insurance, etc., discrimination, segregation, and discriminatory or unfair treatment are also major causes of mental health problems for migrant workers in Shenzhen.

Several studies concerned with rural-to-urban migration in Vietnam report employment prejudices toward migrant workers. They show that there are urban jobs that are implicitly reserved for rural migrants. It means that local residents do not desire to work these jobs, and some people claim that these jobs are more suitable to migrants than local ones. These jobs are often in the informal sector. They are disregarded by society and are underpaid, hard, dusty, and sometimes toxic, sometimes unsafe. Employees taking on these jobs are also often mistreated by their employers (Pham and Tran 2015a, b; Le and Nguyen 2011). Discrimination, however, differs according to the type of employment groups. The group who most usually experiences stigma is construction workers, while the second position belongs to the group of workers in industrial zones. Migrants who work in small business are the less discriminated (Pham and Tran 2015a, b).

Differences in culture and lifestyles and discrimination are barriers to migrants' participation in and integration into the urban social life. There are two aspects of stigma: experienced directly and perceived. Pham and Tran (2015a, b) report that about one-third of the migrants working in the informal sector which they studied have had a direct stigma experience in the form of bullying, threatening, and discrimination. Nearly 20% have been shunned by local people or despised by government and police representatives. In this study, the authors also found out that most of migrants never or rarely visit the local resident's houses, government representative's houses as well as union representative's houses. Besides, these authors found that the majority of workers feel apprehensive about the unfriendly attitude of people at work, and fear of discrimination at work. According to them, these problems are particularly salient among young women, freelance workers, and street vendors. They conclude that over half of the migrants they studied don't feel confident in interacting with local people, avoid being involved in social activities, and are afraid to interact with local authorities and mass organizations. This hampers the expansion of migrants' social networks in cities. In another study from Dang et al. (2012), there is a minority of migrants who are invited to the neighborhood meeting, to local festival, or to meeting of local government. It seems to be a trait of stigmatization.

Domestic work and street vendors are two of the urban job which urban resident considered as "migrant jobs." Street vendors are groups that use public spaces to practice their jobs. They are subject to many prejudices, stemming both from society and state (Tran 2009; Nguyen 2017; Turner and Schoenberger 2012). Regarding society's discrimination, they experience the common stigmatization with other groups of rural migrant workers. Moreover, they are always worried about the legislation of their job (Tran 2009) because of some city government's regulations about migrants workers in general and vendor workers in particular. Nguyen (2017) reports that there were a number of regulation related to street vendor which specify the responsibility of all levels of government in counting, following and arranging the space and training about hygiene and food safety for street vendor. However, those regulations are not implemented by the government. Therefore, the street vendors lack government's supports (Nguyen 2017).

The situation is even more difficult for domestic workers, many of whom live with the host families that control them strictly. This limits communication and integration with urban communities, and reduces the growth of their social networks. Moreover, this work does not receive formal protection from the law (Dao 2009). For their part, factory workers tend to live within a closed environment. Workers often go out of their factory areas and accommodation. They can interact daily with local peoples, neighbors, and landlords but feel a distance between themselves and the local people around them. Their participation with local mass organizations or organizations is very limited (Nghiem 2003; Nguyen 2015).

Research conducted in China shows that prejudices against migrant workers in big cities are also reflected by the media, which conveys negative messages about them (Wong et al. 2007). They are considered a threat to the stability of society and cause of the increase in crime rate in the city. They are also considered to compete with unemployed locals. The public perceives them as stupid and ignorant. As a result, studies conducted in China and Vietnam have shown that migrant workers can feel despised by and alienated from urban people (Wong et al. 2007; Le and Nguyen 2011).

In Vietnam and China, an important institutional factor influences and reinforces the stigmatization and social exclusion of migrants: the household registration policy. This policy prevents migrants from being treated as equal to urban people in accessing public services and integrating into city life (UN Vietnam 2010; Pham and Tran 2015a, b; Houghton 2018). This is echoed in a study of Chinese immigrants by Du et al. (2018) showing that while migrants feel increasingly familiar with the city over time, few feel connected with the city or consider themselves as urbanites. The household registration policy limits the forming of a social cohesion between migrants and non-migrants in cities and is a barrier that holds migrants from asserting themselves as urban people.

4.2 Social Network and Inclusion

At the most general level, social inclusion refers to the ability of society members to participate fully in social processes and get all benefits and opportunities in society, including economic, health, and spiritual benefits, along with access to essential social services such as employment, housing, education, healthcare and legal aid (UNICEF and UNESCO 2013; Dang et al. 2012; Pham and Tran 2016; Tran 2014).

Researches in Vietnam and in other countries have investigated the question of social network and social inclusion by documenting whom migrants make friends within cities (other migrants or others), who they interact with and who they rely on to deal with the difficulties they meet, what activities they participate in at home and in their working place, and how they spend their leisure time. (La and Nguyen 2011; Nguyen and Ta 2013; Pham 2013).

The most important role of social network toward migrants is information and job opportunities. Studies showed that migrants rely on personal and fellowmen relationships to find jobs and settle in cities during the early phases of migration. Besides, network of friends and fellowmen is a pulling factor that encourages individual to make migration decision. It also helps them adapt to their new place (Fan 2003; Nghiem 2003; Pham 2013).

Luu and Nguyen (2009) stated that there is a large amount of migrants who move to urban areas do not register as temporary residents. This makes the government to face many difficulties in managing. This also brings some disadvantages for migrants when they live and work in cities. According to these authors, those migrants who do not register as temporary residents in their urban destination are in a much lower social position in cities. This seems to hamper their capacity to develop a social network in the city which, in turn, negatively affects the overall urban cohesion (Luu and Nguyen 2009).

Using free time is a way to maintain and develop social network and social inclusion. However, studies show that migrants' main activities when they are off work are sleeping and watching television. Other activities include chatting, vising friends, and shopping. Furthermore, migrants spend only limited time participating in social activities such as gathering together to cook, wander, drink coffee (La and Nguyen 2011; Nguyen and Ta 2013). When migrants visit or are visited by friends, these are mainly other migrants. Overall, this population has limited relationship with local people or local authorities (Nghiem 2003; Pham 2013). Reflecting on their low level of social integration, migrants rarely participate in neighborhood or workplace activities. It is worth mentioning that this trend is more salient in Hanoi than in HCMC. Migrants' limited social networks underpin their tendency to seek help from roommates and migrants when faced with difficulties. They rarely reach out to local people or authorities in these situations (Le 2017; Pham and Tran 2015a, b).

4.3 Public Spaces

Drummond (2000) states that: "Public space is an external zone that belongs to the whole community, although it is regulated by prevailing social rules and legal provisions." The concept of public spaces does not only refer to large and majestic spaces such as large parks or spaces of political representation. It also includes smaller and more mundane spaces such as sidewalks and flower gardens (Kim 2012). Many studies also point out that the functions of public space are not limited to entertainment functions but also accommodated commercial activities conducted by a part of the population (Nguyen 2017).

Up to recently, few studies looked at rural migrants' relationship to urban public spaces in Vietnam, but for a few studies about entertainment spaces in industrial zones surrounding cities (where most workers are rural migrants). Research on workers' life in industrial zones points to this population's limited opportunities to take part in cultural and entertainment activities (Hoang 2012; Nguyen 2012; Nguyen and Ta 2013; Pham 2013). Echoing the situation of rural-to-urban migrants belonging to other employment groups, industrial workers work long hours and therefore have very limited leisure time. They report being exhausted after work and tend to use the limited free time that they have to meet basic needs such as eating, bathing, doing laundry, and sleeping. Many reports having no energy to even read newspapers, listen to the radio, or watch TV (Bui 2009; Hoang 2012; Pham 2013).

Yet, another break on industrial workers' ability to take part in cultural and entertainment activities is the lack of adequate and accessible public spaces in and around industrial zones (Bui 2009; Nguyen 2012; Pham 2013). In major industrial parks, export processing zones or industrial clusters, there are few outdoor entertainment places such as sports fields, table tennis, or clubs (Nguyen 2012; Pham 2013). When enterprises have trade unions or mass organizations (such as youth unions), the life of workers may be more diversified but it remains very limited (Nguyen 2012).

As mentioned above, public spaces not only serve purposes of entertainment, relaxation, social interaction. In Vietnam in particular, they are also used for commercial activities, notably those conducted by street vendors who are mainly migrants. Peddlers conduct their business in public spaces like sidewalks, roadways, alleys, vacant spaces near traditional markets, residential areas, and schools. However, in the current context, the use of such public spaces for street vending is often prohibited, banished, or restricted. In order to access and use public spaces, street vendors often have to use their social networks and sometimes "tactics" (de Certeau 1990) to deal with policies, management, and communities' discrimination. The role of these social connections and finesse is well articulated by studies conducted by Vietnamese and foreign scholars (Nguyen 2017; Turner and Schoenberger 2012).

5 Conclusion

Migration is a crucial process, although it is beneficial for individuals, community, and economy. Unskilled labor migrants include young, low-educated, low-skilled, and unmarried people. Their living and working conditions are quite harsh. Studies have shown that migrants in Vietnam and surrounding countries are under pressure, difficulties in living, employment, income, and social inclusion. They face various forms of discrimination (institutional, economic, social) which, like a large part of their lives, are not always visible to the rest of the population, especially in large cities. Stigma puts pressure on migrants and limits their capacity to develop social networks in cities, access urban services, and integrate into urban life. Public spaces are seen as a way, a means to help immigrants relax, reproduce labor power as well as expand social network and social inclusion. However, the lack of public space in areas of cities where migrants concentrate, and aspects of cultural and policy discrimination like household registration, come along with lack of spare time, tough work that has created restrictions on social inclusion of migrants. This reduces the effectiveness of the migration process.

Acknowledgements We would like to thank the Tryspaces project (http://tryspaces.org/presentation-du-projet-tryspaces/) sponsored by the Social Science and Humanities Research Council for allowing us to conduct the research and write this article. We would also like to thank Dr. Prof. Danielle Labbé, School of Urban Planning, University of Montreal for providing advice, support, comments in the research and writing process.

References

Bui, Thi Thanh Ha (2009) Công nhân nhập cư và việc tìm kiếm bạn đời (Migrant workers and finding a partner). Journal of Sociology, 2:41–50

Bui, Dac Thanh Nam, Vu, Thi Hoang Lan & Le, Thi Kim Anh (2013) Những rào cản ảnh hưởng đến tiếp cận và sử dụng dịch vụ y tế của người di cư mùa vụ độ tuổi 18–50 tại phường Phúc Xá, Ba Đình, Hà Nội (Barriers affecting access to and use of health services by seasonal migrants aged 18–50 in Phuc Xa Ward, Ba Dinh District, Hanoi). Journal of Public Health, 27(27):23–28

Bui, Sy Loi (2013) Định hướng hoàn thiện chính sách xã hội trong điều kiện di dân nông thôn - thành thị ở nước ta (Orientation to improve social policy in rural-urban migration conditions in Vietnam). Vietnam Journal of Science and Technology, 10:10–13

Bui, Van Dung (2015) Giải quyết vấn đề nhà ở cho người lao động trong các khu công nghiệp ở các tỉnh Bắc Trung bộ hiện nay (Housing issues for workers in industrial zones in North Central Vietnam). Vietnam Social Sciences Journal, 90(5):30–37

Chu, Manh Hung (2005) Vấn đề trẻ em gái giúp việc gia đình tại các thành phố lớn (The problems of girls working as domestic workers in big cities). Journal of Law, Special issue on Gender equality, pp:17–21

Cirera, X, Lakshman, Rajith WD (2017) The impact of export processing zones on employment, wages and labour conditions in developing countries: systematic review. J Dev Eff 9(3):344–360. https://doi.org/10.1080/19439342.2017.1309448

Cu, Chi Loi (2005) Rural to urban migration in Vietnam. In: Impact of socio-economic changes on the livelihoods of people living in poverty in Vietnam, vol 71, pp 115–143. https://doi.org/10.1037//0033-2909.I26.1.78

Dang, Nguyen Anh (1998) Vai trò của mạng lưới di cư xã hội trong quá trình di cư (The role of social networks in the migration process). Journal of Sociology, 2(62):16–24

Dang, Nguyen Anh (2005) Chiều cạnh giới của di dân lao động thời kỳ CNH. HĐH đất nước (The gender dimension of labor migration in the period of industrialization and modernization of the country). Journal of Sociology, 2(90):23–32

Dang, Nguyen Anh (2012) Di dân con lắc và di dân mùa vụ trong giai đoạn phát triển mới của đất nước (Pendulum and seasonal migration during the country's new stage ofdevelopment). Journal of Sociology, 4(120):40–45

Dao, Bich Ha (2009) Hiˆẹn trạn g công viˆẹc và đ̛ ơ i số ng cu a nˇ̛u nhˆạp cư̛ làm giú p viˆẹc nhà tại Thành phố Hồ Chí Minh. (Female migrants working as Domestic helpers in Ho Chi Minh City). Journal of Sociology, 106(2):51–58

Dao, Mong Diep (2014) Pháp luật về Lao động là người giúp việc gia đình và kiến nghị hoàn thiện (The Labor Law about domestic workers and recommendations). Journal of Law, 12:3–8

Dao, Thi Mai Ngoc (2016) Lao động giúp việc gia đình tại Việt Nam (Domestic workers in Vietnam). Vietnam Social Sciences Journal, 6(103): 56–63

Dinh, Quang Ha (2013) Di dân tự do ở Việt Nam: thực trạng và giải pháp. (Spontaneous migration in Vietnam: Situation and Solution). Journal of Vietnam Social Sciences, 11(72):1–6

Dinh, Van Thong (2010) Di dân ngoạı tinh vào thành phố Hà Nộı : Vấn đề đặt ra và giai pháp. (Migration into the city of Hanoi: Problems and solutions). VNU Journal of Science: Economics and Business, 26(3):173–180

Drummond, Lisa B.W. (2000) Street scenes: practices of public and private space in urban Vietnam. Urban Studies, 37(12): 2377–2391. https://doi.org/10.1080/00420980020002850

Du, Huimin, Li Si-ming, Hao Pu (2018) Anyway, you are an outsider: Temporary migrants in urban China. Urban Studies, 55(14):3185–3201

Elias, Rebecca, & Jean Francois Trinh Tan (2010) Remittance Sending Behavior of Internal Migrants in Vietnam: A Mixed Methods Approach. [available online, but not downloadable apparently)

Fan, Cindy C (2003) Rural–Urban migration and gender division of labor in transitional China. International Journal of Urban and Regional Research, 27(March): 24–47

Hàng rong trong lịch sử kinh tế - xã hội Hà Nội (Street vendors in the socio-economic history of Hanoi). Journal of Sociology, 2(138):57–69

Haughton, Jonathan, Sun Wendi, Le Thi Thanh Loan (2018) Discrimination against migrants in Urban Việt Nam. International Advances in Economic Research 24(3).211–232

Hoang, Thi Nga (2012) Tác động chính sách xã hội tới đời sống nữ công nhân lao động Trong các DN ngoài nhà nước hiện nay (Impact of social policies on the lives of female workers in non-state enterprises today). Journal of Sociology 120(4):46–63

ILO (2012) Action-oriented research on gender equality and the working and living conditions of garment factory workers in Cambodia ILO Regional Office for Asia and the Pacific. http://ilo.org/wcmsp5/groups/public/---asia/---ro-bangkok/---sro-bangkok/documents/publication/wcms_204166.pdf

Jensen, Rolf and Donald M. Peppard (2011) Người bán hàng rong ở Hà Nội - Một cái nhìn về khu vực phi chính thức ở thành phố (Street vendors in Hanoi - A look at the informal sector in the city). Journal of Sociology 4(76):36–47

Jensen, Rolf, Jensen, Donald M. Peppard Jr &, Vu Thi Minh Thang (2009). Di cư tuần hoàn của phụ nữ ở Việt Nam: Một nghiên cứu về người bán hàng rong tại Hà Nội (Circulating migration of women in Vietnam: A study of street vendors in Hanoi). Journal of Sociology 2:59–71

Kim, Annette M. (2012). The mixed-use sidewalk: vending and property rights in public space. Journal of the American Planning Association 78(3):225–238. https://doi.org/10.1080/01944363.2012.715504

La, Thi Thu Thuy, Nguyen Thi Phuong Hoa (2011) Thay đổi tâm lý của thanh niên công nhân xuất thân từ nông thôn (Psychological change of young workers those who are from rural areas). Social Science Publishing house. https://doi.org/10.22144/ctu.jvn.2016.562

Le, Bach Duong, Nguyen Thanh Liem (2011) From rural to cities—The impact on social — economy of migration in Vietnam. Institute of Studies and Social development, Labour Publishing House.

Le, Van Son (2014) Lao động di cư nội địa ở Việt Nam hiện nay. (Internal migrant workers in Vietnam nowaday). Vietnam Social Sciences, 74(1):57–66

Le, Thi Hong Nhung. (2017) (Nghiên cứu trường hợp tại KCN Thăng Long - Hà Nội). (Social integration of workers in the industrial zone (Case study of Thang Long industrial zone - Hanoi)) (Liên kết xã hội của công nhân trong khu công nghiệp hiện nay: Nghiên cứu trường hợp KCN Thăng Long - Hà Nội). (PhD Dissertation in Sociology), Học viện Chính trị quốc gia Hồ Chí Minh (Ho Chi Minh National Academy of Politics), Hanoi.

Luong, Ngoc Thuy (2014) Di dân từ nông thôn ra thành thị trong quá trình công nghiệp hóa và đô thị hóa ở nước ta hiện nay (Migration from rural to urban areas during industrialization and urbanization in our country today). Journal of Human studies, 2(71):49–60

Luu, Bich Ngoc, Nguyen Thi Thieng (2009) Tác động của thanh niên di cư đến khu vực phi chính thức tại Hà Nội trong phát triển Kinh tế - xã hội (The impact of young migrants on the informal sector in Hanoi in socio-economic development). Vietnam Studies Conference - the Fourth (p. 647–658)

Luu, Bich Ngoc &, Nguyen, Thi Thieng (2010) Kiến thức, thái độ và hành vi chăm sóc sức khỏe của thanh niên di cư làm việc trong khu vực phi chính thức tại Hà Nội. Tạp chí Xã hội học, số 4

(112):22–35. (Health care knowledge, attitudes and behaviors of young migrants working in the informal sector in Hanoi). Journal of Sociology, 4(112): 22–35

Luu, Bich Ngoc & Nguyen, Thi Thieng (2011) Thanh niên di cư làm việc trong khu vực phi chính thức tại Hà Nội là ai? (Youth migrants working in the informal sector in Hanoi, who are they?). Journal of Economics and Development, 168, 9–16

Meissonnier J (2006) Marchands de rue à Istanbul - présence urbaine d'une offre commerciale en perpétuelle recomposition spatiale. http://www.cete-nordpicardie.developpementdurable.gouv.fr/IMG/pdf/Marchands_de_rue_cle1d7dc8.pdf, accédé le 13 octobre 2011.

Monnet J (2006) Le commerce de rue, ambulant ou informel et ses rapports avec la métropolisation: une ébauche de modélisation. Presses de Sciences Po. No 39 pages 93 à 109, 46. http://www.cairn.info/revue-autrepart-2006-3-page-93.htm (truy cập ngày 8 tháng 9 năm 2013)

Molisa (Ministry of Labour, Invalides and Social Affairs) (2013) Thực trạng việc làm, đời sống của lao động di cư từ nông thôn ra thành thị và khu công nghiệp (Employment, living conditions of migrant workers from rural to urban areas and industrial zones) . Dự án Tình hình di chuyển lao động từ nông thôn ra thành thị và các khu công nghiệp trong bối cảnh Việt Nam gia nhập WTO (Project of Rural - Urban transition and Industrial Zones in Context Viet Nam Joining WTO)

Mou J, Hillier S, Zhang D (2011) Internal migration and depressive symptoms among migrant factory workers in Shenzhen, China. J Commun Psychol 39(2):212–230. https://doi.org/10.1002/jcop

Nghiem, Lien Huong (2003). Tính liên tục của nông thôn - Thành Thị: Cuộc sống của công nhân may di cư tại Hà Nội, Việt Nam. Trong sách Hiện đại và động thái của truyền thống ở Việt Nam: Những cách tiếp cận nhân học (The rural - urban continuity: life of migrant garment workers in Hanoi, Vietnam. In Modernity and the dynamics of tradition in Vietnam: Anthropological approach). Ho Chi Minh City National University Publishing house, p.289

Nguyen, Ha Dong (2016). Việc làm cho người di cư tự do từ nông thôn ra đô thị: Khó khăn và sự thích ứng (Employment for rural-to-urban migrants: Difficulties and adaptation). Master Thesis. University of Social Sciences and Humanities. https://doi.org/10.22144/ctu.jvn.2016.562

Nguyen, Tuan Anh, Le, Thi Mai Trang, & Tran Van Kham (2016) Hoà nhập xã hội của nhóm lao động nghèo nhập cư ở Hà Nội. (Social inclusion of poor migrant labourers in Hanoi). Journal of Sociology, 134:26–35

Nguyen, Thi Phuong Hoa (2012) Những khó khăn trong cuộc sống của công nhân và tác động của nó tới động cơ thành đạt trong lao động của họ (The difficulties of workers' lives and their impact on the motivation for success in their labor). Journal of Psychology, 11(164):61–72

Nguyễn Thị Khoa (2015) Phát triển dịch vụ xã hội cho người lao động tại các khu công nghiệp trên địa bàn thành phố Hồ Chí Minh (Developing Social Service for Worker in Industrial Zone in Ho Chi Minh city). Science & Technology Development, 18, 40–55

Nguyen, Thi Khoa (2016) Phát triển dịch vụ xã hội cho người lao động tại các khu công nghiệp trên địa bàn thành phố Hồ Chí Minh (Developing social services for workers in industrial zones in Ho Chi Minh City). PhD. Dissertation. Can Tho University

Nguyen, Thuy Diem Huong, Ta Thi Thu Thuy (2013) Đời sống vật chất và tinh thần của công nhân nhập cư tại thành phố Hồ Chí Minh hiện nay (The material and spiritual life of migrant workers in Ho Chi Minh City). Journal of Vietnam Social Sciences 10:46–51

Nguyen, Thi Anh Thu (2009) Tâm trạng lo lắng của người bán hàng rong. (The anxiety of street vendors in Hanoi). Journal of Psychology, 121(4):53–58

Nguyen, Tuan Minh (2017) Le commerce ambulant: Une économie populaire (le cas de la capitale Hanoi au Vietnam) [Street vendors: A popular economy (the case of Hanoi capital in Vietnam)]. Thèse de doctorat de l'Université Sorbonne Paris Cité. https://tel.archives-ouvertes.fr/tel-01920355/document.

Nguyen, Van Chinh (2002) Di dân nội địa ở Việt Nam - Các chiến lược sinh tồn và những khuôn mẫu đang thay đổi (Internal migration in Vietnam - Survival strategies and patterns are changing). In Quynh Tran and Nguyen The Nghia (2002). Sustainable Urban Development. Social Science Publishing house. pp:145.

Nguyen, Van Huy et al (2010) Stress and coping among migrant labourers in urban Vietnam: adaptation cycle and healthy vulnerability. International Journal of Migration, Health and Social Care May 2010

Oxfam (2015) Báo cáo tóm tắt: Rào cản pháp luật và thực tiễn đối với người lao động di cư trong tiếp cận an sinh xã hội. (Brief report: Legal and pratique barriers for migrant workers in access to social protection). Hanoi: Hong Duc publishing house

Pasquier-Doumer, Laure Xavier Oudin, Nguyen Thang (2017) The importance of household businesses and the informal sector for inclusive growth in Vietnam and the informal sector. A co-publication of the Vietnam Academy of Social Sciences and the French National Research Institute for Sustainable Development. NXB Thế giới.

Pham, Phuong Lan (2013) Về chính sách xã hội đối với công nhân trong giai đoạn hiện nay (About the social policies for worker in Vietnam nowadays) . Ho Chi Minh city Journal of Social Sciences, 179(7), 80–84

Pham, Van Quyet & Tran, Van Kham (2016) Hòa nhập xã hội của lao động nhập cư nghèo tại các đô thị Việt Nam: phân tích từ góc độ mạng lưới xã hội. (Social inclusion of the poor migrant workers in Vietnamese urbans: a social network based analysis). Vietnam Journal of Science and Technology, 5-2016, 28–33

Pham, Van Quyet & Tran, Van Kham (2015a). Sự kỳ thị đối với lao động nhập cư nghèo tại các đô thị Việt Nam. Tạp chí Khoa học Xã hội Việt Nam, 11(96):43–50

Pham, Van Quyet & Tran, Van Kham (2015b). Hòa nhập xã hội của người di cư tại các đô thị Việt Nam: Hướng đến một mô hình trợ giúp xã hội (Social inclusion of urban migrants in Vietnam: Towards a social assistance model). Journal of Sociology, 2(130):45–58

Pham, Thanh Thoi (2013) Đời sống xã hội của thanh niên nhập cư lao động phổ thông tại các cơ sở sản xuất nhỏ ở TP Hồ Chí Minh. (Social life of young migrant manual workers at small workshops in Ho Chi Minh city). Ho Chi Minh city Open University Journal of Sciences, 33(5):95–107

Phan, Thanh Nguyet (2017) Sự thích ứng của thanh niên nông thôn di cư làm việc tại các khu công nghiệp, khu chế xuất hiện nay. [Young rural migrants adapting to industrial and export processing zones today]. PhD Dissertation in Sociology. Graduate Academy of Social Sciences, Hanoi.

Tran, Nguyet Minh Thu (2014) Một số đặc điểm di cư lao động ở Hà Nội. (Some characteristics of labour migration in Hanoi). Journal of Family and Gender Studies, 1-2014, 54–65

Truong, Van Tuan (2016) Một số đặc trưng của phụ nữ nhập cư tại Thành phố Hồ Chí Minh giai đoạn 1999-2009 (Some characteristic of female immigrants in Ho Chi Minh city, period 1999–2009. HCMC University of Education - Journal of Science, 83(5):126–134

Turner, Sarah and Schoenberger, Laura (2012) Street vendor livelihoods and everyday politics in Hanoi, Vietnam: the seeds of a diverse economy? Urban Stud 49(5): 1027–1044. Retrieved from: https://doi.org/10.1177/0042098011408934

UNICEF and UNESCO (2013) Social inclusion of internal migrants in India. http://unesdoc.unesco.org/images/0022/002237/223702e.pdf.

United Nations in Viet Nam (2010) Di cư trong nước - Cơ hội và thách thức đối với sự phát triển ở Việt Nam. (Internal Migration: Opportunities and challenges for socio-economic development in Viet Nam). Chu'o'ng trình về chính sách kinh tế và xã h^ọi (Programme Coordination Group on Social and Economic Policies). Report. Hanoi. Retrieved fromhttp://www.un.org.vn/en/publications/doc_details/173-di-cu-trong-nuc-co-hi-va-thach-thc-di-vi-s-phat-trin.html

UNFPA and General Statistic Office (2015) Điều tra di cư nội địa quốc gia 2015: Các kết quả chủ yếu (2015 National Internal Migration Survey: Major results)

Vu, Thi Thao (2013). Making a living in rural Vietnam from (Im)mobile livelihoods: a case of women' s migration. Popul Space Place, 102:87–102

Wong, Daniel Fu Keung; Li, Chang Ying and Sssssong, He Xue (2007) Rural migrant workers in urban China: living a marginalised life. International Journal of Social Welfare, 16 (1), 32–40

ReStructuring Urban Space of Hanoi City on the Basis of Urban Mass Transit Development

Nguyen Thi Thanh Mai and Nguyen Thi Mai Chi

Abstract The appearance of mass rapid transit in megacities like Hanoi is now changing the current urban structure and land-use. According to experiences in many cities in the world, urban structure will transfer from central-core structure with spreading development on the basis of street networks designed for private vehicles to poly-centric structure with central development based on the public transportation framework, around stations in order to efficiently exploit transit services. Development potentials of these areas vary depending on the importance of transportation nodes, quality of connecting infrastructure, and urban facilities, as well as the possibility of utilizing land value and job opportunities there. Classification of stations, transit hubs, and affected areas will help policymakers to orientate the development of land, infrastructure, and urban services conforming current conditions of each area. This paper initially studies about the theory and forecast of restructuring Hanoi's urban space with the support of public transportation.

Keywords Spatial structure · Urban transit hub · Transit-oriented development (TOD)

1 Introduction

Developing public transportation mode in urban areas is considered as one of the most important strategies to reach sustainable urban development. Nowadays, in many megacities all over the world, public transit is on the rising trend and has a higher market share (David 2015) in which urban railway takes a crucial role, increases about 2.7% every year (Berger 2014). In China, about 450 subway stations are constructed every year, and by 2020, there will be about 1,050 and 970 km of subway in Beijing and Shanghai, respectively, making their networks to become one of the longest in the world (Salat and Olliver 2017a).

N. T. T. Mai (✉) · N. T. M. Chi
Department of Urban Infrastructure Planning, National University of Civil Engineering, Hanoi, Vietnam
e-mail: maintt@nuce.edu.vn

L. T. T. Huong and G. M. Pomeroy (eds.), *AUC 2019*, Advances in 21st Century Human Settlements, https://doi.org/10.1007/978-981-15-5608-1_31

Most cities owned private automobile vehicles as a key transportation mode before developing the public transportation framework such as the US, Singapore, or Japan. The transformation of urban space will change in the direction that considers public transit as a key mode, along with the change in travel behavior, the way to exploit and use land, the scheme to connect housing and service facilities. Experiences from the world show that developing residential and economic activities concentrated within a distance of less than 1 km around the stations, transit hubs, and corridors will bring a positive result in environment and economic growth. Some countries, such as the United Kingdom, Australia, and Denmark, have made profits by collecting land value taxes in the affected areas around the stations (Wang et al. 2015). New economic centers that are able to be established with their nuclear will be promising transit hubs. These centers which can be considered as new city centers existing along with core center CBD will complete the poly-centric structure of megacities in the post-industrial period—also known as "multi-polar urban agglomeration" (Banister 2011; Cervero 2013). It is obvious that not all stations and hubs are eligible to become large-scale economic-transportation hubs, or lower scale as residential service centers. This depends on the location, traffic importance of the hub, current conditions, potentials of human, land, and employment, as well as the availability of facilities and services in the neighborhood. Local governments will gain the growth's benefits from the sustainable development model based on public transport, if they can categorize the potential centers and construct suitable land development policies along with a strategy to concentrate higher population and employment, strengthen the multi-modal connection.

The Capital—Hanoi belongs to "special" in Vietnam's city ranking, with a total area of 3324.52 km^2, a population of over 7.3 million people (General Statistics Office 2016), with most of mobility by using private vehicles: about 546 thousands of cars and about 5 million motorbikes (Hanoi City's Public Security 2016). In recent years, Hanoi has begun to deploy a number of mass rapid transit routes such as BRT and light rail (Government 2016). Accordingly, with the support of public transportation, the aim to become a multi-polar urban space as in the Hanoi Master Plan will be achieved (Government 2010). The transformation of an urban space structure based on private vehicles to public transport is an indispensable option to achieve economic growth and sustainability. Therefore, the city needs to grasp the trend of structural transformation to the poly-centric system, indicates the potential areas that can be developed based on transit system, appropriately issues policies to adjust land and establish suitable planning options in both city-scale and neighborhood-scale. Then, the government should develop an appropriate land development mechanism, forecast changes in land value, encourage investors to participate in the process of constructing public transport infrastructure, housing, and service facilities through advanced cross-subsidizing schemes.

This is a complex topic; thus, the paper is considered as the initial research about the theory and analysis of structural change forecasts for Hanoi, thereby opening up the next in-depth researches. The methodologies used in this study are collecting and analyzing past-studied researches, correlating comparative analysis, map analysis, and data aggregation from reliable sources. Two specific objectives of the paper: (1)

Clarify the theoretical basis for the classification of economic-transportation centers based on the characteristics of urban transit hubs; (2) Analyze the case of Hanoi through the author's previous research results on spatial change in Hanoi.

2 Theoretical and Empirical Fundamentals of Spatial Structure Transformation Based on Urban Mass Transit Development

2.1 Public Transit Nodes/Stations Typology

It can be considered that urban public transportation network as a system of lines and nodes. Herein, nodes include stations and public transportation hubs. These nodes are not equivalent depending on the extent of importance of their connection in the network, based on the "centralities" of the nodes. The centralities of the node are assessed by these measures: "closeness centrality"—the average distance from the node to other stations in the network; "betweenness centrality"—the number of the shortest connecting routes between the stations passing through the node; "degree centrality"—the total number of routes linking through the node. There will be a higher score in a station/node if this station/node tends to link to other stations in the network with the shortest distance, while the remaining measurements tend to reach to the maximum value. Besides, the attributes to assess the important level of node also depend on the size of the transit volume, trip frequency, or total number of trips at the station per day (Salat and Olliver 2017a).

Transit nodes in public transport network can be divided into three categories:

- *Type 1—Intermodal hub/highly connective hub)*: connecting many mass rapid transit lines (such as metros, light rail, bus, BRT) at city scale, regional scale, and national scale, including connection to airport and railway stations. The volume of transit passenger here is enormous and diverse. In many cases, it can be a complex transit hub integrating many mass rapid transit modes.
- *Type 2—Transit interchange/core transfer station*: concentrating two or three public transit lines, where passengers can transfer to other types of transportation modes. The number of trips and trip frequency are medium or relatively high.
- *Type 3—Single-line station*: there exists only one public transportation line across, always is the radial line starting from the center on which there are stations to stop and pick-up passengers. The number of trips and trip frequency are not much.

This division is quite similar to the classification of Hung and Phuong (2016). He divided into three categories, including public transportation interchanges at regional scale: connecting inter-regional traffic and urban areas through the inter-provincial buses and railway stations; public transportation interchanges at urban scale: connecting on urban main axes; and public transportation interchanges at neighborhood scale: transit centers with one transit route and connecting with private vehicle.

The Calthorpe team proposed a simpler classification for San Diego. The stations were divided into two categories according to the characteristic of the public transportation route on which the station was located. For the main routes of the regional public transport network, there were light railway stations or inter-city/transit bus stations; other stations were located on high-frequency bus routes (urban links) or feeder buses (connecting neighborhoods) with a 10-min distance to the station on the main inter-regional public transport route.

2.2 *Transportation—Economic Center Typology Based on the Classification of Transit Interchanges*

2.2.1 Classify According to the Value-Assessment Framework

The opportunities for growth in the area generated by the stations are not the same in the city's general public transportation network. It is estimated that only about 15% of the station areas have the potential to develop into a high density and mixed economic center, the remaining is less potential for trade growth (Oliver 2017). Therefore, it is essential to have a tool to classify station areas according to their location, characteristics, potential values, and level of development in each neighborhood.

To assess the development potential and classify station areas/public transportation hubs, Salat proposed a value-based evaluation framework, including node value, place value, and market potential value. The node value refers to the level of traffic concentration at the terminal/hub. The place value refers to the quality of utilities, accessibility to the station, service facilities, and the landscape of station area. The market potential value refers to the level of concentration, job creation, population, as well as land availability and other conditions for economic growth. In this way, the station areas are highly connected, easier to access jobs, residential, commercial, and other service activities. Station areas in Type 1 and Type 2 have more potential than Type 3 to become mixed-use, high-density centers or large-scale urban centers.

In fact, there is a complex interaction between values in different station areas depending on the station type and their location in urban space.

- The area around the highly connected public transportation interchanges is located in the city center with the most promising potential, connecting many lines and serving at the urban scale. It has a high place value because of its diverse urban structure, valuable architectural buildings, and heritage. The formation of additional urban rails here will be catalyst for growth of this area.
- Urban-periphery station areas, adjacent to the urban center, where there are many new construction projects combined with urban infrastructure improvement activities. This is also considered as an area with potential for development near the urban core.

- Suburban station areas, usually single-line stations or railway terminals. These areas are considered to have low traffic value and not high place value, due to their separation from the vibrant economic center, modern utilities, and old infrastructure. These areas are less attractive to investors due to low market value.

2.2.2 TOD Concept

The term TOD—Transit-Oriented Development—has become more and more popular recently. "TOD" refers to the general urban spatial structure developing based on public transportation framework. TOD can also be defined as residential neighborhood or mixed-developed commercial areas around stations, within the walking distance about 400–500 m, and be designed in order to maximize the public transportation's travel demand. TOD can be seen as the proper integration between transportation hubs/public transportation interchanges and land-use planning in affected areas nearby stations to efficiently exploit public transportation service and improve accessibility (Suzuki et al. 2013; Calthorpes Associates 1992).

In the research for San Diego, Calhopre group classified TOD, including Urban TOD and Neighborhood TOD (Calthorpes Associates 1992). Urban TOD is established based on the inter-region link-roads, connecting many railway stations, inter-provincial bus terminal, and on the main urban roads. Neighborhood TOD is established based on neighborhood-scale street networks, and around single-line transit stations. Based on the evaluation of transit stations' location in urban space, Calhopre also proposed three types of TOD: TOD inside the urbanized areas—where construction land was filled up and along with redevelopment activities, TOD in urbanizing areas—where it has unused land interwoven with current residential areas, TOD in urban periphery—where it contains a huge unused land reserve to develop a complete TOD structure (Calthorpes Associates 1992).

2.2.3 Core and Branches Network—Poly-centric Urban Structure

Although many global cities such as Tokyo, Seoul or London, Paris, New York are different in terms of geography and economic potential, their urban structure is based on the full development of public transportation network (mainly subway), representing a nuclear—branches structure (Salat and Olliver 2017b). The core of the network is formed on the basis of dense assemblies of highly accessible stations and high-density public transit lines. Employment and population are highly concentrated at the most accessible transportation hubs. The limit of the cores does not exceed 5 km to ensure the cost is not too high in maintaining the high density of stations. Further from this limit, the density of stations' distribution decreased significantly, thus, only single-line stations existed along the branches of the structure (The Federal Transit Administration 2013).

According to the station categorization, large-scale multi-modal hubs can cluster in the network to develop into a new center. Meanwhile, transportation hubs forming in urban centers are likely to boost the economic growth here by increasing transport connections. These are the elements to form a multi-polar structure, with a CBD center and many secondary centers. According to (Japan International Cooperation Agency 2015), secondary centers can also be established from the stations at the intersection of inner city railways, and at suburban stations, where a huge number of passengers concentrate and have a high developing potential. Besides, creating mixed-development corridors with high employment diversity and social activities along the branches starting from the network core, or in areas adjacent to transit stations, with medium-scale are preferable (Cervero 1994).

At the neighborhood scale, enhancing employment and population concentration within a radius of 1,000 meters from the station is needed in order to create a lively, efficient, and diverse development environment with the support of public transport.

3 Hanoi—Spatial Structure Depending on Private Automobile

3.1 *City of Pedestrian and Other Ancient Means*

At the end of the nineteenth century, before the French arrived in Vietnam, Hanoi's urban structure was typical of Asian feudal urbanism. Urban space was mainly limited within a radius of about 2 km, with walking as the most popular traveling mode. Rural residential areas and towns were clustered and scattered in urban space. There were three main areas including inner city area, townships, interwoven with agricultural and handicraft villages. Hanoi City completely changed its appearance when becoming a Western colonial city since the French construction activities were promoted during the first colonial exploitation period (1888–1920) and expanded built-up areas on the overall scale during the second colonial exploitation period (1921–1945) (Bao et al. 2011). The radius of Hanoi had increased to about 3.5 km, with more diverse urban transport modes such as rickshaws, horse-drawn carriages, tramcars (1900), and trains connecting to nearby regions (1902). Based on the urban planning principles from the modern Western at that time, Hanoi's urban structure was divided into many clear functional areas, with Ho Guom (Hoan Kiem Lake) as an administrative center and the residential area was developed mainly to the west of the Imperial Citadel and to the south of Ho Guom. The population density was spread widely in urban space based on the grid plan and low-rise housing constructions along the streets. A number of routes installed the tramcars which connected the administrative center of the Hoan Kiem lake with the wholesale markets and trade centers, and between rural and urban areas at that period. The wholesale markets are listed as followed: Mo, Dong Xuan, Buoi, Cau Giay, and Ha Dong Market (Thanh Nien Homepage 2019).

3.2 *Prominently Bicycle City in the Tendency of Socialist Urban Planning*

From the time of taking over the capital until the 1980s, Hanoi's space mainly developed in the inner city area and gradually extended towards Ring road (RR) 2 (Vinh Tuy Bridge–Minh Khai–Dai La–Truong Chinh–Lang–Buoi–Vo Chi Cong–Nhat Tan Bridge–Vo Nguyen Giap–Truong Sa–Dong Tru Bridge–Ly Son–Gia Lam Underpass–Nguyen Van Linh–Dam Quang Trung–Vinh Tuy Bridge). Bicycle vehicle was the main transportation mode used by government officials and workers, mean while buses and trams are popular transport mode for unskilled laborers and small-scale shop sellers. An urban narrow road network was formed basically on the basis of grid-street system that existed from the French colonial period. Most of the residential areas were developed along a number of radial roads coming from the inner city such as De La Thanh Street, National Road 6 (Nguyen Trai)… and on the agriculture land among RR1 and RR2. The living model considered as a representative of new lifestyle during that period was collective housing quarter, which was "imported" from the previous Socialist countries during the 1960s. They can be listed as: Nguyen Cong Tru, Kim Lien, Trung Tu, Giang Vo and Thanh Xuan collective houses… Besides, Hanoi's spatial structure was planned with clear functional land divisions: two prominent centers are the political-administrative center in Ba Dinh district and the cultural-administrative center in Hoan Kiem district, adjacent to residential areas and small scale factories… Urban road system was classified into several levels, including main roads, collectors and inner/cul-de-sac roads. Residential areas and service facilities were classified into 3 categories, including city-, district- and neighbourhood-level based on a hierarchical street network. The role of public transportation then was unclear and not crucial enough for promoting market potentials in surrounding areas under the context of a socialist country maintaining the centrally planned subsidized economy.

3.3 *Higher Density-Multi-modal City in the Open Market*

From 1986 to now, Hanoi traffic is a mixture of private vehicles in which motorbike-users share the highest rate. Particularly in Hanoi, as of the first quarter of 2019, the number of registered motorbikes was more than 5 million, increased average of 7.66% per year, accounted for over 86% of the number of vehicles traveling in Hanoi (Homepage 2019). The number of cars has also increased significantly in recent years, on average of 10.2% per year for the period 2011–2016, reaching 739,000 vehicles at the beginning of the first quarter of 2019 (Vietnam Registry Department Homepage 2019). Bus service also improves the quality and quantity. In 2008, the bus modal-share ranked second after motorbikes (about 10%) (Musil 2010). In addition, BRT line No. 01 Cat Linh–Ha Dong has been put into operation since 2016, 02 urban railways No. 2 and 3 are still under construction and completion, in line with other vehicles such as bicycles, electric bicycles, electric motorcycles, taxis are also being used.

The urban transportation system has been expanded, completed, and the number of ring roads and the radial roads, together with the traffic network density has increased significantly. The central city space is developed evenly on the basis of the current street network, with motorbikes, as the main transportation means, under the way of door to door access. People choose to travel by motorbike accounting for 67.4% and by car/taxi accounting for 21.8% (Hanoi Department of Transport and Transport Development and Strategy Institute (Ministry of Transport) 2016). The number of bus routes has increased significantly, covering about 71.7% out of Hanoi's space (Hanoi Urban Transport Development Project 2016), connecting many long-distance route in the urban areas and from center to many satellite cities and towns in the expanded area of Hanoi.

Due to rapid urbanization, the city space is constantly expanding through the Ring road 2, 3 and then reach Ring road 4. It is continuing to cover the entire agricultural space of Ha Tay province and other important cities of this province such as Hoa Lac, Son Tay, and Xuan Mai (according to the overall planning of the capital in 2010). Hundreds of new urban projects appeared around the Ring road 3 toward the Ring road 4 and the north of the Red River, along with many new city centers such as the Western region (Trung Hoa–Nhan Chinh, Kengnam), the Southwest region (Ha Dong), the north of Red River (Long Bien). Residential areas with a hierarchical planning structure tent to shift into a non-hierarchical structure, with the development of land and commercial services along the main roads of the city. Neighborhood unit tends to develop mixed use rather than to remain pure residential function as before. The system of public service centers also changes in perceptions: from the type of social welfare to goods services, operating under the market economy rule.

4 Stronger Returns of Public Transport and Future Poly-centric Structure Based on Mass Transit System

4.1 Development of Public Transport and Mass Transit Network

The Hanoi's Master Construction Plan in 2011 and the Hanoi's General Transportation Plan in 2016 clearly stated that developing public transport system is one of the most important strategies, in which urban railway (UR) plays as a key role. The public transport system including both UR and BRT will contribute to the establishment of an efficient combining traffic network. 4 out of 8 URs are radial links connecting the city center to satellite cities, such as lines 1, 2A, 3 and 5, and the remaining such as lines 4, 6, 7, and 8 principally operate within the inner city. BRT lines include 8 lines and 3 transitional BRT lines will be implemented before having a complete UR network. Some extended BRT routes such as lines 1, 2, and 3, will

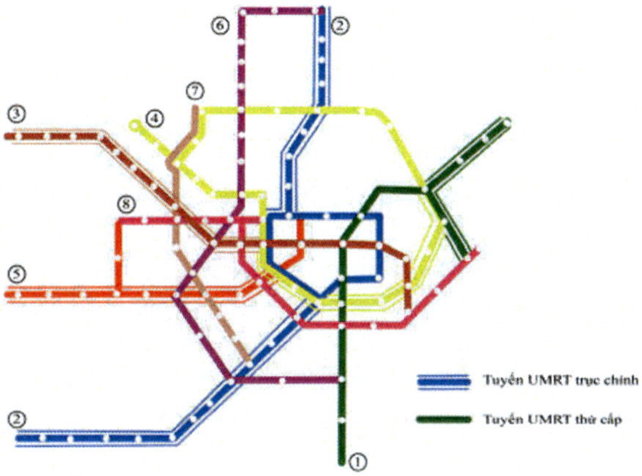

Fig. 1 Classification of main axis and the secondary urban railway lines in Hanoi (*Source* Research team, 2017)

support URs to connect existing city to rural towns and satellite cities in the context of underdeveloped railway infrastructure. In the future, BRT line No. 9 will become UR No. 4: Me Linh–Sai Dong–Lien Ha; line 10 will transform to be UR No. 8: Son Dong–Mai Dich–Duong Xa. BRT No. 11 will be studied to become a monorail line. Similarly, some railway lines will apply monorail mode if they respond fully to the technical demands (cross section, travel demand, etc.). According to our research, the urban railway network will be classified into two types: main axis connecting to the expanded urban area and the secondary lines serving in the existing city.

According to our research, the urban railway network will be classified into two types: main axis connecting to the expanded city area and the secondary line served in the existing city areas (Fig. 1).

4.2 Urban Transportation Interchanges and Future Transit Centers

In current, public transportation interchanges in Hanoi mainly support for bus, one BRT line, and two urban railways which are under construction, Hanoi bus network consists of 75 routes, with 2210 stops, 370 bus shelters, 5 interchanges, 77 terminals including two large-scale bus interchange Long Bien and Cau Giay. BRT route No. 1 is with a length of 14.8 km with 21 bus shelters, including 2 terminals at Cat Linh and Yen Nghia. Urban railway 2A is 13.1 km long with 12 stations, connecting with BRT 01 at Yen Nghia, Van Khe, La Khe, and Cat Linh stations. The route connects with metro line 3 at Cat Linh station, Metro line No. 2 at Thuong Dinh station and connects with Yen Nghia bus station (the Southwest of the capital). Urban railway 3 is 12.5 km long with 12 stations. In particular, Cat Linh station becomes an interchange connecting Line 2 A with BRT01 and the national railway. Besides, the Line 3 also connects to Line 1 (Yen Vien–Ngoc Hoi) at Hanoi Station. When the urban railway and BRT network are completely constructed, it will create more high-density multi-modal transit interchanges (Japan International Cooperation Agency 2007).

Hanoi's bus terminal plan was approved in 2018, intended to move four existing inter-provincial bus terminal outside of the center and convert their functions into public transit interchanges, incorporating with public parking spaces (Hanoi People Committee 2019). 07 inter-provincial bus terminals will be newly planned in the periphery of Hanoi's center, and 05 inter-provincial bus terminals will be located in satellite cities according to the master planning of these cities. In addition, there are urban bus stations located in urban centers of district towns and ecological towns. Theoretically, the system of bus terminals will effectively connect with urban railways and BRT to become a major multi-modal transportation hub in Hanoi in the future.

4.3 Forecast for the City of a Poly-centric Model and Corridor Development Based on Mass Transit

4.3.1 Legal Factors Affecting Hanoi's Overall Urban Structure

Construction Plan for Hanoi Capital Region approved under Decision 490/QD-TTg signed on May 05, 2008 and the adjustment plan of the capital approved under Decision 768/QD-TTg signed on May 06, 2016 had confirmed the organic relations between the capital and adjacent areas (nine provinces) in socioeconomic development. The satellite cities of Hanoi have been affirmed as a gateway center of regional connection: Xuan Mai, Hoa Lac, Son Tay, Soc Son, Phu Xuyen–Phu Minh.

Hanoi's Master Construction Plan to 2030, vision to 2050 (under Decision 1259/QD-TTg signed in 2010) clearly pointed out the viewpoint of developing a model of urban agglomeration consisting of a CBD, five satellite cities, district towns, and ecological towns in the green corridors. Meanwhile, the plan also identified the boundary of Hanoi's core center by the green belt and established a center fringe along the Ring road 4 which consists of many emerging centers such as Dan Phuong, Hoai Duc, Ha Dong, Thuong Tin and Me Linh, Dong Anh, Gia Lam.

Hanoi's Transport planning approved under Decision 519/QD-TTg signed on March 31, 2016 proposed to strengthen the connection of Hanoi center by the UR network and BRT which will support the urban road network. Some radial lines will connect to satellite cities such as Line 1, 2A, Line 5, Ring Road 3 and 4, with the support of some extended BRT lines. The terminal of these railway lines ends at satellite cities and runs through towns in the green corridor (Fig. 2).

4.3.2 Urban Center Typology Following Poly-centric Model Based on Public Transportation in Hanoi

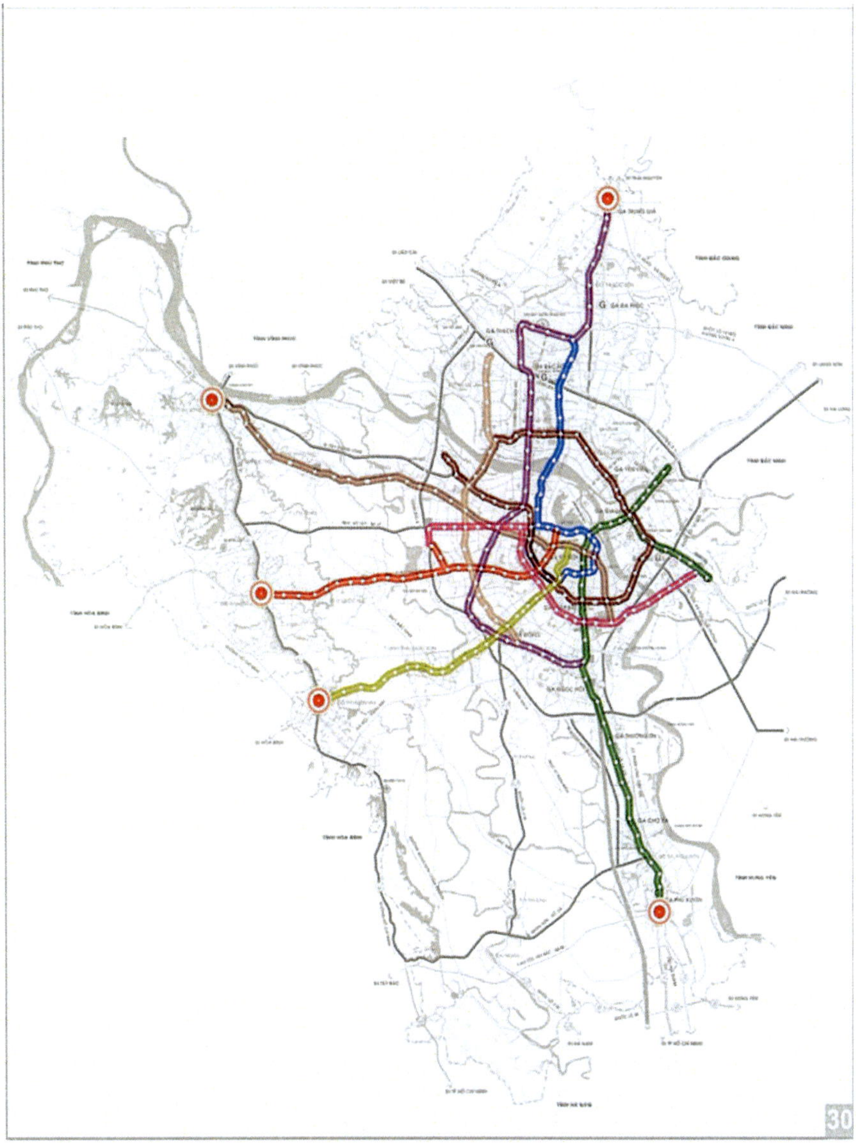

Fig. 2 Proposal for economic-transport development corridor serviced by UR framework (by Research team, 2017)

High-density economic transport core: located in the inner city, within Ring road 4, where UMRT, BRT lines densely concentrate on a high number of transit hubs. These areas own an enormous potential growth to continue being the most important CBD of the capital.

Inter-regional secondary center: developed in satellite cities where transport infrastructure, economic, employment, and population potential are available. These are also planned with an estimated population from 210,000 to 750,000 people (equivalent to urban areas of categories II and I), playing an important role in the development of economic, cultural, and scientific technology and service industry. In the future, these cities will become the inter-regional transportation hubs which locate the transit terminal to Hanoi center and connect to their future own specialized transit network. For instance, Soc Son, which consists of Noi Bai international airport and the public transport network from the core city, will become an economic-transit hub at regional and urban level.

City-level urban center: Established in newly emerging urban centers (along Ring road 4). Herein, there are key transit interchanges, which are the intersections of UMRT, BRT, and monorails and P + R transit parking spaces, such as Yen Nghia, Ngoc Hoi, and Dong Anh station. In addition, a number of existing city-level centers within the core city will become economic-transport hubs based on current economic and market potentials, and with the support of public transport, such as Cat Linh, Long Bien, and Cau Giay stations.

Neighborhood-level urban center: Established in existing urban centers and towns in the green corridor of Hanoi with the support of radial urban railways. The TODs here will enhance connectivity with existing transportation network, and boost economic activities, jobs creation, and services related to agriculture, handicrafts, or tourism. Some examples are Phung, Quoc Oai, Phuc Tho, and Chuc Son. Besides, the neighborhood-level urban centers also appear in the center of current residential in the core city. They will be planned as future stations, which are expected to have high market potential by attracting large numbers of commuters, such as My Dinh, Giap Bat–Nuoc Ngam, and Yen So.

In addition, many single-line stations on the main radial and secondary ring lines are only considered as pick-up and drop-off stops (Tables 1, 2 and 3).

4.3.3 Forecasting the Spatial Impact on Station Areas in Urban Space

The formation of new urban centers and the classification of urban centers under the impact of public transport can be considered as the transformations in Hanoi's urban structure. In addition, depending on the characteristics and location of the stations in the network, the development areas surrounding them will have different impacts on the status of population, land-use, infrastructure, and development potential in the future. Based on the urbanized spatial belts, it can be divided into three types of TOD by location: TOD located in the historic inner city with redevelopment activities, TOD in the areas under strong urbanization process, and TOD in a new development area.

Table 1 Theoretical attributes to evaluate the important role of node/transportation hub-based on its center level

Theoretical attributes	Value (high score)	Center level	Node's important role
Average distance between the node and other nodes in the network (Closeness centrality)	Shortest	Highest	Highest
Total number of connecting lines across the node (Degree centrality)	Most	Highest	Highest
Total number of shortest lines connecting stations across the node (Betweenness centrality)	Most	Highest	Highest
Volume of transit passenger (Intensity of node activities)	Large	High	High
Trip frequency	Large	High	High
Total number of trips	Large	High	High

Source Aggregated by research group (2019)

A series of residential TODs along the public transport routes will become the TOD corridor. In Hanoi, some noticeable TOD corridors along the lines with large transit volumes such as line No. 2, No. 4, No. 3, and No. 5 have been already formed the service-commercial axis.

5 Conclusion

It can be said that the orientation of transforming the urban spatial model toward transit development is an indispensable option of megacities with high population concentration and economic activities. The positive effect of this model is to save time and money, limit greenhouse gas emissions, and environmental pollution through concentrating high population and employment density around stations and along transit corridors, and utilizing transit services.

The basic feature of this model is to avoid dispersed planning across the entire urban space, but to centrally develop in potential locations, within the service area of the stations, transit interchanges to ensure passenger trips—from/to work/housing within 30 min. In areas with high market value potential, high connectivity will allow for the development of high density, mixed-use, strengthen accessibility, while reducing and controlling growth is needed in those areas having less potential.

The station categorization not only determines its importance of traffic connection in the network, but also indicates the areas with potential for economic growth with high level of employment and population concentration. Stations with lots of high-quality public transport routes, diversely connecting many transportation modes, and with huge travel demand, will have higher potentials to attract jobs, housing and commercial services. In fact, there exists a typology of economic-transport hubs in

Table 2 Forecast about the classification of transit interchanges based on current conditions and early-stage planning (Research team, 2019)

	Stations/Transit interchanges	Location	Current/Planning	Connecting transportation routes
Level 1—regional transit interchange				
	Son tay	Satellite cities	According to master plan of satellite cities; Hanoi's bus terminal plan	Inter-provincial road network, BRT, UMRT 2
	Hoa Lac			
	Phu Xuyen			
	Soc Son			
	Xuan Mai			
Level 2—major urban transit interchange				
	Cat Linh	Hanoi center	Current	UMRT, BRT, city bus stations
	Yen Nghia			
	Cau Giay			
	Long Bien			
	Inter-provincial Bus Terminal (IBT) Gia Lam	Hanoi center	Current will change the function into transit interchanges according to future plan	
	IBT My Dinh			
	IBT Giap Bat			
	IBT Nuoc Ngam			
	Bus Terminal (BT) Yen So-Transit interchange	Hanoi center	Constructing based on the plan and will change the functions in the future	
	BT Đong Anh	Periphery of Hanoi center	Planning	City bus stations, UMRT, BRT
	BT Yen Nghia			
	BT Northeast Co Bi			
	Southern bus terminal (Ngoc Hoi, Ring road 4)			
	BT Northern (Noi Bai)	Satellite cities, towns		
	BT Western (Son tay 1)			
	BT Northwest (Phung)			
Level 3—neighborhood transit interchange				
	Interchanges established in district towns and ecological towns	Urban centers, ecological towns, district towns	Hanoi's master plan, Hanoi's plan for parking spaces	Bus terminal, connecting to railways and radial BRT

(continued)

Table 2 (continued)

	Stations/Transit interchanges	Location	Current/Planning	Connecting transportation routes
	Other transit interchanges	The development belt of central Hanoi	Hanoi's urban transportation Plan and Hanoi's plan for park and ride parking spaces (13 parking spaces)	Ring roads, radial-axis roads near transportation hubs and transit interchanges

Table 3 Classification of TOD based on its location in Hanoi urban spatial structure (Research team, 2017)

Classification of TOD based on location	Location	Place characteristics	Types of transit	Actions
Historic inner city with redevelopment activities	On the right bank of the Red River, within RR2, expected population: 0.8 million people	Long time development, store many cultural values, architectural works, infrastructure; less vacant land; high land value, great economic potential Difficult to form a complete TOD	Metros, urban BRT	Controlling density and economic activities; enhancing walking and bicycles activities to stations by current pedestrian spaces or converting the function of some motorized roads
Areas under rapid urbanizing process	From RR2 to Nhue river, urban areas locating in the Northern of red River, expected population: 0.8–0.9 million people	Some vacant and unused land areas are stuck in urban areas	UMRT, radial BRT, ring BRT	Developing TOD in current sites, dense functions, supporting urban services, commercials, and transportation reaching current nearby residential and villages
Newly developing areas	Satellites cities, megalopolis in the east of RR4 (estimated population: 1.2–1.4 million people), old urban areas, towns in green corridor, expanded urban areas	Land is available, low construction and population density, having potential economic and infrastructure developing factors. Be able to renovate and develop into TOD	UMRT, radial BRT	Developing TOD around stations, transforming land-use, function, and encouraging economic activities, urban services, and multi-modal transportation

the city's transit network. Fully understanding the typology will help policymakers and governors can define the areas having developing potentials which are suitable to its current conditions and also discover the deficiencies and improve to reach the location's stature.

As above mentioned, an urban structure based on public transport will tend to focus developing in the centers, along the transit corridors, and toward the multi-polar city model. The distribution of economic activities tends to decrease from the station area where it can easily access traffic, quality facilities, urban infrastructure, and high market potential. The space between the high-density concentrated development areas will be strictly controlled, not over-developed to ensure efficient economic development, infrastructure exploitation, energy consumption reduction, and environmental improvement.

Hanoi urban areas are formed as a complex structure, through many periods. Hanoi's urban space gradually expands from the core city beyond the Ring road 2, 3 and reaches to Ring road 4. Its space basically develops outstretched and be a symbol of automobile-dependent city. Since the early 2000s, the structure and urban space of Hanoi have some transformations due to the impact of the market economy: residential and commercial activities tend to locate along the main roads. This process takes place quite spontaneously due to limitations of urban management in reality (the incomplete detailed planning).

In the future, Hanoi aims to become a multi-polar city structure including a CBD and a system of secondary centers located in satellite cities, suburban towns, and areas with potential on transit connections, economic development, high population density in central Hanoi. The centers will be classified according to the characteristics of the stations, transit interchanges, their place value, and the potential of land exploitation, including Regional transit interchanges, key public transportation hubs, and single-line transit stations. Besides, there will be formations of economic-transport corridors which connect the station areas along transit lines. Especially, transportation interchanges at the intersections of the ring roads, and the main radial routes should be highlighted because new and potential nuclear centers tend to be formed here, where several high-complexity transit hubs are concentrated.

The paper provides initial and theoretical analysis of Hanoi's urban structural transformation due to the impact of public transit systems in overall scale. In order to more accurately forecast the impacts, more in-depth studies, based on the stations which have already been designed on detailed planning, are needed. The classification of the transit stations and the development centers from the transit cores also requires more in-depth investigation about current status. Assessment tools, computer technology, specialized programs, and software should be studied to assist in accurately quantifying the complex impacts of transit interchanges/stations. It will support to set up more reliable database for governors and policymakers to make appropriate decisions to efficiently utilize land exploitation.

References

Banister D (2011) Cities, mobility and climate change. J Transp Geogr 19–6:1538–1546

Bao TQ, Mai NTT et al (2011) In: Bao TQ (ed) Hanoi architecture and planning in French Colonial period. Construction Publishing House

Berger R (2014) UNIFE world rail market study–forecast 2018 to 2023

Calthorpes Associates (1992) City of San Diego land guidance system: transit—oriented development design guidelines. The City Council, California

Cervero R (1994) Rail-oriented office development in California: how successful? Transp Q 48:33–44

Cervero R (2013) Linking urban transport and land use in developing countries. J Transp Land Use 6–1:7–24

David S (2015) Vale: transit oriented development, integration of land use and transport, and pedestrian accessibility: combining node-place model with pedestrian shed ratio to evaluate and classify station areas in Lisbon. J Transp Geogr 45:70–80

General Statistics Office (2016) Vietnam statistical book

Government (2010) Decision 1259/QĐ-TTg Hanoi on master plan to 2030 and vision to 2050. Approved by Prime Minister

Government (2016) Decision 517/QĐ-TTg on Hanoi transportation plan approved by Prime Minister, Hanoi

Hanoi City's Public Security (2016)

Hanoi Department of Transport and Transport Development and Strategy Institute (Ministry of Transport) (2016). Project Strengthening management of private vehicles to reduce traffic congestion in Hanoi City

Hanoi People Committee (2019) Planning for parking spaces, logistics centrés and roadside stations in Hanoi to 2030 and vision to 2050 under Decision No. 15/NQ-HĐND signed in 05/12/2018

Hanoi Urban Transport Development Project (2016). Mid-term report—feasibility study for BRT (HUTDP—Feasibility)

Hung HN, Phuong HT (2016) Mass rapid transit in urban development. Construction Publishing House, Hanoi

Japan International Cooperation Agency (2007) The comprehensive urban development programme in Hanoi capital city of the Socialist Republic of Vietnam (HAIDEP). Final report. Hanoi

Japan International Cooperation Agency (2015) Project for studying the implementation of integrated UMRT and urban development for Hanoi in Vietnam

Mai NTT (ed) (2016–2017) Restructuring urban structure based on the development of public transport network in Hanoi city. Science and Technology Research Project of Ministry of Education. No: B2016-XDA-06, Hanoi

Musil C (2010) Building a public transportation on system in Hanoi: between emergency and constraints. Conference paper

Oliver G (2017) Presentation in MDTF on sustainable urbanization. The China—World Bank Trust Fund

Salat S, Olliver G (2017a) Transforming the urban space through transit oriented development. The 3V approach. The World Bank

Salat S, Olliver G (2017b) Transforming the urban space through transit oriented development. The 3V approach. The World Bank, pp 21–24

Suzuki H, Cervero R, Iuchi K (2013) Transforming cities with transit: transit and land use integration for sustainable urban development. World Bank, Washington, DC

Thanh Nien Homepage (2019). https://thanhnien.vn/doi-song/giao-thong-cong-cong-xua-o-hanoi-819164.html. Accessed 30 Sept 2019

The Federal Transit Administration (2013) Transit-oriented development and joint development in the United States: a literature review

Vietnam Registry Department Homepage (2019). http://www.vr.org.vn/thong-ke/Pages/tong-hopso-lieu-phuong-tien-giao-thong-trong-ca-nuoc.aspx. Accessed 30 Sept 2019

VTV Homepage (2019). https://vtv.vn/trong-nuoc/xe-may-dang-chiem-86-luong-phuong-tien-giaothong-dang-tham-gia-tai-ha-noi-20190404142702196.htm. Accessed 30 Sept 2019

Wang Y, Potoglou D, Orford S, Gong Y (2015) Bus stop, property price and land value tax: a multilevel hedonic analysis with quantile calibration. Land Use Policy 42:381–391

Urban Slums and Affordable Housing

Informality in the Southern City; An Enquiry into Informal Practices in Housing

Ruhamah Thejus

Abstract As the Southern City continues to expand, much of the housing space created in it is termed informal. However, there is still much to understand about the range of informal housing—both in and outside the slum—in terms of housing tenure, how informality is defined and how access to housing is negotiated. This paper aims to look at the ways in which informal housing is created and understood in the megacities of the Global South. It uncovers some of the actual practices followed to gain access to informal housing, with special reference to Bangalore in India. This access to housing is viewed in terms of its location across the city, the arrangements entered in to access it and the connectedness between housing and economic activity, enabling an understanding of the dynamics that create informal housing arrangements. A bifocal view of informal housing from academic and empirical perspectives is used to shed light on three important nuances in the understanding of informality of housing. The findings from the study shed light on the wide variety of legal and illegal arrangements that fall under the bracket of *informal* housing. It is hoped that these findings will contribute to a shift in the understanding of informality of housing in the Southern City, which can contribute to a shift in the legal and contractual environment, calling for more flexibility and creativity in the housing arrangements it allows for. Ultimately this paper will enhance the understanding of cities and will highlight the fact that the city is not created by planners and managers only, but by every resident who inhabits it.

Keywords Informal · Formal · Housing · Location · Arrangements · Southern

1 Informality in the Southern City

As the Megacity of the Global South continues to burst at its seams, housing—ranging from tall apartment blocks to villas and gated communities—dominates the landscape, comprising what is commonly known as formal housing, while the slum is

R. Thejus (✉)
Dalai Lama Institute, Bangalore 562109, India
e-mail: rthejus@dalailamainstitute.edu.in

L. T. T. Huong and G. M. Pomeroy (eds.), *AUC 2019*, Advances in 21st Century Human Settlements, https://doi.org/10.1007/978-981-15-5608-1_32

viewed as the ultimate form of informal housing. Residential space becomes a pursuit for all in the city, whether its purpose is to increase prestige and social standing or simply to lay a weary head at night. However, a closer look at informality in housing arrangements reveals a more complex story.

While informal settlements appear in all large cities—and house 'one-third of the urban population in the global South' (King et al. 2017, p. 2)—the elements that constitute informality are varied, contested and open to modification. In addition, the changing world order—where the globalisation of the 1990s is giving way to regionalism—has led to each part of the world developing its own unique flavour of urban informality with regard to housing, none more pronounced than that of India. Professor of Geography, Majid Hussain, describes the Indian Slum as a cluster of twenty or more households, lacking drinking water, toilet and drainage facilities (Hussain 2011, p. 379). This basic description sheds light on the issue of definition in the first place—to classify housing as slum housing involves addressing the complexity of the elements that make up housing, and to achieve uniformity of definition across an entire nation is impossible. Such definitions of informal housing also focus on what Ananya Roy labels 'subaltern urbanism'—an informality that exists particularly among the poor in the city—while failing to address the many wealthy and elite forms of informality in housing that are 'valorised' (Roy 2011a, b). Hussain puts the proportion of Indian urbanites living in slums at 30% in the year 2001 (Hussain 2011, p. 379)—reflective of the situation in the Global South. This classic textbook definition and enumeration of informality in housing terms is widely used both in the study of informality and in policy action.

This paper seeks to scrutinise informal housing—looking at Bangalore as an empirical point of departure—and to understand the processes that create access to the range of informal housing arrangements that exist. A focus on the poor seeking access to housing is present here, considering that current modes of planning in many cities of the South serve to exclude the poor in systematic ways (Watson 2009). This analysis calls for a look at informal housing beyond its mere structure and form—to understand the assemblage of residential space as a dynamic process. In contrast to considering the land value and analysing urban housing as a direct outcome of the prevailing political economy, allowing for an assemblage view of housing enables an understanding that reflects the process of gaining access to housing. Colin McFarlane suggests that 'by paying detailed, ethnographic attention to processes of assemblage, urbanists may better understand how actually existing urban situations are constituted and, on this basis, may be better equipped to imagine alternatives to those situations' (Brenner et al. 2011, p. 228). In order to craft policies that envision access to housing for all, it is vital to first understand what informal housing is, how it has come about and how it serves the purpose of generating a right to the city for millions in the Global South.

This paper responds to the need to continually recalibrate an understanding of the informal as a set of practices, rather than as a physically confined space with certain identifiable features. It also calls to question the notion that formality, defined in terms of legality and planning, is the better way to inhabit the city, while the informal, which is more practical and is characterised by self-planned, auto-constructed arrangements, which are negotiated on a case-by-case basis—where 'deviations' from a legal perspective have become the norm (van Gelder 2013, p. 518)—is to be disapproved and removed at all cost.

To do this, the experiences of five diverse households currently accessing residential space in an informal manner in Bangalore city will be studied in order to highlight the complexity and the multiplicity that needs to be grasped to understand informality in the area of housing. Further, three unique characteristics of informality in housing, which are evident in these five cases, will be drawn out from the wide literature available on informal housing in the Global South. This bifocal reading into informality will provide the means to understand the day-to-day reality of accessing housing, following the principle proposed by Gautam Bhan to speak of 'practice…rooted in an empirical specificity' (Bhan 2019, p. 3).

2 Inhabiting the City

Five households accessing urban housing space in varied ways have been selected as case studies for this paper. Their stories are outlined below in terms of the space they occupy, how they gained access to this space and how they maintain this access on an on-going basis.[1] These cases will shed some light on the diversity of arrangements people enter into to obtain housing.

2.1 The Private Slum

The first housing situation comprises a group of 70 households—200 individuals—located beside a wide, open drain on a large plot of land in a typically marginalised location within a Bangalore Development Authority (BDA) approved residential layout in Bangalore. At first glance, the settlement appears to be an archetypal slum, where the auto-construction of informal structures made of thatch and iconic blue plastic sheets—particularly vital in the rainy season—on public land has grown over time. However, a closer look at this settlement reveals that the land is indeed a formally defined housing plot owned by an individual living at another location in Bangalore city. For a sum of one thousand rupees per month, a family can mark themselves a spot on this land and live there. A municipal water connection runs into the settlement, and the homes are lit by solar lanterns, lacking a supply of electricity. Conditions for residence in this informal settlement include engagement in work in the city, keeping out of trouble with the police, and maintaining good relations with the owner as well as the other tenants on the plot. Of course, this form of sub-letting

[1] All responses included here have been provided voluntarily and with the knowledge of their purpose for this paper.

of housing plots is not legal, but surveillance in this part of the city is minimal, which, coupled with relations between the owner of the plot and the city authorities, has allowed the settlement to grow unchecked over the last six years.

Dulliya is a resident of this settlement. He arrived in Bangalore from Raichur district in North Karnataka—all the other residents of the settlement are also from Raichur district, having arrived in Bangalore at different times. Dulliya's wife and two young children live with him in their tent, which started out as a basic structure three years ago, but now accommodates a light-weight cot, kerosene stove and trunks to contain their belongings. They purchased all the material required to set this space up from their own savings, and the small additions were made gradually as their income in the city increased. Both Dulliya and his wife work at construction sites as daily-wage labourers, which means, they find themselves at different locations across the city depending on where work is available. While this situation is not ideal, they appreciate the stability that this housing arrangement offers them—for now—and in any case, their livelihood is more secure than what their rural farm offered them.

2.2 *Building Whilst Dwelling*

A construction site in a formally defined residential area forms the location of the second housing arrangement. A household of twelve people—all members of one extended family—work at the site during the day and double up as the night watch when construction work ceases. This arrangement began five years ago, when the construction of the building commenced, and the household initially lived in a plastic tent, guarding the precious construction materials that were stockpiled on site. As the ground-floor parking lot was completed, this family moved in and continued to work on the rest of the building, which is now nearing completion.

Meenakshi, the lady of the home, lives here with her husband and three children, along with her brother with his family, as well as their ageing mother. One of the engineers working on this building is a relative of hers, and by his recommendation, they were recruited directly from their village in Raichur district in the north of Karnataka. Electricity and water connections have now been provided to the building, and Meenakshi's family is allowed their use by the building contractor, although this is a relatively recent development. Anything else the family needs has to be bought through their own means—they have created temporary partitions and a kitchen area to provide a semblance of privacy in the otherwise open parking lot. They are happy with this arrangement and with the long way they have come from that plastic shed five years ago.

In addition to working on construction during the day, Meenakshi works as domestic help in neighbouring homes in the evenings to help make ends meet so they can comfortably send their children to school. Living amidst wealthy homes provides her access to plenty of opportunities, and her income from domestic help exceeds that from construction.

2.3 Watching and Waiting

The third housing arrangement is again found within the walls of a formally defined, privately owned, BDA approved housing plot in a residential layout in Bangalore. The owner of this plot is seldom around, and has arranged for a woman, with her husband, both in their late fifties, to live here to guard the plot, with no written agreement and with rare checks to confirm that all is well.

The agreed resident is Povatha, and the housing she has been provided with consists of one room measuring 10 feet by 10 feet. However, the plot is occupied by Povatha, her husband, their adult son with his family and their daughter with four children, totalling 10 people, which has caused displeasure to the owner when he has come by on spot-visits. Subsequent expansion has yielded a shed for storage, cooking and washing, a separate bathroom and a large water tank. A self-constructed connection into the city sewage system has been created and is well-used. While a kind neighbour allowed Povatha the use of electricity for minimal lighting, their favour ran out when she began to use the electricity for additional appliances, and she is now dependent on solar or kerosene lanterns. The plot lacks municipal water supply, but a tanker of water can be bought to fill the tank on site, which lasts the household for a week.

Living here has enabled Povatha to take a formal loan to buy her son an auto-rickshaw, while she herself operates a fruit stall in the wealthy neighbourhood that surrounds her home. She views the bribes she has to pay the police in order to keep her fruit stall in a popular location and to allow the supplier to deliver to her location as routine payment—something she budgets for every month.

Povatha has lived in Bangalore for 38 years under similar arrangements at various locations, this ten-year engagement being of the longest tenure. As long as the owner does not want to build on his plot, she knows she is secure here. Once the inevitable construction of a house on this plot begins, she will seek work at the site, or will offer her services to guard some other vacant plot in this area—there are plenty of opportunities.

2.4 Continuation Housing

A single room measuring ten feet by eight feet in the corner of the ground-floor parking lot of an apartment block in a residential layout in Bangalore, housing a family of five, could almost go unnoticed as a part of the apartment. Bhagiamma and her husband have lived here for fifteen years, since the laying of the foundations of the building. Both Bhagiamma and her husband were employed in construction work here, where her husband began as an unskilled worker, and picked up the skills required for masonry, tile-laying and electric wiring. Their family grew here to include their three daughters. While the construction of this building was completed five years ago, Bhagiamma has managed to retain this housing space, as she continues to work as a cleaner in the building. She holds no document in writing to underscore this arrangement, pays no rent and receives no payment for her work. Her husband works in construction at various locations around the city, with his current assignment being fifteen kilometres from their home.

The home shows clear evidences of incremental development in the form of a kitchen and a washing area outside. Bhagiamma was asked to use liquefied petroleum gas for cooking in place of the firewood that she previously used in order to reduce the accumulation of soot on the walls of the building. To obtain a gas connection, she was provided with the required documentation from the owner, marking this location as her address.

The family feels privileged to have access to this space, though it is small, and the growing children clearly need more space. They do not doubt the security of tenure they have here and say they consider it their own house.

2.5 Living on the Edge

The pavement—a common location for informal housing in many Southern cities—provides the final scene of residence for this study. The pavement dweller in question is Santhosh, a young man, who has lived in Bangalore for the last seven years, leaving his wife, children and elderly parents back in Kanpur district of Uttar Pradesh—his trade: the sale of potted plants. He follows the market wherever it goes, and for now, he has found lucrative business on the pavement of this wealthy residential neighbourhood.

His housing has improved incrementally, beginning with the humble plastic tent, and now including a second, slightly more stable tent, cooking stove, bed, chair and of course the beautiful front garden that comes with the job. The bamboo, tarpaulin and tin sheet, as well as the furniture were all bought over time, using money from his income or borrowed from friends—Santhosh knows many other nursery workers from Uttar Pradesh in Bangalore. His dependence on the goodwill of neighbours has only grown over the last three years he has lived at this location—a neighbouring home provides him with water for bathing, while a bakery supplies his drinking water. For the plants in his nursery, Santhosh has to buy water, which is a sizeable expense. However, he is happy that the Bruhat Bengaluru Mahanagara Palike (BBMP)—which is the municipal administrative body that provides civic amenities—allows his residence at this location for a small *fee*, as he gets excellent business here. The location comes with the understanding that Santhosh will discourage people from dumping garbage on the neighbouring BBMP-owned land. For now Santhosh is satisfied, but he knows it is only a matter of time before the pavement will need to be cleared and he will be required to move—this is an undesirable, yet necessary aspect of his life that he has come to terms with.

3 Understanding Informal Housing

Informality in terms of land tenure, access to housing and a right to the city—simply understood as the informal comprising the local and the everyday, and the formal involving a high level of state planning (McFarlane and Waibel 2012, p. 1)—is most pronounced in the large cities of India, owing to the rapid influx of rural migrants, people from towns and smaller cities and people moving between as well as within megacities, creating a so-called shortage of housing stock—their quest: the best possible combination of livelihood, living conditions and income. The five case studies outlined here shed important light on informality in housing in particular, since all five homes possess the character of lacking documentation to confirm an arrangement that allows the residents to inhabit the home, have been built incrementally, and lack the amenities required.

A wide array of households—ranging from those who have recently arrived from remote rural areas, to chain migrants who have settled their families into the city, right up to families that have lived for generations in the same house, now located in a reclassified area—constitute the residents of informal housing in India. The households which participated in this study have all migrated into Bangalore for economic purposes, as is the case with a large proportion of informal settlers in the city—Ramachandra Guha explains this as the increasing likelihood that Indians—skilled as well as unskilled—will travel long distances for employment Guha 2008, p. 711). Jean-Louis van Gelder names rural-to-urban migration and poverty as two important sources of informality (van Gelder 2013, p. 494). Some households benefit from high levels of capital—financial and human as well as social—while others find a poor fit between what they possess and what is required to access housing space in the city. This results in significant variation in the ability of the poor to negotiate and access housing. In addition, as evidence from other parts of the world indicates, increases in incomes do not guarantee a reduction in housing informality (van Gelder 2013, p. 494). So while all the participants in this study earn significantly higher incomes than their rural agricultural pursuits yielded, this increased income has not generated access to the stable and permanent housing they envisaged for themselves.

Against this backdrop, viewing housing as informal and formal in a binary manner becomes problematic, owing to the fact that it is found in a variety of locations within the city, and not in particular, geographically defined sections alone; in areas of the city meant for housing and those meant for other functions. So, while informal settlements—commonly known as slums—deserve attention and action, not all informal housing arrangements are found in geographically bound areas that resemble the slums of the city, but rather are spread in large or small pockets all over the city, and are also found scattered throughout more formally recognised neighbourhoods. This is illustrated by the five households considered in this paper, where housing space is found in planned and approved areas of the city, with a few formal elements, but where the spaces are inhabited in self-defining and ambiguous ways. These households find themselves in an assisted squatting arrangement of sorts—where the practice of squatting is described by Alexander Vasudevan as the occupation of residential spaces that are unauthorised, dwellings are make-shift and assembled, and occupancy remains uncertain and precarious (Vasudevan 2015). In addition, these five arrangements occur in isolation from other informal settlements, leaving the residents entirely at the mercy of the owner, and dependent on the goodwill of neighbours, city authorities and others who directly or indirectly control access to these spaces.

It would make sense to begin to apply Andy Pratt's suggestion for a shift in thinking about informality in general, to housing: 'we should consider informality as a hybrid of what we have hitherto termed formal and informal' (Pratt 2019, p. 614). Pratt also dissuades the use of blanket terminology, where the formal is viewed as planned, legal and government-controlled in contrast to informal, which then becomes other to formal, and encompasses everything that is created, self-regulated, local and stands outside of the limited framework of the state (Pratt 2019). 'Binary thinking, by opposing two distant poles, making them exclusionary, and separating them with an artificial divide, places informality as that which needs to be repaired, fixed and corrected' (Acuto et al. 2019, p. 482). Whereas, the situation on the ground reveals a formal system that cannot possibly keep up with the housing needs of the people it seeks to serve—or prefers to refrain from meeting these needs—alongside a self-created and regulated system that accommodates in a more inclusive manner, and functions especially in favour of the poor.

Here lies the first nuance in understanding the informality of an Indian city—not all residents who access residential space in an informal manner are residents of slums, or informal settlements, neither do they all occupy spaces that are conventionally considered informal. This facilitates the study of urban housing without over-emphasising the formal or informal nature of its location, but rather focusing on the right to the city it creates and the enabling context it offers to the people it houses.

The range in location of auto-regulated (informal) housing within the city leads on to an appreciation of the variety of arrangements that occur within the informal realm under which residents occupy housing space. The binary set of arrangements wherein house ownership or non-ownership determines access to housing indicates an insufficient understanding of the ways in which housing space is accessed. Theoretically, it could be assumed that the presence of a clear property title on paper would result in the right to urban space, and the lack thereof would mean no access to urban space—specifically housing. In reality, however, 'legitimacy of land occupation is not systematically coupled with the existence of formal property rights, especially in the developing world' (Garland 2013, p. 51), and a number of arrangements along the spectra of official to unofficial and legal to illegal give rise to the occupation of what is known as urban housing in general—but more specifically, informal housing. Ananya Roy highlights the complexity of the informal realm through what she terms subaltern urbanism, where neither the ingenious resourcefulness used to survive—celebrated by many—nor the lack of political agency required to put together a decent living—bemoaned by others—are the predominant themes in understanding the factors at play in creating and sustaining informal housing (Roy 2011, p. 226–228). Rather, she highlights what is referred to as *Occupancy Urbanism* by Solomon

Benjamin where the informal resident develops the autonomy required to get around the proper real estate, legal and financial procedures to create and maintain residential space (Roy 2011, p. 229). The variety of arrangements required to access housing is reflected in the practices followed by the five households used in this study. All agreements with owners of the space are undocumented and could be terminated at any time, making them impermanent, and leaving the residents with no means of redressal in the case that the flimsy agreement is ended for whatever reason. This feature of evanescence is well understood by all occupants, and they consider it a matter of fact in the process of habiting the city. The legality of the occupancy is also questionable in all five cases, whether it is the dwelling on the pavement or the guarding of space and material in exchange for its use, or the occupancy through undocumented financial exchange, highlighting the paradox that accessing space in an informal manner can lead to either penalty or legitimization (van Gelder 2013, p. 495). However, this does not prevent the successful assembling and incremental improvement of residential space, and indeed the use of this space and mode of occupancy, not as an end in itself, but as a scene for the development of an active urban life.

This second nuance considers the varied modes of putting together a space to live, which are as diverse as the residents themselves, and sheds light on the different levels of tenure security that residents of informal housing experience. Many housing arrangements are, in fact, a combination of formal and informal practices, resulting in what Roy and AlSayyad term 'hybrid arrangements' or a 'formal-informal continuum' (McFarlane and Waibel 2012, p. 2). Although the residents occupy spaces with the permission of the owners of the properties, this is undocumented at any level, again leaving the residents reliant on the benevolence of those involved in the other side of the agreement to occupy the space in question.

In addition to this complex diversity with regard to the kinds of residential space accessed by residents of Indian informal housing, and to the arrangements they enter to access it, is the range of activities they engage in to survive—some evanescent, and others lasting. Abdou Maliq Simone describes the process of 'the majority of urban residents of the South' as 'constantly putting together some workable form of income and habitation.' He goes on to state that any semblance of stability requires a constant recalibration of edges, boundaries and interfaces' (Bhan et al. 2018, p. 264). This coexistence and inextricable link between housing and livelihood is illustrated by the five households in the study. Two scenarios are evident: In one case, a form of employment has led to the provision of the current housing arrangement—which is true in the cases of the construction workers' families—and in addition to which, the households must look further afield to supplement the incomes they earn from construction, in the form of domestic help, or work at multiple construction sites, using varied skills. In a second case, the housing facilitates the practice of a livelihood, making it successful in a particular area of the city, which is evident in the cases of the nursery and potted plant and the fruit vending and auto-rickshaw businesses. Any change in the economic climate of the area occupied will necessarily lead to a suitable recalibration of the income-habitation amalgamation. In all cases, housing is not an end in itself, but has to give rise to a workable context, within which a household can survive the city.

The commonly conjured-up image is that of the Dharavi Slum in Mumbai, as portrayed by Kevin McCloud in the popular documentary *Slumming It*, in which the innovation and enthusiasm of informal residents are not only celebrated, but also indeed rendered desirable. The Slum here is a self-contained area, where, in addition to housing and amenities, all livelihoods and enterprises are available for practice within (McCloud 2011). Contrary to this image, however, the residents of informal housing are required to participate in economic processes and to access resources from multiple sources and locations across the megacity.

This complexity in the area required to operate in and negotiate outcomes in the city provide the third nuance to understanding informality in the Indian city: not all operations of informal residents are centred within a small geographical area—they are required to spread their operations across a city or even out of it in order to gain access to enough resources to survive (Bhan et al. 2018. p. 271). In fact, residents of informal housing, especially those living in alternative housing arrangements, are more likely to seek housing that allows for an amalgamation of function—as residential space and as an area of economic activity, thus leading them to seek out housing arrangements that facilitate a supplementary or primary source of income in addition to residential space.

4 The Practice of Housing in the City

The understanding of informality in housing thus requires a study of day-to-day practices employed by the poor to access and occupy housing spaces through a variety of informal arrangements. This paper has highlighted three of the nuances in informality found in the practice of accessing housing: (a) diversity in location of informal housing across the city, (b) diversity in arrangements or negotiations entered in to gain access to housing space, and (c) diversity in the spread of operations and human interactions required to maintain a workable combination of livelihood and residence.

In all cases, either the households seek out employment that brings a housing arrangement with it or they seek to live at their employment site. Here the importance of the concept of the amalgamation of livelihood and habitation in the informal realm is underscored. Its effectiveness in reducing commuting times and costs, creating access to markets, cutting across social stratification common in the Southern city and creating the most efficient patterns for productivity is vital to understand in the context of housing. Compounding this are the social relations and contracts that undergird the arrangements.

While arrangements that enable free access to housing with certain jobs are common in the formal sector—residence on campus in educational institutions, access to parsonage housing for faith-based jobs or live-in arrangements for health-care-providers being common situations in Indian cities—there is evidence through this study that workers living in informal housing also seek out such housing situations, and are drawn towards arrangements that provide a degree of housing security. Work in the construction industry is one such area where the possibility of housing being accessible to workers through informal arrangements is high, increasing the prospects for the livelihood-habitation nexus. This feature of informality must be taken into consideration when considering the planning of the city.

5 The Collaborative Southern City

This paper by no means seeks to condone the apathy of formal planners of the city towards the poor, neither does it promote the view that the legitimization of informal access would solve the problems of informal housing as proposed by the likes of Hernando do Soto (van Gelder 2013, p. 495), but rather calls for a closer look at the varied modes of inhabiting the city in order to create an urbanism wherein flexibility between the formal and informal can result in more workable solutions for all, especially for the poor. Evidence shows that new ways of approaching and dealing with informal housing in the Southern City are required, where the disconnect between planning, policy and practice can be addressed by bringing them to the same table, and where solutions arise from a collaborative process. 'An approach that integrates legal and social perspectives' is required (van Gelder 2013, p. 496).

Finally, I would like to conclude that the variety of informal practices described in this paper, highlighting the geographical spread of housing informality, the range of arrangements involved in accessing housing informally and the livelihood-habitation amalgamation required to make informal housing effective for the poor, can be used to indicate areas where policy can incorporate forms of housing practice that would otherwise be considered outside of standard planning practice. 'This results in new forms of planning practice from within the state apparatus' (Bhan 2019, p. 7), but also calls for a legal environment within which a variety of arrangements are permissible and a sense of creative practice is fostered, enabling the poor to occupy housing space in a more secure manner in less conventional locations, and through less conventional arrangements, allowing for the effective amalgamation of livelihoods and habitation in the Southern City.

References

Acuto M, Dinardi C, Marx C (2019) Transcending in(formal) urbanism. Urban Stud 56(3):475–487

Bhan G (2019) Notes on a southern urban practice. Environ Urban 1–16. Sage Publications, International Institute for Environment and Development (IIED). https://doi.org/10.1177/0956247818815792

Brenner N, Madden D, Wachsmuth D (2011) Assemblage urbanism and the challenges of critical urban theory. City Anal Urban Trends Culture Theory Policy Action 15(2):225–240. https://doi.org/10.1080/13604813.2011.568717

Bhan G, Srinivas S, Watson V (eds) (2018) Companion to planning in the global south. Orient Black Swan, India
Garland AM (ed) (2013) Innovation in urban development. Incremental housing, big data and gender. Wilson Center, Washington D.C.
Guha R (2008) India after Gandhi: the history of the world's largest democracy. Picador, India
Hussain M (2011) Human geography, 4th edn. Rawat Publications, Jaipur
King R, Orloff M, Virsilas T, Pande T (2017) Confronting the urban housing crisis in the global south: adequate, secure, and affordable housing. Working paper. Washington, DC. World Resources Institute. www.citiesforall.org
McFarlane C, Waibel M (eds) (2012) Urban informalities: reflections on formal and informal. Routeledge, Abingdon
McCloud K, Simpson H (2011) Slumming It. Roadshow Entertainment, Pymont N.S.W
Pratt AC (2019) Formality as exception. Urban Stud 56(3):612–615. https://doi.org/10.1177/0042098018810600
Roy A (2011a) Urbanisms, worlding practices and the theory of planning. Plan Theory 10(1):6–15
Roy A (2011b) Slumdog cities: rethinking subaltern urbanism. Int J Urban Reg Res 35(2):223–238
UN-Habitat (2009) Planning sustainable cities: global report on human settlements. Earthscan, London
van Gelder J-L (2013) Paradoxes of urban housing informality in the developing world. Law Soc Rev 47(3):493–522
Vasudevan A (2015) The makeshift city: towards a global geography of squatting. Prog Hum Geogr 39(3):338–359
Watson V (2009) The planned city sweeps the poor away…: urban planning and 21st century urbanization. Prog Plan 72(3):151–193

Upgrading Slums in India

Braj Raj Kumar Sinha

Abstract Slums, the vulnerable spaces in the urban centers, lack fundamental resources and capabilities and result due to failure of policies, bad governance, corruption, inappropriate regulation, dysfunctional land markets, unresponsive financial systems, a fundamental lack of political will, rural-urban migration, urbanization, poor housing planning, poor infrastructure, social exclusion and economic stagnation, informal economy, poverty, politics, social conflicts, natural disasters, etc. Upgrading slums is a process of improving basic municipal infrastructure services such as access to potable water, sanitation, toilets, waste collection, access roads, paved foot paths, storm drainage, electricity, street lighting, public telephones, regularizing security of land tenure, and affordable housing improvements, as well as improving access to health, education, training, food and nutrition, child care, transportation, and other municipal services. Objective of this paper is to assess an improvement in various aspects relating to upgrading slums. This paper is based on the data collected from the Census of India, 2001 and 2011 and National Sample Survey Sixty-ninth Round conducted during July–December 2012. For the analysis of upgradation of slums, a Composite Z score was calculated by employing important demographic and economic variables. Findings show some improvement in respect of road, water supply, street lights, electricity, latrine facility, sewerage, drainage, garbage disposal, educational facility at primary level, and medical facilities over a period between 2007 and 2012. Spatial analysis shows a sharp regional difference in the level of slum upgradation in India. This paper is supported by the relevant cartographic representations.

Keywords Upgradation · Slum · Urban centers · Municipal infrastructure services · Demographic and economic variables

B. R. K. Sinha (✉)
Department of Geography, Institute of Science, Banaras Hindu University, Varanasi 221005, U.P., India
e-mail: sinha_brk@yahoo.co.in

L. T. T. Huong and G. M. Pomeroy (eds.), *AUC 2019*, Advances in 21st Century Human Settlements, https://doi.org/10.1007/978-981-15-5608-1_33

1 Introduction

Slum Upgrading or slum improvement means many things, but at its simplest it has come to mean a package of basic services: clean water supply and adequate sewage disposal to improve the well-being of the community. But fundamental is legalizing and "regularizing" the properties in situations of insecure or unclear tenure (Retnaraj 1998). Upgrading customarily provides a package of improvements in streets, footpaths, and drainage as well. Solid waste collection is frequently included with its positive impact on health, along with street lights for security and night activity. Slum upgrading consists of physical, social, economic, organizational, and environmental improvements to slums undertaken cooperatively and locally among citizens, community groups, businesses, and local authorities (Madhusoodhanan 2006). The main aim of upgrading slums is to improve the lower standard of living of slum dwellers. Most of the slum areas face scarcity of basic services such as provision of safe drinking water, sanitation, and wastewater and solid waste management. The responsibility of arranging and supplying such services is of the local bodies of the government. "Slum upgrading" is used mainly for projects developed by or engaged by the Government, NGOs, and other similar agencies. Most of the scholars and practitioners of the developing countries considered it an essential and significant component of urban development. Though, many people do not believe that slum upgrading is successful. They point to the difficulties in providing the necessary resources either in a way that is beneficial to the slum dwellers or in a way that has long-term effectiveness (https://en.wikipedia.org/wiki/Slum_upgrading). Alternatives to slum upgrading include the construction of alternative tenements for people living in slums (rather than fixing the infrastructure itself) or the forced removal of slum dwellers from the land (Sufaira 2013). The benefits are simply that people obtain an improved, healthy, and secure living environment without being displaced. The investments they have already made to their properties remain and are enhanced—this is significantly better than removing them to costlier alternatives that are less acceptable to them. Recognizing title and security of tenure makes a positive contribution to both the economic prospects of the poor, as well as to the national economy. Experience has shown that slum-upgrading projects are associated with social and economic benefits that are particularly high. For example, in a recently upgraded area of El Mezquital, Guatemala, infant mortality rates fell by 90% and crime by 43%. The possible slum-upgrading approaches or strategies or schemes (the result of poverty alleviation and National draft slum policy of Government of India) are as follows:

Slum Upgradation: This is the strategy for dealing with slums in the city in which the required facilities or the basic infrastructure would be provided to most of the existing slums, thereby upgrading the living environment.

Slum Redevelopment: It intends to redevelop the slum which is located on local body and private land at the same place where they are found existing by recognition and re-densification. The land belonging to local body will be developed by local body itself and the private plots will be developed by the concerned party.

Slum Rehabilitation: As a part of rehabilitation, government has targeted to relocate the slums which are situated in flood-prone area, reservation plots, and on road alignment for the rehabilitation.

2 Objective

This paper aims at assessing an improvement in various aspects relating to upgrading slum in India.

3 Data Source and Methodology

The data on slums in India is based on the Census of India, 2001 and 2011 and National Sample Survey Sixty-ninth Round, Schedule 0.21, conducted from July to December 2012. Composite Z score has been calculated from various demographic and economic variables such as literacy rate, female literacy rate, sex ratio, child sex ratio, work participation rate, and female work participation rate. The final Z score has been used to examine the Level of Demographic and Economic Development of slum population in India. Here, this is worth mentioning that the data of social amenities of slum population was collected only in 2011 Census and not in 2001. Due to such limitation in the availability of data, the investigator has taken into consideration only those variables which were commonly available in both the censuses. Hence, the assessment of the levels of Demographic and Economic Development of slum population in India is based on the final Z score relating to the above variables, and change in the status of the respective state between 2001 and 2011 indirectly indicates the slum upgrading and capability building in India. The National Sample Survey data relating to improvement in various aspects such as road approach, road within, water supply, street lights, electricity, latrine facility, sewerage, garbage disposal, drainage, and educational and health facilities have been used to assess slum upgrading and capability building. This paper is supported by the relevant cartographic representations.

4 Analysis

4.1 Spatial Variation in the Level of Demographic and Economic Development in India

Slum demography had been presented on the basis of actual count in Census 2001 for the first time in the history of the Population Census in the country. Detailed demographic data about slum areas across the cities/towns in the country having population more than 50,000 in 1991 Census were enumerated. The slum population was reported from 26 to 31 states/union territories in India in 2001 and 2011, respectively. Nine states/union territories namely Himachal Pradesh, Nagaland, Mizoram, Sikkim, Arunachal Pradesh, Manipur, Dadra and Nagar Haveli, Daman and Diu, and Lakshadweep did not report any slum population in their cities/towns in 2001, while four states/union territories namely Manipur, Dadra and Nagar Haveli, Daman and Diu, and Lakshadweep did not report any slum population in their cities/towns in 2011.

Table 1 and Fig. 1 show the state-wise variation in the level of demographic and economic development in the years 2001 and 2011. Five States/UTs namely Meghalaya, Kerala, Pondicherry, Tamil Nadu, and Tripura were demographically and economically well developed in the year 2001 in comparison to other states of India. Three out of these five States/UTs namely Kerala, Tamil Nadu, and Tripura remained in the category of high development even in 2011. Two states namely Mizoram and Sikkim joined this group in 2011, whereas two states namely Pondicherry and Meghalaya slumped down to the category of high developed states from 2001 to 2011. Three states namely Rajasthan, Bihar, and Uttar Pradesh were in the category of very low developed states in comparison to the other States/UTs. These three states maintained their status quo in 2011 also by showing the very low level development. In addition to these three states, four new States/UTs namely Chandigarh, Jammu and Kashmir, Haryana, and Uttarakhand slumped down to the level of very low developed states from 2001 to 2011. Apart from the above-mentioned States/UTs, the other States/UTs were in the category of high or low level of demographic and economic development.

4.2 *Status of Social Amenities in Slums in India Between 2007 and 2012*

Assessment of the status of social amenities is based on the information obtained from the Sixty-ninth Round National Sample Survey of slums. In this round of NSS, an attempt was made to find the dynamics of the conditions of slums by assessing—whether certain undesirable conditions were improved or not over time. For this purpose, about a dozen facilities were taken into consideration for evaluation to know the proportion of slums reporting improvement in the concerned facility during the last five (from 2007 to 2012) years, the proportion of slums reporting deterioration in the facility, the proportion of slums reporting no change in the condition, and the proportion of slums reporting that the concerned facility was neither existing earlier nor existing now. For several aspects of living conditions in slums, including water supply, availability of electricity, drainage system, garbage disposal system, road used as main thoroughfare within the slum, approach road to the slum, and so on, it was ascertained whether the slums had experienced an improvement, a

Table 1 State-wise level of demographic and economic development

Composite Z score of demographic and social indicators	2001		2011	
	Name of states	Number of states	Name of states	Number of states
Very high above 05	Meghalaya, Kerala, Pondicherry, Tamil Nadu, Tripura	05	Mizoram, Kerala, Sikkim, Tripura, Tamil Nadu	05
High 00–05	Karnataka, Andaman and Nicobar Islands, Andhra Pradesh, West Bengal, Maharashtra, Assam, Chhattisgarh	07	Andaman and Nicobar Islands, Karnataka, Meghalaya, Nagaland, Pondicherry, Himachal Pradesh, Goa, West Bengal, Assam, Andhra Pradesh, Maharashtra	11
Low 00 to −05	Punjab, Odisha, Uttarakhand, Delhi, Goa, Madhya Pradesh, Gujarat, Jharkhand, Haryana, Jammu and Kashmir, Chandigarh	11	Chhattisgarh, Odisha, Madhya Pradesh, Punjab, Delhi, Arunachal Pradesh, Gujarat, Jharkhand	08
Very low below −5	Rajasthan, Bihar, Uttar Pradesh	03	Bihar, Chandigarh, Jammu and Kashmir, Haryana, Uttar Pradesh, Rajasthan, Uttarakhand	07
Data not available	Himachal Pradesh, Sikkim, Arunachal Pradesh, Nagaland, Manipur, Mizoram, Daman and Diu, Dadra and Nagar Haveli, Lakshadweep	09	Manipur, Daman and Diu, Dadra and Nagar Haveli, Lakshadweep	04

Source Census of India 2001 and 2011

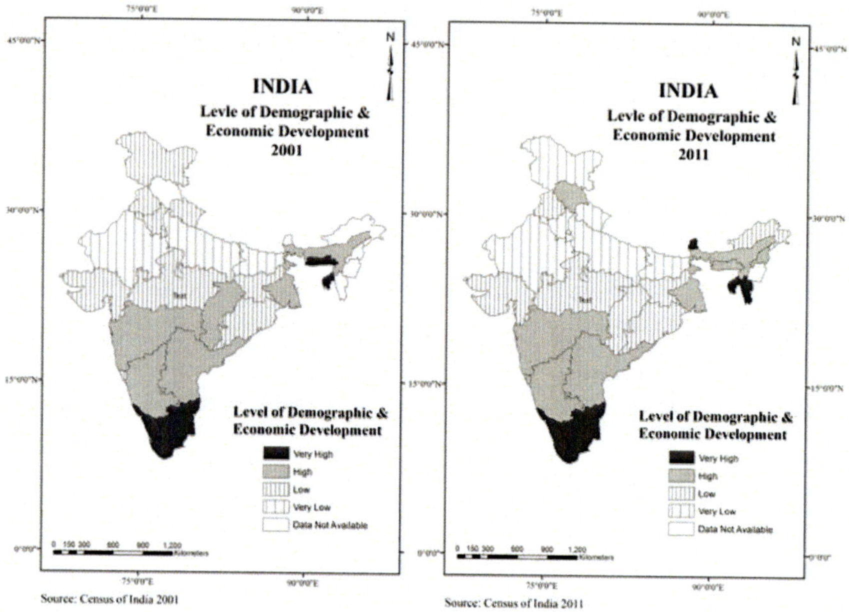

Fig. 1 Spatial variation in the level of demographic and economic development in India

deterioration, or no change in the concerned facility during the last five years (from 2007 to 2012). For detailed description, different social amenities and its status such as improvement, no change, deterioration, and neither existing earlier nor existing now are clearly shown in Table 2. By adding the percentage reporting improvement in a specific facility and the percentage reporting deterioration, and subtracting the sum from 100, one would obtain the percentage of slums that reported no change, plus the percentage of slums in which the facility was neither existing on the date of the survey or nor it was existing five years earlier.

Table 2 and Fig. 2 show the improvement in various social amenities in the last five years (from 2007 to 2012). This indicates a type of improvement in the condition of slums in India. At all-India level, 36.23% of slums were in the improvement category of the status of social amenities; however, different social amenities in improvement category were found varying between 17.24 and 53.24%. While 49.11% of the total slums showed no change/no improvement in the status of different social amenities during the period of 2007–2012. 1.29% of slums registered even deterioration in the condition of above-mentioned social amenities, and 12.75% of slums reported that the above-mentioned social amenities were neither existing in 2007 nor existing in 2012. Slightly more than 53% of slums showed improvement in respect of road approach and 47.56% in respect of roads within during the last 5 years. The condition of water supply, street lights, electricity, and garbage disposal improved between 39.20 and 45.45%, whereas other amenities such as latrine facilities, sewerage, drainage, educational facility at primary level, and medical facility exhibited improvement

Table 2 Status of social amenities in slums in India over the period of five years between 2007 and 2012

Social amenities	Status				Total no. of slums
	Improvement	No change	Deterioration	Neither existing earlier nor existing now	
Road: approach	469 53.23[a]	377 42.79	22 2.50	13 1.48	881 100.00
Road: within	419 47.56	402 45.63	24 2.72	36 4.09	881 100.00
Water supply	400 45.45	404 45.91	34 3.86	42 4.77	880 100.00
Street lights	366 41.59	389 44.20	14 1.59	111 12.61	880 100.00
Electricity	345 39.20	500 56.82	8 0.91	27 3.07	880 100.00
Latrine facility	267 30.44	431 49.14	20 2.28	159 18.13	877 100.00
Sewerage	151 17.24	418 47.72	13 1.48	294 33.56	876 100.00
Drainage	291 33.22	407 46.46	18 2.05	160 18.26	876 100.00
Garbage disposal	345 39.43	371 42.40	17 1.94	142 16.23	875 100.00
Educational facility at primary level	256 29.26	515 58.86	7 0.80	97 11.09	875 100.00
Medical facility	190 21.66	529 60.32	8 0.91	150 17.10	877 100.00
Average	318 36.23	431 49.11	17 1.92	112 12.75	878 100.00

Source NSS sixty-ninth round, schedule 0.21, July–Dec 2012
[a]Decimal Figures in the Cells are the percentages to total number of slums in respective social amenity

in the range of 21.66–33.22% during the period of 2007–2012. Besides the above-mentioned facilities, private and community toilets are also found; however, these are not mentioned in the table. Such toilets do not exist in the majority of the slum areas. Where there are no toilets, people defecate in the open space, and it creates several health-related problems to the people in slum areas.

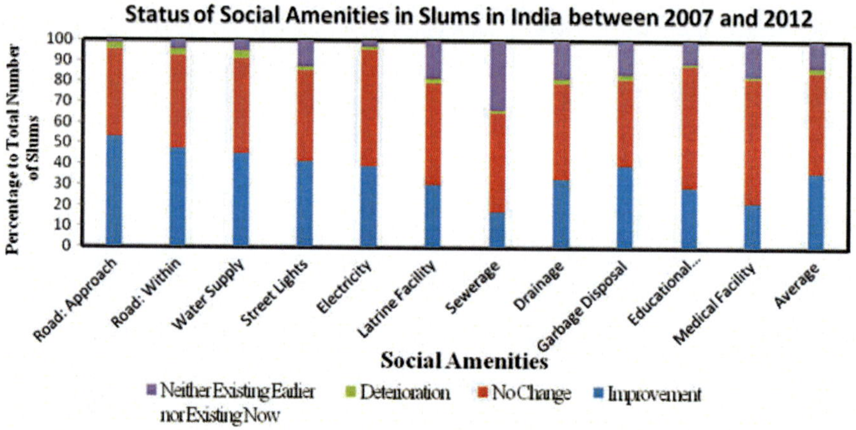

Fig. 2 Status of social amenities in slums in India between 2007 and 2012

4.3 *Participatory Groups and Social Amenities in Slums in India During 2007–2012*

The Government of India and various states have taken certain initiatives for the development of infrastructure as well as improvement of environment in urban settlements with a view to improve the quality of life. Apart from government, various Non-Governmental Organizations (NGOs) and Residents of the slums have also played some roles in slum upgradation or slum improvement. From the slums that reported improvement in any of the facilities discussed in the previous paragraphs, it was ascertained whether the source of improvement was Government (including municipal bodies), NGOs, the residents themselves, or any other agency. The information has been mentioned and shown in the following Table.

Table 3 and Fig. 3 show the percentage distribution of slums reported improvement in facilities by source. From the table, it is clear that more than 93% of urban slums reported improvement in social amenities through the efforts of the Government. This is because it is the sole responsibility of states to provide road, water supply, street lights, electricity, educational institutions, and medical facilities. The residents of the urban slums had contributed very little toward the improvement of various facilities available except the latrine facility, which was above 28.84%; this is mainly due to the increasing number of private toilets in slum areas. The other sources including non-governmental organizations were coming forward to some extent to improve the condition of urban slums in respect of latrine, sewerage, water supply, and garbage disposal, etc. The NGOs can play very important role in medical and educational facilities at primary levels, and residents of slum can improve latrine facilities considerably in comparison with other sources.

Table 3 Participatory groups and social amenities in slums in India during 2007–2012

Social amenities	Participatory groups				Total no. of groups
	Government	Non-govt. organization	Residents	Others	
Road: approach	456 97.23[a]	2 0.43	6 1.28	5 1.07	469 100
Road: within	407 97.14	3 0.72	7 1.67	2 0.48	419 100
Water supply	380 95	2 0.5	11 2.75	7 1.75	400 100
Street lights	355 96.99	7 1.91	2 0.55	2 0.55	366 100
Electricity	330 95.65	8 2.32	5 1.45	2 0.58	345 100
Latrine facility	185 69.29	2 0.75	77 28.84	3 1.12	267 100
Sewerage	143 94.70	0 0.00	6 3.97	2 1.32	151 100
Drainage	276 94.85	5 1.72	7 2.41	3 1.03	291 100
Garbage disposal	328 95.07	5 1.45	9 2.61	3 0.87	345 100
Educational facility at primary level	232 90.63	18 7.03	0 0.00	6 2.34	256 100
Medical facility	164 86.32	16 8.42	1 0.53	9 4.74	190 100
Average	543 93.06	12 1.94	24 3.74	7 1.26	583 100.00

Source NSS sixty-ninth round, schedule 0.21, July–Dec 2012
[a]Decimal Figures in the Cells are the percentages to total number of slums in respective social amenity

4.4 *Spatial Variation in the Level of Improvement of Slum Condition in India*

Table 4 and Fig. 4 show the state-wise status of level of improvement of slums in India. For this, first of all the percentage of number of slums reported improvement in the concerned facility to the total number of slums belonging to the same facility in respective state was calculated. Secondly, the average value of calculated percentages of various facilities concerning improvement in respective state was worked out and four categories of levels of slum upgradation were made for analyzing spatial variation in the level of improvement of slum conditions in India. Four states namely Tripura, Pondicherry, Karnataka, and Chhattisgarh out of the total 35 states and

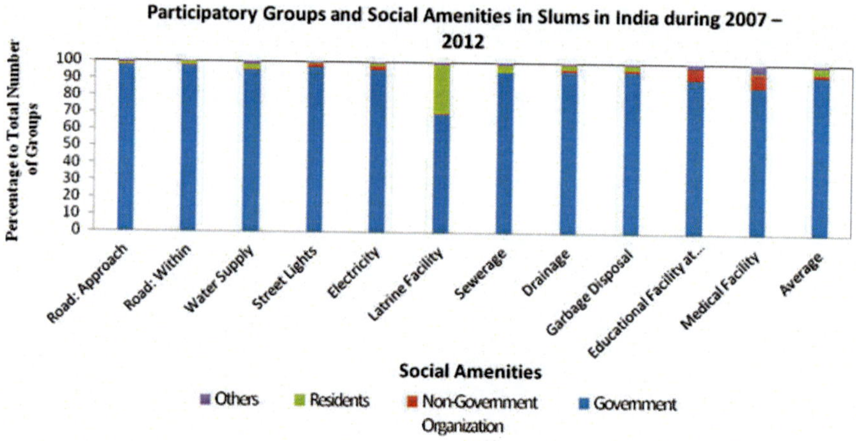

Fig. 3 Participatory groups and social amenities in slums in India during 2007–2012

Table 4 State-wise status of level of improvement in condition of slums in India over the period of five years (2007–2012)

Level of slum upgradation	Name of states	Number of states/UTs
Very high (above 45.22)[a]	Tripura, Pondicherry, Karnataka, Chhattisgarh	04
High (31.05–45.22)	Madhya Pradesh, Rajasthan, Meghalaya, West Bengal, Punjab, Kerala, Andhra Pradesh, Maharashtra, Jammu and Kashmir, Arunachal Pradesh	10
Average (31.05)	India	
Low (16.87–31.05)	Jharkhand, Tamil Nadu, Odisha, Uttar Pradesh, Himachal Pradesh, Assam, Bihar, D and N Haveli, Gujarat, Delhi, Uttaranchal	11
Very low (Below 16.87)	Goa, Chandigarh, Haryana, Mizoram, Daman and Diu	05
Data not available	Sikkim, Nagaland, Manipur, Lakshadweep, Andaman and Nicobar Islands	05

Source NSS Sixty-ninth Round, Schedule 0.21, July–Dec 2012
[a]Based on the average value of the percentages of slums reported improvement in different facilities to the total number of slums of concerning facility in respective state

union territories experienced very high level of upgradation in comparison to other states and union territories of India. On an average, more than 45% of slums in these states and union territories reported improvements in various aspects of the condition of slums. Ten states/UTs demonstrated high level improvement in various social amenities in the range of 31.05–45.22% to total slums. Eleven states and union territories exhibited low level improvement indicated by the percentage range of 16.87–31.05.

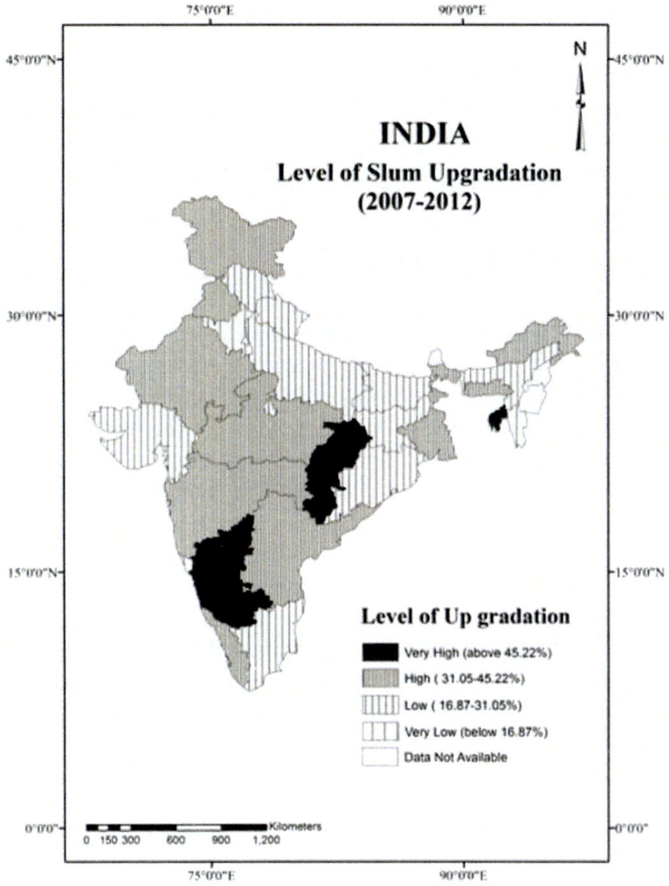

Fig. 4 Level of slum upgradation

In five States/UTs namely Goa, Chandigarh, Haryana, Mizoram, and Daman and Diu, the level of improvement was very low in the percentage range of below 16.87. On an average, below 16% of the total slums in these states registered improvement during the period of 2007–2012. Therefore, these states need special attention as far as slum upgradation is concerned. In the remaining five states and union territories, the data concerning improvement in slums was not available.

Figure 4 reveals sharp regional differences in the level of slum upgradation in India. Northern, Eastern, and few pockets of Northeastern, Southern, and Western part of the country experienced low level of upgradation. In these states, improvement was registered in between 16 and 31% of slums. While in Central, Western, Southern, Eastern, and Northeastern states, the improvement was in the category of high level. Only in few pockets of Northeastern and Northern parts of the country, improvement was very low. While in few pockets of southern, Central, and Northeastern portions, the improvement in various aspects of slum condition was very high. Overall, it can be said that there is a sharp regional difference, which exists in the level of slum upgradation in India. These regional variations may be due to the considerable variation in government's funding in developing quality of physical infrastructure in slum areas.

5 Findings and Suggestions

Findings show that there has been an improvement in respect of Road: Approach (53.23%), Road: Within (47.56%), Water Supply (45.45%), Street Lights (41.59%), Electricity (39.20%), Latrine Facility (30.44%), Sewerage (17.24%), Drainage (33.22%), Garbage Disposal (39.43%), Educational Facility at Primary Level (29.26%), and Medical facilities (21.66%) over the period of five years (i.e., from 2007 to 2012). It has also been found that the Government of India as one of the participatory groups has played a very crucial role in improving the above facilities from the point of view of upgrading slums in India. Spatial analysis of the level of improvement of slum condition in India demonstrates a sharp regional variation. All the states and union territories do not show the same level of upgradation. The majority of states and union territories experienced improvement in the range of 16.87–45.22%. Such regional variations may be due to the considerable variation in the allocation of government's funding as well as variation in the level of participation in enhancing quality of physical infrastructure in slum areas.

The following strategy may be taken into consideration from the point of view of improving living conditions in Indian slums:

In situ *development*—Wherever there is feasibility of in situ development, the sites of such slums should be identified and taken up under a systematic plan of action with a view to make provisions of basic facilities/services so that such slums may be habitable.

Tenement Scheme—The slums located in congested and unhygienic areas of the urban centers wherein equitable distribution of space is not possible should be cleared and new houses should be made and distributed to slum dwellers.

Rehabilitation Approach—The slums where neither the in situ development nor the scheme of tenements is feasible, slum dwellers should be rehabilitated and resettled in tenements in nearby locations to spend their life.

In addition to the above schemes, the development of cooperatives or multipurpose cooperatives as alternative approach should be formed for motivating and educating the slum dwellers about the philosophy and ideology of cooperation and self-help. The local level NGO should be given the responsibility to organize such cooperatives. When the cooperative society becomes experienced, self-reliant, and in a position to run their affairs independently, the concerned NGO should be withdrawn and slum dwellers should be free to function from their own resources. With the experience so gained and training and skills so acquired, the NGO may shift to such other areas that require their assistance/guidance in the process of upgrading slums.

References

Madhusoodhanan V (2006) Rehabilitation measures for slum dwellers in Thiruvananthapuram city. In: Nair KN, Gopikuttan G (eds) Housing in Kerala, New Delhi, pp 16–62
Retnaraj D (1998) Rapid urbanization in Kerala: exploration of its causes. J South Econ 13–14
Sufaira C (2013) Socio economic conditions of urban slum dwellers in Kannur Municipality. IOSR J Humanit Soc Sci (IOSR-JHSS) 10(5):12–24
Wikipedia. The free encyclopedia. https://en.wikipedia.org/wiki/Slum_upgrading

Status of Elementary Educational Facilities in Slums Across Different States of India

Prabhakar Nishad and Braj Raj Kumar Sinha

Abstract Education is an important ingredient of human resource development. It contributes to well-being of individuals by improving income and standard of living. Elementary education in India, in general, has shown considerable improvement over the past few decades, but the educational level of slum residents in India is very low, and there is an urgent need to improve the educational achievement of urban slums. The objectives of this paper are to: find out the availability of the government primary school in slums across different states of India, analyze the improvement in the condition of primary level educational facilities during 2007–2012, and demonstrate the sources of improvement in the condition of educational facilities at primary level in slums of India. The results show that 59.40% of slums were within the distance of half kilometer from the government primary schools. Improvement in the educational facilities at primary level during 2007–2012 was reported by 30% of slums, and "no improvement" by 57% of slums. At India Level, less than 1% and in West Bengal 4% of the total slums reported decline in the elementary level educational facilities. Out of the total improvement at the primary level educational facilities in slums in India has been found by the effort of government of India and 17% by the NGOs; however, some variations have also been noted across different states of India.

Keywords Slum · Elementary education · Educational facilities · Source of improvement

1 Introduction

Education is an important ingredient of human resource development. It contributes to well-being of individuals by improving income and standard of living. It plays a

P. Nishad (✉)
Department of Geography, D.A.V. Post Graduate College (J.P. University, Chapra), Siwan 841226, Bihar, India
e-mail: prabhakaragna@gmail.com

B. R. K. Sinha
Department of Geography, Institute of Science, Banaras Hindu University, Varanasi 221005, Uttar Pradesh, India

L. T. T. Huong and G. M. Pomeroy (eds.), *AUC 2019*, Advances in 21st Century Human Settlements, https://doi.org/10.1007/978-981-15-5608-1_34

crucial role in making a suitable foundation for the overall socio-economic development of a country. For every developing nation, elementary education is a means to success. India as a developing nation also gives maximum importance to elementary education. In India, elementary education means the education given to the children of 6–14 years. The Government of India has made the elementary education free and compulsory for 100% literacy as it is necessary for the nation's progress (Goswamee and Kumar 2017, p. 103). India's achievement in the field of education has not been very inspiring even after independence. Statistical data show that India is even behind the countries like Sri Lanka. The most basic measurement of educational status of the people is literacy. High level of physical and mental status of population of a country is a prerequisite for its economic, social, and political advancement. That is why, literacy rate is considered to be a good indicator of development in any given society (Hannan and Munir 2015, p. 211). Illiteracy is one of the major hurdles in bringing any positive change in the society. On the one side, India is developing very fast in the field of technology and trying to establish the status of "Digital India"; but on the other hand, 26% of the country's total population are illiterate (2011 Census). Illiterate population is mainly from the economically weaker section of the society.

In Indian cities, the two classes of people can be easily distinguished; one is from weaker section who are living below the poverty line and other who are enjoying most of the facilities so called above the poverty line. The people who are living below the poverty line are mostly from the slum areas and are characterized by pitiable builtup area with compact setting, overcrowding and by the lack of drinking water, electricity and hygienic condition. A slum, as defined by the UNHABITED Agency of the United Nations, is a run-down area of a city characterized by substandard housing, squalor and tenure unsecurity. Since most of the slum dwellers suffer from the livelihood and the basic needs rather than receiving education. As a result, we are unable to achieve the target of 100% literacy. Recognizing the importance of elementary education, Indian Government is spending a lot of money in the field of elementary education. The Right to Education Act 2009 provides free and compulsory education to each and every child up to the age of 14 years. This act also provides free and compulsory admission as well as attendance in the nearest school till the completion of their elementary education. To take care of malnutrition of the children Mid-Day meal is also provided by schools; but all these attempts are not able to provide 100% enrollment of children in the elementary education. The rate of drop out children from the schools is also high. Slums have universal existence as most of the cities in the world have slums. The underprivileged children living in the slums have no basic facilities in comparison with those children living out of the slum areas and enjoying some special attention from their parents and society. According to the Census 2011, in India, approximately 13.7 million households or 17.4% of the total urban households are living in slum areas. Inaccessibility of safe drinking water and lack of sanitation in slum areas are negatively affecting the population and specially children in slum area. The rate of population growth in slum areas is also high, and the literacy rate in slum area is lower with higher number of out-of-school children in comparison with other areas. Apart from these, the ill-health, poverty, and lack of basic infrastructures are the basic problems of the slum areas, specially in the developing nations like India. These are the basic reasons behind the stagnation and slow development of the country and their citizens (Nath I et al. 2014). Due to the above-mentioned reasons, it is important to focus on the development of slum population.

World Bank rightly recognized poverty as multi-dimensional problem, which includes less accessibility of opportunities for developing human capital as well as education rather than problem of only low-income people (World Bank 1994). A different perspective of cost-benefits of education is created with the combination of economic factors and labor market uncertainties (Hussain 2005 p. 134). The basic problem in widening the coverage of elementary education in India is that there is a huge number of students coming from challenging backgrounds of both the rural as well as urban areas. As the economic returns from the elementary education are low, the parents of the students belonging specially to the underprivileged families do not give much value to the elementary education (Khasnabis and Chaterjee 2007, p. 2091). Higher infant mortality rates, lower literacy rates, and low accessibility of safe drinking water and sanitation have been observed in non-notified slums in comparison with notified slums (Vaijayanti and Subramanian 2015, p. 55). Large family size, poor health, substandard living condition, poor economic condition, low parental educational background, unfavorable housing situation, language problem, unbalanced occupation and schooling conditions are the major constraints in attaining elementary education by the slum children in general and specially by the children of 6–14 year age group in particular (Kapur 2016, p. 38). Education in general and elementary education in particular plays an important role in improving economic productivity, lowering the population growth, and improving the health and nutrition. All the above-mentioned circumstances make the present study noteworthy.

2 Out-of-School Children in India

Out-of-school children means the children who are neither registered nor are they going to school. If the number of out-of-school children is higher, it means there is more need to focus on achieving 100% literacy rate. There are few children from the age group of 6–14 years who may not have been enrolled in any school. There are few other children who may have initially been enrolled but later on dropped out of school without completing their elementary education. The special study of out-of-school children helps us in identifying the geographical locations which require the greater attention for achieving higher literacy. Similarly, different policies can be framed on the priority basis for various social groups and particular gender.

Approximately 37% of the total children of urban areas were found living in the slums as per the estimate of the National Survey on Estimation of out-of-School Children 2014. This means roughly 19.8 million children between the age of 6 and 14 year were living in slum areas. Out of the total elementary education age group children living in the slum areas, 2.38% (0.473 million) were neither registered nor they were going to the school. Gender wise analysis shows that the number of out-of-school female children (2.73%) is more in comparison to that of the male children amounting to approximately 2.14%.

Figure 1 shows the state wise regional variation in the pattern of out-of-school slum children in India. The proportion of the out-of-school children in the states like Kerala, Goa, Sikkim, Himachal Pradesh and Tamil Nadu was below the national average. In Punjab, Uttaranchal, Bihar and Jharkhand states more than 10% of the total slum children from the age group of 6–14 years were not attending the schools. Such a large number of out-of-school children in these states is a matter of serious concern even after the elementary education became a fundamental right for each and every child belonging to the age group of 6–14 years as par eighty-sixth constitutional amendment 2002 and the enactment of the Right of Children for Free and Compulsory Education Act 2009. The main reason behind higher proportion of out-of-school

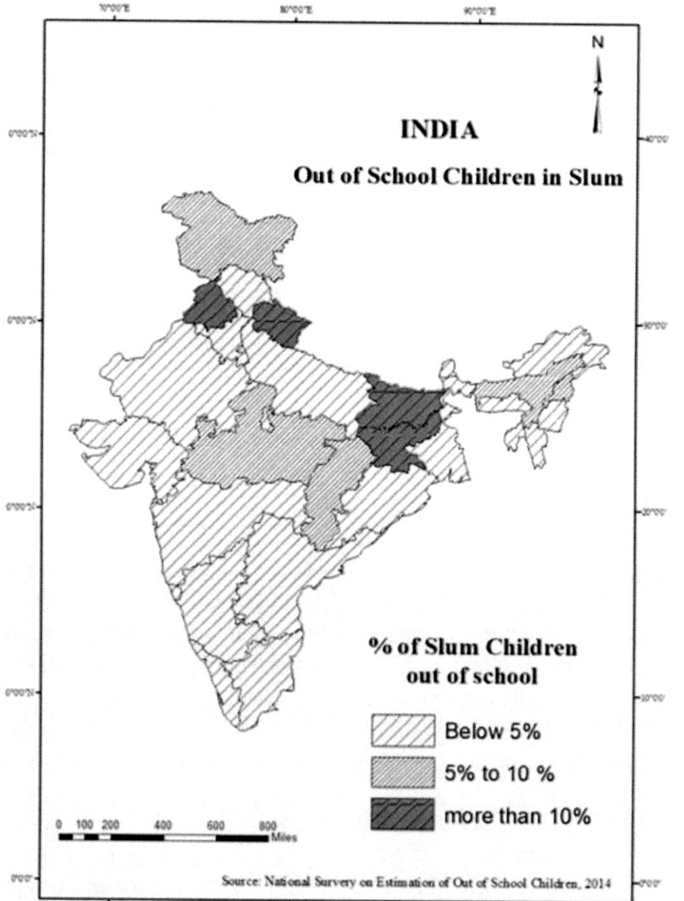

Fig. 1 Spatial Pattern of Out of School Children from Slums in India

children is poverty. According to the National Survey on Estimation of Out-of-School Children, 23.76% respondents reported poverty or poor economic condition as the main reason for children not going to school. In the case of dropouts from the primary school, economic conditions and the poverty were reported as the major reasons by 28.52% respondents; while 23.45% respondents reported the same reasons in case of those children who were never enrolled in the primary schools. 27.70% slum respondents reported "Child not interested in studies," as one of the reasons in causing dropouts from the primay schools. 20.74% respondents reported "children needed to support household income" as a major reason in the case of those children who were either not attending or not enrolled in the elementary education schools in the slums.

3 Availability of Government Primary School in Slums Across Different States of India

Universal access to primary education demands schooling facilities within reasonable distance from the children's residence. Distance is an important factor as far as elementary education is concerned. This is especially more important in case of those deprived children who reside in the slum areas. So, if schools are not situated within their reachable distance, they may not be able to complete their schooling even if they may have formally been enrolled in the schools. Therefore, the Right to Education Act, 2009 makes it mandatory for appropriate authorities of central, state, and local governments to provide primary schools within the defined area/limits of neighborhood. But, at the same time, this act does not clarify the distance or neighborhood limits as centralized norm. The central government leaves this definition to the appropriate State or local Governments to notify aforesaid distance limits or area in the RTE Rules. This is because India has very diversed geography and the requirements of each and every region are different. So, the local or state government would be in better position to define the distance limit, considering the local conditions of different children.

However, in the Right to Education Act circulated to different states, Central Government has made an attempt to illustrate the distance limit of elementary school. According to the circular, the school for the children up to the fifth class should be in the radius of one kilometer from their residence, and for the children of sixth to eighth class, it should be within three kilometers. Keeping in mind the safety of children from the point of view of risks like difficult terrain, lack of transport infrastructure, landslides in mountainous region, the local or state government may define the distance limit of elementary school at its own. This act also provides accessibility to schools to those children who were earlier not allowed to take admission on the social, cultural, caste religious grounds. Such rules should be decided by the local or state government by considering the local contexts and requirements. The

Table 1 Distance of nearest primary school from the slum residence across major states of India

State	Distance range and percentage of nearest primary school				
	<0.5 km	0.5–1 km	1–2 km	2–5 km	>5 km
Andhra Pradesh	71.2	20.4	3.9	4.6	0
Bihar	94.4	3.6	1.2	0.8	0
Chhattisgarh	79.2	15	1.1	4.7	0
Gujarat	61.7	20.6	17.7	0	0
Karnataka	76.8	23.1	0	0	0.2
Madhya Pradesh	65.6	24.3	9	1.1	0
Maharashtra	46.4	38.1	15.1	0.4	0
Orissa	76.8	10.1	13.1	0	0
Rajasthan	52.7	37.9	8.8	0	0.5
Tamil Nadu	57.7	33.6	8.1	0.6	0
Uttar Pradesh	25.8	30.3	33.3	10.7	0
West Bengal	83.7	15.4	0.9	0	0
All-India[a]	59.3	28.4	10.3	2	0

Source Ministry of Statistics and Programme Implementation, Govt. of India
[a]Based on all States and UTs, including States and UTs not shown in this table

above-mentioned rules have been included in the Right to Education Act to apply by the Union Territories also without legislature.

In respect of the availability of educational facilities at the primary level, it is found from the Table 1 that 59.3% of the slums had at least one Government primary school located within half kilometer; it was also observed that 28.4% of the total slums had at least one primary school stablished within 0.5–1 km. Table 1 shows that the percentage of slums having a Government primary school within half kilometer was highest in Bihar (94%), followed by West Bengal (83%) and Karnataka (76%). It is interesting note that Uttar Pradesh is such a state where the percentage of total primary schools coming within the range of 1 km was found lowest.

4 Improvement in the Conditions of Primary Level Educational Facilities

Section 19 of RTE Act, 2009 has established various norms and standards for schools. The private or government schools not maintaining the prescribed norms and standards should fulfill the established criteria within the period of three years from the date of enactment of the aforesaid act. Therefore, this is the responsibility of the governments and local bodies to improve the conditions of the primary level educational facilities to achieve and maintain the minimum standard of educational facilities in each and every school in the defined jurisdictional area.

Table 2 Percentage of slums by improvement in the condition of primary level educational facilities during 2007–2012 across major states of India

State	Improved	No change	Deteriorated	Not existed
Andhra Pradesh	20.8	57.3	1.4	20.5
Bihar	84.6	2.6	0	0.8
Chhattisgarh	64.4	35.2	0	0.3
Gujarat	10.1	67.4	0	22.4
Karnataka	46.8	51.8	0	1.3
Madhya Pradesh	57.6	36.6	0	5.8
Maharashtra	22.5	72	0	5.5
Orissa	41.7	53.4	0	4.9
Rajasthan	45	37.9	0	17.1
Tamil Nadu	1.8	75	0	23.2
Uttar Pradesh	21.8	43.5	0	22.4
West Bengal	32.9	58.6	4.3	4
All-India[a]	30.1	57.1	0.8	11.2

Source Ministry of Statistics and Programme Implementation, Govt. of India
[a]Calculated by including those States and UTs not shown in this table

Table 2 shows the improvement in the conditions of elementary level educational facilities across major states of India from 2007 to 2012. The statistical data shows that 30% of the total slum schools improved the conditions of education facilities at primary level during 2007–2012. 57% slum schools reported no change, 11% slum schools showed that such facilities did not exist even today, and less than 1% of slum schools at all-India level exhibited the condition of deterioration in the educational facilities at primary level.

At the state level, the picture of improvement in the educational facilities at primary level widely varies across the major states of India. 84% of all slums in Bihar, 64% in Chhattisgarh, 45% in Rajasthan and 47% in Karnataka show the improvement in terms of the primary level educational facilities whereas Tamil Nadu and Gujarat demonstrate only 2% and 10% improvement respectively. Similarly, 4% of the total slums in West Bengal and 1.4% in Andhra Pradesh reported deterioration in the elementary level educational facilities during the period of 2007–2012. The above-mentioned state wise variation in the improvement of the conditions of elementary level educational facilities generally determined by the deferential geographical settings and socio-economic development levels across different states of India.

5 Sources of Improvement

This section explain the sources of improvement in the educational facilities at primary level as discussed above, it is necessary to find out the sources of improvement that include the Governments (including state/central/local) or various NGOs or local residents themselves or any other source. The government has been a major source of improvement in the primary level educational facilities in the slums.

Table 3 shows that more than 80% of the improvements was due to government interventions. State wise data also shows that in the states like Bihar, West Bengal, Chhattisgarh, Andhra Pradesh, Gujarat, Karnataka, Orissa and Tamil Nadu almost 100% of the improvements was due to the efforts of the governments only. NGOs contributed 17% of the total improvements in the educational facilities at primary level, though in this respect state wise variations have also been observed, for example, the role of NGOs in Uttar Pradesh remains the highest (63%) as compared to other states of India. The role of residents during the period of 2007–2012 was negligible in almost all the major states as far as the source of improvement in the educational facilities at primary level is concerned.

Table 3 Sources of improvement in the elementary educational facilities across major states of India

State	Sources of improvement			
	Government	NGO	Residents	Others
Andhra Pradesh	98.6	1.4	0	0
Bihar	100	0	0	0
Chhattisgarh	99.4	0	0	0.6
Gujarat	100	0	0	0
Karnataka	100	0	0	0
Madhya Pradesh	97.4	1.9	0	0.7
Maharashtra	79.9	15.2	0	4.9
Orissa	100	0	0	0
Rajasthan	76.6	4.1	0	19.3
Tamil Nadu	100	0	0	0
Uttar Pradesh	36.2	63.8	0	0
West Bengal	100	0	0	0
All-India[a]	80.6	17	0	2.4

Source Ministry of Statistics and Programme Implementation, Govt. of India
[a]Based on all States and UTs, including States and UTs not shown in this table

6 Conclusion and Suggestions

Education is an important ingredient of human resource development as well as birthright of every child of the country. The deprived children living in slum areas also come within this fold. Poor economic condition has adversely affected the accessibility of elementary education, especially in slums. In respect of the availability of educational facilities at primary level, it is found that about 59% of the slums had at least one Government primary school situated within half kilometer and about 28% of slums had at least one primary school located within 0.5–1 km. Improvement in the condition of educational facilities at primary level during 2007–2012 was reported by 30% of the total slums and 57% slums reported no change. 11% slums reported that such facilities did not exist even today, and less than 1% slums at all-India level reported deterioration in the educational facilities at primary level. The government has been a major source of improvement in the educational facilities at primary level in slums. The state wise variation in the improvement of the conditions of elementary level educational facilities may be attributed to the diversed geographical settings, varied levels of development, and different sets of socio-economic conditions among the various states of India. Following are the suggestions for the improvement in the quality of primary level education in the slums of India:

- Teachers and various authorities should develop a positive attitude toward the underprivileged children living in slum areas. This will develop confidence among such children and will act as a motivation to overcome various challenges faced by them in day-to-day life. Their negligence may accelerate dropout rate from the schools, which will adversely affect their human capital development.
- Basic facilities such as separate toilet facilities for girls and boys, at least one classroom for each class, playing ground with the required equipments, and one library should be made available for all class students.
- Schools should be kept open even afternoons, in the evenings, weekends, and during the summer vocation for informal classes, for reading and writing in the library as well as for the art, laboratory and the like activities.
- The nursery schools (Aganbari) should be opened for the children aged three—six within the range of 0.5 km in every slum settlement.
- Schools should provide more attractive and joyful learning environment. Computer-based learning system should also be made available to the children for practice. This will make learning system more enjoyable and interesting.
- Dependency on the private tuition should be minimised by reducing the amount of homework, revising the syllabus and improving the learning system only within the school hours.

References

Goswamee G, Kumar RA (2017) Problems of primary education among the slum children of Guwahati. Clarion 6(2):103–107

Hannan A, Munir A (2015) Changing dimensions of literacy scenario and their determinants in India: a geographical perspective. Forum geografic. Studii și cercetări de geografie și protecția mediului XIV(2):211–220

Hussain Z (2005) Analysing demand for primary education: Muslim slum dwellers of Kolkata. Econ Polit Wkly 40(2):137–147

Kapur R (2016) Education of the slum dwellers. Int J Transform Bus Manag (IJTBM) 6(II):37–43

Khasnabis R, Chatterjee T (2007) Enrolling and retaining slum children in formal schools: a field survey in eastern slums of Kolkata. Econ Polit Wkly 42(22):2091–2098

Nath I, Maiti NC, Halder MK (2014) Elementary education of slum children: an attempt to reach the unreached. Available http://www.caluniv.ac.in/academic/education_journal/8%20Elementary%20Education%20of%20Slum%20Children%20%20An%20Attempt%20to%20Reach%20the%20Un-reached.pdf. Accessed 09 May 2016

Vaijayanti K, Subramanian M (2015) Educating the urban poor: a case study of running preschools in non-notified slums of India. Int J Early Years Educ 23(1):54–66

World Bank (1994) World Development Report 1994: Infrastructure for Development. New York: Oxford University

Concept, Status, and Progress of Affordable Housing in Urban India

Braj Raj Kumar Sinha and Prabhakar Nishad

Abstract In urban area, the prices of real estate and land are very high. This forced the economically weaker section of the society to live in a marginal land with pitiable housing condition in combination with congestion and lack of basic amenities. The poor people can only afford cost of house up to five times of their annual income, and Equated Monthly Installment (EMI) or rent should be less than 30% of their monthly income. The high land price and cost of building materials are rising year by year especially in urban areas. So, making affordable housing and providing it to poor becomes a daunting task. The objectives of this paper are to i. throw light on the concept of affordable housing with special reference to India, ii. find out shortage of affordable urban housing among different economic categories, iii. demonstrate variation in the spatial pattern of shortage of affordable urban housing across different states in India, iv. map out progress of "Pradhan Mantri Awas Yojana (Urban)—Housing for All" across different states in India, and v. point out problem and its possible solution relating to affordable housing for urban poor in India. Findings show that 88% of total shortage of affordable urban housing pertains to Economically Weaker Sections (EWS) and another 11% for Lower-Income Groups (LIG). There is state-wise variation in the shortage of affordable housing in urban India. The present rate of growth of PMAY-U is very less. Therefore, there is need for accelerating the present growth rate to achieve the target of building 18 million houses.

Keywords Pradhan Mantri Awas Yojana · Urban poor · Affordable housing · Economic categories · Growth rate

B. R. K. Sinha (✉)
Department of Geography, Institute of Science, Banaras Hindu University, Varanasi 221005, India
e-mail: sinha_brk@yahoo.co.in

P. Nishad
Department of Geography, D.A.V. Post Graduate College (J.P. University, Chapra), Siwan 841226, Bihar, India

L. T. T. Huong and G. M. Pomeroy (eds.), *AUC 2019*, Advances in 21st Century Human Settlements, https://doi.org/10.1007/978-981-15-5608-1_35

1 Introduction

Globally human population is continuously rising. The total population of the world in the mid-2017 reached about 7.6 billion. This means that almost 1 billion populations were added in a span of 12 years (WPP 2017). As per current estimate, every year world's population is being added by around 83 million people. According to the Department of Economic and Social Affairs of United Nation, global population will be 8.6 billion in 2030 and 9.8 billion in 2050 and 11.2 billion in 2100. The major contribution to this fast growing world population is mostly from developing nations. China having population of 1.4 billion (19% of world population) and India having 1.21 billion (18% of world population) are the two most populous countries of the world. Overall, approximately 60% of global population lives in Asia.

According to the Census of India in 2011, around 31% of its population lives in urban areas. It was estimated by FCCI (Federation of Indian Chambers of Commerce) that urban population of India will witness net growth of 900 million people. The rate of urbanization in India would be at a Compound Annual Growth Rate (CAGR) of 2.1% by 2012–2050, which means almost double than that of China (TG-12 2012). The large number of people living in Indian cities is under very severe pressure. Indian cities are struggling to accommodate the growing number of urban population caused due to rural to urban migration. Mumbai with 12.5 million population followed by Delhi with 11 million population are the two largest metropolitan cities by population in India. Due to increasing rate of urbanization, Indian cities are facing the problem of slum and squatter settlement. These settlements are characterized by poor health and education facilities, scarcity of land and housing, lack of basic infrastructure facilities like safe drinking water, electricity, sewerage, drainage system, etc. All these adversely affect the quality of life of slum dwellers. Kudva (2009) and UN-Habitat (2014) considered slums as temporary phenomena and emphasized that the slums would disappear with the passage of time owing to the development of affordable housing. But at present slums seem to be a permanent feature among the cities of the low and middle income countries.

Due to unavoidable rural to urban migration, population of Indian cities is growing very fast. So, the demand of affordable housing is also growing very rapidly day by day. In cities, because of scarcity of land, real price of real estate is very high, and due to unplanned urbanization in developing countries like India, slums are growing very rapidly. Cost of building material rose very much in recent years and real estate developers are developing luxury houses to maximize their profit (Sethi 2017). The shortage of affordable houses in India is mainly because the real estate builders are providing houses mainly to Middle-Income Group (MIG) and Higher-Income Group (HIG), whereas Economically Weaker Sections (EWS) and Lower-Income Group (LIG) remain unnoticed (Gupta 2018). Demographic and economic factors together make India a large and growing market for housing development for years to come. Low-income rental housing will continue to be in high demand on many reservations where high poverty levels persist (Pierson 2010). Affordable housing is in national agenda in developing countries like India. Housing is now considered as a basic need, so every government tries to find out the ways to provide affordable house to their citizens (Gopalan 2015). In the present world, especially developing countries like India, housing is a big problem. Housing is one of the basic needs considered for living; it also helps prevent diseases and improves health, provides social security, and helps in increasing national and household income (Maulik et al. 2016). The increasing urban population in developing countries like India is facing various socio-economic, institutional, and financial challenges in providing adequate affordable housing. So, for effective planning and its implementation in providing affordable housing to their citizens, it becomes very crucial to understand the situation of dynamics of demand side (Kumar 2015).

2 Concept of Affordable Housing

The term "housing" in developed nations refers to a dwelling unit of standard type, mostly an independent house or apartment, whereas in developing countries like India, it has different notion. For example, the census of India refers it to sheltering taken either on pavement or in a thatched hut or in a bungalow. Affordable is a relative term, so it has no straightforward meaning. It has various meanings in various contexts.

In the context of housing in urban area, RICS in a report on Making Urban Housing Work in India elaborates the meaning of affordability as "adequate shelter" on the basis of sustainability. That means providing security of tenure to the common household living in urban area. According to Practice Standard Guidance Notes (GN 59 2010) of RICS, affordable housing means providing house to those people who needed it but not able to afford it in open market.

KPMG in its report entitled as "Affordable Housing—A Key Growth Driver in the Real Estate Sector" defines affordable housing on the basis of three parameters: level of income, dwelling size, and affordability (Table 1). The two parameters, namely, income level and size of dwelling are not dependent on each other but the third parameter, namely, affordability is interrelated with the income and prices of property.

Table 1 Definition of affordable housing—KPMG

Economic categories	Income level	Size of dwelling unit	Affordability
Economically weaker section	Below INR 1.5 lakhs per annum	Up to 300 sq ft	**EMI to monthly income**: 30–40% **House price to annual income ratio**: Less than 5:1
Lower-income group	Between INR 1.5 and 3 lakhs per annum	300–600 sq ft	
Middle-income group	Between INR 3 and 10 lakhs per annum	600–1200 sq ft	

According to US Department of Housing and Urban Development, affordable housing means those houses which can be bought by paying less than 30% of their annual income. It will be considered as cost burden to a family who is paying more than 30% of their annual income on housing. This may also create difficulties for affording other necessary items like medical care, clothing, food, and transportation.

The Task Force on Affordable Housing formed in 2008 by the Ministry of Housing and Urban Poverty Alleviation (MHUPA), Government of India also defines affordable housing. According to this task force, affordable house for two economic categories, namely, economically weaker section and middle-income group may be defined on the basis of size of the dwelling and its cost (Table 2).

The Mission Directorate of the Jawaharlal Nehru National Urban Renewal Mission (JNNURM) also defined affordable housing on the basis of size of dwelling for three economic categories, namely, economically weaker section, lower-income group, and middle-income group. According to this definition, the EMI or rent for housing should not be more than 30–40% of buyer's gross monthly income (Table 3).

Table 2 Definition of affordable housing—MHUPA (2008)

Economic categories	Size of dwelling unit	Cost	EMI or rent
Economically weaker section	300–600 sq ft carpet area	Not exceeding four times the household gross annual income	Not exceeding 30% of gross monthly income of buyer
Middle-Income Group	Not exceeding 1,200 sq ft carpet area	Not exceeding five times the household gross annual income	Not exceeding 40% of gross monthly income of buyer

Source Task Force on Affordable Housing, MHUPA, 2008

Table 3 Concept of affordable housing by MHUPA

Economic categories	Size	EMI or rent
EWS	Minimum of 300 sq ft super built-up area Minimum of 269 sq ft (25 sq m) carpet area	Not exceeding 30–40% of gross monthly income of buyer
LIG	Minimum of 500 sq ft super built-up area Maximum of 517 sq ft (48 sq m) carpet area	
MIG	600–1,200 sq ft super built-up area maximum of 861 sq ft (80 sq m) carpet area	

Source Guidelines for Affordable Housing in Partnership (Amended), MHUPA, 2011

3 Shortage of Affordable Urban Housing Among Different Economic Categories

The urban households are classified into three economic categories in India on the basis of their monthly income. Those families whose monthly income is less than Rs. 5000/- are known as EWS, and the families whose monthly income is between Rs. 5001/- and Rs. 10,000/- are classified as LIG and those families whose monthly income is more than Rs. 10,000/- are called as MIG. According to MHUPA, in 2012, almost 95% of the total shortage is from EWS and LIG in India.

Table 4 shows the Urban Housing Shortage in India among different economic categories, as estimated by the Technical Group constituted by the Ministry of Housing and Urban Poverty Alleviation (MHUPA). Total housing shortage in Urban India was 18.78 million units in 2012. The table further shows that 56.18% of this shortage belongs to the category of Economically Weaker Sections (EWS) and 39.44% of this shortage pertains to Lower-Income Groups (LIG). The estimated shortage for Middle-Income-Groups is only 4.38%. So, we can say that housing shortage in urban India is mainly from the EWS and LIG categories. But bulk of the urban housing that has been built in India is beyond the bearable cost of the EWS and LIG of the society. Due to high profit gained from luxury and high-end housing, real estate developers, especially, private developers are targeting MIG and HIG. Because of scarcity of land, outdated laws, unnecessary delay in approval, high land cost, and adverse banking policies, low-cost housing becomes less profitable or sometimes unprofitable. So, the low-cost housing to EWS and LIG are provided by the government for the purpose of social welfare. But these low housing projects are not enough to accomplish the existing and ever increasing urban housing shortage in India. So, the affordable housing need of the EWS and LIG are neglected. This results into massive shortage of affordable housing, especially, demanded by EWS and LIG in India.

Table 4 Economic category-wise housing shortage in urban India (2012–2017)

Category	Shortage (In Millions)	Percentage to total shortage
Economically weaker sections (EWS)	10.55	56.18
Lower-income group (LIG)	7.41	39.44
Middle-income group (MIG)	0.82	4.38
Total	18.78	100

Source Report of the Technical Urban Group (TG-12) on Urban Housing Shortage 2012–2017, Ministry of Housing and Urban Poverty Alleviation, September 2012

4 Shortage of Affordable Urban Housing Across Different States in India

Table 5 gives clear idea about variation in the pattern of urban housing shortage in India. The highest number of housing shortage is in Uttar Pradesh (3.07 million) followed by Maharashtra (1.97 million). The other major contributing states of urban housing are Andhra Pradesh (1.27 million), West Bengal (1.33 million), and Tamil Nadu (1.25 million). From Table 5, we can see that 76% of the total shortage of housing in urban areas is from top ten states starting from Uttar Pradesh to Gujarat. It also shows a mixed picture where less developed as well as more developed states

Table 5 State-wise estimation of urban housing shortage in India

Name of state	Housing shortage in millions	Name of state	Housing shortage in millions
Uttar Pradesh	3.07	Nagaland	0.21
Maharashtra	1.94	Uttarakhand	0.16
West Bengal	1.33	Jammu and Kashmir	0.13
Andhra Pradesh	1.27	Manipur	0.08
Tamil Nadu	1.25	Puducherry	0.07
Bihar	1.19	Goa	0.06
Rajasthan	1.15	Dadra and Nagar Haveli	0.05
Madhya Pradesh	1.1	Himachal Pradesh	0.04
Karnataka	1.02	Arunachal Pradesh	0.03
Gujarat	0.99	Meghalaya	0.03
Jharkhand	0.63	Tripura	0.03
Kerala	0.54	Mizoram	0.02
Delhi	0.49	Chandigarh	0.02
Haryana	0.42	Sikkim	0.01
Odisha	0.41	Daman and Diu	0.01
Punjab	0.39	Lakshadweep	0.01
Chhattisgarh	0.35	Andaman and Nicobar Islands	0
Assam	0.28	**India**	**18.78**

Source Report of the Technical Urban Group (TG-12) on Urban Housing Shortage 2012–2017, Ministry of Housing and Urban Poverty Alleviation, September 2012
Source MoHUA, 2019

are having the shortage of urban housing. The people of these states are forced to live in pitiable housing conditions.

Variation in the land costs, building bye laws, high regional variation in the pattern of urbanization and regional variation in other facilities like power, technology, banking, and insurance, etc., are considered very important factors for regional variation in the pattern of affordable housing shortage in India. Due to variation in the availability of as such facilities some regions get special favor in respect of development of affordable housing projects; whereas, some other regions of the country face high scarcity of affordable housing.

5 Progress of "Pradhan Mantri Awas Yojana (Urban)—Housing for All" Across Different States in India

In June 2015, Pradhan Mantri Shahari Awaas Yojana (Prime Minister Urban Housing Scheme) was launched. The main objective of this scheme was to provide low-cost housing for urban poor in India. Under this scheme, to provide affordable housing to EWS and LIG, it was planned to build 22 million affordable houses by the year 2022. For this scheme, a financial assistance of Rs. 2 trillion (US$29 billion) was provided by the center to various state governments. PMAY-U has the following four components:

1. *In situ Slum Redevelopment (ISSR)*: The main objective of this scheme is to provide affordable housing to slum dwellers through redeveloping the slum areas existing on public or private land. The land is used as a resource in this scheme. Under this scheme the central government provides financial assistance of Rs. 0.1 million per house to the implementing authorities.
2. *Credit-Linked Subsidy Scheme (CLSS)*: To reduce housing loan effectively, this scheme provides financial assistance to EWS, LIF, and MIG families directly to their account.
3. *Affordable Housing in Partnership (AFP)*: The objective of this scheme is to promote private developers by providing them financial assistance. Under this scheme, if any private real estate developer is developing at least 35% of the total houses of their project to EWS, then the central government will provide financial assistance of Rs. 0.15 million per EWS house.
4. *Beneficiary Led House Construction (BCL)*: In this scheme, for the construction of the new house or for extension of any existing house of the family belonging to EWS/LIG, an assistance of Rs. 0.15 million will be provided by the central government.

In PMAY, financial assistance will be between US$1,400 to US$3,300. The main feature of this scheme is that it will give interest subsidy of 3–4% for MIG and 6.5% for EWS and LIG on housing loan. This subsidy will be for the period of

20 years from the date of start of loan. The houses under this scheme will be preferably constructed through eco-friendly technology. The preference would be given to physically handicapped and senior citizens in the process of allotment of ground floor in any housing project under this scheme.

Figure 1 shows the progress of PMAY-U across different states in India. States of Rajasthan, Uttar Pradesh, Madhya Pradesh, Punjab, Sikkim, Tripura, and West Bengal have achieved significantly higher house completions under the PMAY-U than other states. The success can be attributed to the proactiveness of the respective state government in establishing Public–Private Participation practices and their focus on streamlining execution and maintaining efficiency in timelines. These mainly include provision of single-window clearance for building approvals and quick validation of applications. The progress of PMAY-U in the states of Kerala, Andhra Pradesh, Bihar, Jammu and Kashmir, Arunachal Pradesh, Nagaland, and Mizoram is very low.

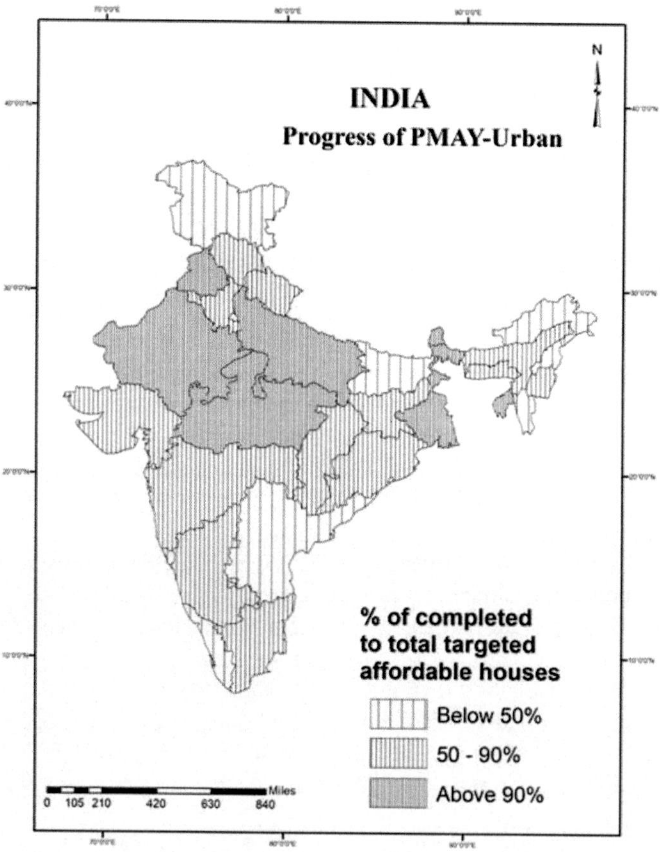

Fig. 1 Progress of PMAY-Urban in India (2015–2019)

5.1 Challenges Toward Affordable Housing for Urban Poor in India

India is facing several challenges in the process of developing low-cost housing project. Supply of affordable housing is facing the problems like scarcity of land, rising cost of building material and urban regulatory issues. On the other hand, inaccessibility of home finance is adversely affecting the demand of affordable housing in urban area. As a result, buying capacity of affordable housing in organized sector is considerably reduced for EWS and LIG. Though, some of these challenges are gradually being palliated, still multi-faceted efforts are required for mass development of low-cost housing in India. Following are the few challenges regarding affordable housing, especially, for urban poor in India:

1. *Scarcity of Land in Urban Area*: Due to rural to urban migration, the density of urban population in India is very high. As a result, the demand of land for housing purpose in urban India is also very high. So, there is huge shortage of land in urban area, and this scarcity is further aggravated by the old and irrelevant laws of the central/state/local authorities.
2. *Rising Costs of Construction*: Construction cost is an important factor in deciding selling price of any affordable housing because it constitutes 50–60% of selling price. In the luxury housing project construction, cost constitutes only 18–20% of its selling price. So, the minimization of construction cost is an important aspect of making low-cost housing feasible. In recent years, price of raw materials for housing project like cement, sand, bricks, and steel have witnessed 20–50% rise. The shortage of labor in urban areas due to success of Mahatma Gandhi National Rural Employment Guarantee Act (MGNAREGA) results into the rise of wage rate. This further raises the construction cost of affordable housing projects.
3. *Inaccessibility of Home Finance*: India has a widespread network of financial institutions like banks, etc. But due to the fear of ensuring repayment, inaccessibility of home finance still prevails for EWS and LIG.
4. *Unnecessary delay in approval of Affordable Housing Projects*: According to one estimate, for any housing project in India, developers have to take approval from 40 departments at various levels. Cost of housing projects increases between 20 and 30% due to delay in approval from various departments. After finalization of the agreement of land purchase, it takes at least 2–3 years for the commencement of construction.
5. *Scarcity of Skilled Labor*: The real estates of India are struggling due to the shortage of skilled labor. The cost and delivery of low-cost housing projects are adversely affected by this shortage.
6. *Economic Constraints for EWS and LIG*: The present financial system in India is skewed toward MIG and HIG. So, EWS and LIG are facing difficulties in securing formal financial assistance. Banks and other financial institution are serving mainly to HIG and MIG. Since the EWS and LIG is unable to provide required official documents of housing loan, they are rarely served by official financial institution. One another reason, why the financial institutions are not

serving EWS and LIG is because the income of these households varies from time to time. This makes them vulnerable for repayment of secure installment of housing loan.

7. *Old and ambiguous building bye laws*: In India, the building bye laws are very old and in several aspects lack clarity. So, the acquisition of land becomes very tedious process in India. Sometimes, set of laws get modified significantly after the beginning of housing projects. This further exacerbates the existing problems of private developers in the sector of real estate.

6 Conclusion and Suggestions

Those houses which are affordable to the Economically Weaker Section and Lower-Income Group are known as affordable houses. Primarily, EWS and LIG categories are facing the problems of housing shortage in urban area. However, the private real estate builders are providing houses mainly to Middle-Income Group (MIG) and Higher-Income Group (HIG). High price of majority of the houses in urban regions makes them beyond the buying capacity of EWS and LIG segments. Because of scarcity of land, outdated laws, unnecessary delay in approval, high land cost, and adverse banking policies, Low-cost housing becomes less profitable or in sometime unprofitable. State-wise analysis shows that 76% of the total shortage of housing in urban areas is from top ten states. States of Rajasthan, Uttar Pradesh, Madhya Pradesh, Punjab, Sikkim, Tripura, and West Bengal have achieved significantly higher house completions under the PMAY-U than other states. The success can be attributed to the proactiveness of the respective state government in establishing Public–Private Participation practices and their focus on streamlining execution and maintaining efficiency in timelines. Following are some suggestions through which the construction cost can be reduced and large-scale development of affordable housing may become possible:

Identification of right beneficiaries: To identify right beneficiaries, there is urgency to devise procedure for it. The role of AADHAAR card (biometric identity card provided by Unique Identification Authority of India) becomes important in this aspect of identifying right beneficiary if the person's income is also added to it.

Use of advanced technologies: For the development of affordable housing at mass scale, the role of technology becomes crucial. With the use of innovative technology, the cost of housing project can be reduced considerably. For example, with the use of few techniques like prefabricated construction technology and portable modular housing unit, we can fulfill the demand of affordable housing.

Ease of Land availability: The identification of easily available land and bringing peripheral land around the city could fulfill the scarcity of land to some extent. For this, various authorities must act in coordination so that regular supply of land parcel to low cost housing project can be maintained.

Ease of Development Norms: In the development of affordable housing, old and rigid norms should be relaxed. There is need for revision of urban development plans on regular basis so that only relevant norms should be modified on the basis of contemporary need.

Harmonize State and Central policies: The central government policies should be synchronized with the state and local government policies. This will help in reducing the shortage of affordable housing in India.

Development of Housing microfinance: There is urgent need for development of housing micro-financial institutions for EWS and LIG. This will help them in getting financial assistance provided by government.

Single-window approval system: The unnecessary delay in the approval of affordable housing project results in significant increases in construction cost. So, the single-window approval system will save time as well as money.

Relaxing tax and duty charge: The government should exempt tax and duty charges on construction materials for affordable housing projects. This will considerably reduce construction cost of the affordable housing. Government should also provide subsidy in Research and Development of new technology and low-cost materials. This will motivate various agencies and institutions for the development of affordable housing projects in India.

References

Gangani MG, Suthar HN, Pitroda J, Singh AR (2016) A critical review on making low cost urban housing in India. Int J Constr Res Civil Eng (IJCRCE) 2(5):21–25

Gopalan K (2015) Affordable Housing: Policy and practice in India. IIMB Manag Rev 27(2):129–140

Gupta N (July, 2018) The challenges in "Affordable Housing" for urban poor in India—deciphering the "Housing for All by 2022"—a comprehensive approach to decrement housing shortage—a case study of Delhi. Int J Sci Eng Res 9(7):341–372

Kudva N (2009) The everyday and the episodic: the spatial and political impacts of urban informality. Environ Plan A 41(7):1614–1628

Kumar A (2015) Housing shortages in urban India and socio-economic facets. J Infrastruct Dev 7(1):19–34

Pierson BL (2010) Developing affordable housing in Indian Country. J Afford Hous Commun Dev Law 19(3/4):367–390. 10842268. Spring/Summer

Report of the Technical Urban Group (TG-12) on Urban housing shortage 2012–17. Ministry of Housing and Urban Poverty Alleviation, Sept 2012

Sethi HK (2017) Affordable housing in India. Int J Eng Res Technol (IJERT) 6(6)

United Nations Human Settlements Program (UN-Habitat) (2014) A practical guide to designing, planning, and executing citywide slum upgrading programmes

World Population Prospects (2017) The 2017 revision. United Nations, New York

Slums, Squatter Settlements and Affordable Housing in the Dhaka Metropolitan Area

Nurul Islam Nazem and Shahana Sultana

Abstract Dhaka Metropolitan Area accommodates about ten million people of whom at least four million live in about 3500 slums and squatter settlements of the city. About 80% of these settlements are owned by private owners who rent out their slum houses to the poor. Most of the slum dwellers are renters and pay very high house rent as there is no institutional mechanism to control rent hike. Various studies show that the slum dwellers pay comparatively a higher amount as house rent per unit area than the middle-income people. Even they pay more for the utility services than those are paid by the middle-income households. This is a paradox that the poor spend more than the middle and even the high-income people for housing, while their affordability is low. The paper examines this paradox taking example of poor workers' housing provision in the slum areas of Dhaka. The poor workers experience severe housing crisis, as their affordability is low. At present they spend about Taka 4,000 to 5,000 for about one hundred square feet of floor space in slum areas, which is equivalent to 20–40% of their total household income. Although they pay so high for housing, cannot afford to own even a small unit due to high land price and lack of credit facilities. The housing market is totally controlled by the private sector, where the government or institutional arrangements for making provisions for housing the poor are conspicuously absent. In such situation, the poor face increasingly tougher challenges to own accommodation and survive in Dhaka Metropolitan Area.

Keywords Slums · Housing affordability · Rental market · Institutional arrangements · Urban poor

N. I. Nazem (✉)
Department of Geography and Environment, University of Dhaka, Dhaka, Bangladesh
e-mail: nazem.info@yahoo.com

S. Sultana
Department of Geography and Environment, Jagannath University, Dhaka, Bangladesh
e-mail: sspapri52@gmail.com

L. T. T. Huong and G. M. Pomeroy (eds.), *AUC 2019*, Advances in 21st Century Human Settlements, https://doi.org/10.1007/978-981-15-5608-1_36

1 Introduction

It is not very easy to conceptualize the poor and poverty. In urban areas, understanding poverty is even complex. The manifestation of poverty in urban areas is reflected in many ways. Ownership of land and other key resources and the level of income are important indicators. However, these cannot be easily traced. Housing and housing-related problems such as water supply, sanitation condition, electricity and cooking materials, access to education and health facilities, etc. are easy identifiers and thus can be considered as appropriate indicators to measure poverty. Access to all these services provided by the state and society is another important issue. We need to understand the nature of accessibility of the urban poor to facilities like health, education, infrastructure services and recreation, etc. being provided through the market economy. The poor lack power and have limited access to legal provision and little ability to raise their voices along with many other deprivations.

Poverty should be understood in a combination of all these variables. Urban poverty is, in fact, one of the serious problems in urban areas, particularly in big cities. However, our city authorities hardly consider this issue seriously for meaningful interventions. On the contrary, city authorities and local administration tend to consider the poor as a burden for cities; and most often publicly express that the poor should go back to their places of origin instead of living inhuman life in cities. The presence of the poor in cities and their arrival from rural areas is considered negatively. This is perhaps the reason why the authorities are reluctant to provide services and manage them for their better living in cities. The authorities hardly try to realize as to why and under what circumstances the poor come to cities and struggle in cities to survive. As a result, the problem spiralled and gradually turned into a serious one to be solved easily. Meanwhile the number of the poor has been increased, their living circumstances have become further complicated. Thus, the cities in Bangladesh could not be poverty-free.

2 Poor Settlements in Urban Areas: Number and Slum Dwellers

There are about 530 towns and cities in the urban systems of Bangladesh. Census 2011 shows that nearly 40.2 million people lived in these towns and cities (BBS 2014). By now, perhaps the number of population in cities and towns has crossed 50 million. However, the actual number of the poor is not known. The definition of the poor is even not clear. In 2005, Centre for Urban Studies (CUS) Dhaka defined urban slum areas where the poor live. The definition includes five indicators such as monthly household income to be less than Taka 5000; very poor structural quality of housing: very high density of population or room crowding; insufficient municipal utility services, especially water and sanitation and lack of housing security of tenure

(Islam et al. 2006). All these indicators together make a household poor with low socioeconomic status.

Using this definition, the study identified over 4000 slums and squatter settlements in the largest six cities of Bangladesh. In 2005, the total population of these six cities was 15.4 million of which about 5.4 million were found to be slum dwellers (Islam et al. 2006). Thus, 35% of all dwellers live in slums in these cities. The highest proportion of slum dwellers, 37% was found in Dhaka city, followed by 35.4% in Chattogram, 32% in Rajshahi, 30% in Barishal, 27% in Sylhet and 19% in Khulna.

An overwhelming majority of the slum dwellers are urban poor. However, there are poor who live outside the slum areas. Another study of CUS (2010–2011), which mapped poor settlements of 27 towns and cities of Bangladesh, found about 44,000 poor settlements (Nazem and Hossain 2012). Many of these settlements were not slums as defined above, yet the dwellers were considered to be the poor. These settlements contained 55% of the total households of 27 study cities (Nazem and Hossain 2012). Among the identified 1.1 million households in poor settlements 36.3% lived in extreme poverty situation. About 25% were very poor, 23% were poor and only 16% were poor marginally. Dhaka Metropolitan Area was not included in this survey.

In 2014, Bangladesh Bureau of Statistics (BBS) conducted a countrywide census and survey on urban slums. This census shows that there were 2.23 million slum dwellers in urban areas in thirteen thousand nine hundred slums (BBS 2015). The number of slums and the slum population as recorded by the BBS census seem too low if compared with the previous studies. The reason behind such a low figure is perhaps the definition of slum and the method survey. At this stage, it is very difficult to say that the number of poor in cities has drastically reduced. However, poverty has decreased countrywide. The urban areas, however, continue to receive poor rural migrants, and that influences the increase of urban poor. There are evidences that some of the indicators of poverty such as income and housing condition in slum areas have improved. This is due to overall development of the country. There is hardly any slum development programme in urban areas visible in recent years.

3 Housing Situation in Slum Areas

Housing condition has never been satisfactory in slum areas of urban Bangladesh. Although, over the years, some physical characteristics of slum houses show improvements, the shortage of housing units has remained central to the housing problem. Demand side for low cost and cheaper housing units in urban areas is as high as ever. On the other hand, supply is low as usual to meet the high demand for housing. In 1991, housing deficiency in urban areas was only 0.65 million, which reached 4.6 million in 2010 (World Bank 2012). At present (2019), this deficiency seems to have reached 8.5 million. This trend of increasing housing demand is largely due to influx of low-income people in cities. Thus, there exists a huge gap between housing demand and supply, particularly for low-income people. To address the situation, housing supply for low-income people should be at the centre of development programmes in urban areas.

Land is very crucial for housing supply. A limited supply of land and high price along with lack of necessary infrastructure prevent required supply of land for housing and to meet the increasing demand. The high price of land is considered to be main obstacle for smooth housing production in the densely populated cities of Bangladesh. Particularly in urban areas, high price of land, which prevents the production of low-income housing, is considered to be the main constraint in building the buying capacity of citizens. In city like Dhaka, more than 70% of its households do not own land or house, and they are compelled to live in rented houses (BBS 2012). At least one-third of households of Dhaka belong to the category of very low-income group. A decent and livable house is beyond their capacity. Most of them live in congested slum areas. A study by CUS (2005) shows that about 37% of the people of Dhaka city live in only 5% of land, and 35% of the people of Chattogram city live in 3% of its areas (Islam et al. 2006). Other cities also show almost similar condition. This indicates that urban housing condition in Bangladesh remained in an acute imbalanced condition between the poor and the middle and high-income groups. No one can expect such an imbalanced situation in the pattern of our national development.

Housing is the key problem of the urban poor. The poor come to cities for earning income, not for occupying a house. When they get a job or any opportunity to earn income, need shelter first to continue their earnings. However, a large portion of these earnings goes to meet house rent. Several studies show that the poor pay more as house rent than those paid by the middle-income people for housing (Islam and Shafi 2008). This is a very hard reality.

One example can be cited. Monthly house rent for a one-room accommodation in slum areas in Mirpur (North-west Dhaka) is about Tk. 3000–4000. The room is approximately about 100 square feet in size. Bathing, toilet and cooking facilities are common, which is usually shared by 4–5 similar households (Average size of household in urban areas of Bangladesh is 4.1). If this is compared with a housing unit of about 1500 square feet in middle-income areas like Kalabagan or Maghbazar in central Dhaka, with an average house rent of Tk. 25,000 to 30,000 (maximum); it would be surprising to observe that what the poor people pay for 100 square feet unit. If a 100 square feet slum housing unit costs Tk. 3,000 per month, then a 1500 square feet unit in middle-income areas should be at least Tk. 45,000 per month. In reality, such a unit costs much less. This shows that the poor pay more for a fragile unit when compared to those rented by the middle-income people. The poor pay more not only for housing but also for utility services. Moreover, there is no control or monitoring of the situation. This can happen only in unregulatory market economy, where demand side for poor's housing is so high (Mowla and Hossain 2007).

Various studies show that over the years the quality of housing in slum areas has improved slightly in large cities (Baker 2007). Most of the thatch and *katcha* (Non-concrete) houses were replaced by semi-*pucca* units. Water supply, sanitation and electricity supply have also improved, though the costs of utilities are very high in slum areas. Most of the slum dwellers now drink pure water. Most of them have electricity connections. Waste collection condition, however, is not yet satisfactory. There is hardly any road network except walkways in slum areas (Islam and Shafi 2008).

In the absence of any control, house rent is increasing day by day, which influences the slum dwellers in shifting their houses frequently. The rate of home shifting is the highest among the slum dwellers. They change their rented accommodation due to frequent rent hike and this happens due to the fact that there is no legal document between the homeowners and renters in slum areas. Fixing of house rent and terms and conditions are always verbal. It gives houseowners an edge over the renters, which the owners use to force the renters to leave their houses at any time. Frequent movement creates serious problems for the slum dwellers. Changes in children's schools, disruption in the health care system, borrowing microcredit, etc. are some of the serious problems for them. Some NGOs provide primary health care services, but the slum dwellers are really helpless during serious diseases.

It is to be noted here that the slums are usually controlled by local political power holders. They even have influence on the newcomers to the slum areas. The slum dwellers also accept their illegal dominance to avoid harassment. It is really difficult to do anything in the slum areas bypassing the *mastaans* (Hooligans) and local power elites. Slum dwellers are often used politically, such as for attending rallies and public meetings, etc.

The following are some reality of urban poor based on research findings.

- Room/house rent is very high in slum areas, and there is no control over this rent hike. Slum dwellers pay this rent as they have no other ways to protest and no other places to go. They usually pay a higher amount for accommodation and utility services when compared to those living in the middle-income areas.
- Mobility of slum dweller within the city in terms of changing their residences is very high. It is due to rent hike and uncontrolled rental market.
- Slum dwellers' spending on housing does not match with their household income. They pay maximum for minimum services, which are also not matching with their income.
- Over the years, the size of slums (in terms of area) is getting smaller and smaller. It is due to the fact that large slums and squatters have been gradually evicted as these were mostly located on public land. The trend now is the privatization of slums in small areas. This leads to an increase in the number of slums, while their size in terms of area and population is decreasing. At present, the number of small-sized slums is more than the number of big-sized slums. The large-sized slums are disappearing, and the small-sized slums are increasing.
- Most of the large-sized slums are located in the government-owned land or institutional land or in *khas* land. Small-sized slums are located on privately owned land.
- It has been observed that the slums located on privately owned land show incremental improvement. Physical infrastructure is good and utility services are better

than earlier. Most of these slum houses are semi-pucca. In the case of large-sized slums/squatter settlements which are located mostly on public land illegally, such as squatters, physical development is either slow or absent.

- Slum dwellers are comparatively more vulnerable due to social and environmental problems rather than economic problems. Even, this urban poor are more vulnerable than the rural poor.
- It is necessary to realize as to why the poor come to cities. It is also necessary to evaluate their contribution to the city economy. Housing can be an indicator for a healthy city.

4 Importance of Housing Sector to Reduce Poverty

4.1 Housing and GDP

In developing countries, 3–8% of GDP comes from construction sector, of which one-third is contributed by housing sector. On the other hand, construction sector contributes to 40–70% of the national investment. Informal sector housing market provides about 80% of total housing supply. GDP of the country would have been much enriched if we could consider and assess the informal sector housing properly.

4.2 Employment in the Housing Sector

Housing sector provides employment and contributes directly to the economy. Building and construction industry is one of the major employment providers for skilled and unskilled labour force of cities. Examples of various countries indicate that investment in housing industries and employment can play important role to improve the economy. Construction and related industries of the country contribute to at least 10% of the employment sector. Such contributions can be increased further by creating employment in specialized building and construction material sector. Especially, demands for products such as rod, cement, tin, tally, ceramic materials and its export demand are also rising. This indicates that housing can be a strong economic sector in future. However, these products should be made available for the poor.

4.3 Housing as Source of Income

Housing sector not only provides accommodation or shelter, but the sector can be considered to be an important source of income. The sector plays an important role in generating employment of the poor. This is quite a vibrant employment sector, where skilled and unskilled workers can be absorbed for income-earning activities. For this, the housing industry should be developed in such a manner that it can contribute to the national economy.

5 Conclusions

Slums are settlements where the urban poor live. The poor cannot be separated from our cities. Thus, for our national development, it is necessary to consider the poor and their development positively. One of the key issues is the provision of housing for the poor in cities. Considering their contribution, it is a stronger demand that the poor have the right to own a house. To ensure the positive benefits of urbanization in the country, it is necessary to provide shelter to the poor and this should be an essential goal of urban planning and development in the country. Housing sector contributes to the employment and economic growth of cities. Thus, housing sector should be considered with its due importance. This will help achieve livable environment, good health, proper education, employment, economic growth as well as the overall development of cities. Most importantly, this will help achieve many of the SDG goals.

However, the city authorities' attitude and the approach to development in cities do not help the development of the poor. The approaches are hardly inclusive. Authorities need to change their attitude and approaches to development particularly of the poor and homeless, by recognizing their contribution. Their contribution, especially, in production and labour market both in formal and informal sectors is very much remarkable. To reach the desired goal of inclusive development, proper investment in low-income housing sector is necessary. And also, a change in the existing policy and planning system is required along with good governance.

The proposed Urban Policy and Housing Policy of Bangladesh have been expressed regarding the rights of poor. But the solution to the housing problem of the urban poor seems difficult through the existing market economy, where the role of the State is gradually shrinking. If the poor do not get loan facilities without interest or less interest, it is not possible to solve housing problem of the individual poor. Secondly, land should have been kept for poor people in various areas through zoning so that they can buy land with reduced price. Thirdly, government can facilitate rental housing system for the poor, which if they wish can buy and own through instalment. The most important thing is the state's assistance in the low-income housing for the poor.

References

Baker JL (2007) Dhaka: improving living conditions for the urban poor. Bangladesh development series paper, 17. World Bank

BBS (2012) Population and Housing Census 2011. National Series, Volume 4: Socio-economic and Demographic Report. Bangladesh Bureau of Statistics

BBS (2014) Population and Housing Census 2011. National Volume 3: Urban Area Report. Bangladesh Bureau of Statistics

BBS (2015) Census of slum areas and floating population 2014. Bangladesh Bureau of Statistics

Islam N, Shafi SA (2008) A proposal for a housing development programme in Dhaka city. Centre for Urban Studies

Islam N, Angeles G, Mahbub A, Lance P, Nazem NI (2006) Slums of urban Bangladesh: mapping and census 2005

Mowla QA, Hossain MS (2007) Living environment for Dhaka's urban poor–an analysis. In: Architecture for the Economically Disadvantaged: Proceedings of the International Seminar. Dhaka, BUET

Nazem NI, Hossain MA (2012) Mapping poor settlements in 27 cities and towns of Bangladesh: UPPR experience. CUS Bulletin on Urbanization and Development. 60–62:16–20

World Bank (2012) World Bank indicators: Bangladesh urbanization

Evaluation of the Residential Satisfaction in Affordable Housing for Low-Income People and Its Social Impact on Urban Planning in Hanoi, Vietnam

Trinh Thi Kieu Trang and Bui Ngoc Tu

Abstract Providing appropriate housing for the people is one the responsibilities of the governments. In recent years, the government of Vietnam decided to implement the affordable housing for low-income households with many projects in Hanoi and Ho Chi Minh city. The aim of this research was to evaluate the satisfaction rate of affordable housing residents in Hanoi, where the population density is extremely high and the average income is low, compared to the housing price. This research used survey methods, data analysis, statistical methods, comparison and expert method. A field survey was implemented for data collection. The main fieldwork was undertaken during June 2019 and August 2019 in Linh Dam, Hanoi. Findings revealed that the overall level of satisfaction of the residential environment, including housing features and services, is moderate. However, this level of satisfaction is not constant; rather, it fluctuates across the variables of housing features, service quality, and environmental policies. This paper also examined the determinants of residential satisfaction and its social impact on urban planning in Hanoi. The overall conclusion is that attempts to improve low-income housing and the surrounding services may result in improving overall satisfaction.

Keywords Residential satisfaction · Affordable housing · Low-income · Services · Hanoi

1 Introduction

Residential satisfaction is an assessment of the extent to which the present dwelling of residents and the quality of the environment is close to aspiration of their favorite one (Galster 1985). Or more specifically, Residential satisfaction, defined as the feeling of contentment when one has or achieves what one needs or desires in a house, is an important indicator, and planners, architects, developers, and policymakers use it in a number of ways (Mohammad and Adel 2014). Residential satisfaction

T. T. K. Trang (✉) · B. N. Tu
Faculty of Geography, VNU University of Science, Thanh Xuan, 334 Nguyen Trai, Hanoi, Vietnam
e-mail: trinhthikieutrang@gmail.com

L. T. T. Huong and G. M. Pomeroy (eds.), *AUC 2019*, Advances in 21st Century Human Settlements, https://doi.org/10.1007/978-981-15-5608-1_37

studies are quite vital among low-income resident; most of the studies on affordable housing have been conducted in the area of planning to increase affordable housing supply and less to evaluate the results of planning policies on residential satisfaction (Crook and Whitehead 2010; Karki 2015; van den Nouwelant et al. 2015). In reality, dissatisfaction in most cases motivates families to relocate (Rossi 1955), and it affects regional development policies in the future.

The production of affordable housing has three stages, which include policy-making, design, and post-occupancy evaluation. In this process, three major groups of professionals including policymakers, designers, and researchers are involved. There are a lot of research on this area (Checa and Arjona 2010; Huang et al. 2015; Mohit et al. 2010; Amérigo and Aragones 1997; Etminani-Ghasrodashti et al. 2017; Ibem and Aduwo 2013). However, there is no studies on evaluation of the residential satisfaction in affordable housing for low-income people in Vietnam. This article offers a new category for residential satisfaction which can be more useful for policy-makers, designers, and researchers. In developing countries such as Vietnam, where they still do not have vast experiences in low-income housing projects, this category can help relevant experts in the development process of affordable housing.

Started in 1997 with over 200 ha, including two residential areas (North Linh Dam residential and Linh Dam peninsula and general service area), Linh Dam new urban area is one of the first new urban areas in Hanoi. It contributes to create a driving force for the "real estate economy" with the appearance of number of new urban areas and development until today. The Hanoi People's Committee had planned to turn Linh Dam area into one of the best model urban area in 2009. However, this urban area seems to be overloaded by changes in land use purpose, lack of investment in essential life services, and faces excessive population growth problems. The appearance of affordable housing for low-income people with 12 apartment buildings with high-density construction and excessive population growth has totally changed the urban planning architecture of this area. Residential satisfaction in affordable housing is one of the most severe challenges of policymakers around the world. However, there were insufficient studies on assessment of residential satisfaction in Vietnam. Hence, this article attempts to take the first step in this area. This paper contributes to the residential satisfaction literature with affordable housing for low-income people by assessing the survey to improve residential satisfaction. The main purpose of the research is to find out the reasons for (dis)satisfaction in Linh Dam HH affordable housing. Another purpose is to examine the factors that affect the residential satisfaction the most. The findings will help understand the housing needs of low-income families as well as their housing expectations and reflect it in planning systems and design principles.

2 Literature Review

The concept of satisfaction refers to the wide domain of utilities and demands to meet the essential human needs. Satisfaction is a subjective evaluation of the performance of the products in meeting customers' needs and expectations (Parker andMathews 2001). Thus, when we tend to choose and achieve what we want; our satisfaction is obtained (Gifford 1999). Housing factor is one of the important components of life satisfaction. In fact, residential satisfaction is considered as one of the most studied topics in the field of residential environment (Ge and Hokao 2006).

Residential satisfaction refers not only to physical aspects of residential setting such as dwelling, housing developments, and neighborhoods, but also social, economic, and organizational or institutional aspects of such settings. Therefore, we can define housing satisfaction as the feeling of contentment when one person achieved to his needs or desires in a house (Francescato 1987).

The issue of assessing residents' satisfaction with the residential environments can play a great role in achieving the goals of urban projects. It is essential to have a multidimensional approach to understand the complexity of emergence and development of residential satisfaction. The studies on residential mobility of Rossi (1955) realized that households in life cycle face lack of fit regarding their housing needs. These problems are caused by household size, age, prestige, etc., and lead to dissatisfaction. While Wolpert (1966) combined two forgoing theories (adjustment theory and housing needs) and created an ecological approach to deal with environmental stress. He belived that families adjust themselves to some environmental factors decide to migrate.

Residential satisfaction is also presented as a concept with multiple features. Overall, residential satisfaction depends on the satisfaction level of two general components comprising satisfaction of neighborhood and residential units directly and also personal, cultural, and social features indirectly.

3 HH Linh Dam Urban Area Scheme

With about 160 ha area of which nearly 50% is water surface, Linh Dam peninsula has had outstanding advantages in landscape, and it has been planned to be a model of new urban area in Hanoi. But, new urban areas such as VP3, VP5, and HH Linh Dam have disrupted the spatial structure of this area. Especially, the 5 ha planned for the general service center was replaced by 12 apartment buildings named HH Linh Dam which has completely devastated the landscape of this area. HH Linh Dam is located in the southwest of Linh Dam peninsula. It is an affordable housing project of Mr. Le Thanh Than for low-income people. In 12 buildings, each building has 36–42 floors, and each floor has about 20 apartments. This project solved housing problems for low-income people, creating opportunities for people to have an apartment in the capital.

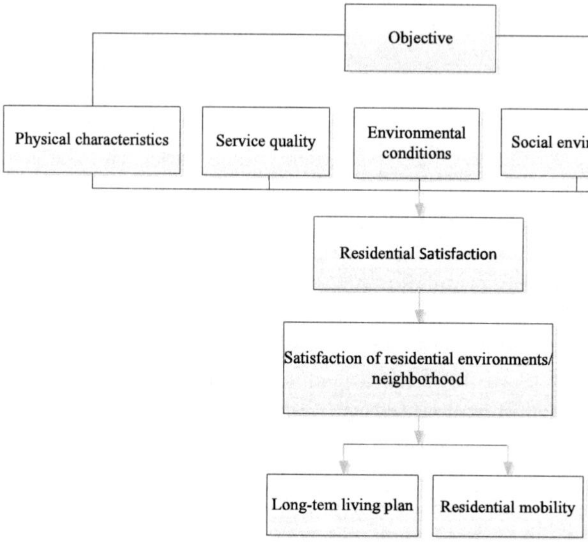

Fig. 1 Model to measure the satisfaction level of residential research

After being completed, this HH Linh Dam urban area will add about 60,000 people to Linh Dam peninsula, making big problems in this area. The project broken the original urban planning for Linh Dam area. The construction density is too dense. Population density is too high. Infrastructure and social services do not meet the needs of people living in the area. Environmental landscape is affected. There are no parks, schools, play areas for children. Traffic is often congested. Owners of apartments could not get the house ownership certificate because the investor have built exceeded the number of floors permitted. (Fig. 1).

4 Methodology and Fieldwork

A field survey methodology was considered the most appropriate data collection approach for this paper. The research population for the study is the inhabitants of the HH Linh Dam urban area. Low-income housing in HH Linh Dam is comprised of a wide range of housing styles (type, area, level) and services. Furthermore, the satisfaction of inhabitants with their housing in this region has not been investigated previously. A questionnaire was used as the main method of data collection; 150 questionnaires were personally distributed by a small team of trained research assistants during June 2019 and August 2019, of which 112 were considered as useable.

Reliability was tested through the “Alpha” test, and all scales used in the questionnaire were considered reliable. Descriptive statistics in terms of means and frequency were used to describe the respondents’ characteristics. Factor analysis was used to

Table 1 Resident satisfaction with the physical characteristics

	N	Mean	S. D
Sunlight, air into the apartment	112	3.36	0.721
Aesthetics of the apartment	112	3.43	0.625
Soundproof, anti-noise	112	2.93	0.835
Views from the apartment to the surroundings	112	3.65	0.768
Apartment location	112	2.52	0.847
Quality of interior and interior materials	112	2.67	0.842
Layout of the room space	112	2.70	0.847
Apartment security system	112	2.81	0.729
Fire protection system of the apartment	112	3.58	0.755
Construction quality and interior design	112	3.00	0.805
Total	112		

answer the research questions. Finally, the principal components of factors affecting residential satisfaction were identified using spider web diagram of five groups of factors.

5 Findings Analysis

To measure the satisfaction level of dwellers under each category of satisfaction attributes, the 32 selected attributes of satisfaction were subclassified into five groups of elements.

5.1 Satisfaction with the Physical Characteristics of the Apartment

There are ten elements for physical services. Table 1 demonstrates that they tend to be rated as middle–low level; it can therefore be construed that physical infrastructure is in poor conditions. Especially the groups of factors of fire protection system, apartment's views, and the aesthetics have the lowest satisfaction level.

5.2 Satisfaction with Service Quality

With about 60,000 people living in 12 buildings of HH Linh Dam urban area (density of 9 people/m^2), this place becomes the most "crowded" residential area in Hanoi.

Table 2 Resident satisfaction with service quality

	N	Mean	S. D
Children's play area and welfare facilities in the block	112	2.95	0.837
Public institutions, hospitals, and shops	112	3.02	0.890
School and educational environment	112	3.04	0.853
Privacy protection	112	2.86	0.868
Infrastructure (electricity, water, elevators)	112	3.84	0.778
Convenient parking	112	4.12	0.836
Distance from home to work	112	3.40	0.650
Public transport and road	112	3.68	0.633
Ensuring security in the housing area	112	2.87	0.691
Total	112		

Infrastructure group, parking space, and public transport become serious problems. Some reasons are due to dense population but with no parking space; elevators do not meet the demand of residents, especially, during rush hour, it takes about half an hour to get in. Lobby and playground for residents are encroached as parking lot, shops, market. Traffic congestion occurs regularly in the urban area. Table 2 shows that residents are not satisfied with service quality in HH Linh Dam apartment area.

5.3 *Satisfaction with Environmental Conditions*

Linh Dam urban area was considered as a model urban area of Hanoi in 2009. It was surrounded by a large natural lake and an artificial lake in the middle, with a canal designed and constructed to connect with the lake. But most of the evaluation factors about environmental conditions group tend to approach dissatisfaction. The canal in Linh Dam has long had no water to circulate; plus the people discharging domestic waste have seriously polluted the canal. This canal is located along the row of apartment buildings HH2B, HH2C, HH3B, and HH3C. Since the canal is heavily polluted, there are almost no visitors anymore. Proposing local authorities and residents should start dredging and cleaning the canal soon, returning the freshness to the urban area. Recently, HH Linh Dam apartment complex has been continuously witnessing fire and explosion which has left people extremely worried about the quality and fire safety in the apartment (Table 3).

Table 3 Resident satisfaction with apartment environmental condition

	N	Mean	S. D
Green space in the housing area	112	3.36	0.598
Visibility and good landscape	112	3.23	0.644
The freshness of the air in the housing	112	3.25	0.637
Waste management	112	3.41	0.637
Fire extinguishing system in the building	112	3.10	0.553
Public spaces and clean streets	112	3.20	0.598
Total	112		

Table 4 Resident satisfaction with social service

	N	Mean	S. D
There are relatives living in the house	112	2.58	0.706
Relationship with neighbors	112	2.62	0.762
The meetings to collect opinions of the people	112	2.92	0.673
Total	112		

5.4 Satisfaction with the Social Environment

The Table 4 showed that most of the residents chose HH Linh Dam urban area because of the affordable criteria; so, the criteria for social environment are rated at an average level, not focusing much on these elements.

5.5 Satisfied with the Economic Environment

HH Linh Dam is affordable housing for low-income people; so, the criteria for house price, rents, and expect value of housing are rated at an average level, but the brand of HH Linh Dam urban area is not appreciated (Table 5).

Table 5 Residents' satisfaction with economic environment

	N	Mean	S. D
House prices, rents	112	2.55	0.757
Cost of maintenance and service	112	2.67	0.821
Brand of the urban area	112	3.00	0.794
Expect value of housing and investment to increase	112	2.74	0.791
Total	112		

Table 6 Summary factor analysis

	N	Min	Max	Mean	S.D
Satisfaction with the apartment	112	2.00	4.40	3.0643	0.44069
Satisfaction with the service	112	2.22	4.89	3.3075	0.38141
Satisfaction with environmental conditions	112	2.50	4.67	3.2574	0.31551
Satisfaction with the social environment	112	1.00	4.00	2.7054	0.58295
Satisfaction with the economic environment	112	1.00	4.25	2.7411	0.60676
A	112				

5.6 General Assessment

According to the heads of households' reporting, lack of unit facilities and lack of area for vehicle parking are the reasons of their dissatisfaction. These mentioned factors that lead to residents' dissatisfaction are derived from failing to follow standards in complex building designing. Residents had no idea about physical features factor and declared neutrality about this issue.

Majority of the respondents were not satisfied with the living environment quality and service quality. The reason of this dissatisfaction can be found in (1) Poor performance of housing cooperatives in the delivery to households (2) Condominiums are degraded, people do not enjoy basic utilities, and security has arisen recently. In addition to these factors, most of the residents' complaints are related to traffic, public transport and road, social service. The high-density resident in narrow area is the main cause of this problem (Table 6).

According to studies, interviewees are not satisfied with the service quality and living environment. The low environment quality, lack of fire safety, poor quality of life, lack of necessary utilities, and dozens of other shortcomings make many families consider selling apartments in HH Linh Dam to find better new places to live. In contrast, more than 60% were satisfied with the security and participation of residents in social activities.

General aspects of HH Linh Dam housing residents' satisfaction was examined by measuring axes such as apartment feature, service quality, the quality of the local living environment, the social environment and economic environment; resolving the problem of housing by providing this kind of low-income homes, and adaptation delivered housing for low-income families (Fig. 2).

More than 30% have plan to move to another area in future. One reason is Linh Dam is quite far from the workplace or school; the system is not yet synchronized for good living conditions (Tables 7 and 8).

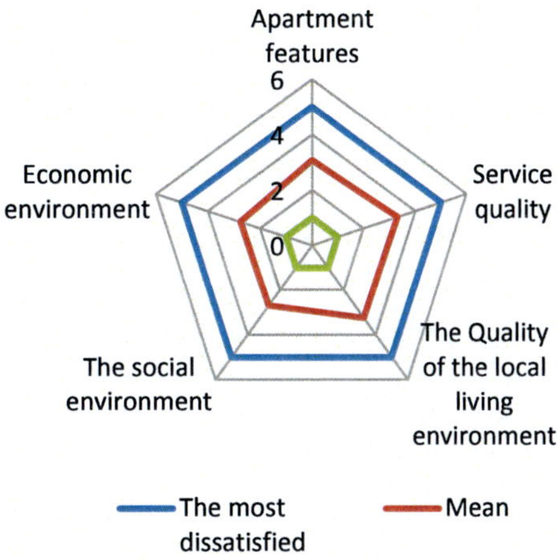

Fig. 2 Spider web diagram

Table 7 Residents' plan in future

	Frequency	Percent	Valid percent	Cumulative percent
Long-term living plan	80	71.4	71.4	71.4
Moving plan	32	28.6	28.6	100.0
Total	112	100.0	100.0	

Table 8 Desired accommodation when moving

	Frequency	Percent	Valid percent	Cumulative percent
My Dinh or Trung Hoa-Nhan Chinh	23	20.5	20.5	20.5
Royal City	16	14.3	14.3	34.8
Times City	6	5.4	5.4	40.2
Other urban areas in downtown areas	59	52.7	52.7	92.9
Hanoi suburb	8	7.1	7.1	100.0
Total	112	100.0	100.0	

6 Conclusion

HH Linh Dam urban area is consistent with macroeconomic policies of Vietnam government in providing housing for low-income households, but it suffers from a number of disadvantages that deserve special attention. The level of resident satisfaction factors can assist to understand the strengths and weaknesses of these projects. This study can measure the satisfaction level of HH Linh Dam housing residents.

This study showed that the housing problem of majority of residents of HH Linh Dam housing has not been solved by this kind of affordable housing. Therefore, the hypothesis of this study that is HH Linh Dam buildings' residents are not satisfied with affordable housing that was provided for low-income families was supported. Also, HH Linh Dam housing household complained of the living environment, service quality, access to public transport, lack of facilities out of unit, lack of area for vehicle parking, and costs. In contrast, internal conditions, security, and social participation of residents were reported as the main advantages of the sites. It can be claimed HH Linh Dam housing project has not been able to meet the demands of low-income people, and residents are not satisfied with their affordable houses.

To increase satisfaction of HH Linh Dam housing, the following strategies can be suggested:

- Monitoring and quality controlling the performance of cooperatives that build the affordable housing for low-income people;
- Government's commitment to approved quality of affordable housing;
- Designing of complexes in the metropolitan area;
- Increasing public facilities and services in the area of town;
- Establishing consult centers to reduce social problems of the residents;
- Providing good public transport facilities to access the city center, working places, school, hospital…
- Constructing pre-designed parks and green spaces in the complex;
- Reducing construction density in affordable housing projects and issue house ownership certificates for apartment owners.
- Attention to car spaces in the design of residential units.

References

Amérigo M, Aragones JI (1997) A theoretical and methodological approach tothe study of residential satisfaction. J Environ Psychol 17(1):47–57

Andersen HS (2008) Why do residents want to leave deprived neighbourhoods? The importance of residents' subjective evaluations of their neighbourhood and its reputation. J Housing Built Environ 23(2):79–101. https://doi.org/10.1007/s10901-008-9109-x

Checa J-C, Arjona Á (2010) Residential satisfaction in a context of urban exclusion. Psyecology 1(2):221–237. https://doi.org/10.1174/217119710791175632

Crook T, Whitehead C (2010) Intermediate housing and the planning system. The role of intermediate tenures, Making housing more affordable, pp 101–119

Etminani-Ghasrodashti R, Majedi H, Paydar M (2017) Assessment of residential satisfaction in Mehr housing scheme: a case study of Sadra New Town, Iran. Housing, Theo Soc 34(3):323–342. http://doi.org/10.1080/14036096.2017.1298536

Francescato G (1987) Residential satisfaction: its uses and limitation in housing research. In: Willem et al (ed) Housing and neighborhoods, theoretical and empirical contribution. Greenwood Press, New York

Galster GC (1985) Evaluating indicators for housing policy: residential satisfaction vs marginal improvement priorities. Soc Indic Res 16(4):415–448

Galster GC, Hesser GW (1981) Residential satisfaction: compositional and contextual correlates. Environ Behav 13(6):735–758
Garde A (2016) Affordable by design? Inclusionary housing insights from Southern California. J Planning Edu Res 36(1):16–31
Ge J, Hokao K (2006) Research on residential lifestyles in Japanese cities from the viewpoints of residential preference, residential choice and residential satisfaction. J Landsc Urban Planning 78:165–178
Gifford R (1999) Perception and recognition environmental translated dehbashy. Architect Cult J 2:21–29
Huang Z, Du X, Yu X (2015) Home ownership and residential satisfaction: evidence from Hangzhou, China. Habitat Int 49:74–83
Ibem EO, Aduwo EB (2013) Assessment of residential satisfaction in public housing in Ogun State, Nigeria. Habitat Int 40:163–175. http://doi.org/10.1016/j.habitatint.2013.04.001
Karki TK (2015) Mandatory versus incentive-based state zoning reform policiesfor affordable housing in the United States: a comparative assessment. Housing Policy Debate 25(2):234–262
Mohammad AM, Adel Mahfoud MAK (2014) Residential satisfaction-concept, theories and empirical studies. Urban Planning Local Govern 3:47–66
Mohit MA, Ibrahim M, Rashid YR (2010) Assessment of residential satisfaction in newly designed public low-cost housing in Kuala Lumpur, Malaysia. Habitat Int 34(1):18–27.http://doi.org/10.1016/j.habitatint.2009.04.002
Parker C, Mathews BP (2001) Customer satisfaction: contrasting academic and consumers' interpretations. Market Intell Planning 19:38–46
Rossi PH (1955) Why families move: a study in the social psychology of urban residential mobility. Free Press
Salleh AG (2008) Neighbourhood factors in private low-cost housing in Malaysia. Habitat Int 32(4):485–493. https://doi.org/10.1016/j.habitatint.2008.01.002
van den Nouwelant R, Davison G, Gurran N, Pinnegar S, Randolph B (2015) Delivering affordable housing through the planning system in urban renewal contexts: converging government roles in Queensland, South Australia and New South Wales. Aust Planner 52(2):77–89
Wolpert J (1966) Migration as an adjustment to environmental stress. J Soc Issues 22(4):92–102

Intergenerational Perspective on Right to the City in Relation to Changing Nature of Social Contracts in the Slums of South Delhi

Swagata Basu and Gloria Kuzur

Abstract Slums are an outcome of the dependent urbanisation process in most of the post-colonial countries. The nature of dependence was such that it siphoned off indigenous resources to subsidise the development of the metropolis at the cost of its own population. Deteriorating economic conditions along with growing population pressure compelled the urban settlers to compromise on the quality of environment, comforts, dignity and security that became precarious over time. Drawing from the Lefebvrian concept of 'Right to the City', this paper attempts to chart the inter-generational shifting goalposts of expectations of new social contract among slum dwellers regarding their rights to affordable housing, privacy and security of the most vulnerable sections of inhabitants (children, women, senior citizens and differently abled persons) of Kusumpur Pahari, New Delhi. The study area is a spatially compact, yet socioculturally layered slum juxtaposed with affluent neighbourhoods, state-of-the-art shopping malls dotted with ultra-modern housing colonies of foreign diplomats and the bureaucrats in its vicinity. The concept of 'congestion' expressed in terms of the availability of exclusive rooms for married couples within the question of 'housing inadequacy' is explored through in-depth interviews of the residents of the slum. The first-generation settlers' struggles to create a 'socially just city' find articulation in the aspirations of the formally educated younger generation of slum dwellers who express their desire to be a part of the decision-making process of city building just as they create the 'oeuvre' of the city through their quotidian practices of the production of urban space.

Keywords Urban slums · Housing inadequacy · Congestion · Right to the city · Socially just city

S. Basu (✉)
Department of Geography, SSV (PG) College, Hapur, India
e-mail: swagatab@rediffmail.com

G. Kuzur
Centre for the Study of Regional Development, Jawaharlal Nehru University, New Delhi, India

L. T. T. Huong and G. M. Pomeroy (eds.), *AUC 2019*, Advances in 21st Century Human Settlements, https://doi.org/10.1007/978-981-15-5608-1_38

1 Introduction

> Any city, however small, is in fact divided into two, one the city of the poor, the other of the rich; these are at war with one another.—Plato.

The processes associated with 'Dependent Urbanisation' in the post-colonial nations come with a natural corollary of the creation of slums and squatter settlements in urban centres. Cities in India bear testimony to this historical process of slum creation and the issues of governance and aesthetics associated with it continue to plague urban planners. D'Souza (1979) observed that slums are found in cities irrespective of its size, or the time period of its existence. Vaddiraju (2016) referring to the Census of India 2011 data documents that 377 million people in India live in urban areas who accounts for 31.6% of the country's population. There are 53 million cities in India, and more than 43% of the urban population inhabit in those cities. Kundu (2011) terms India's urbanisation process as 'exclusionary' since the local elites have had a head start in appropriating the city-space. The result of this exclusionary process leads the way to the gentrification of cities coupled with congestion and other aspects of social pathology. The slum population in India account for 5.41% of the total population but in the case of the capital city, namely New Delhi, the slum population accounts for 10.63% of the population. The Directorate of Economics and Statistics, NCT Delhi, defines slums as ' a compact settlement of at least 20 households with a collection of poorly built tenements, mostly of temporary nature, crowded together usually with inadequate sanitary and drinking water facilities in unhygienic conditions'. The large mass of people living in slums of Delhi often elicit questions of 'Right to the City', derived from Lefebvrian concept of the right to reclaim public spaces, habitation and housing and looking for ways and means to secure the future of the slum dwellers from the onslaught of 'top-down' beautification process undertaken by the civic authorities of the cities. Bhan (2016) reiterates the 11th goal of the United Nation's Sustainable Development Goals (SDGs) which aims to make 'cities and human settlements inclusive, safe, resilient and sustainable' for housing 'to ensure access for all to adequate, safe and affordable housing and basic services and upgrade slums'.

The new SDGs see housing as an integrated issue with adequacy, safety and affordability. The means of measuring 'adequacy' involves ascertaining the materiality of different parts of the house like the roof, wall and the material used for construction, also included are the householder's access to water and sanitation, estimates of serviceable and non-serviceable housing. MoHUPA (2012) draws attention to the Kundu Committee Report on the question of congestion within the scheme of slum dwelling. His analysis strides beyond the physical and social conditions of living in the slums and defines the phenomenon of congestion as the condition where married couples do not have exclusive rooms for themselves. Incorporation of this aspect of congestion explains the need to critically study the conditions of the living of the urban dwellers.

Harvey (2008) clarifies that 'Right to the City' as not only concerned with the right to access the city, something migrants already possess. The right must encompass the power to alter the city according to their needs. One of the most significant human rights ensures the right and ability to create a qualitatively different kind of 'urban sociality'.

In order to construct a 'socially just city' in the times of existence of inalienable right to private property, one needs to address the questions of alienation and injustice. Harvey also claims that in the times of endless capital accumulation; the existing rights to the city should be reviewed and it needs to be asserted politically. Other derivative rights like the right to be treated with dignity should become a fundamental right.

Purcell (2003) described the right to the city as the right to appropriate urban space and to participate centrally in the production of urban space. Purcell draws from Lefebvre's writings and mentions the city's oeuvre, which is created by the quotidian practices of its inhabitants. Participation of the city dwellers in producing city space must ensure their participation in decision-making.

Rousseau (1998) writes that civil society should be so constituted, that the members never have to relinquish their birthright of being free. He distinguishes the will for personal benefits against the will for all in the best interest of the group. Only by following general will can humans escape the breakdown of social life.

Lefebvre (1972) in Marxist thought and the City re-reads the situation of the working class of England from the works of Engles. The urban reality in its 'horror' emerges from the readings. The stark juxtaposition of splendor and squalor or wealth and poverty that existed in the nineteenth century can be seen to replicate itself in the present cities too. Lefebvre asks whether all classes, despite having the same abilities, possessed the same interest in happiness. The concealment of sight of poverty too has a long history. The processes of pushing away the producers of city-space from visible zones foment feelings of alienation and it becomes a natural corollary to the life of a slum dweller.

Questions on 'alienation' have been examined by Butola (2019), where he interprets the Marxian idea of alienation vis-a-vis ownership of private property which is considered the 'product, the result, the necessary consequence of alienated labour of the external relations of the worker to nature and to himself.' All revolutions have been set up to protect one's property at the cost of another. All private properties are nothing but alienated labour. Possession of property by one most often means loss of property of another or possession of the property by one entails non-possession of property by others. Though Hegel analysed property to be the basis of realising self-actualisation, Marx interpreted that not only are the propertyless alienated, but, even those with the property are alienated too since property reduces all human qualities to quantities leading to extreme alienation. In the context of commodity being a reason for alienation, Butola (2019) shows the way for geographers to comprehend the Marxian tenet with ease since it is they who interpret man–nature relationship to be symbiotic and not one alienated being fighting against the forces of nature. This leads to the understanding that commodity relations lead to alienation since labour power itself changes into commodity.

Policymakers, aware of the physical conditions of the households and neighbourhoods of the residents of the slums attempt to formulate policy changes to ameliorate the living conditions of the urban poor/slum dwellers. In Indian cities, the Jawaharlal Nehru National Urban Renewal Mission (JNNURM), 2005, and The National Urban Housing and Habitat Policy, 2007, are significant policy documents that addressed the plight of the urban dwellers. The report on the Technical Group on Urban Housing Shortage (2012–2017) was prepared by the Ministry of Housing and Urban Poverty Alleviation, National Buildings Organisation, Government of India. The Group prepared a scheme named Rajiv AwasYojna (RAY) for providing affordable housing for the poor. This scheme aims to provide homes to the homeless, support existing households with incremental housing and upgrading their living conditions and in this way attempting to ease the demand and supply and imbalance of the urban housing market.

In Delhi, during 2012–13, there was an attempt to build low-priced housing units for the poor that were priced between 5 and 10 lakhs. The construction sites in Delhi for these units are Bhiwadi and Bawal. The issues that plague the real estate developers also determine the successes of the schemes. Issues that hinder the successful implementation of such schemes are unavailability of urban land, delay in approvals, rising construction costs, lack of skilled manpower, financing constraints of residents of low-income groups, limited financing avenues of developers, archaic government laws, disputable tax regimes, construction costs and lack of manpower. The way forward was devised by ensuring the availability of land via Urban Local Bodies (ULBs) and Urban Development Departments (UDDs). The Delhi Development Authority (DDA) plans to start in situ re-development as per the Delhi Slum and JJ Rehabilitation and Relocation Policy, 2015, prepared by the Delhi Urban Shelter Improvement Board. Despite the assurances offered by the State for improvement of the quality of housing in cities, the urban poor too live in distress regarding the possibility of eviction even if the eviction holds the promise of resettlement. The aspects of slum living discussed in the foregoing part may be documented in the study area chosen for this paper. The following section introduces the study area.

2 Introducing the Study Area

The study area chosen for this paper is a three-decade-old slum named Kusumpur Pahari, located in the southwest region of the capital city of New Delhi. The slum was established on the ridge of Aravali Hills during the 1970s when a new university named Jawaharlal Nehru University was being built which attracted migrants as construction workers at the site. The migrant workers working on the site of the construction built their tent-like hutments and drew groundwater from the shallow acquirers of the quartzite hills and continued to live on the upland site as newer streams of migrants joined them. In 2011, the Census of India recorded a population of 17,000 living in 3782 households. The growth of the slum in terms of population size has been spectacular over the three decades. From a meagre number of 24 persons in 1971, the slum has grown into a full-grown colony with the establishment of commercial and governmental institutions along with socio-religious structures that vouch for the vibrant community life of the inhabitants.

Kusumpur Pahari houses multiple streams of migrants originating from different states of Haryana, Rajasthan, Uttar Pradesh, Bengal, Bihar, Tamil Nadu among others. The duration of the stay of migrants in the colony ranged from 40 years to recent settlers. Pradhan and Sriram (2013) record that the slum has also been spatially segregated on the basis of the ethnicity of the migrants. The blocks, namely A, B and C, are home to migrants from Haryana, Uttar Pradesh and Rajasthan, respectively, while blocks D and E, which are newer blocks, are home to mixed groups of settlers.

The other interesting feature of the settlers of Kusumpur Pahari is that many households house three generations of migrants who continue to stay in the colony, and this fact offers the opportunity to study their changing perspectives on city life, opportunities and means of establishing social contract as viewed from the gaze of inhabitants of the older generation who undertook the migratory journey to settle in the colony vis-a-vis younger generations who were born and brought up in the colony. Kusumpur Pahari is also one of the six slums of Delhi which have been earmarked for in situ resettlement scheme elaborated in the section above.

Apart from all the above reasons, Kusumpur Pahari fits into the classic schema of the rich versus poor neighbourhood of the city. The slum area is located between two rich neigbourhoods, namely Vasant Vihar and Vasant Kunj. The area surrounding the slum is also the zone of prosperous shopping area, where state-of-the-art malls have been constructed that caters to consumers who belong to the affluent sections of the city. The study area thus holds great possibilities of analysing the juxtaposition of wealth and prosperity occurring within the vicinity of areas drowned in squalor and grime. The study area also elicits the urban processes that hinder visibilisation of the poor quarters of the city by keeping the slum residents away from the public eye.

For analysis of the nature of social contracts that were historically laid down in the slum area among the residents, and the newer contracts that are continuously getting established and altered, a variety of database and methodology were adopted to bring out the perceptions of the residents of the slum regarding their collective right to the city. The following section briefly documents the process of collection of information for the present study.

3 Database and Methodology

The paper is based on empirical information for which, extensive field visits were made to the study area and micro spatial units within the slum were selected for interviewing the residents. Block B and C were chosen for the study for the discerning difference in the perceptions of migrants originating from different states. Semi-structured interviews were used which subsequently turned into biographic interviews of the respondents. All exchanges were recorded with prior permission from the interviewee. To eschew issues of 'positionality' and 'reflexivity'; strict adherence to 'no consoling, no giving advice, no interpretation of situations, not intruding oneself into the life history and not suggest what the interviewee might want to say next' was followed. Only when the interviewee "dried up" did the researcher proceed to the next section of the interview. Serendipitous and Convenient mode of sample selection technique was adopted for the study; followed by control quota sampling to represent the diversity of regional, religious composition of the complainants. The present paper is a compilation of the responses of 100 inhabitants of Kusumpur Pahari, 20% of the respondents were first-generation migrants, 37% second-generation, and finally, 43% of the sample was third-generation inhabitants of the slum. Forty-six per cent of the respondents were males, while the rest were females and 41 of the respondents were chosen from Block B indicating that they had migrated from Uttar Pradesh and the rest were from Block C and had migrated from Rajasthan. The method of analysis adopted for the study is the use of qualitative analysis of the interview schedules for culling out feelings of alienation from land, property, and commodities. In the following section, the experiences of establishing social contract in the slum and the process of being alienated from land, property and commodities have been drawn from the life histories of the inhabitants.

4 Analysis

The location of the slum Kusumpur Pahari sandwiched by two affluent colonies, namely Vasant Vihar and Vasant Kunj other than being in the vicinity of a group of shopping malls of which one of the malls, named DLF Emporio happens to be the most expensive mall of India (Ghertner 2015). The location and situation of the slum bring almost similar visual imageries that Engels lucidly presented in 'The Conditions of Working Class in England' (Lefebvre 1972). The description of the poor quarters of the city was repugnant, the brilliance of the aristocratic quarter contrasted strongly with the foulness and wretchedness of the old town, which housed the working class.

Lefebvre further explains that the idea of concealing of poverty from sight was undertaken by 'unconscious segregation' whereby, the town is built in a way that a person may live in it for years yet never come in contact with the working quarters or even the workers. This completely holds true for the residents of Kusumpur Pahari, who live amidst affluent neighbourhoods, where many of them provide services as gardeners, domestic help, cooks, security guards, electricians, plumbers among other professions; yet, their employers have little or no clue about their living conditions.

Nandy (2013) explains the need to make slums invisible. Slums according to him are loathed as they constantly remind the upper-middle class of city dwellers of poverty and exploitation on which prosperity is built. Simply planning resettlement schemes can never eradicate slums as long as large scale deprivation of the migrants by urban middle class and till rural folk believes that 'proletarianisation' is better than being employed in the primary sector. To save on commuting costs to work, the workers would live close to the place of work, and that in turn would lead to the proliferation of slums. Nandy also warns that any attempt to free the city of slums would only make slums invisible. In this context, the process leading to the migration of rural folk to the city is discussed below.

4.1 Migration

Rural agrarian distress and its role in sending forth steady streams of migrants to cities all over India is a well-documented phenomenon (Lal 2013). The residents who migrated from their villages to Kusumpur Pahari were individuals who had experienced not only rural distress for them to migrate to cities, but some also moved to the city in search of freedom from the mundane social oppression they experienced in their native villages. Women residents migrated primarily due to their marriage and their shifting base to their husband's household. The journeys undertaken by the migrants to Kusumpur Pahari have been documented in this study.

In the new environment, the residents made efforts to establish social contract with other fellow residents. Initially, the basis of contract would invariably be linked with their occupation. Women often made companions during their water collection ritual. In the past, women travelled long distances to draw groundwater, but in the present times, they meet at the zones where water tankers supply water. With time the residents began organising religious festivals prompting them to meet at the temples in the colony or churches. For the younger generation, the basis of establishing social contracts is their association from their schools, friends they grew up with and their colleagues they work with.

Once the individual and his family settle in the slum region, their sense of identity also gets altered and associated with not just their residential area but also the city as a whole. The residents in the next section have articulated the aspirations linked to living in a big city.

4.2 Aspirations—Right to the City

The residents of Kusumpur Pahari have a strong sense of belonging to their slum; yet, most of them were not in a position to articulate their rights to the city. The participants of the present research were probed to elucidate their feelings about whether they possessed any sense of entitlement from their life in the city.

Ramvati, 70 years old, who had migrated to Kusumpur Pahari in 1979, expressed her deep-seated anguish stemming from the fact that it was she along with the first stream of migrants who had constructed hutments with rocks and leaves on the barren uninhabited zone of the city. She left her village home in Devaria in Uttar Pradesh with her husband and children in search of livelihood. She dismissingly speaks of Vasant Vihar. "...this colony didn't even exist. Only a few houses were being built then. The market at A Block was the only commercial site with tiny shops selling daily needs…" Her assertion indicates the role of the residents of Kusumpur Pahari in developing the affluent neighbourhoods seen in the vicinity. Her husband was an employee in one of those shops. She couldn't get her children educated because of lack of funds but she has made sure that her grandchildren study in a private school by availing the Economically Weaker Sections reservation criterion. She attributes their inability to rise up the socio-economic ladder despite living in Delhi for almost four decades to her husband being an alcoholic. Her fears and anxieties related to the impending in situ resettlement scheme are palpable. Ramvati is wary of the Government's move for resettlement as she feels each resident of the colony has his/her livelihood linked with the contiguous neighbourhoods where they work as service providers. She also expressed her concern about the children of the slum who study in the local schools. Any shifting process of households during the time of construction of the multi-storied resettlement colonies would disrupt the lives of the residents. All women and men who were over 50 years old and had toiled for establishing their lives in the city shared this feeling and did not wish to be relocated.

There appeared a difference in the opinion of younger residents; those who were born in the slum, formally educated in schools and work as blue-collar workers. Their formal education and social networks have developed a different sense of 'good life' and aesthetic normativity (Ghertner 2015). The youngsters wanted the slum to get demolished and aspired to live in a clean environment since their cognitive faculties understand the reason for the need to demolish slums for making Delhi a world-class city. According to them, slums are a nuisance. 32 years old Raju, who migrated with his father at the age of 10, was educated in a government school and has worked as a gardener, labourer, painter and, finally, found a job at the International Airport terminal in the city. Raju is one of the newer generations who have a completely different notion of his right to the city and his aspirations of living in a slum-free city has been carefully constructed by the discourse which 'conducts the conduct' of the subjects. Raju and his generation support the in situ resettlement scheme though there appears lurking fears in their minds too of being left out while the possession of the new buildings take place.

As the idea of their rights to the city varies, their feeling of contentment and dissatisfaction with their quality of living in Kusumpur Pahari showed divergent opinions. The factors that foment satisfaction and dissatisfaction have been discussed in the following section.

4.3 Congestion and Adequacy

The number of family members dwelling in a single room measures the concept of congestion. This aspect is critically analysed by re-assessing whether married couples had separate rooms of their own (Bhan et al. 2019). It was documented that most couples living in joint family structures where their in-laws' family also lived along with them, did have separate rooms for themselves. Rooms were usually constructed as an extension of their existing household. The site of these rooms was often on the terrace. In the case of nuclear family structure, where husband and wife lived with their young children or adolescent children, had no private space for themselves. Most often it was observed that in rooms of 10 feet by 12 feet size, couples slept on the floor along with their children. In some households, which owned beds, the couples made arrangement for the children to sleep on the bed while they themselves slept on the floor. While all women interviewed were asked whether they experienced stress from lack of privacy as a couple, all of the interviewees denied lack of space being a cause of stress, yet in the same breath informed that they know of couples who undergo distress arising from access to private space.

In connection with questions of adequacy, access to water emerged as the prime factor that weighed heavily on the residents. Each and every resident complained about not having piped water and were dependent on water tankers to fill water for their entire household needs. Women were the ones who waited in queues with two to three huge water containers, lifted them on bicycles and paddled back to their homes. Most of the households did not have their own toilets and bathing cubicles and had to travel to one of the mobile toilets located in the colony. Yet, it was their uncertain water supply that disrupted their lives. Women described leaving crying babies or food in order to fill water from the tankers. Many women also felt that had it not been for the uncertainty about water, many of them would have been in the position to take up paid employment. Even within the colony, there emerged varied responses regarding cleanliness and sanitation. Some blocks had wider and cleaner roads, while others un metaled roads with overflowing open drains. Despite the visible discomforts, the residents shared a feeling of oneness with their neighbourhood. Women respondents pointed out to 'safety' as one of the reasons people chose Kusumpur Pahari over other colonies. Respondents were aware of the fact their residential area would soon get demolished for making way for low-cost housing. The impending loss of home-space appeared to have created a deep sense of anxiety and restlessness among the residents. In the light of losing claim to their only possessions, the aspect of urbanism-induced alienation has been examined in the following section.

4.4 Alienation

Parker (1978) observed that when people move from small groups to large urbanised complex societies, their sense of alienation heightens. This sense is accompanied by feelings of isolation, feelings of meaninglessness and self-estrangement. Drawing from Butola (2019), the respondents of this study expressed a deep sense of alienation which arose from their loss of land at the place of their origin. The feeling of alienation was not as strong for being disposed off from their rural base, but it deeply affected the residents who owned property in the slum and may have to lose their homestead for the government provided tenement. The younger generation, who did not associate much with their villages, shared feelings of alienation arising from their labour being converted into a commodity. Each person was reduced to the value of his or her labour. One of the residents wailed and asked for help from the residents of the rich localities where they invest their labour. Her pleading appears rhetorical since it is the upper-middle class wishing to do away with the squalor and filth surrounding their neighbourhoods. The plight of the residents of Kusumpur Pahari captures the essence of the Lefebvrian concept of 'atomisation' of the worker. The class that appropriates their labour fails to recognise that all classes possess the same interest for happiness.

5 Summary of Conclusions

Kusumpur Pahari offers ample scope to study class-based residential segregation with stark differences in living standards with its neighbouring colonies. In a bid to turn cities free of slums, the residents who offer their productive labour to the upper classes are uprooted from their only possessions. D'Souza's (2019) reasons for slum dwellers to remain rooted to their colonies is primarily because they value their social networks and prefer remaining in slums over newly constructed homes. Such social networks are a source of informal credit and insurance that support the slum dwellers during financial difficulties. Fetish for aesthetics becomes one of the driving factors for urban governments to attempt to demolish slums. The understanding of the Right to the City varies among the residents and this makes it difficult to become a pressure group and resist the planning schemes that undermine their existence. The Sustainable Development Goals may be pursued and achieved in a better way if the human face is incorporated in urban development schemes.

References

Bhan A, Deb A, Harish S (2019) Understanding inadequacy: a view from urban India. https://iihs.co.in/knowledge-gateway/wp-content/uploads/2017/10/Understanding-Inadequacy.pdf. Last accessed 29 Sept 2019

Bhan G (2016) In the public's interest: evictions, citizenship and inequality in contemporary Delhi. New Delhi: Orient Blackswan

Butola B (2019) Concept of alienation in Marx. Human Geograp 12(2):73–79

D'Souza VS (1979) Socio cultural marginality: a theory of urban slums and poverty in India. Sociol Bull 28(1/2):9–24

D'Souza R (2019) Housing poverty in urban India: the failures of past and current strategies and need for a new blueprint. ORF Occasional Papers

Ghertner AD (2015) Rule by aesthetics: world class city making in Delhi. Oxford University Press, New Delhi

Harvey D (2008) The right to the city. New Left Rev 53:23–40

Kundu A (2011) Politics and economics of urban growth. Econ Politi Weekly 46(20):10–12

Lal V (2013) Introduction. In: Lal V (ed) The Oxford anthology of modern Indian city: politics, culture and life forms. Oxford University Press, New Delhi, pp i–xlviii

Lefebvre H (1972) Marxist thought and the city. University of Minnesota Press, Minneapolis

MoHUPA (2012) Report of the technical group on urban housing shortage. New Delhi

Nandy A (2013) Slums as self confrontation. In: Lal V (ed) The Oxford anthology of modern Indian city: politics, culture and life forms. Oxford University Press, New Delhi, pp 234–237

Parker JH (1978) The urbanism-alienation hypothesis: a critique. Internat Rev Modern Soc 8(2):239–244

Pradhan K, Sriram J (2013) Delhi's ugly underbelly: caught between shiny malls and sprawling bunglows, the serving class lives and dies in a cesspool of deprivation. https://www.indiatoday.in/magazine/special-report/story/20130128-slums-in-south-and-central-delhi-761488-2013-01-19

Purcell M (2003) Citizenship and the Right to the Global City: Reimaining the Capitalist World Order. International Journal of urban and Regional Research. Vol.27.3. 564–590

Rousseau J (1998) Social Contract: or Principles of Political Right. Wordsworth Editions Limited. Hertfordshire

Vaddiraju AK (2016) Urban governance and the right to the city. Econ Polit Weekly LI93221-25

Urban Renewal and Redevelopment

From Welfare to Participation—The Changing Narratives of Slum Rehabilitation and Housing for the Urban Poor in Delhi

Ushosee Pal

Abstract Slum evictions and resettlement colonies have been a part of Delhi's urban history since the late 1950s. As the central space of political power and administration, Delhi's history of political contestations has shaped its housing practices pertaining to the urban poor. Post-colonial Delhi prior to liberalization of the Indian market saw housing for the urban poor as a 'welfare benefit' to be provided to the low-income groups with the help of the civil society. Neoliberal Delhi views the same benefit as an opportunity for a 'slum-free' city with 'world-class' infrastructure by way of public–private participation. What were the processes which brought about this shift in the narrative and approach of housing and urban redevelopment in Delhi? The advent of the market in housing for the urban poor is a fairly recent phenomenon with its own peculiarities. The present work analyses the neoliberal turn in housing for the urban poor in Delhi through its housing policies, plans and an ethnographic case study of its pilot in situ slum rehabilitation project under public–private participation in Kathputli Colony. It attempts to review the stronghold of the state–market nexus in the development of lower income groups and consequent problems of inclusivity and right to the city.

Keywords In- situ slum rehabilitation · Housing for the urban poor · Urban redevelopment

1 Introduction

Development in India has largely witnessed a change of hands since the liberalization of her economy in 1991. In post-colonial India, welfare was seen as a benefit to the helpless poor, an idea which has ever since dominated the development discourse albeit with some narrative changes. According to T.K. Oommen, the post-colonial policymakers 'viewed development as a non-political, techno-managerial activity'.

U. Pal (✉)
Department of Sociology, Jamia Millia Islamia University, New Delhi 110025, India
e-mail: ushosee@gmail.com

L. T. T. Huong and G. M. Pomeroy (eds.), *AUC 2019*, Advances in 21st Century Human Settlements, https://doi.org/10.1007/978-981-15-5608-1_39

In this model of development, people's participation had little or no space; the beneficiaries of development were defined as clients and targets'(Gill 2005, 28). Often, welfare schemes could not sustain the economy of the newly formed nation states. Forced to open up their markets in order to stabilize the economy, global capitalism began dictating policy norms of developing nations often without enough checks and balances, and the welfare state became even harder to sustain (Rudra 2002).

Policies in India in recent times have increasingly invited the global market to coalesce with the democratic state in managing developmental work. The ideas of participation or participatory policies entail heavy public sector disinvestments from erstwhile welfare schemes and entry of the market in developmental activities making consumers or clients out of citizens. (Noronha and Sharma 1999; Leitner et al. 2007).

The transition of India from welfare to public–private participation (PPP) is embedded in a neoliberal process where populist politics is funded by the elite market forces making the latter dominant in decision making once elections are won. The urban poor of Delhi is a powerful vote bank living in the slums of the city. 90% of these slums lie on state-owned land.[1] With urban land increasingly gaining value as a commodity for the real estate market, urban policies have integrated slum redevelopment with PPP promoting mixed land use over erstwhile land clearance for infrastructure and resettlement policies. The loss of livelihoods of urban poor and re-entry into slums making resettlement housing a hub for middlemen was given as the logic behind in situ slum rehabilitation which is the latest policy adopted in Delhi.

This study is an outcome of doctoral research conducted from July 2016 to November 2017 on the in situ slum rehabilitation project of Kathputli Colony. The present work is divided into three parts: (1) a review of the ways in which housing for the urban poor has been addressed in urban governance of Delhi from the 1950–90s by way of narratives from laws, plans and policies of land and housing; (2) a description of the neoliberal turn in urban redevelopment through the post-millennial focus of the government on housing for the urban poor in Delhi; (3) a discussion on the implications of this change through accounts and official paper trails from the Kathputli Project procured through ethnographic fieldwork.

The urban poor of Delhi largely engaged in informal employment—ignored or underestimated in their resilience for long enough create makeshift self-made housing on vacant city lands they do not have ownership of. The space created by planners in cities of the global south often comes from emulation of models already in use in the global north not factoring in the unique conditions posed by urban informality (Roy 2005; Watson 2009). Nonetheless, planners have time and again tried to address slums that seen to threaten the symmetry and the orderliness of the planned city. The following section explores the presence and the treatment of slums in the official documents stating acts, policies and plans pertaining to housing in Delhi from the 1950s to the 90s.

---

[1]According to the 69th round survey by the National Sample Survey (NSS) in 2012, 'About 90% of slums were built on public land, owned mostly by local bodies (46%), railways (28%) and state government (16%), etc.' (GNCTD 2015, ii).

2 1950–1990—a Review of Land, Development and Housing in Delhi

The starting point in this work is the late 1950s, as this was one of the most eventful times for the urban development of Delhi, the capital city of newly independent India. From 1956 to 1957, three important legal events followed the reorganization of states and consolidated the Central Government's control over Delhi's land, development and governance, namely

- Slum (Improvement and Clearance) Act of 1956 for clearing out encroached land in the Union Territories of the time[2]
- Municipal Corporation Act of 1957 which led to the creation of the Municipal Corporation of Delhi (MCD), an apex body to govern all civic services in Delhi headed by the Lt. Governor;
- Delhi Development Act 1957 which mandated the Delhi Development Authority (DDA), also headed by the Lt. Governor, specifically created for acquiring land and land use planning.

It is in this politico-legal backdrop of control that the first Master Plan of Delhi (MPD—1 hereafter) was formulated by the DDA. Consequently, the DDA monopolized the ownership and management of land, putting Delhi's land under the direct control of the central government (Pugh 1991; Durand-Lasserve and Royston 2002; Sivam 2003).

Land acquisition by the Central Government was reified by MPD-1 as essential to the planning and development of Delhi. The justification to the large scale and largely arbitrary acquisitions was that,

> The ownership of land by the Government makes planning and the implementation of plans easier and is imperative if slum clearance, redevelopment and subsidised housing and provision of community facilities according to accepted standards have to be undertaken as indeed, they must be in Delhi, in a determined way. (DDA 1962, 7)

By the 1980s, DDA had become the largest land-owning body in Delhi due to massive land acquisition and was also responsible for all developmental work. For the housing for urban poor, the policies up until the 80s were used contextually and somewhat arbitrarily.[3] Furthermore, the slow processing of acquired land and the inability of civic and municipal bodies in implementing developmental goals derailed the planning of Delhi (Chatterjee 1984). From the 50s up until the 80s, Delhi's main policies on slums were plot resettlement on outskirts and environmental upgradation

[2]Incidentally, Delhi went from being a full state to Union Territory under the States Reorganisation Act (1956) passed just a month prior to the Slum Act.

[3]In a historical study on a resettlement colony in the outskirts of Delhi, Emma Tarlo discussed how Delhi's city space and its resettlement colonies are spatial products of historical events and political changes which affected its governance. Citing instances such as the 1965 war, Emergency, and the Sikh riots of 1984, Tarlo showed how 'urban spaces like resettlement colonies are structured as much through a succession of historical events at national level as through local politics and urban planning.' (Tarlo 2000, 52).

with basic services. Housing for the employees of the newly independent central government and for the partition displaced people from 1947 to 1971 was prioritized. Housing for the middle-income groups was seen to deserve more attention.

The JJ[4] removal scheme in 1960 gave the slum residents plots of 80 square metres on 99 years lease outside of the city (Dupont and Ramanathan 2008; Kaur 2019). Around 47 resettlement colonies came up by 1977 under this scheme which eventually became large slums lacking basic services due to an 'anti-urban bias' amongst the planners of the time. Urban poverty and slums were seen as a consequence of lack of rural development and hence housing for the urban poor was not prioritized (Kaur 2019). It is pertinent to note that this was also the time when the government lead by Indira Gandhi had declared an emergency and was removing squatter settlements rapidly. It was often remarked that her election slogan '*garibi hatao*' (remove poverty) eventually turned into '*garib hatao*' (remove the poor) (Kaur 2019) due to the extensive slum demolitions which occurred in the city during this time. Over time, the regulations and the size of the resettlement plots kept changing. At one point, the plot size became as small as 12.5 square metres (Dupont and Ramanathan 2008). Housing for the urban poor was largely arbitrary and neglected in the implementation of the first plan. The second master plan (MPD 2 hereafter) was no better in this regard.

3 Post-90s—Transformation and the Neoliberal Turn

There was a delay of almost a decade in the termination of the MPD 1 and coming up of the MPD 2 in 1990. This was due to more politico-legal changes that Delhi underwent at this time. The city was fast expanding and the DDA had underestimated the rate of urban growth in Delhi. Migrants from rural areas settled on public land acquired but unused with the help of contractors and councillors for whom this population was a large vote bank. The decade of a gap between the two master plans led to further unplanned growth of Delhi and coming up of more slums (Nath 1995).

MPD 2 came after prolonged deliberations in sync with the Regional Plan for the newly formed National Capital Region (NCR). Both the plans were projected for the year 2001 and carried over more or less the same ideas which formed the first plan.

Land acquisition was still seen as vital although the justification of land prices and terms of exchange had complicated by this time. On one hand, there was a heavy emphasis on freezing the land acquisition rates to a back date as a higher price of land due to development in areas surrounding was seen as 'unearned increase' (Chatterjee 1984, 33). On the other hand, alternative forms of compensation for land were suggested, such as returning one-fifth or one-tenth of land to the owners with usage regulations after acquisition and development as the value of developed land was held to be far greater than raw land (Chatterjee 1984, 36). Moreover, it was

[4] JJ refers to '*jhuggi jhompri*' which is a Hindi term for makeshift homes in squatter settlements in Delhi.

believed by senior planners of the time that heavy focus on resettlement can lure more people into slums in hopes of acquiring planned housing. There were talks amongst planners on regulated migration in order to prevent squatting. It was also mentioned in the deliberation prior to the second master plan that there needs to be a greater focus on rural development in order to prevent migration. While poverty alleviation was mentioned and discussed in planning, the anti-urban bias remained intact in the second plan. The welfare state of the time, in its policies, thus found urban poverty only a consequence of rural backwardness, and the dominant narrative was to prevent the poor from settling permanently within the city.

By this time, the DDA had transferred the task of slum demolitions and improvement to the Slum Wing of the MCD which came up with a three-pronged redevelopment policy for slums that mandated either relocation (most prevalent), or in situ upgradation or environmental improvement.

The slum relocation drives through the 90s did not help in reducing the slums or urban poverty in Delhi. The dominant trend in housing for urban poor was slum demolition and resettlement to the outskirts in the form of either housing or plots. This was exacerbated by the elite and the middle classes in the city which often found slums to be a threat to the environment (Baviskar 2003), or as spaces of nuisance (Ghertner 2012) to the general public life, collectively moving courts for demolition orders (Bhan 2016). The demolitions were rarely ever successful in housing the urban poor. Relocation to the outskirts of the city would often lead to the loss of livelihood due to heightened proximity to the workplace. Meanwhile, land acquired for public use was still underutilized. This often led to the re-establishment of slums within the city. Consequently, by 1998, 27% of Delhi's urban population was residing in JJ colonies (Dupont 2008).

With the Commonwealth Games 2010 around the corner, slum demolitions continued with a significant amount of violence till the early years of the millennium. However, some significant legal cases and a decade of research on the urban poor established that housing for the urban poor required to be undertaken within the city.

The policy narrative on slum redevelopment shifted in 2007 from slum resettlement or relocation to in situ *slum rehabilitation* with the coming up of the third Master Plan of Delhi (MPD-3 hereafter) for the year 2021. Under the policy of the Master Plan, Kathputli Colony, a slum in the heart of Delhi was chosen as the pilot project for this form of slum redevelopment. Interestingly, it was combined with the ideas of mixed land use through Public–Private Participation (PPP). Land was seen as a resource that could be utilized to lure market investments for a 'world-class city' on one hand, and make the city 'slum-free' on the other. Mixed land use under PPP gave the real estate corporations cheap prime city land to develop for their elite urban customers in lieu of developing a relatively larger part of the land for the slum residents who have 'encroached' on it over some decades due to a failure of planning by the DDA (Bhan 2013).

In the meantime, the Slum Wing of the MCD was restructured and named Delhi Urban Shelter Improvement Board (DUSIB) through the DUSIB Act 2010 under the state government which came up with its own policy framework on slum relocation and rehabilitation and directed the use of the same model of the redevelopment as DDA has used in the case of Kathputli Colony. The Delhi Development Act had mandated the DDA, a central government body to be the nodal authority to acquire and develop all land in Delhi. The state government of Delhi, on the other hand, stated that DUSIB was the nodal authority responsible for all slum redevelopment in the city. With most land owned by the DDA, and the great potential for generating revenues through private investments in land development projects, a conflict of interest emerged as to which organization was entitled to carry out development agreements with real estate conglomerates. From a policy afterthought to a category of utmost importance, slums and their residents in Delhi today find themselves caught up between a political tug of war between the central and the state governments over who would be the harbingers of development in their area. Incidentally, this was the post-2010 era which saw a new populist socialist political opponent increasingly come to power in Delhi's state government. With antagonistically opposed political parties at the state and the centre sharing the jurisdiction of land, development and governance in Delhi, there are often forthright overriding and attempts to capture power over these areas through brute force, as was seen in the case of Kathputli Colony. Today, while construction has begun at a rapid pace in Kathputli Colony which is now a 'site', a case still goes on in court between activists representing and vouching for the DUSIB and officials representing the DDA. The Chief Minister of Delhi was conspicuously absent in the foundation laying ceremony of the project held in April 2018. Ministers and officeholders of the central government have become the faceless faces of the project.

On a busy December afternoon in 2016, a convoy of men from DDA and Raheja Developers, navigated the narrow and congested gullies of the slum, guarded by hundreds of police and paramilitary personnel, were distributing pamphlets promising the residents, *sapnon ka ghar*—'dream homes', on loudspeakers. In less than a year, the land was cleaned out with bulldozers, an elaborate and strategically timed foundation laying ceremony was held and the construction of Delhi's first in situ slum rehabilitation project began at the 'site'. The slum homes were razed to the ground to make space for squatter verticalization (Budds et al. 2005).

The following section is an attempt to provide a timeline of the Kathputli Project up until the demolition of the slum in October 2017 and through it, the journey of its residents.

4 Housing the Poor Within the City—A Case Study of Kathputli Colony[5]

Kathputli Colony is a slum that is believed to have been created in the 70s when the Indira Gandhi government notoriously razed large chunks of the Old City. Nomadic folk performers often called *ghumantu* or *khanabadosh* found shelter in the fallow lands (erstwhile stone quarries) of Shadipur, back then, away from the city center devoid of any urban development.

From the late 1980s, the DDA has made attempts to relocate Kathputli Colony—once in 1986 to Vasant Kunj and once in 1996 to Mehrauli. Both these projects did not materialize as both the relocation areas were not as central as Shadipur. Meanwhile, from the 1980s migrants from rural Bihar, Uttar Pradesh, West Bengal and Maharashtra, began to settle alongside the artisan communities, starting off an internal power struggle for politically distributed resources (water, electricity, gas connection, ration card, etc.) amongst the residents. During the talks about earlier resettlements, non-artisan communities claimed equal entitlement to housing, consequently clashing with the artisan communities. Till the time, the land was inhabited by the two groups, the slum was physically divided into two parts by the single main street it had. The homes closer to the Shadipur bus depot which was also better built with more floors and rooms belonged largely to the artisans while relatively smaller homes closer to the railway tracks in north-western periphery of the slum mostly belonged to the non-artisan communities. Old residents and newer residents could easily be distinguished from the type of materials used to construct the dwellings.

Despite plans and collaboration with civil society organizations of the time, the DDA had not been keen on building a separate artists' village in Delhi. Citing paucity of funds and the need to distribute land equitably amongst other urban squatters as well, separate plans for the artisans never materialized (Dubey 2016).

By the millennium, the Delhi Metro began to alter the transport map of Delhi, roads were bettered and near Shadipur, high-income residential areas nearby such as East Patel Nagar and Rajinder Nagar bringing it within the ambit of the city centre. From a fallow piece of land lying on the outskirts, Kathputli Colony turned into an invaluable commodity with its real estate potential unused.

Kathputli Colony was now a large slum with water and electricity connection. The residents had ration and voter identity cards; other migrant populations of semi-skilled or unskilled labour employed in the informal sectors began to share the space with the artisan communities; and most importantly, the slum enjoyed political immunity from local MLAs and party leaders due to vote bank politics.

On September 4, 2009, the PPP agreement of Kathputli Colony was signed between the DDA and Raheja Developers after a 2-year-long process of tender

[5]The data for this section was collected in the duration between July 2016 and October 2017, from ethnographic fieldwork and acquisition of the official records of the entire correspondence between the DDA, Raheja Developers and other stakeholders for over a decade in the housing project up until October 2017 using the Right to Information Act, 2005. This was done due to reluctance of office holders in speaking about the project.

bidding. The developers were to be handed over the land of 5.22 hectares owned by the DDA at a price of 61 million Indian INR for mixed-use redevelopment.

It is fairly simple to understand the nature of PPP in the Kathputli Project. On one hand, a residential tower complex based on the EWS[6] housing guidelines of the DDA is to be built by the Raheja Developers for 2800 'eligible' residents on 3.42 hectares of land (65.5% of the total land). On the other hand, to compensate for the cost of EWS housing, the real estate company has been given the remaining land as a 'remunerative component'. This land of 1.8 hectares which supposedly compensates the cost of housing for the urban poor can be used for free sale housing and commercial real estate, the latter not exceeding 10% of the Floor Area Ratio of the total residential component.

Thus, on the remaining land of 1.8 hectares, the developer is building the Navin Minar, a luxury residential skyscraper, and beside it the Raheja Mall, a multistoried centre of retail and commerce. Structures of opulence and grandeur catering to the wealthiest in the country with promises of a helipad, glass-bottomed swimming pool, gold souk, multilevel car park and more.

It took 8 years from the signing of the agreement (which was supposed to be fulfilled within 2 years), and 10 years from the inception of the project, to clear out the land.

Firstly, while the project was supposed to be completed by 2011, it was only in February 2014 that the first 500 families comprising almost the entirety of the Maharashtrian Dalit Buddhist communities and some people from the Muslim migrant communities of Bihar shifted to the Transit Camp. This was due to the DDA's inability to find land to build Transit Camps required to temporarily shift the population while the in situ redevelopment is carried out. In one instance, an official letter shows that there was resistance from the residents of Basai Darapur to the construction of the camps in their vicinity.

Meanwhile, a second delay was in execution of the survey of the population which was carried out *after* the signing of the PPP agreement, or more precisely, after it had already been legally decided that no more than 2800 dwelling units would be built for the slum dwellers on the land of Kathputli Colony.

It was only after the first notice of eviction on February 2, 2014, that the residents and the local leaders of Kathputli Colony realized that the project was not really hearsay. This was the first time the project was met with resistance from the residents with the help of local leaders and political action groups. It was around this time according to some resident accounts, when the police are said to have entered Kathputli Colony, dragging men associated with the resistance out of their homes in the middle of the night and assaulting women who tried to protect them. They were placed with criminal charges which some of them have been trying to clear to date.

[6]EWS (Economically Weaker Sections) housing according to the third Master Plan of Delhi comprises of apartment blocks of approximately 30 square metres (30.5 in case of Kathputli Colony). These are flats built in low cost and allotted on lease at a subsidized down payment are usually constructed by the DDA in order to provide housing to slum dwellers. These are also called Janta flats in some localities of Delhi.

Cut-off dates determined the eligibility of the residents to the allotment of apartments in Kathputli Colony. This date was revised twice—first from September 4, 2009, to January 30, 2011, to incentivize voluntary shifts to the Transit Camp; and again later, to January 1, 2015, in order to accommodate the increase in the population of the land over time.

Through 2014 until October 2015, the residents of Kathputli Colony tied up with more and more activists and organizations to resist their evictions from the land. Letters were written to the DDA signed by some local slum leaders in collaboration with these organizations to stop the project from citing non-adherence to the MPD-3 guidelines by the DDA. Men belonging to the Bhat community of nomadic folk artists from Rajasthan, a dominant group in the slum, were at the forefront of this movement. The colony came into the international media lens through documentaries and reports.[7] Meanwhile, local leaders from the groups which had already shifted to the Transit Camp and the Raheja Developers wrote counter letters to higher authorities and the DDA respectively, citing that non-governmental organizations (NGOs) were misleading the slum dwellers and preventing them from shifting to the Transit Camp. On October 31, 2015, the Raheja Developers, for the first time held a meeting with the residents to explain the projects to them. It was at this point that the resistance started to weaken. Some activists working for the resistance against the project maintain that it was around this time that some influential local men at the forefront of the movement were hired by the real estate giant to advocate in the interest of the project. However, some local leaders level the same accusation on the activists and organizations which were (and are) working to stop and enable transfer of the project to the DUSIB.

On December 19, 2016, door to door eviction drives were conducted by the DDA and the Raheja Developers, making residents sign the eviction agreements on the spot and immediate transfer to the Transit Camp. In the prolonged tug of war between authorities and political activists mediated and fuelled by informal local leaders for their own politico-economic gains, the residents began to be divided into their opinions and actions which reflected in the fragmentation of the physical space of the slum. Individual residents showed anxieties regarding their everyday lives which were increasingly governed by uncertainty. After December 2016, the process of vacating Kathputli Colony picked up pace despite mobilizing resistance. More and more people began to leave Kathputli Colony for the Transit Camps despite being participants in the resistance. In August 2017, the final cut-off date was revised to January 1, 2015. This was followed by a fresh survey conducted in July 2017 based on the new cut-off date. In October 2017, a series of events took place leading to the entry of bulldozers and armed police personnel to transform the slum into a site. Between October 20 and 25, a 1000 newly constructed EWS apartments in Narela, a locality near the north-west borders of Delhi, were approved for relocating the

[7]The film 'Tomorrow We Disappear' came out in April 2014 features artist communities of Kathputli Colony which equated the impending commencement of the project with the death of traditional art forms. One of the key subjects in the documentary was leading the resistance along with the leader of the Bhat community against the DDA, later himself shifted to the Transit Camp, motivating several others of his community to follow suit, leaving the leader and his crumbling support behind.

excess eligible residents of Kathputli Colony. Three new enumeration lists surfaced showing 2800 eligible residents at the Transit Camp, 492 eligible for Narela and 771 rejected applications.

On October 30, 2017, the DDA entered Kathputli Colony with bulldozers while several residents were still staying within its premises. Police personnel barred entry of activists inside and also at the Transit Camp. Although the officials present at the site claimed that eviction and demolition notices were displayed at several locations in the slum five days prior to the demolitions, none of the residents seemed to be aware of it. Several residents accused the DDA officials and workers of starting the demolition work without even waiting for people to vacate the houses. Several people were sleeping in the open air for the next few days until a temporary shelter guarded by police officials was built for those who were suddenly displaced. Some of the evicted residents also claimed that at the Transit Camp, the dwellings allotted to them had been broken into and occupied by other residents and thus they didn't have a place to go to.

5 Concluding Remarks

The price of a single apartment in Navin Minar starts at approximately 60 million INR, almost the same as the cost of the entire land at the time of the agreement. With 125 apartments, the cost of a single flat reaches up to 340 million INR which means, a lower estimate of the income from Navin Minar alone will be roughly 7.5 billion INR.

The cost of building 3000 EWS flats in A-14 Kalkaji Extension under in situ mode, has been approximately 2.5 billion INR. This means that building around 2800 dwelling units in a similar amount of land in Kathputli Colony would cost roughly the same amount if not less which shows a basic excess income from Navin Minar alone, at a minimum of 5 billion INR. With penthouse properties advertised in Navin Minar of roughly 340 million INR each, this projected income is likely to be much higher than 5 billion in actuality. Even if the cost of construction and the time lag is considered, it is for sure that the income from Navin Minar and Raheja Mall will not merely reimburse the costs incurred by the developer for slum rehabilitation but also bring a great amount of capitalistic gain to them. The Navin Minar and the Raheja Mall are being built to cater to the top elites of the country so much so that a panel of experts will determine the eligibility of a person to buy a house in Naveen Minar. In such a scenario, the gentrification of the entire surrounding locality of Kathputli Colony comprising of other low-income residences cannot be ruled out.

Finally, the largely urban poor residents of Kathputli Colony must make a down payment of 142,000 INR to the DDA to receive allotments of the EWS apartments they are entitled to. Hence, another 396.7 million INR shall be earned by the DDA, 313.6 million out of which would just be revenue from the allotment of the EWS houses to the urban poor who are currently residing in the Transit Camp, and the rest has been charged as the cost of maintenance for 5 years. This down payment is one of the many mandatory conditions imposed on these low-income groups to avail of the benefit of housing within the city. Further, the apartments will not be freehold property for at least 10 years after they are allotted to the residents generating limited legal titles to their property.

The Kathputli Project boasts of incentivized in situ redevelopment of the slum by realtors instead of the government where what is claimed to be participatory is actually a process of unchecked privatization. The benefit model of welfare only crystallizes in this kind of participatory development where communities at large are kept misinformed and fragmented systematically for the gain of those in power positions. For the urban poor in Delhi, housing within the city is a negotiable matter with relocation to the outskirts still in practice despite new laws and policies. The dominant narrative thus seems to suggest that, planned housing within the city comes at a price, especially for those who can least afford it within the city.

References

Baviskar A (2003) Between violence and desire: space, power, and identity in the making of metropolitan Delhi. Int Soc Sci J 55(175):89–98

Bhan G (2013) Planned illegalities: housing and the'failure'of planning in Delhi: 1947–2010. Econ Polit Wkly 15:58–70

Bhan G (2016) In the public's interest: evictions, citizenship, and inequality in contemporary Delhi. University of Georgia Press

Budds J, Teixeira P, Sehab (2005) Ensuring the right to the city: pro-poor housing, urban development and tenure legalization in Sao Paulo, Brazil. Environ Urban 17(1):89–114

Chatterjee M (1984) Viewpoints on Delhi's master plan report on a panel discussion. India Int Cent Q 11(1):27–55

Delhi Development Authority (DDA) (1962) Master plan for Delhi. New Delhi

Dupont V (2008) Slum demolitions in Delhi since the 1990s: an appraisal. Econ Polit Wkly 12:79–87

Dubey S (2016) Understanding participation in a heterogenous community: the resettlement of Kathputli Colony. In: Space planning and everyday contestations in Delhi. Springer, New Delhi, pp 35–36

Dupont V, Ramanathan U (2008) The courts and the squatter settlements in Delhi–Or the intervention of the judiciary in urban 'Governance'. New forms of urban governance in India. Shifts, models, networks and contestations. New Delhi, pp 312–43

Durand-Lasserve A, Royston L (eds) (2002) Holding their ground: secure land tenure for the urban poor in developing countries. Earthscan

Garibi Hatao or Garib Hatao?: India's journey through family planning and poverty 'Removal' (2019). https://www.epw.in/engage/article/family-planning-emergency-indira-gandhi, 28 Aug 2019

Ghertner DA (2012) Nuisance talk and the propriety of property: middle class discourses of a slum-free Delhi. Antipode 44(4):1161–1187

Gill R (ed) (2005) State, market, and civil society: issues and interface. Rawat Publications

Government of National Capital Territory Delhi (GNCTD) (2015) NSS 69th round survey. Urban Slums in India. July 2012–December 2012

Kaur K (2019) Shifting urban redevelopment policies and the aspirations for a slum free India: experiences of urban poor in Delhi. Man India 99(1–2):157–167

Leitner H, Peck J, Sheppard ES (eds) (2007) Contesting neoliberalism: urban frontiers. Guilford Press

Nath V (1995) Planning for Delhi and national capital region: review of plan formulation and implementation. Econ Polit Wkly 2:2191–2202
Noronha E, Sharma RN (1999) Displaced workers and withering of welfare state. Econ Polit Wkly 5:1454–1460
Pugh C (1991) Housing and land policies in Delhi. J Urban Aff 13(3):367–382
Roy A (2005) Urban informality: toward an epistemology of planning. J Am Plan Assoc 71(2):147–158
Rudra N (2002) Globalization and the decline of the welfare state in less-developed countries. Int Org 56(2):411–445
Sivam A (2003) Housing supply in Delhi. Cities 20(2):135–141
Tarlo E (2000) Welcome to history: a resettlement colony in the making. In: Dupont V, Tarlo E, Vidal D (eds) Welcome to history. Delhi: urban space and human destinies
Watson V (2009) Seeing from the South: refocusing urban planning on the globe's central urban issues. Urban Stud 46(11):2259–2275

Information Economy, Employment Vulnerability, and the Emergence of New Urban Marginality in Dhaka City, Bangladesh

Rasel Hussain

Abstract The twenty-first century is witnessing huge remarks in the informationalization of the world economy and the restructuration of the capitalist economy as a side effect. The paper aims to explore the new urban marginality due to the rise of information-based economy in the city of Dhaka, Bangladesh. The study is based on the theoretical frameworks of Castells, Sassen, Fainstein, and Friedman and Wolff. The study reveals that most of the poor people living in these neighborhoods are involved in the informal sectors of the economy where they have to work long hours and they receive lower income. It further reveals that the poor are marginalized due to their poverty and illiteracy about technology and thus they are denied the access to the urban services as well. The study argues that due to the intervention of information as well as technology-based economy economic restructuration taken place and changes becoming visible in the economic sphere particularly in the job market which discriminates the urban poor and illiterate mostly.

Keywords Information economy · Economic restructuration · Marginality and Dhaka city

1 Introduction

In the era of neoliberalism, a new economy emerged mostly in the last quarter of the twentieth century on a worldwide scale, which is termed as informational, global, and networked economy where everything is integrated through a network. The global integration of financial markets since the early 1980s, made possible by new information technologies that are taking place to a greater extent day by day with the acceleration of informationalization. The new technological paradigm changed first the scope and dynamics of the industrial economy, creating a global economy and fostering a new wave of competition between existing economic agents as well as between them

R. Hussain (✉)
Senior Lecturer, Department of Sociology, East West University, Dhaka, Bangladesh
e-mail: raseldu1971@ewubd.edu

L. T. T. Huong and G. M. Pomeroy (eds.), *AUC 2019*, Advances in 21st Century Human Settlements, https://doi.org/10.1007/978-981-15-5608-1_40

and a legion of newcomers. The information technology, which brings information-based economy is a mixed blessing for economic progress (Castells 2000).

The flexible management and production system along with all the other things are becoming integrated into the information-based economy. There is informational agriculture, manufacturing, and service activities that produce and distribute based on information and knowledge embodied in the work process by the increasing power of information technologies (Castells 2000).

The economic structure of the global city leads to social polarization, as the leading sectors, on the one hand, employ a group of extraordinarily high-earning individuals and on the other hand create a demand for low-paid, low-skilled service workers like informal activities (Sassen 1991). Information-based economy and the vulnerable work situation of the poor are interrelated due to the establishment of information-based economy in the world economic system. Due to their poor wages, they are not capable to meet their daily basic needs on a regular basis, thus they are facing multidimensional problems like housing, food, water and sanitation, access to basic health services, and others. Now the society is differentiated between the skilled and unskilled workers. The citadel and ghettos exist in a closer proximity in the world cities and in this case the citadel is the most benefited areas, whereas the ghettos are becoming the most polarized areas of the city.

In the period of neoliberal globalization, the cities of Bangladesh are representing a concentration of surplus population. Knowledge-based economy is dominating the world that actually promotes the highly skilled and literate working force thus ignore the huge amount of surplus unskilled labor. For example, the apps-based transportation system like Uber, Pathao, Sohoj, Obhai is welcoming a huge number of technologically skilled manpower and thus creating employment for many which has an impact on the unskilled labor force like rickshaw puller and so on. The unskilled vendors have been replaced by the online service providers. Hotel and restaurants replaced the hotel boy by a robot. As a result, most of these ignored and backward surplus populations are now working in informal sectors of the economy. This surplus urban population is a specific characteristic of dependent urbanization occurring in this country. Dependent urbanization incorporates over-urbanization and parasitic cities with a low standard of living and high rate of unemployment. It is the informal sector backed by economic liberalization that has given the unemployed urban population unprotected, unregulated, and self-employed informal jobs for maintaining their survival.

Dhaka, the capital of Bangladesh, is also the dumping ground of surplus population and most of them are migrants from rural areas. They come to the city with a hope for better livelihood and soon they understood that many things have been changed with the passing of time. Now, the city is for the skilled and literate workforce not for the unskilled laborers. Thus, without finding no other option most of them got involved in informal sectors of the economy like street vending, rickshaw pulling, begging, and so on (Hossain 2011, 2013). The urban marginal groups who are continuously fighting for their means of subsistence are the real victims of social polarization. These groups of marginal people are mostly residing in slum areas in the urban peripheries of the city. Information-based economy created through information and communication technology is ultimately responsible for the new urban marginality in Dhaka.

It is in this context; this paper aims to explore the marginality of the lower segment of city dwellers. It highlights the impact of information-based economy upon the marginal groups and their vulnerabilities to employment which is taking place due to the information-based economy.

2 Data and Methods

The study has been conducted in the city of Dhaka which experiences a rapid urban transformation in recent decades while millions of rural poor moving here for employment. The study was conducted in five major neighborhoods, namely Kamrangichar, Hajaribagh, Kawranbazar, Mohammadpur, and Mirpur, having large informal sector involvement and the concentration of the new urban poor. Different major informal economic sectors including manufacturing, garments industry, transportation, and services, day labor, street vending have been taken emphasize in the survey.

The study is based on a mixed methodology that is the compilation of both qualitative and quantitative techniques of data collection. The data have been collected through surveys of 120 households along with in-depth interviews with the most relevant respondents. The sample ensures the inclusion of main informal activities and the geographic distribution of the poor neighborhood areas of Dhaka City Corporation. The primary data were supplemented by secondary data collected from different sources on informal sector workers in global, regional, and national arenas by urban scholars.

3 Theorizing Economic Restructuring and Urban Marginality

In the period of neoliberalism, capitalism stands in its highest peak and automated the whole world with its own economic process. Now, industrialization focus on production, and for this, vertical integration is required in different layers like from labor to capital, manager to producer, and then to the consumer. The development of technology, especially communication technology makes the world a global village. Due to time–space compression, now people are moving freely from one place to another and flexible process of accumulation taken place around the world which brings with its flexi workers. New markets of products established due to the new innovation in commercial, technological, and organizational sectors. Digitalization makes possible by time–space compression which also accelerates the decision-making process. Organizational shift toward vertical disintegration started because of the firm's care establishing its local networks all over the world and concentrate on them from the center. Production became information- and knowledge-based which requires skilled manpower those who are technologically sound, thus it creates employment opportunity and at the same time a huge number of people became jobless because of technological illiteracy. All these people became socially polarized and, with the loss of the job, their standard of living falling down.

Scholars have studied some of the issues of economic restructuring, transformation of work, social polarization, and marginal condition of the workers. But very a few of them focus on the information-based economy. New information technologies are facilitating more complex interactions that are organized by networks. Castells (1996) argued that new information technologies, such as the internet, allow such structural types to pervade social and economic processes. He describes how the fundamental aspects of networks allow for changes that are leading to a variety of transformations, such as decentralization within firms, telecommuting of workers, interactions in the virtual community, and economic globalization. Networks can expand without limits by simply integrating new nodes that share the same means of communication with other nodes. According to Castells (2000) restructuring slowed largely from relative demand shocks and those prominent levels of permanent job destruction and permanent discharges were distinguishing features of industries subject to restructuring. In addition, most workers displaced in restructuring industries relocated to other sectors. While this process of reallocation led to large increases in productivity in industries cracking workers, it also resulted in prolonged periods of unemployment for displaced workers. Moreover, relocating workers suffered ample reductions in earnings, consistent with substantial losses in their specific human capital.

Fainstein (2001) reports that cities show signs of increasing inequalities during the 1990s driven by growth at the top of the occupational ladder and by the exclusion of the marginal from the labor force at the bottom. Whereas Butler et al. (2008) argued that ‘skilled manual workers’ and ‘intermediate non-manual workers’ as forming part of the middle classes, which comprise, inter alia, security guards, traffic wardens, sales assistants, telephone operators, and routine clerical workers. Whereas less qualified members of the working class tend to circulate between low-skilled jobs in different sectors, alternating periods of work with episodes of unemployment.

Pratschke and Morlicchio (2012) showed that the process of economic restructuring makes employment for a large segment of the population and at the same time unemployed the non-skilled workers worldwide. Bina and Davis (2008) argued that a by-product of the increase in productivity over the last 20 years has been a massive surplus population—a gigantic reserve army of unemployed—at the global level.

One of the most important facets of the restructuring of production in both manufacturing and services is that it tends to rely on new social and information technologies, which require highly skilled employees. Now the urban peripheries become the dumping grounds of the urban poor in the era of surplus humanity. The nature of life and labor in the peripheries clearly indicate their poverty and vulnerability. Most of the urban poor are involved in low-paid economic activities. They mostly work as rickshaw pullers, street vendors, construction workers, transport workers, garments workers, and low-grade employees in the government and private sectors and other informal activities in the peripheral areas of the city (Hossain 2013). Women are the worst sufferers in this era of surplus humanity. The poor women working in the garment factories are getting low wages although they work for long hours. The condition of their workplace is very unsafe that is a threat to their reproductive health and they often become the victims of accidents.

4 The New Urban Marginality in Dhaka City

4.1 Demography, Households, and Residence Pattern

The people living in the poor settlements are to some extent different from other sections of urban dwellers in terms of their demographic features. Among the urban poor surveyed from this study, one-fourth of them are less than 20 years of age, whereas 10.8% of respondents are more than 50 years of age. In this study, 67.5% male and 32.5% female respondents were interviewed. Out of total respondents, 44.2% are married, 40% are unmarried, and 10.8% are widowed. The study further reveals that among the respondents 26.7% completed their primary education, whereas a large segment (17.5%) of the respondents never attended school and only 9.2% can sign only. An insignificant of the respondents (5%) attended a higher secondary level and above. Indeed, overall educational status of the poor people was not satisfactory at all which ultimately make them marginal in the city.

Almost half of the respondents opined that in their households at least two members involved in earning activities another 22.5% of the respondents argued that they are the only earner in their families. However, there is a relationship between the number of family members involved in the earning activities and the economic condition of the family.

The surveyed respondents are proportionately distributed as Kamrangichar (20.8%), Hajaribagh (16.7%), Mohammadpur (20.8%), Mirpur (20.8%), and Kawaranbazar (20.8%) and more than half (57%) of the respondents are living in slums, whereas only 42.5% of them are living in the non-slum areas which are often considered as low-cost housing areas in the city. They mostly migrated to the city for unemployment and underemployment in their place of origin and opportunities for new jobs in the informal economy in the Dhaka city. The survey reveals that 62.5% of the respondents are living in Dhaka city for more than 6 years and above. The study reveals that, recently, migrants are vulnerable in the city to housing and employment than their long-term counterparts.

4.2 Employment Vulnerability

The poor people who migrated from rural areas to the city are mostly engaged in informal activities due to scarcity of the formal jobs. Moreover, they are lacking skills to get jobs in these formal sectors of the economy. They often engage themselves in the employments which usually require small capital and need not require a higher level of skills. The survey shows that 30% of the respondents are involved in street vending, about 22% are involved in lower grade professions like peons, 18.3% of them who are mostly females are working in garments, 11.7% respondents are working as salesmen, and another 11.7% of them are working as day laborers.

The poor people surveyed responded differently to the reasons for being involved in their present employment. About 57.5% of the respondents involved in the present employment only for subsistence and 43% of them are involved in their present jobs as they could not manage other jobs. It further reveals that 26.67% are involved in their present jobs, as they did not possess skills to engage in the higher grade of professions.

Robiul, a 19 years old young man, working in a boutique shop in Hajaribagh commented,

> I am not literate enough to get a good job with better salaries. Nowadays one has to be skilled to get good job and I have nothing of that, that's why without finding any other option I got involved in this job.

It is observed that people in urban sectors often change their employment which is also valid for the poor sections of the city dwellers. This study also reveals that significant portions of them were involved in other jobs prior to moving to the present jobs. More than 22% of the respondents were involved in the day labor activities and almost 16% of them were students and housewives before moving to the present employment. In fact, the poor have limited access to employment training which could play an important role in their promotions in employment. The survey shows that 75% of the respondents do not have any training opportunities in their present job.

The vulnerability to employment is expressed through the difficulties they experience at the workplace. Among the respondents, more than 65% need to go 1–2 km far from their residence to attend a job. Another 17.5 and 12.5% have to pass 3–4 and 5–6 km, respectively, to reach their destinations. They experienced different forms of harassment in the workplace. The survey shows that more than half (57.5%) of the respondents talked about long working hours; 30.83% talked about lower income, whereas 24.7% were the victims of eviction and have to work in an unhealthy environment. Moreover, poor people mostly working as day laborers are unable to manage the job which is the real challenge for them. Anwar Hossen, a 55 years old man, working as a day laborer in Mohammadpur commented

> I came to this place at 7 a.m. after having a cup of tea and a piece of bread in search of work and waited for the last 5 h but nobody came here to hire me today, and it happens sometimes. I have to starve on the day when I fail to manage any work.

4.3 Restructuration of the Labor Market and New Forms of Poverty

Due to the development of information and communication technology many changes occurred in the job market which actually open the door for the skilled and limit access for the unskilled marginal group of people. Flexi workers become available in the cities. The survey reveals that 74.5% of the respondents mentioned the changes in the labor market especially the part-time jobs are increasing to a greater extent. Almost 40% of respondents opined that due to the lack of skill and formal job now more and more people are engaged in informal activities.

About 31% of respondents viewed that job for women is increasing day by day, and 28.33% mentioned the increase of computer-oriented jobs. Sharmin, a 24 years old married woman, living in Basantekh slum, argued that

> The present Dhaka is not the same as it was before her childhood. Now everything has been changed. Many women are currently working in the garments and other offices, the shops at the footpath increases in a greater extent, offices became computerized, cost of living also increased more than five times and marginalize the new urban poor.

As poor people are mostly involved in low-paid jobs due to their low level of skills and employment training, they manage their lives with limited earnings. The study reveals that 39.2% of the poor households have income between 10,001 and 15,000 Tk per month, whereas 20% of the households have a monthly income between 15,001 and 20,000 Tk. It is difficult for the people involved in lower grade employments to save some money after meeting their daily requirements.

The income is insufficient to manage their livelihood in the city. This study reveals that a large segment of them (82%) are not satisfied with their present income as they failed to meet their basic needs. Many of them are not getting adequate food needed for their survival. The survey shows that 61.7% of the respondents eat two times a day and only 30% of the respondents eat three times a day. Food they are eating is insufficient for marinating their good health. The study reveals that 83.3% of the respondents replied that food is not enough to keep them healthy.

Moreover, the price of the daily necessary goods is increasing to a greater extent day by day. Due to the lower level of income, almost 90% of respondent's income doesn't support to get and to have the necessary things for their family members. When the respondents were asked whether they have any plan to change their current job, more than 88% of the respondents replied positively. Shahajahan Molla, a 65 years old man, living in Begunbari slum is currently working as a peon in a private bank replied

> I want to continue the present job until retirement as I have no option and opportunities to change the job.

4.4 The New Urban Marginality Is the Byproduct of Information Economy

In the era of neoliberalism, nothing is static and everything is taking new shape and the process which makes it possible is the expansion of information and communication technology. In the globalized world, many new features of the economy are substituting the old one which, in one sense, is good for a definite group of people and, for others, it fetches so many misplaces. Groshen and Potter (2003) argued that the distinguishing feature of restructuring industries was a relatively heavy reliance on job destruction, although a reduced pace of job creation was also evident in these industries. And this statement has been strongly supported by Friedman and Wolf (1982). They argued that the world city today is in transition. Structural instability manifests itself in a variety of ways such as the dramatic changes in the distribution of employment, the polarization of class divisions, physical expansions, and decaying older areas and political conflict.

As a result, many people without finding any other option they are getting involved in informal activities and became lower grade professionals where skill is less important. This study reveals that 30% of the respondents involved in street vending activities; 22.5% as lower grade professions; 18.3% in garments activities; 11.7% of respondents worked as salesmen; and another 11.7% are working as day laborers.

In this economic transformation, the service-oriented industry places greater attention in the global cities. Sassen (2002) argued that the global city economies contain a broad complex of specialized service industries that enable transnational corporations to coordinate production, investment, and finance on a world scale. Economic restructuring also changes the physical sphere of the countries inside and outside of their own territories. In the past, the main economic base of Bangladesh was agriculture which is loosening its position now to a greater extent as new sectors like garments and lather industry have taken its place and playing the main agent of the economy.

In the present urban economic system of Bangladesh, it has been witnessed that the persons who came to the city with a hope to leading their life in a charming way by appealing themselves in both formal and informal sectors, more than half of them are running their life by working in the informal activities in the city areas thus leading a negligible life. They became the new urban marginalized group in the city which becomes the dumping ground of these surplus populations (Hossain 2013). Most of the time they have been distraught and offended by both the general and civil forces as their activities are considered as unlawful. Friedman and Wolf (1982) argued that the primary social fact about world city formation is the polarization of its social class divisions in which the transnational elites are the dominant class in the world city and the city is arranged to cater to their lifestyle and occupational necessities. On the other side, the permanent underclass of the world city could scarcely be more striking and the victims of the system and also fails them in the nerve centers of the world economy where they are queuing for a job. They are actually serving the transnational elites and provided personal services to them. It has been also observed that these two hostile groups of people often living in a close proximity but of them especially the upper class always maintain the distance.

Increasing inequalities during the 1990s, driven by growth at the top of the occupational ladder and by exclusion from the labor force at the bottom (Fainstein 2001). The unfortunate group of people lives in the ghettos in most of the cases and thus hypotheses a subculture of their own where their needs never achieved.

The present form of urbanization is the reflection of economic restructuring and rearrangement of the marginal group of people in the urban ghettos. Hossain (2013) argued that neoliberal urbanization is characterized by economic restructuring and relocation of the poor communities in the urban peripheries. It is playing an important role in the formation of new urban peripheries in Dhaka. The impact of economic restructuring led to a particular focus on the nature of 'ghetto unemployment' and structural forms of exclusion from work. To be involved in the information-based as well as knowledge-based economy someone has to be literate enough and should have skills. At first, they were used to get the opportunity to be involved in the manual work in the industries but now this is not possible for them due to the computerization of the industries. The present economic structure plays a crucial role in the development of unemployment as well as a lower level of employment opportunities in the world economic arena. The present study reveals that almost most of the respondents are not happy with their present job as they do not see the prosperity in the present job so they opined that they want to make a stable and prosperous life for themselves and for their family as well.

The subculture that has been settled by the new urban marginal group of people once started continued through generation to generation and thus they mingle themselves in the cycle of social polarization. The information and communication technology makes possible the network-based society. The newly developed economic structure opens the door for extraordinarily high-earning individuals and just shut the door for the unskilled labor forces. Due to the involvement in the lower grade jobs, respondents did not get enough money to run their livelihood. And in some cases, they are the only ones who involved in the earning activities in their family. According to the survey, one-fourth of the respondent's family income is between 10,001 and 15,000 Tk in a month which is not enough to lead a well-mannered life. The marginal groups have to somehow manage their lives with their present income either sinking in loan or not.

Moreover, the poor communities are often excluded from many jobs. Castells (2010) viewed that social exclusion is the process by which certain individuals and groups are systematically barred from access to positions that would enable them to in autonomous livelihood within the social standards framed by institutions and values in a given context. Institutionally polarize the soul and conceal the will and individually mental illness, nervous breakdown, psychiatric repression, exclusion from the society. In Bangladesh, it has been observed that a large segment of the urban negligible group of people living in the slums leading a reckonable life where most of the services are not obtainable to those people and they have no access to most of the urban services at all and thus foremost a polarized livelihood.

5 Conclusion

In the information economy only, the people having computer literacy have access to the new forms of employment. As a result, huge numbers of poor people migrated to the city with hope and prosperity are often excluded from the new information-based economy. These vulnerable groups do not meet the basic needs of themselves and their family members throughout the years and they basically lived in settlements located in poor neighborhoods in the peripheries of the city. Furthermore, their livelihood initiative exacerbates when they are unable to find out jobs in the city. These new poor often lead a marginal life in the city where neoliberal globalization has created huge opportunities for the newly rich. In the city social polarization occurs due to income inequality, economic displacement, unequal distribution of wealth, and unequal access to the labor market.

The economy of Bangladesh has been restructured over the decades due to the intervention of information and communication technologies in the economic sectors. Now lower grade jobs, part-time work, jobs for women are mostly available in the cities of Bangladesh. Gradually, full-time jobs in the public sectors are disappearing. Although the poor have the right to employments as citizens, they are denied the access to the new employment sectors. By creating more formal employments in the cities, their vulnerabilities to work could be to some extent resolved. But this is very difficult to strengthen the public sectors while every single sector is being privatized under the new wave of privatization. Employment training could be another option for the marginalized groups for increasing their access to the new information-based economy. In fact, proper governance and planning could ensure sustainable livelihood for this greater segment of the urban population.

References

Bina C, Davis C (2008) Contingent labor and omnipotent capital: the open secret of political economy. Polit Econ Quart 4(15):166–211

Butler T, Hamnett C, Ramsden M (2008) Inward and upward: marking out social class change in London. Urban Stud 45:67–88

Castells M (1996) The Rise of the network society (the information age: economy, society and culture, vol 1). Blackwell, Malden, MA

Castells M (2000) End of millennium, the information age: economy, society and culture vol III. Wiley Blackwell, Cambridge, MA; Oxford, UK

Fainstein S (2001) Inequality in global city regions. In: Scott AJ (ed) Global city-regions: trends, theory, Policy. Oxford University Press, Oxford

Friedman J, Wolf G (1982) World city formation: an agenda for research and action. Int J Urban Reg Res 6(3):309–344

Groshen EL, Potter S (2003) Has structural change contributed to a jobless recovery? Current issues in economics and finance. Research and Market Analysis Group of the Federal Reserve Bank of New York

Hossain S (2011) Urban poverty in Bangladesh: slum communities, migration and social integration. IB Tauris, London

Hossain S (2013) Migration, urbanization and poverty in Dhaka, Bangladesh. J Asiatic Soc Bangladesh (Hum.) 58(2):369–382
Pratschke J, Morlicchio E (2012) Social polarization, the labour market and the economic restructuring in Europe: an Urban perspective. Sage, New York
Sassen S (1991) The global city: New York, London, Tokyo. Princeton University Press, Princeton
Sassen S (2002) Locating cities on global circuits. J Environ Urban 14(1)

Breakthrough to Promote the Urban Economy of Vietnam Urban System in the Forthcoming Period

Quoc Toan Nguyen and Thi Nhu Dao

Abstract Nowadays, urban areas with a high concentration of economic activities are considered as the main driving forces that have contributed up to 70–80% GDP growth for countries having a high rate of urbanisation. In Viet Nam, this contribution also has witnessed up to 70% of gross domestic product (GDP). Therefore, it is extremely important to set up urbanization policies and strategies to maximize the economic promoting role of urban areas. Recently, Viet Nam's urban development model mainly focusing on spatial organization but paying less attention to economic space arrangement and management driven by the market has constrained cities' GDP contribution as well as strong economic growth. By reviewing the situation, characteristic, and process of Viet Nam urbanization recently, the article synthesizes lessons learned about the relationship between the urbanization process and the economic contribution role of the urban areas. The article then comes up with breakthrough suggestions to promote the role of the urban economy for Vietnamese urban system in the forthcoming period.

Keywords Breakthrough · Urban economic · Vietnam urban system

1 Introduction

Urban areas are considered as main driving forces that have contributed up to 70–80% to economic sector and GDP growth of countries with a high rate of urbanisation. In Viet Nam, the contribution from urban areas has witnessed up to 70% of gross domestic product (GDP). In 2012, five cities directly under the central government—Ha Noi and Ho Chi Minh Cities (special cities), Hai Phong, Da Nang, and Can Tho (first class cities)—provided more than 35% of the national GDP, creating more than

Q. T. Nguyen (✉)
National University of Civil Engineering, Ha Noi, Viet Nam
e-mail: toannq@nuce.edu.vn

T. N. Dao
Urban Development Agency, Ha Noi, Viet Nam
e-mail: nhuxd@yahoo.com

L. T. T. Huong and G. M. Pomeroy (eds.), *AUC 2019*, Advances in 21st Century Human Settlements, https://doi.org/10.1007/978-981-15-5608-1_41

37% of export revenue, and more than 56% of total national budget revenue, although these cities only accounted for about 20% of the national population and 2.9% of the land area (Canh et al. 2013). Therefore, it is extremely important to set up urbanization policies and strategies in order to give a reasonable direction which maximizes the promoting role of urban areas. From the economic renovation (from 1986), generally, the rapid and strong urbanization process has boosted the economic development as well as has improved the quality of life of the people. However, besides the obvious achievements, urban system development which was characterized by sprawl, fragmented development (World Bank and Ministry of Planning and Investment 2016) also has led to negative consequences such as environmental pollution, saturation of public facilities and urban transport, traffic congestion, housing pressure, flooding, depletion of natural resources, loss of identity of urban landscape, and so on. In addition, the impact of climate change and the global development commitments that Viet Nam has participated in are also an important context that creates additional challenges on urban development toward quality, strong economic growth, and sustainability.

By reviewing the situation, characteristic and process of Viet Nam, the article synthesizes lessons learned about the relationship between the urbanization process and the economic contribution role of the urban areas. The article then comes up with breakthrough suggestions to promote the role of urban economy for Vietnamese urban system in the forthcoming period.

2 Vietnamese Urban Sector

2.1 *Vietnamese Urban Identity (Urban–Rural Relations)*

According to the Planning Law (30/2009/QH12), urban areas are defined as "densely populated areas and mainly operating in non-agricultural economic sectors; being the political, administrative, economic, cultural or specialized centre, with the role of promoting the socio-economic development of a country or a territory or a locality; including the inner city and suburbs of the city, urban centres and suburban areas of towns and townships."

According to the administrative hierarchy (Fig. 1), a typical characteristic of Vietnamese urban areas is: *urban in rural area and rural in urban areas*. As a result, while there are many "urban area" reflect a high economic and population agglomeration, some are predominant with rural characteristic. The urban and suburban administrative scope does not completely provide urban's or rural's characteristics because the fact that the part identified as suburban may have a high accumulation point in terms of population, urban economic activities as well as built-up density. For example, in recent years, urban development in Ha Noi and Ho Chi Minh Cities mainly has taken place in suburban areas and outside urban core areas (World Bank and Ministry of Planning and Investment 2016, p. 236).

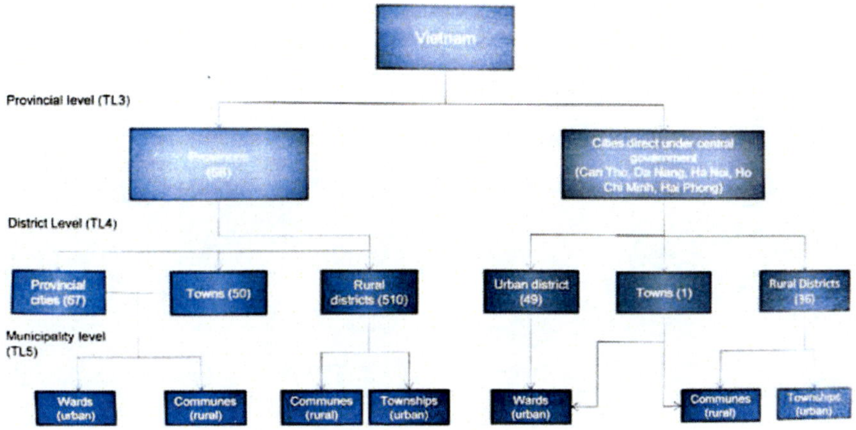

Fig. 1 Administractive structure of Vietnam. *Source* Ministry of Construction of Viet Nam 2017, unpublished and General statistic office of Viet Nam 2015

According to Resolution 1210 of Standing Committee of the National Assembly, urban classification is based on administrative area. The Resolution also refers to the case of "expected area of future urban formation is classified by corresponding urban type criteria". Accordingly, the recognition of the expected urban area is followed by two cases: The first one is "***standards-satisfied***" urban (based on the approved master plan and corresponding current urban infrastructure evaluation for the given area) and the second one is recognized as a corresponding urban type (if the given area also has a certain approved administrative boundary (for example, a commune centre satisfies urban technical standards and is recognized as a township, it can be official ranked as a IV class city)). In other words, the recognition of "***urban-being***" (in the sense of opposition to the countryside) according to the Vietnamese law is always associated with the role of an administrative unit (such as township, town, city).

Besides, it can be said that intertwining urban–rural is an important nature of Vietnamese urban (Urban Development Agency 2019; Consulting Eptisa 2019; Quyen 2015). This relationship is not only a historical feature of urban development, it is also considered as a positive point to trigger improvement in rural areas as well as a mechanism to control over-development, harmonize landscape, and ensure food security in urban areas. However, this fact also creates confusion in Viet Nam's urban management due to the differences in economic accumulation and functions between these parts. For example, from cities of class IV to upper classes, management activities always are based on both urban planning and rural construction planning. Many experts argue that the existence of rural land in urban areas can cause farmers' discontent about land prices, force farmers to keep their agricultural land, while under the market pressure, the annexation of development investment projects on these areas, as being fertile areas with cheap prices is inevitable. Master planning for Ha Noi

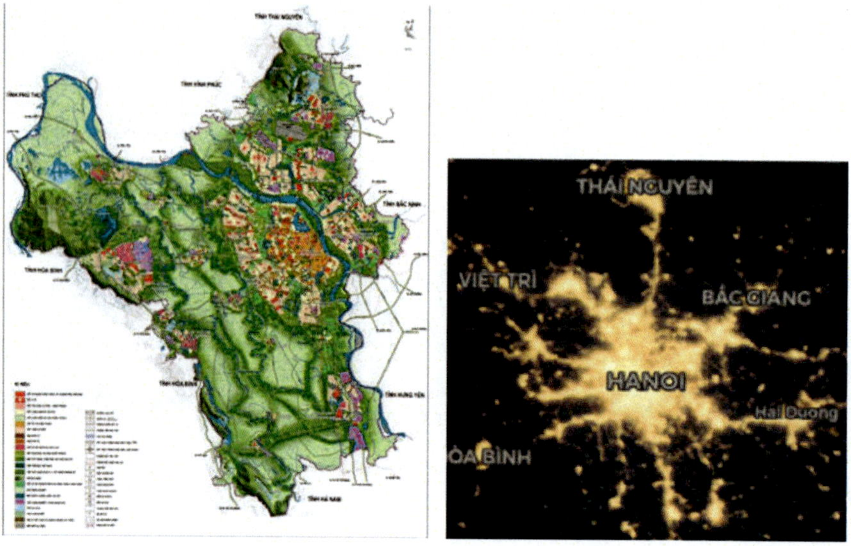

Fig. 2 Ha Noi master plan by PPJ with green belt (left) and the actual development out of local control based on earthnight map (right)

by PPJ [1] planed the agricultural green belt area in the city with an explanation to maintain the ecological region, landscape, and food security for the capital and urban–rural relationship. But the reality shows that this area has developed out of the local authority control (Fig. 2). The Ministry of Construction is preparing urban–rural system planning as an important national sector plan to guide this important relationship with an aim to create connection ability and maintain economic and value chains between them creating effective harmonious coordination.

Rate of urbanization[2]: In line with the aforementioned urban concept, Viet Nam's urbanization rates are rather different from the world's. While the world normally relies on indicators representing economic growth reflected by the night light model or determining the built-up area on satellite imagery (in other words, urbanization is defined very close to the nature of economic accumulation and economic transition as well as the improvement of life's quality). The rate of urbanization in Viet Nam is, in fact, more influenced and controlled by the size (in terms of area and population) of urban administrative units rather than the effect of economic and associated population accumulation. Actually, urbanization areas in Viet Nam conceal many questions such as sprawl development, fragmentation, unfocused, and disconnected (World Bank and Ministry of Planning and Investment 2016; Urban Development

[1]PPJ: PERKINS EASTMAN, POSCO E&C and JINA.

[2]The urbanization rate is calculated by dividing the urban population by the natural population. Accordingly, the urban population (according to the General Statistics Office) is "people living in urban areas, towns and townships. All residents of other administrative units (in rural communes) are considered to be the rural population.".

Agency 2019; Consulting Eptisa 2019). The urbanization rate of a municipality in Viet Nam is sometimes higher than the actual agglomeration of economy or urban population in that area (as the fact of low density). While this figure is also likely to be lower than that of urban development in general, since there are indeed many areas reflecting "urban characteristics" but are not inclusively accounted for (because their location did not belong to the administrative unit of a township, town, or city. That situation is so-called "hidden" urbanization in the report of WB, for example, during 1990–2012 in the Ho Chi Minh City, 48.5% of the population increased, and 62% of the urban land area increased were in suburban communes (Kontgis 2014). According to VIUP,[3] the figures of urbanization reflected in many documents of other organizations are significantly different: according to GSO it was 34–35%, WB 36%, UDA was around 38%, while to VIUP it could be up to 52%. The difference of urbanization situation and nature may lead unreasonable or asynchronous strategy and policy for urban growth.

2.2 *Current Situation of Urbanization and Urban Issues*

It can be said that if the city is seen as a picture, the Vietnamese urban picture is a collection of many scattered, discrete, and unconnected development areas. This situation is also the cause of many problems of inefficient use of urban land, resources as well as urban operation.

Main core, secondary core, and undeveloped areas in region: The accumulation of non-agricultural economic sectors, employment, services, and urban population has been mainly concentrated in the centre of the city (the main core) and pulled out some locations around suburban industrial areas (secondary core). While in the middle of these development cores, low density was still common with the lack of urban infrastructure connection, creating a big gap and a significant disparity (World Bank and Ministry of Planning and Investment 2016).

The existence of these "undeveloped points/areas" in municipalities can be considered as isolated oases lagging behind joint development. Lacking be connected and well planned for urban standard services have limited the development of high-density housing, industry and commerce. These areas are also often "hot" points for the poor and informal classes with limited economic capacity but keeping the willingness to accept the consequences of a marginal area such as the lack of urban infrastructure and services, inadequate amenities, inefficient land use, and maintaining the "rural" character in the city (such as lifestyle, way of thinking, way of expression, small-scale economy…). Meanwhile, the overload development in the cores also has led a serious pressure on this area, typically the issues of overpopulation, housing, infrastructure, services, and the environment. As the result, frequent traffic congestion, inundation, degraded environmental quality, threatening urban identity heritage spaces were seen. It can be said that both development areas (main,

[3]VIUP: Vietnam institute for urban and rural planning.

secondary cores) and "undeveloped" areas have had their own problems of life quality and the degradation in different levels and characteristics. In particular, the lack of connectivity between them has disrupted the synergy of urban hierarchical structure, thereby limiting the competitiveness of urban and region.

Widespread urbanization is also a result of emerging movement of urban upgrading and re-classifying[4] to take advantage of the state budget allocation mechanism. The process involves the situation that local authorities only focus on their own direct interests rather than the regional development and cooperation [because they could benefit directly from expanding their businesses within their boundaries, so they ignored the cluster development potential and the economic advantage of regional infrastructure investments (World Bank and Ministry of Planning and Investment 2016, p. 258)]. This situation significantly widens the gap for low density and weak connectivity, while the efficiency of the accumulation of urban population, economic sectors, services, and labor tended to be reduced.

2.3 *Assessment of the Urbanization Process in Viet Nam*

The relationship between urbanization (in terms of population and land) and urban economic growth

From population urbanization to land urbanization: In the period of 1992–2005, Viet Nam witnessed faster urbanization of the population than this of land which was accompanied with a higher economic density in urban areas. However, recent years have shown that urbanization of land has been faster than residential urbanization and not be associated with substantive economic development activities. This change recorded a decline in GDP growth due to the reduction of labor productivity growth rate (World Bank and Ministry of Planning and Investment 2016, p. 233).

The relationship between population accumulation and urban growth

In the past, since the 1990s, the accumulation of urban population in Viet Nam was mainly due to the flow of rural–urban migration. In the 2000s, liberalization movement law made it easier for people to migrate within country. From 2004 to 2009, the inter-regional migration flow increased in parallel with the industrialization process, contributing 42% of the population growth in urban areas in Viet Nam. After the 2008 economic crisis, the accumulation due to this inter-regional migration flow has no longer played many roles due to the looses in industrial and construction

[4]Urban Classification: In order to qualify for one of six urban levels, cities must achieve a minimum of 70/100 points. The population density accounts for a maximum of five points and only equals to the ratio of non-agricultural labor, which is 2–11 times lower than other indicators. A municipality can easily achieve a minimum score by building more infrastructure, expanding administrative boundaries to increase population and inflating local gross product (this may be a reason for local GDP growth is often one and a half times higher than the overall growth of the country) while ignoring more substantive criteria such as population density (World Bank and Ministry of Planning and Investment 2016, p. 258).

sector. From 2009 to 2014, population agglomeration was highly dependent on urban boundary expansion, of which almost half of population growth in urban areas was related to administrative boundary changes, 21% due to migration from rural areas, 32% due to natural growth (Kontgis et al. 2014, OECD 2018). This shows that population accumulation accompanied by industrial development, economic, and service accumulation more contributed to urban economic growth. In addition, actual research data shows that this phenomenon was also directly related to a higher level of average income per capita (Figs. 3 and 4).

The relationship between urbanization and industrialization

The process of urbanization has been also associated with the policy of industrializing the country. However, ironically, in the last period, the strong increase in industrial

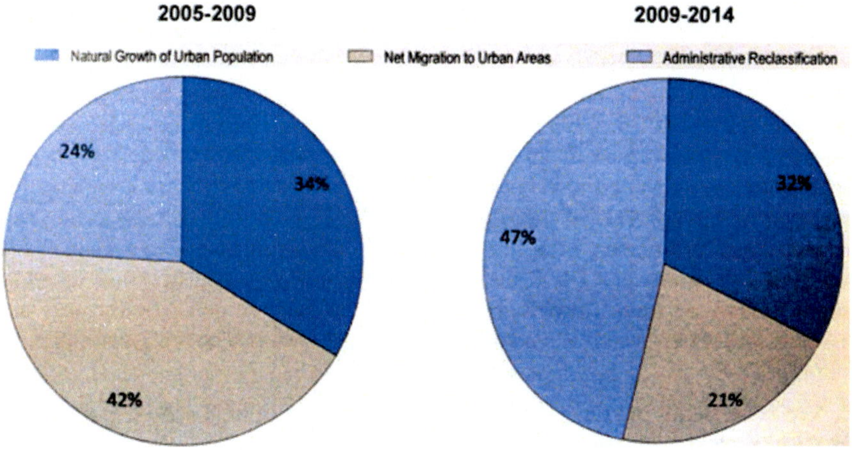

Fig. 3 Contribution to urban population growth. *Source* (Quyen 2015)

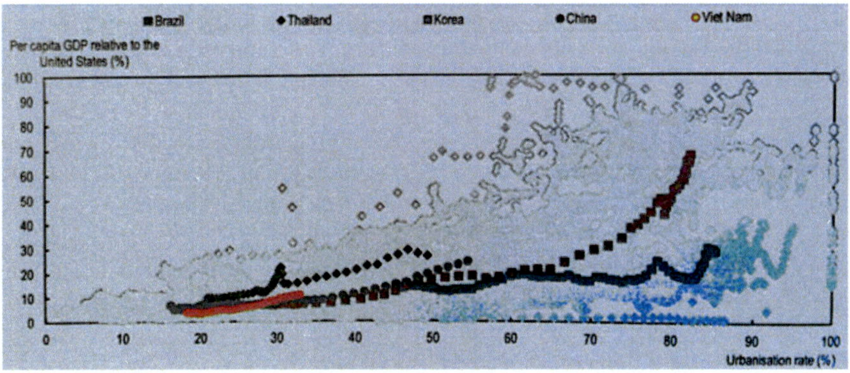

Fig. 4 Economic coverage and urbanization (1970–2015). *Source* (Kontgis 2014)

zones was more like the expansion of urban areas, which was more influenced by administrative decisions than by the needs of the urban development and the market demand, led to no more benefits for economic growth. In fact, many concentrated industrial parks have operated ineffectively, even becoming abandoned industrial zones. Statistics show that there are 70,810 hectares of vacant industrial land in the areas of Ha Noi capital, Da Nang and Ho Chi Minh Cities spending an investment of US $20 billion. According to the Ministry of Planning and Investment's estimation, in the next 10 years nearly US $40 billion should be required just to fulfill building infrastructure structure for approved industrial and economic zones (World Bank and Ministry of Planning and Investment 2016).

Regarding the urban spatial development orientation, the urbanized region

Viet Nam policies have been oriented to focus on growth poles and regions [Decision 445 in 2009[5] provided a development roadmap, by 2015 promoting urbanization in priority economic regions; by 2025, promoting urbanized areas, and by 2050 promoting urban networks]. In fact, the Vietnamese urban system and urban region have formed clearly with growth poles, the central cities acting as the driving force and development corridors which have largely contributed to urban economic development, growth poles such as Ha Noi and Ho Chi Minh Cities (in the Red River Delta and Southeast-including a part of the vicinity in the Mekong Delta, respectively), followed by the Central Coast region]. It can be said that the urbanization trend of Viet Nam keeps going along the global urbanization roadmap, forming urbanized region in the trend of "regional urbanization" and "regionalization" to increase competitiveness.

However, if we zoom into the growth poles, dynamics centers and these corridors to explore the urban functions of the city, we may find more about the policy of " concentration" which mostly was characterized by a set of discrete, separate, unconnected development points, so the operation of the city differs greatly from the roles of dynamic urban areas in real effective urban region (Fig. 5).

The lack of infrastructural connectivity and especially this of economic linkages and value chain in these poles are remarkable points. There are still many "empty" points about population and economic agglomeration, expressed in low-density and distributed/fragmented development. In other words, the concentration policy, in reality, is still horizontal and physical expansions. These growth poles, urban centers, or development corridors were established by a "driven will" rather than by a market vision focusing on real urban value agglomeration. In other words, this "concentration" phenomenon has not yet related to a need or orientation of building an effective urban economy, creating an urban value chain, forming an smooth transport connection but reflecting an "individual plan" to exploit land resource, convert agricultural land use into urban areas, meet the criterion for upgrading urban types to benefit the state budget allocation [the decisions on urban administrative expansion in the period of 2000–2010 were not the main factors responsible for

[5]Decision 445/QD-TTg of the Prime Minister in 2009 approving the adjustment of the Master Plan for Vietnam's urban system development to 2025 and vision to 2050.

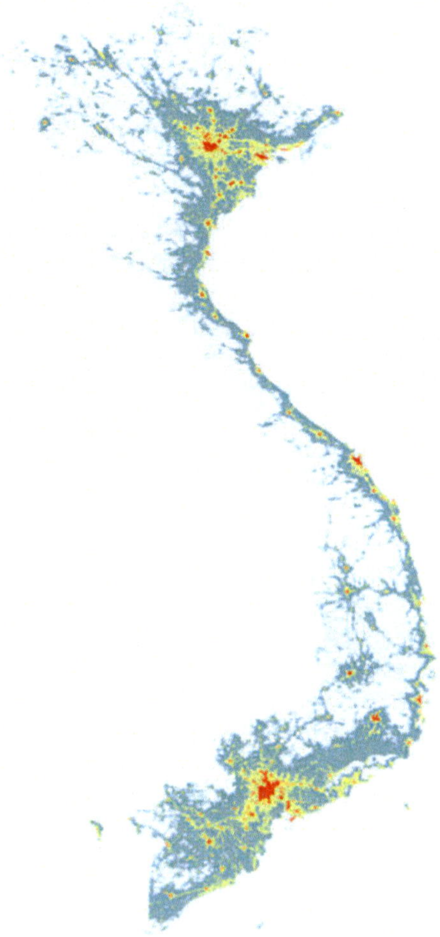

Fig. 5 Urban regions in Viet Nam based on map of night light intensity, 2017. *Source* WB

the urban economic growth in the period of 2011–2012 thereafter (World Bank and Ministry of Planning and Investment 2016, p. 241)].

3 Breakthrough Suggestions of Urban Economic Growth

Smart financial allocation for urban growth poles and to prepare a supporting roadmap for development urban centres and other potential fast-growing urban areas in the region.

The current policy of state budget allocation seems to follow an equal model for all urban areas (although in reality, more or less allocation may be related to the ability of mobilization and management of each locality). It can be said that this policy has not discouraged the responsible, active and creative role of local governments in small and medium-sized cities so it consequently has not stimulated economic growth and a significant life quality change for these kind of cities, while it even has not provided enough growth engine or the proper investment resources for outstanding growth in big cities (poles). Although many experts consider that reliance on state resources is not the primary source of urban development as private sector normally play important role in this field, this mechanism creates a less fair game, or a "discontent" for large cities (poles) that are contributing more to the national budget but only receiving less allocation, or create the "reliance and lack of autonomy" of small and medium local governments.

Experiences from China and other countries shows that concentrating resources for growth poles and potential developed cities will help to boost the urban economy and increase urban GDP much more than the dispersion of resources and no smart focusing investment. The principle is that developing a city (pole, particularly) requires a huge of investment resources which should be abundant enough to help the city to carry out not only its smart technical and social infrastructure projects, to solve and stimulate economic agglomeration needs, but also to foresee resolutions for a enormous range of other associated urban problems. Therefore, the prioritizing policy of resources concentrating for growth poles always needs the sacrifice of small and medium-sized cities for a while. Then, a growth pole with strongly enough development in return invests potential cities by order to ensure that each urban area once be invested will be gained the most effective concentrated investment, ensuring the sufficient accumulation of economic and human resources.

The process of urbanization in recent years in Viet Nam has clearly shaped two metropolitan areas,[6] showing that Viet Nam's urbanization trend is gradually coming in line with the world's one forming large urbanized regions and areas. Increasing intra-regional and inter-regional connectivity and promoting small urban systems (network developing) especially urbans belongs to supply and value chain should be considered to invest at the same time or in their next steps.[7]

Frame transportation infrastructure connection, especially for supply chain and logistic chains

This policy is to meet not only the needs of the residents' life but also this of goods circulation to support and consolidate value chains, supply chains, as well

---

[6]Ho Chi Minh growth pole (in the Southeast region), Hanoi city (in the Red River Delta). Besides, Da Nang city is a vital and remarkable development centre in the coastal corridor.

[7]From a regional construction perspective, the role of large and small cities as well as their functional decentralization is very important. International experiences show that each type and size of urban area has its own function and often has the nature of supporting each other. A big city function could not be replaced by a bunch of small ones but other the same-sized one. Economic activities create a large city, including manufacturing and high-class services, intersections, administrative and financial centers. While small-town operations are more related to demand from surrounding agricultural areas and more based on economies of scale at the factory level.

as logistics chains (connecting service area with storage, loading, unloading, transportation, payment). This is even more important for the structure of the urbanized region. Recently, the infrastructural system has only paid attention to synchronous investment, while fast-connected transportation has not been focused, particularly to some strategic inter-regional axes, in railway systems and highways to connect the economy between growth poles, development center cities, and large production areas (e.g. connecting between large-scale agricultural commodity production areas in the Mekong Delta, Central highlands with major markets, ports and border gates (World Bank and Ministry of Planning and Investment 2016)).

Also, in terms of building global value chains, currently, Viet Nam is participating in global value chains mainly in the final assembly stage, the domestic economy is just consumption but not stimulating growth for auxiliary production. However, in the coming period, the increase of domestic rate of input materials for the production of exports will bring opportunity to secondary cities where meet development requirement of business clusters in sectors that are related to each other. International experience indicates that medium-sized secondary cities can support the local economy by linking buyers and sellers in the same industry together by spatial cluster (World Bank and Ministry of Planning and Investment 2016, p. 249). Therefore, if there is a lack of infrastructure network, especially fast link thanks to high-speed railway, it may hinder the logistics industry.

More effective urban land-use, particularly in existing urban areas

Optimizing existing urban areas by increasing its value at the highest level rather than focusing resources in new urbanized areas is important. The policy of "value filling and adding" is implemented by strengthening priority policies to attract private sector invest to urban re-development projects or by completing infrastructure connection, providing urban utilities or securing employment sources and opportunities for workers… Increasing land use efficiency through urban redevelopment project tools (Land Readjustment LR and land renewal UR are valuable and successful experiences from Japan (De Souza et al. 2018)). Increasing built-up density (by adding heights or floor areas) by investing more capital on the same land to synchronize with transport investment to ensure good circulation and not add pressure on urban infrastructure.

To improve urban productivity and to boost economic growth by ensuring urban utilities and jobs for labor flows

Labor productivity in urban areas is always much higher than this in rural areas. Therefore, the accumulation of labor in urban areas is a good way to increase productivity and contribute to the urban economy. In fact, the recent population agglomeration in urban areas is a result of the race of expanding urban administration and urbanizing land. In other words, the high concentration of people in urban areas has not been accompanied by the goal of utilizing and promoting labor productivity in urban areas. So that, after becoming city dwellers, these farmers, who lacked qualified skills in the non-agricultural economic sector and were not familiar with urban lifestyles, easily fell into the "trap" of "getting rich accidentally "(by receiving land compensation or selling land) and also quickly became unemployed, or took part in social evils, creating more social burden on the city. China's proactive policy of labor mobility to growth poles has created a momentum for urban economy, which can be said to be a valuable experience for Viet Nam. Accordingly, proactive policies on labor resources for urban areas is always accompanied with a series of sub-policies such as planning for affordable accommodation and employment areas, addressing social welfare regimes, infrastructure services provision, proactive training of skilled resources, etc., so that the immigrant flows to the growth poles can be assured to contribute their all efforts. These "job-ready areas" planned for this migration flow are also expected to become attractive places that trigger economic development, create a spillover growth effect, and gradually fill the economic growth (effective agglomeration formation) for the growth poles and in existing urban areas.

To create and innovate urban resources

In the world, many countries have succeeded in changing their perceptions of urban development resources from their challenges and difficulties (turning challenges into resources) or from their strengths, such as urban space, urban waste, urban cultural heritage, urban landscape, urban culture (lifestyle), characteristics of local people… In addition, the creativity in the use of available resources is also a way to promote the financial capacity for the city. It is not possible to more and more access to limit land resources, and to depend on (raw) land selling. Exploiting the surplus value of land based on smart investment create more benefits, such as investing in infrastructure, providing essential urban services to add value to the land (for example, Land Value Capture (LVC) is a effective tool to help recover added value from land). At present, there are no effective tools in Viet Nam to exploit value added from land in the development and provision of urban services projects. For example, redevelopment projects or road expansion significantly increases the value for houses located deeply in alleys or had poor road access, so it is unreasonable if these households are not responsible to pay for this adding value as well as do not contribute anything. In addition, the evaluation of land value according to market price also can create new revenue sources. According to the World Bank's assessment, Viet Nam's land tax only accounts for more than 0.07% of GDP, nearly ten times lower than the average in developing countries, 30 times lower than the average of OECD countries. The lack of a land and property taxation system restricts the stable and regular revenue earned by local governments (Hoa and Pherson 2010).

Also based on exploiting land resources, the world has applied TDR-transfer of development rights-attached to the land. TDR is a tool that allows purchase, sale, and transfer of "air rights" to protect or develop space on land between landlords who do not need development and those who need it, usually for two purposes to control space for heritage conservation projects and landscape space (by selling air rights) while still meeting market demand of further development in case of being restricted by FAR (through the acquisition of unused "rights" in the surrounding areas or from the planned sending areas. Since 2018, nearly 100 heritage sites in Sydney have been selling "air rights". In New York City, billionaire Donald Trump-when he was not the president of the United States—legally bought TDR above at least 7 low-rise real estate around Trump Tower, in order to make sure a nice view for this building which will not be covered by any other buildings (https://www.abc.net; http://kientrucvietnam.org.vn/australia-thanh-pho-sydney-ban-quyen-khong-gian-de-co-tien-bao-ton-di-san/).

4 Conclusion

The process of urbanization in Viet Nam in recent years has brought positive changes in urban economic development, contributed to the country's GDP, improved quality of life but it also left many challenges for urban management. However, this process has also followed the rules and stages of world urbanization. After a period of sprawl development, the current urbanization period more focuses on city quality and promotion the role of urban economic growth. These challenges require great efforts to renovate development-oriented policies and strategies that concerns about cooperation and further promotion of the role of urban economic chains. In this effort, the responsibility of the central government is to create guidelines and policy frameworks, institutions, and tools to support effective implementation, while local levels should be more proactive and creative.

References

ABC new homepage. https://www.abc.net.au/news/2019-05-02/private-developers-purchasing-air-rights-sydney-heritage-sites/11063642. Accessed 25 Sept 2019

Canh NT, Dien NN, Tinh DPT, Phat NT, Phu HT, Khoa PC, An CQ, Hieu HH, Tri NM, Minh HTH, Tuan NQ (2013) Municipal government in Vietnam: case studies in Ho Chi Minh City and Da Nang. The Asia Foundation, University of Economics and Law, Hanoi, 76 p 2

Consulting Eptisa (2019) Draft of urban development strategy, period 2021–2030 and vision to 2030, Vietnam

De Souza FF, Ochi T, Hosono A (2018) land readjustment: solving urban problems through innovative approach. Japan International Cooperation Agency Research Institute

Hoa HD, Pherson M (2010) Land policy for socio-economic development in Vietnam. Kennedy School of Public Policy, Vietnam

Kontgis et al (2014) Supervise peri-urbanization in Ho Chi Minh City area. J Appl Geogr 53:377–388

Organisation for Economic Co-operation and Development (OECD) (2018) Vietnam urban policy report

Quyen LT (2015) Symbiotic relationship between urban and rural areas in Vietnam. J Plan 89 and 90. https://www.viup.vn/vn/Lien-ket-Do-thi-Nong-thon-n114-Moi-quan-he-cong-born-between-the-future-and-the-thon-Viet-Nam-d6702.html

Urban Development Agency (2019) Urbanization report, Ministry of Construction, Vietnam

Vietnamese architecture homepage. http://kientrucvietnam.org.vn/australia-thanh-pho-sydney-ban-quyen-khong-gian-de-co-tien-bao-ton-di-san/. Accessed 25 Sept 2019

World Bank and Ministry of Planning and Investment (WB & MPI) (2016) Vietnam 2035, towards prosperity, equity creation and democracy. Hong Duc Publisher, Hanoi

Research Proposal for Orientation for Urban Opening Planning Area in Ha Dong District, Ha Noi City

Vu Khac Hung and Tran Van Tuan

Abstract In the process of industrialization–modernization of the country, the urban environment in Vietnam is very lacking in open space areas. Hanoi is one of the two cities with the highest urbanization rate in the country. In 2017, Hanoi had an additional 11 million m2 of houses, 100 times higher than the record of 0.11 million m2 of housing construction in Hanoi in 1978 (Nguyen in Vietnam Architect Mag 05, 2018). However, the area of public space in residential areas has been encroached as a place of business, parking, housing construction to serve the purpose of economics (Nguyen in Vietnam Architect Mag 05, 2018). Among the districts of Hanoi, Ha Dong is considered one of the districts with the largest urban construction and development in recent years. Thereby, the research uses the method of forecasting the land use demand of people in Ha Dong District, identifying the open space needed to meet that demand in the next planning period 2020–2030. In addition, the research results present an overview of open space research in the world as well as in Vietnam, assess the current situation of open space and identify open space planning options in Ha Dong District, Hanoi City.

Keywords Open space · Ha Dong District

1 Introduction

In each country, open space is defined and classified according to different approaches. Studies of urban open spaces appeared quite early, mainly in European countries. The industrial network began from England then spread to the whole world in the late eighteenth century to the first half of the nineteenth century, causing the process of urbanization to take place strongly in European countries. With the explosion of commodity production based on industrial machinery, new European cities were gradually formed, changed, and considered as industrial cities with the

V. K. Hung (✉) · T. Van Tuan
VNU University of Science, Ha Noi, 334 Nguyen Trai, Thanh Xuan, Ha Noi, Viet Nam
e-mail: vukhachung@hus.edu.vn

L. T. T. Huong and G. M. Pomeroy (eds.), *AUC 2019*, Advances in 21st Century Human Settlements, https://doi.org/10.1007/978-981-15-5608-1_42

main functions as a place of exchange and production goods. By the end of the nineteenth century, rapid urbanization and the pressure of hot growth gave the appearance of uncontrolled spatiality in European cities (Wouter 2016). Against this backdrop, urban managers and researchers propose the use of integrated urban planning and design tools with technological advances accumulated from the construction industry, the environment, and economics to systematize public spaces in the city. With urban esthetics as the focus, managers and academics hope to address the spatial clutter caused by urbanization, enhance the image of the city, improve transportation systems, and provide providing green space, entertainment services, promoting intra-urban trade, while improving the quality of life in old urban areas for workers.

In Vietnam, open space emerged from the feudal era in the form of playgrounds, wharves, banyan trees, communal yards, village gates, village markets, village roads, temple temples, meeting halls, houses, worship of ancestors, etc. Through different stages, open space areas influenced by many planning ideas have created unique characteristics for the Vietnamese urban areas. In general, the Vietnamese open space can be divided into three main phases including the feudal period, the French colonial period (1858–1945), and the period of socialist state construction (1945–now).

In the feudal period, the feudal government decided on common issues and had its space of power. During this period, public spaces existed in the form of community spaces, those areas served the needs of a group of subjects such as a royal court, villagers, villages, and were limited to use by the government, a group that objects. Everyone who wants to access and use it requires the consent of the community using the spaces. By the French colonial period (1984–1945), the French had first introduced to Vietnam but the principle of the Western urban planning with a grid of square squares, majestic spatial axes, squares before the big projects such as government, banks, and theaters to show off their physical and cultural power and strength. During this period, the French created the public spaces properly. Outstanding works such as Long Bien Bridge, Hanoi Cathedral, and Bach Thao Park are one of the French architectural heritage brought to urban Vietnam. Since the period of socialist state construction from 1945 to the present, open space constructions have been influenced by two typical social development periods: the subsidy period (1976–1985) and the renovation period (1985–now).

Public space, people's ability to live, and approach are three common phrases in the study of urban open space and it has a mutual impact. This is the result of the research project named "Public space in a livable city" in Singapore (Kien 2018). Currently, with the trend of developing smart cities, the application of information technology in urban management is gradually popular in every country. However, in poor or developing cities, research on open space, livability, and human approach is still the main research directions to improve the quality of life in urban areas.

2 Definition of Term

2.1 Open Space

Open space (OS) itself is a complex and multidimensional concept and does not have a universal, universal definition of it. Open space is created, used, assigned, managed, and regenerated by the political–economic–social needs of different social institutions, at different spaces, and at different times. It is governed by various political, economic, and social systems.

In the area of urban planning, the term "open space" refers to areas reserved for parks, green spaces, water surfaces, and other public areas (Hemakumara and Ranasinghe 2018). Architecturally, the concept of open space is a broad term that can be used to describe all lands that do not contain buildings and constructions, possibly including public and private lands (Nochian et al. 2015). Urban open space objects must be designed to ensure service and toward the use of the people and the community, everyone can access and benefit without paying fees. Areas outside the city limits, such as parks or rural open spaces, are not considered urban open spaces.

The definition of open space is also used in many other studies, the Miami Valley Regional Planning Commission in 1993. The definition of open space in this study is very broad and spans from areas of nature activity, such as parks and soccer fields, to receptor areas. Natural activities such as wetland conservation and flood control, in other words, these are places built and created to serve the community (Miami Valley Regional Planning Commission 2006). However, there is no specific definition of people's accessibility in this definition. Therefore, those who are built to serve the people, they have to pay fees such as health centers, schools are still classified as urban open space objects.

The concept of open space has not been officially mentioned in current legal documents in Vietnam. Some documents of the Ministry of Construction and the Hanoi People's Committee such as QCXDVN 01: 2008/Ministry of construction and Decision No. 1640/QD-UBND Hanoi and other similar decisions have mentioned many works that are considered open space objects including public works, parks, lakes, and gardens.

Therefore, open space is understood as urban void areas, interspersed between buildings, construction works where all people have access and use rights, either free or paid. Depending on different subjects, participate in recreational activities, social, cultural, and health training (Miami Valley Regional Planning Commission 2006; Nochian et al. 2015; Hemakumara and Ranasinghe 2018).

2.2 Function

– ***Open space is the gathering place*** including Squares, pedestrian streets. This is a space where people meet and communicate, so it needs to be designed in accordance with the proportion of people, amenities, dining areas, utilities in close proximity, and not affected by the transport. Squares and street spaces are an important part of public spaces. It is an open space or surrounded by buildings. Different from the street space with the displayed traffic function, the function of the square is difficult to identify when it is just empty space.

- ***Open space is a place to rest and relax*** such as parks, scenic spots, lakes, and playgrounds. It is considered as a place to escape the stresses brought about by urban life where people can enjoy the pleasures of escaping from everyday life such as running, playing, walking, or soliciting. Public spaces of this type require walking paths, biking paths, greenery, and less utility services.
- ***Entertainment and sports space*** provide a space for people to participate in and participate in physical activities, relaxation, social interaction. It can be joined by everyone for playing, interacting, exercising, celebrating, or participating in other activities that bring personal satisfaction or a group of subjects or the community. Public spaces of this group usually include playgrounds in urban areas, dormitories, residential areas, etc.
- ***Ecological green space and concentrated agricultural production areas*** are greenery areas including public-use greenery, street greenery, and civil greenery. In addition, the areas also have a portion of agricultural land forming concentrated agricultural production areas. In order to create beauty, urban greening, trees are also a very important component in preserving the urban atmosphere to avoid pollution by industrial agents.

3 Methodology

The study was conducted through document analysis and comparison of different open space classification systems that have been applied in some developed and developing countries to propose a fully open space classification system for Vietnamese cities today.

In addition, the research results are consulted to evaluate public opinion on the quality and needs of each different OS object. This is a premise to help better orient the planning of urban and ensure use of land reaches its most effective purpose.

The study conducted community surveys through 50 questionnaires with different age groups. The purpose of the survey is to assess the needs and opinions of people about the open space in Ha Dong District in particular and Hanoi capital in general.

4 Research Area

Ha Dong District is located on the junction of the Nhue and the La Khe rivers, between the intersection of National Road 6 from Hanoi to Hoa Binh and Highway 70, Ha Dong is also the starting point of Highway 21B connecting the center of

Fig. 1 Location Ha Dong District. Source: Ha Dong District People's Committee

Hanoi to the southern districts and provinces of Ha Nam and Ninh Binh (People's committee of Ha Dong District 2018) (Fig. 1).

Ha Dong District is considered an attractive attraction for the migration process of people from different provinces to work and live in the capital. With a huge demand for population growth, Ha Dong District is facing a lot of pressure in solving urban planning issues, both to meet people's living quarters and to design reasonable Open space areas. In order to solve the above needs, the study assesses people's satisfaction with open space works in Ha Dong District to make appropriate recommendations.

5 Open Space Classification in Other Countries

In each country and territory, urban open space is defined and classified differently according to the space and user. This difference comes from the approach or characteristics of natural and social conditions in each region. This is a description of the open space system classification of some developed countries with a high level of urbanization (more than 70%).

5.1 Miami, United States

Depending on the research goals, urban open spaces are divided in different ways. According to specific subjects, the Miami Valley Regional Planning Committee of

Table 1 Open space classification in Miami Valley, USA

General Outdoor Recreation Areas		Utility		Landfills / Mineral Extraction	
Swimming Pool	Picnic Grounds	Well field	Wastewater Plant	Landfill	Quarry
Active Recreation	Stadium	Open Space Link		Sand & Gravel Extraction Site	
Plaza/Commons	Horse Racing	Hiking Trail	Large Institution	Cemeteries	
Fairground	Auto Racing	Bikeway	Flood Control	Cemetery	
Ball Field	Golf Course	Pedestrian Walkway		Historical Sites / Museums	
Outstanding Regional Amenity		Natural Environment Recreation Area		Historic Site	Museum
Scenic River	Waterfall	Campground	Fishing Lake	Covered Bridge	
Natural Environment Protection Area		Fish &Game Club	Scout Camp	Airfield	
Wildlife Preserve	Natural Area	School		Airport	Sky Diving
Park Preserve	Wetland Preserve	Public School	College		
Storm Basin	Hunting Area	Private School	University		

Source MVRPC 2005 open space database (Miami Valley Regional Planning Commission 2006)

USA has released the classification of open space objects, shown in Table 1 (Miami Valley Regional Planning Commission 2006).

This is considered the most detailed classification system of OS objects (including urban and rural areas). Some open space objects in this classification require costs when people participate such as stadiums, schools, golf courses, and other similar places.

5.2 *Michigan, United States*

The park and open space classification is proposed by The Michigan Department of Natural Resources, and the objects are divided into different groups (Michigan Department of Natural Resources Grants Management 2009).

Based on Table 2, it can be seen that the classification of open space objects is not based on specific criteria. This leads to incomplete and cluttered object groups due to irrespective of the classification by function or hierarchy.

5.3 *Singapore*

The Urban Redevelopment Authority in Singapore has a classification system for open space objects. Open spaces with parks, beaches, sports, and entertainment create

Table 2 Parks and open space classifications

Classification
Mini-parks
Neighbourhood park
School-park
Community park
Large urban park
Natural resource areas
Greenways
Sports complex
Special use
*Private park/recreation facility
*Parks and recreation facilities that are privately owned yet contribute to the public park and recreation system

Source: Michigan management deparment of natural resources 2009

great places for urban Singapore (Urban Redevelopment Authority of Singapore 2014) (Table 3).

It can be seen that the OS objects are listed quite fully. The difference from the classification system of other countries is that these objects are listed based on the main function of recreation and relaxation. In particular, the object of urban services is limited.

5.4 *London, United Kingdom*

OS objects are divided in the following manner based on the space and area of open space areas (London Borough of Tower Hamlets 2011).

Table 4 demonstrates that the OS classification is only based on 'catchment hierarchy' except for 'Linear Open Space' that could be an item of 'function' classification.

Table 3 OS classification by functional groups

Zoning	Example of developments
Open space	1. Wooded area
	2. Swamp area
	3. Natural open space
	4. Public promenades
	5. Outdoor pedestrian malls
	6. Landscaped plazas
Park	1. National park
	2. Regional park
	3. Community/neighbourhood park
	4. Park Connectors
	5. Zoological gardens, botanic gardens
Beach area	Nil
Sport and recreation	1. Sports complex/indoor stadium
	2. Swimming complex
	3. Golf course
	4. Golf driving range
	5. Recreation club
	6. Campsite
	7. Chalet
	8. Marina
	9. Water sports centre
	10. Outward bound school
	11. Theme park

Source: Urban Redevelopment Authority, Singapore 2014

Table 4 Classification of open spaces in London

Open space category
Regional
Metropolitan
District/major parks
Local parks
Small local parks
Pocket parks
Linear open spaces

Source: London Borough of Tower Hamlets (2011)

6 Result and Discussion

6.1 *Comprehensive Categorization of Open Space Based on Classification Method*

Through the process of comparing open space systems in different territories and countries around the world, in combination with legal documents on urban construction in Vietnam, the study offers a specific classification system of open space objects according to hierarchy, function, and landscape (Ministry of Construction 2008) (Table 5).

6.2 *Assessment of the Appropriateness and Necessity of Open Space Objects, Testing in Ha Dong District*

The study surveyed people on the quality of current public space services in Ha Dong District in particular, as well as Hanoi City in general. Thereby, there have been received some positive results. The survey results have decentralized the priority in selecting open space objects of people by age. The study also determined the level of satisfaction and desire of the people for the planning of open space objects in Ha Dong District.

- *Frequency of using OS objects by age groups*

Research age groups by criteria related to work and needs. Thereby, respondents were divided into four groups including: under 20 years old, from 20 to 39 years old, from 40 to 59 years old, and over 60 years old. Associated with age groups, the general need for using open space objects is also different (Fig. 2).

The chart showed that in the age group under 20 years old, learning space is most used. While other age groups often choose open space objects to entertain, relax, and practice sports. In particular, in the age group of over 60 years, the use of hospital space for medical examination and treatment is used regularly due to the need to protect health.

- *Assessment the level of satisfaction of the quality of OS Ha Dong*

In nearly a decade, with an increase of about 120,000 people, Ha Dong has become one of the districts with the fastest population growth in Hanoi. Ha Dong District is currently one of the two districts of the capital with the fastest urbanization rate. As part of the general planning of the capital until 2030, with a vision to 2050, Ha Dong is focusing on building infrastructure and public utilities, and closely managing the construction works in the area. county table. This is considered an attractive place for real estate and apartment projects. Therefore, the number of works built to ensure the needs of people is increasing.

Table 5 Proposing open space classification system in Vietnam

Hierarchy	Landscape/environmental character	Function
National open space Region open space Provincial open space District open space Urban open space	1. Urban public service works • Education – Nursery school – Primary school – Junior high school – High school, vocational training • Hospital – Clinics – Clinic – General hospital – Maternity homes • Sports – Training ground – Basic sports ground – Stadium – Sports center • Cultural – Library – Museum – Exhibition – Theatre – Cultural house – The circus – Supply children • Market 2. System of urban greenery • Trees for public use • Street greenery • Specialized green trees 3. Urban transportation system 4. The park 5. Urban flower garden 6. The lake 7. Square	• Space for schools, education, and academies • Health protection/examination and treatment • Exercise/physical exercise • Document storage • Display • Historical/cultural heritage • Community gardens/urban agriculture/agriculture • Bicycle path or footpath/entrance • Meeting • Travel • Cemetery/memorial/memorial • Urban landscape • Ornamental gardens, plants, zoos • Relaxing/reflecting/escaping from urban stress

Besides high-rise buildings, condominiums and construction works, open space areas are planned in accordance with the general planning of Hanoi capital until 2050 (70% of green space, 30% for urban development).

However, the state's open space projects are always facing many difficulties in implementation. This is due to the construction cost; there is no compromise between the profit of the investor and the state, typically Ha Dong District green park with an area of over 50 ha.

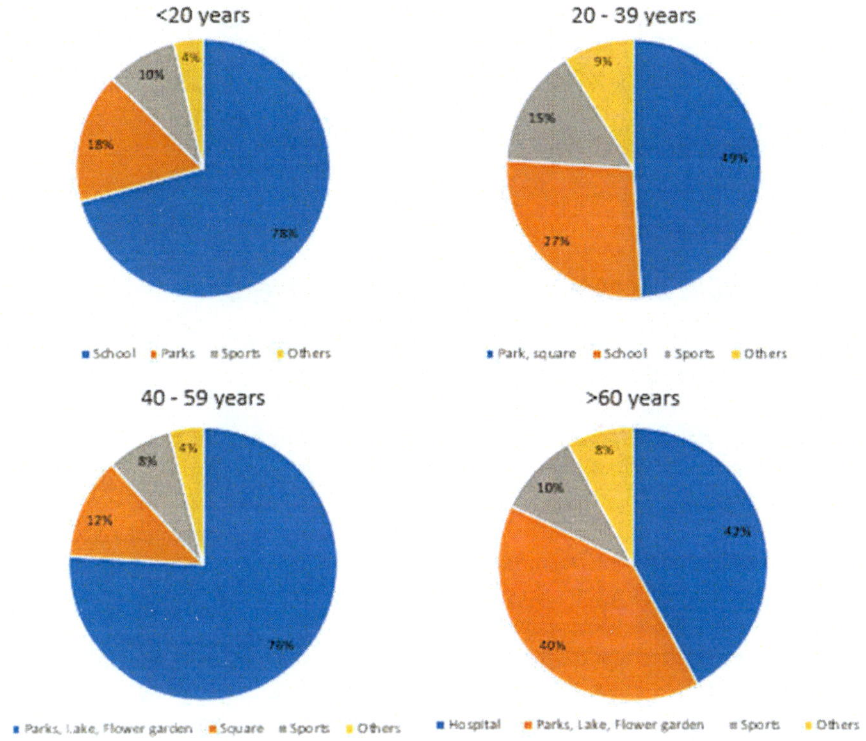

Fig. 2 Differences in the priority level of selecting open space objects by age group. Source: Summary of sociological survey results in 2019

With a population of about 352,000 at the end of 2018, Ha Dong currently lacks green park areas to serve the needs of the people. The Thanh Ha Water Park project, launched in 2019, has addressed a large number of amusement needs in the district. However, the district still needs to accelerate the construction of open space areas to ensure the overall planning of the city (Table 6).

Table 6 Classification people's satisfaction level of open space in Ha Dong District

Classification	Proportion (%)
Satisfaction	**38**
Dissatisfaction (reasons below)	**62**
Incompliance	*15*
Rar from accommodation	*48*
Pollution	*25*
Incompliance	*12*

Source: Summary of sociological survey results in 2019

Through surveys in open space areas, most people are not satisfied with the quality of open space services in Ha Dong district today. Some of the main reasons listed in the survey include the lack of open space, design away from housing, pollution, and other reasons.

Therefore, planning should better meet the needs of the people to ensure the best use of land.

6.3 Solutions for Urban Open Spaces Development

– *Changing management mechanism to attract investment*

The Ha Dong District Green Park project was approved in 2007 with a scale of more than 50 ha, in the list of calling for investment. However, until now, the work has not been implemented.

In February 2015, the People's Committee of Ha Dong District issued a document No. 320 of the Hanoi People's Committee to propose a plan for temporary exploitation and use management for the cleared land area (52.8 ha) of the land area, planning construction of a green sports park in Ha Dong District (Fig. 3).

This is the first time that the phrase "temporary planning" has appeared in land use works in Vietnamese cities. Basically, the temporary planning helps take advantage of vacant land, bringing economic benefits to both investors and the state. However, as a whole, projects that need temporary planning have not brought real value to the people compared to the planning. The population of the district in the 10 years from 2008 to 2018 has nearly doubled while the district's open space project to serve the people has not been implemented.

The cause of the failure to call for investment capital for the green park project is said to be unable to balance the profit between the investor and the state in the

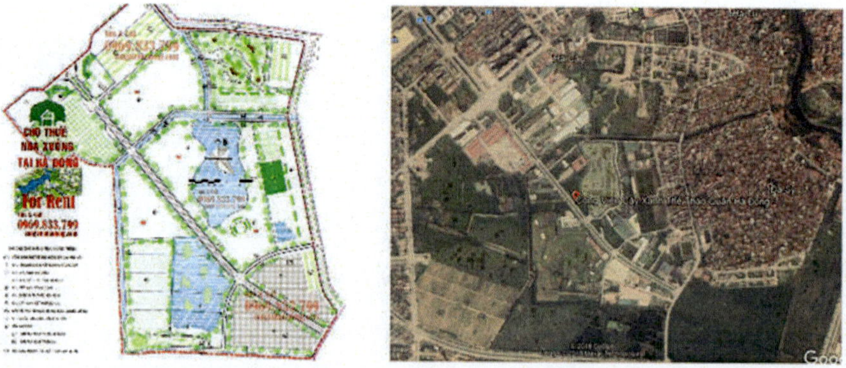

Fig. 3 Example of "temporary planning" in Ha Dong District. Source: Ha Dong District People's Committee and Actual photo taken from Google Earth

construction and exploitation of this open space project. Therefore, in public works built with the capital of the state and businesses, the interests between the two sides should be balanced. It is even necessary to adjust the planning to attract businesses to invest in the project, allowing businesses to build some business items but still ensure the spatial reasonable for planning works.

– *Changing people's habits in participating in public space*

The process of urbanization makes the number of buildings, buildings, and commercial centers increasingly higher. People gradually become lazy in communicating with the environment and the people around them. Therefore, calling and attracting people in more access to public space facilities is considered necessary for human development.

In order to do that, open space objects need to be designed more attractive with many activities to attract the attention of everyone. Ho Guom established in 2016 walking street is considered a cultural—entertainment center of the people of the entire capital. Through promoting and encouraging cultural activities on the streets around the Hoan Kiem Lake on weekends, participating in the opening space of Ho Guom walking street has become a habit of many people to promote the engagement. cultural exchange between people (Fig. 4).

Fig. 4 A number of cultural and entertainment activities at Ho Guom walking street. Source: Photographs of activities collected around Hoan Kiem Lake area, Hanoi

7 Conclusion

Hanoi is one of the two largest urban areas in Vietnam with a high rate of urbanization. With hundreds of projects built every year, Hanoi has changed its appearance quickly, becoming the economic and political center of the country. In particular, Ha Dong District is considered a typical area for the urbanization process of the capital in particular and the whole country in general.

In this context, studies of urban open space design are more necessary than ever. The topic presents an overview of open space objects from the past to the present in Vietnam. In addition, the research results also provide a specific definition and classification system for open space subjects in Vietnam based on the analysis of documents, synthesis of legal provisions on urban construction. This is the premise for the process of approaching Vietnam's OS system in subsequent studies.

The solutions presented above were developed from the process of analyzing real conditions, consulting the community to improve the quality of OS subjects. The design should be done more in urban areas and meet the needs of the people.

References

Daamen T, van Bueren E, Chen Y, Franzen A, Heurkens E, Hobma F, Wouter, VJ (2016) Urban development management: past, present and future

Hemakumara GPTS, Ranasinghe WC (2018) Spatial modelling of the householders' perception and assessment of the potentiality to improve the urban green coverage in residential areas: a case study from Issadeen Town Matara, Sri Lanka. Ruhuna J Sci 9:44–56

Kien To (2018) Public space in livable and humane city 30 + 31:76–83

London Borough of Tower Hamlets (2011) An open spaces strategy for the London borough of tower hamlets

Miami Valley Regional Planning Commission (2006) Miami Valley open space assessment

Michigan Department of Natural Resources Grants Management (2009) Guidelines for the development of community park, recreation, open space, and greenway plans

Ministry of Construction (2008) QCXDVN 01: 2008/BXD Vietnam construction standards construction planning

Nguyen Q (2018) Public spaces in landscape planning and sustainable development. Vietnam Architect Mag 05

Nochian A, Tahir OM, Maulan S, Rakhshandehroo M (2015) A comprehensive public open space categorization using classification system for sustainable development of public open spaces. Alam Cipta (Int J Sustain Trop Des Res Pract) 8:29–40

People's committee of Ha Dong District (2018) Report of land statistics in 2018 in Ha Dong

Urban Redevelopment Authority of Singapore (2014) The planning act master plan written statement

Comparative Study of the Patterns and Characteristics Urban Morphology of the Old City, Bengkulu, and Singapore that Has Relation to Historical Background

Fitrianty Wardhani and Samsul Bahri

Abstract Urban morphology is an applied science that learns about the history of the spatial patterns of a city and learns about the development of a city. The old town area of Bengkulu was an English-built city from 1719, while Singapore was a British formation city as well, due to the London Treaty agreement occurring between the two regions. This paper aims to be a comparison study between two cities with historical linkages in terms of identifying patterns and characteristics of its city morphology. This research is quantitative descriptive research using two analytical techniques, namely diachronic reading and typo-morphology. The indicators used to acquire patterns and characteristics of the urban morphology are Bengkulu and Singapore, which are aspects of detail consisting of buildings and materials, road and block patterns, land use activities, and open spaces. The second indicator is a city layout aspect consisting of spatial patterns and environmental compositions awakened. Through this research is expected to get similarities and differences in the patterns and characteristics the morphological of both cities. This study will provide an overview and input for the development planning of the more advanced city of Bengkulu.

Keywords Comparative studies · Urban morphology · History

1 Preliminary

1.1 Theory of Urban Morphology

Urban morphology is "an approach in understanding the city as a geometric collection of buildings and artifacts with the configuration of the unity of certain physical spaces the product of socio-spatial changes." Our understanding of "urban morphology" cannot be separated from the physical form of the city which is formed primarily by

F. Wardhani (✉) · S. Bahri
University of Bengkulu, Bengkulu, Indonesia
e-mail: fwardhani@unib.ac.id

L. T. T. Huong and G. M. Pomeroy (eds.), *AUC 2019*, Advances in 21st Century Human Settlements, https://doi.org/10.1007/978-981-15-5608-1_43

the physical conditions of the environment as well as the dynamic socioeconomic interactions of the people.

Herbert (1932) in Conzen (2003), the scope of the study of urban morphology is emphasized on the physical form of the urban environment which can be observed from its appearance which includes elements of:

(1) the existing road system,
(2) building blocks both residential and non-residential areas (trade/industry),
(3) individual buildings.

Smailes (1955) emphasizes the scope of morphological studies including:

(1) land use
(2) street patterns and
(3) building types.

Morphology means the science of form. In the urban context, morphology is the study of the forms and shapes of residential environments. Form means an observable form and is a configuration of several objects, while a shape is a geometric feature or an external shape and outline of an object. Although they have almost the same meaning, these two words (form and shape) have a different basic understanding, where the form emphasizes the form consisting of various elements and each element can be clearly observed its characteristics, and visually each element is in one unit (configuration). For example, a road corridor is visually formed from a row of buildings with a certain height and arranged within a certain distance from the road border. Shape emphasizes the external form of the form, or in other words, the silhouette in the context of townscape is often referred to as the skyline. A collection of objects located above the ground surface will form certain patterns (shapes), such as linear, grid, concentric, radial, cluster, and so forth.

Smailes (1955) introduces three elements of city morphology, namely land use, road patterns, and building types or characteristics. Meanwhile, Conzen (1962) also suggested elements similar to Smailes, namely plan, architectural style, and land use.

1.2 Morphological Components

M.G.R. Conzen considers that it is necessary to pay attention to the four components of morphology (Carmona et al. 2003: 61), including:

1. Land use

Land uses (land uses) is a principal component in the growth of the region. This component is considered as an activity system generator that determines the pattern and direction of regional growth. This component has the level of temporality is very high in terms of being able to easily change literature, especially in relation to the economic value it holds. Land use greatly influences the physical appearance of the area, especially in determining the development of built and non-built areas.

Several studies and literature explain how the level of mixing (land use) mixture greatly affects the vitality of the region, economic value, and several other environmental quality components.

2. Building structure

This component is a representation of typology in morphological analysis and can be discussed in two aspects, including mass structuring and building architecture. The structuring of the masses is related to how buildings are scattered within the site along with their density and intensity while building architecture is more a physical manifestation of spaces and buildings that represent the culture, history, and creativity of a community.

3. Pattern

This component can be discussed in terms of size (dimensions) and distribution. The size of the plot will affect the intensity of land use, while the distribution of the plot will affect the establishment of a connecting network. In general, the pattern of this plot is strongly influenced by natural potential, especially contours and geological conditions. Legally, the plot is limited by ownership boundaries which greatly affect the pattern of control, use, and management of space.

4. Road network

This component is a derivative function of land use. As a connecting line, the road network greatly affects the efficiency and effectiveness of regional functions. The road network as a representation of public space is considered as the core generator of the vitality of the region as explained.

Therefore, in this paper, the identification of patterns and characteristics of urban morphology is done by using only two indicators namely building details and materials as well as road and block patterns.

1.3 Bengkulu City

Bengkulu City is one of the colonial cities of the former British and Dutch colonies which until now its legacy can be found in the city of Bengkulu, especially the old city of Bengkulu. The old city of Bengkulu was an area in Bengkulu which was used as a trading center in the sixteenth to eighteenth centuries and built a fort called Fort Marlborough as its center of defense.

The British began to build the city of Bengkulu starting in 1719. As the British colonial authorities to form and design the city of Bengkulu with a good and unique. In addition, the UK also adopted the form of cities with different characteristics in each city under their control. As the ruler at that time, the British placed Bengkulu city based on its function, namely the city whose economy was focused on shipping and trade. The focus of this city arrangement adjusts to the location of Bengkulu city which is located along the coast.

1.4 Singapore City

The geographical position of Singapore to the south of the Malay Peninsula is a country in the form of a small island on the Malacca Peninsula. Astronomically, at latitude 1 ′15 ″LU–1 ′26″ LU and longitude 103 ′40 ″BT–104 ′BT, the area of this country is 818 km^2 with 54 islands around it.

Singapore's national borders are

- North: Johor Strait
- East: South China Sea
- South: Singapore Strait
- West: Malacca Strait.

Modern Singapore occurred in 1819 by Sir Stamford Raffles; at that time, he established a port for the colonies from England. Under British colonial rule. Singapore has turned into a very strategic port, given its location in the middle of the trade route between India and China which has finally become one of the most important ports in the world to this day. Raffles arrived in Singapore on January 29, 1819. He encountered a small Malay village at the mouth of the Singapore River, chaired by a Temenggung Johor. Because of the abundant freshwater at that time, Raffles made the place for his colonists to live. After signing the agreement on February 6 of the same year, Raffles left the settlement, and handed over responsibility to Colonel William Farquhar and at the end of May, Raffles returned, because of the rapid development of the city, he realized the need for a formal urban plan to guide physical expansion. Then Raffles went back and in October 1822 Raffles returned to Singapore and formed a City Committee led by Lieutenant Philip Jackson to formulate a formal plan for the colony which came to be known as the Jackson Plan, namely by laying the foundations of city roads and Zonal layout. For example, the allocation of civic institutions on the north bank of the Singapore River and creating a major commercial area that came to be known as "Commercial Square" and on the southern edge developed into civic and CBD districts on both sides of the river in a grid pattern. While the settlement zone is in China town, Little India, and Kampong Glam, which can attract the attention of tourists. But the city committee plan led by Jackson only lasted eight years because of the rapidly growing population and severe population density, eventually, many residents moved to the suburbs of eastern Singapore. This growth also caused suburban roads to become congested with traffic, especially along Geylang Road which leads to the East Coast.

In 1927, the government tried to capture the situation by establishing the Singapore Improvement Trust (SIT) with the main objective of reducing urban congestion and the provision and improvement of public infrastructure, particularly in widening roads to accommodate and modernize traffic. Their efforts are only in the local area because they do not have the legislative power to produce comprehensive plans or to control urban development. And because the Second World War caused disruption in efforts to reduce the problems of the city, this lasted for 1941–1945. In 1947, the SIT effort did not succeed in overcoming the overcrowding in Singapore. And under the auspices of the ruling People's Action Party at the time, the SIT (Singapore Improvement Trust) was replaced by the HDB (Housing Development Board) which was founded in 1960. This proved to be Singapore's turning point in dealing with population density. HBD has built 5000 housing units that have been built in a span of more than 20 years. In the 1970s, a large proportion of the population had found adequate housing. Most of the current urban planning policies originate from the HDB practices. And Singapore's Regional and municipal planning policies are under the Urban Redevelopment Authority which is to create independent cities and districts that are served by regional centers that have their respective functions. The purpose of urban plans is to maximize land use efficiently but comfortably and to serve as many residents as possible with their respective functions, such as infrastructure, environmental preservation, space for water catchment, and land for military use all require consideration and national urban planners.

2 Research Methods

Using qualitative methods, the research technique is carried out by analyzing data archives originating from the old maps of Bengkulu and Singapore, observations, and evaluating detailed aspects including buildings and materials, road patterns, and blocks. All of them are intended to identify patterns and characteristics of the old cities of Bengkulu and Singapore (Table 1).

Table 1 Data analysis process scheme

Target	Analysis of	Analysis techniques	Results obtained
Identify the pattern and characteristics of urban morphological of the old city of Bengkulu and the old city of Singapore	Detailed analysis Identification – Buildings and materials – Road and block patterns	Diachronic reading and morphological typologies and figure-ground techniques	Get patterns and characteristics urban morphology of the old city of Bengkulu and the old city of Singapore Knowing the similarities and differences in urban morphological between the two cities

Source Whardhani (2019)

3 Results and Discussion

This study is to identify the patterns and morphological characteristics of the city, namely the old city of Bengkulu and the old city of Singapore. The two cities have a historical background in which the two cities are the result of the London Treaty Exchange between England and the Netherlands. In the beginning, Bengkulu was formed by the British then switched to forming and developing Singapore. This comparative study begins by analyzing the detailed identification of buildings and materials as well as the patterns of roads and blocks. Identification is done by using diachronic reading and morphological typography analysis techniques. From the identification results, it is expected to get similarities and differences in patterns and characteristics of the morphology of the two cities, namely between the old city of Bengkulu and the old city of Singapore (Whardhani 2019).

3.1 Patterns and Characteristics Urban Morphological of Bengkulu Old City

Identify building details and materials using typo-morphological analysis

Building details 1895–1945

Bengkulu Port is a component of the city that functions as an economic zone, Fort Marlborough as a defense area and Kampung China, Kebon Keling, Jitra Makan Complex as a residential area and its facilities, while for the government area it is located southeast of Fort Marlborough. The colonial influence government building is a two-story building with a curved entrance decorated with pseudo-masts while the Chinese Village, which had been established since the British Colonial period, was home in the village (Whardhani 2019) (Fig. 1).

It has Chinese architecture. There are 20 houses with Chinese architecture in this area. The houses generally extend toward the back, two stories, and have a curved roof. Also seen the house was decorated with oversized which is above the window that serves as ventilation as generally in Chinese home architecture. Housing in the Keeling Garden Complex adopts home architecture in the United Kingdom which features a large yard, a red roof, and a sharp shape (Figs. 2, 3 and 4).

Building and Material Details from 1945 to present

The British colonial architectural style which is a modern architectural style has increasingly been abandoned. Government buildings are now shifting to using the current architectural style that is suitable for tropical areas that use shield or saddle roofs. Nevertheless, the British heritage building that used to be the Office of the British Governor is now converted into a residence of the Governor. But at this time, the Chinatown area which was once a trading center in the city of Bengkulu is now sinking, and almost becoming a dead area. The area around the port, which is well known as the site of the Paderi now functions as a beach and water tourism area in the city of Bengkulu which is crowded.

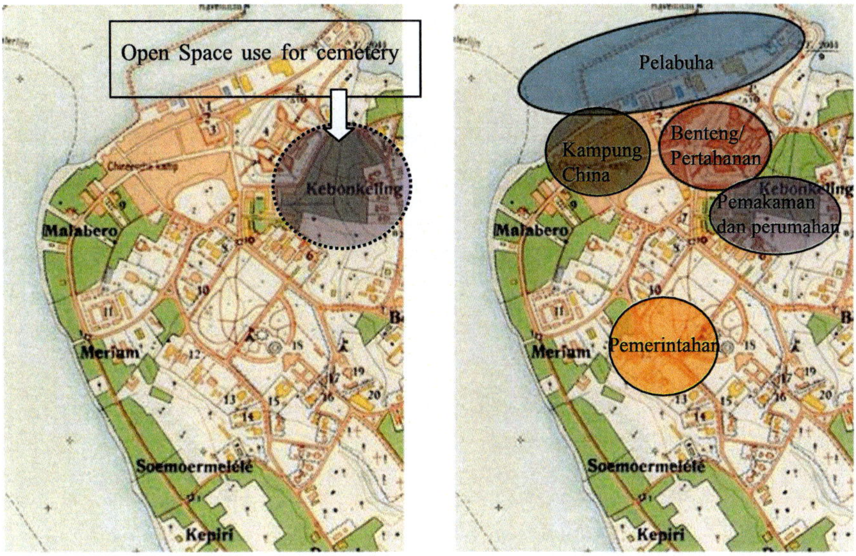

Fig. 1 Structuring of the dutch map source 1924

Fig. 2 Kampung China

Fig. 3 Housing in Bengkulu, 1900

Fig. 4 British government office

Identification of road and block patterns, by using typo-morphology, and figure-ground analysis techniques

The pattern of roads and blocks is the second identification carried out to determine the pattern and morphological characteristics of the old city of Bengkulu. The analysis is done using figure and ground analysis. This analysis is used to identify road and block patterns by looking at the texture and patterns of an urban fabric.

Identification of the city morphology is done by using the figure and ground analysis technique, it can be known the pattern of existing roads and blocks by looking at the existing ones and historical information of the past. This analysis is carried out on the map of the city of Bengkulu in 1867–208. Then it will be known the changes and characteristics of the existing road and block patterns in the old city area (Table 2).

So based on the analysis of road and block patterns, it can be concluded that, in the early 1800s, the pattern of roads and rounded cities (round) and radials concentric centered on the fortress area as a center of defense and trade, later in its development, the form was abandoned where the pattern of roads and blocks becoming increasingly complex and heterogeneous; and the trade and government center have shifted toward the middle of the city of Bengkulu, while the fortress area has been abandoned and turned into a tourist attraction for cultural heritage and beaches.

3.2 Patterns and Characteristics Urban Morphological the Old City of Singapore

Identify building details and materials using typo-morphology analysis

The old city of Singapore consists of China town, Little India, Arab Quarter, Kampung Glam, Geylang Road, Tanjung Pagar, Clark Quan. Just as it was in Bengkulu, the city of Singapore was formed by emphasizing the Port of Singapore (Marina Bay) as a component of the city that functioned as an economic zone, which developed into the CBD District Civic. While the settlements are in the Colonial District, Chinatown, Little India, Arab Quarter, and Kampung Glam.

Table 2 Analysis of road and block patterns

No	Map	Identification
1	**Road and block patterns in 1894** 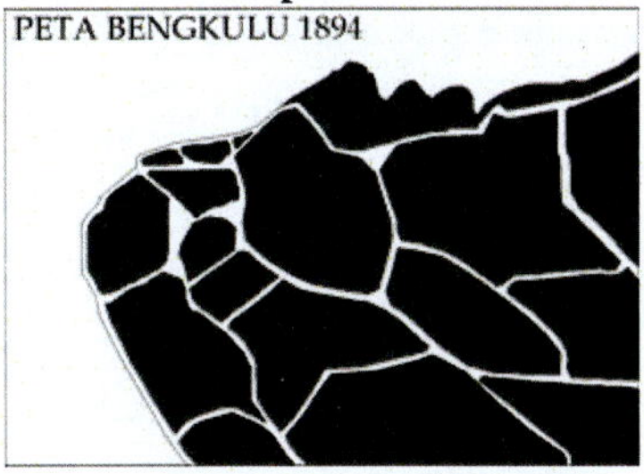 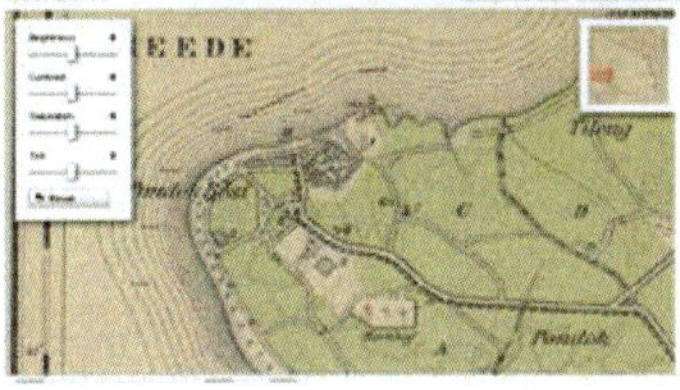	Based on the figure ground analysis, it can be seen that the pattern of the road is a Rounded Cities (round) form. This is the most ideal form for the city, because it can develop into all direction and has balanced, as the advantages Beside that, the pattern of roads and mass blocks formed in 1894 is a concentric radial pattern. The old city of Bengkulu is a city formed by the British as a place for trade and defense that focusing the city on the fort area 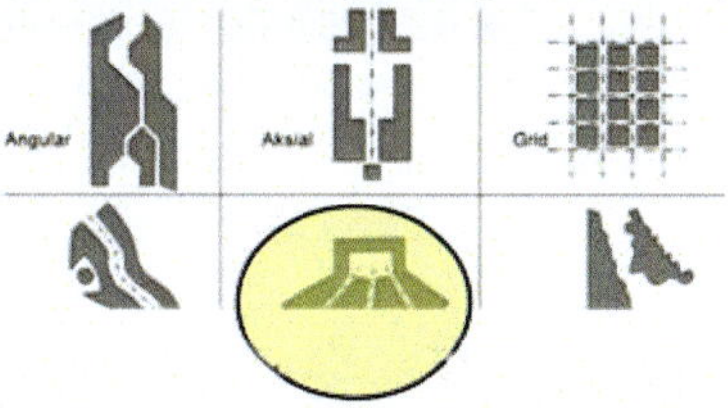 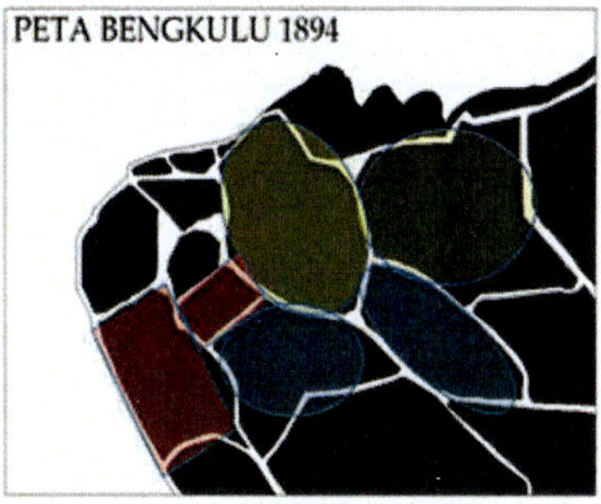 Form Rounded Cities 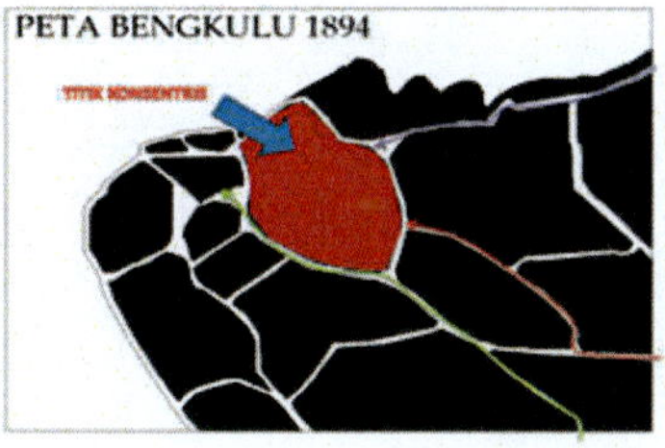

(continued)

Table 2 (continued)

No	Map	Identification
2	**Road and Block Patterns in 1914** 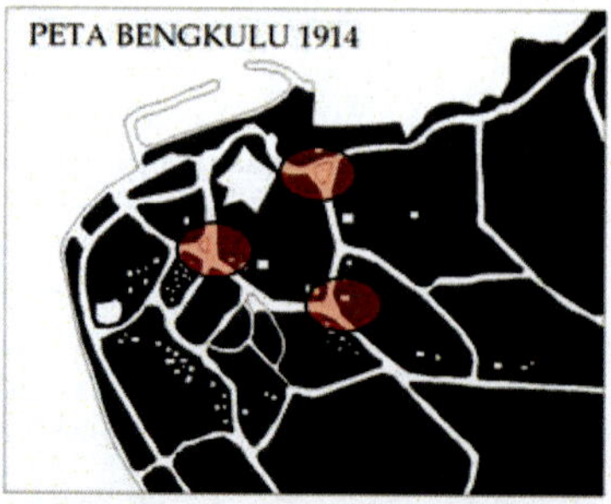 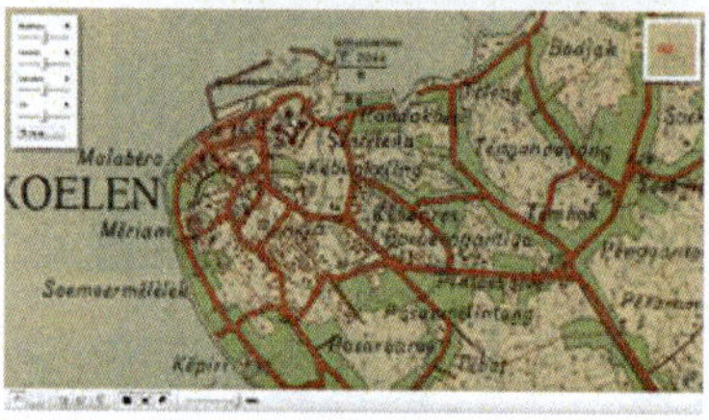	In 1914, the road and block patterns were still the same but the distribution of mass blocks was growth and there were several additional roads, but the pattern and initial shape of the road were still maintained. Radial concentric patterns have been changed. The wider distribution of the mass made current pattern tends to be heterogeneous The center of Bengkulu city in the British colonial period was located on the beach, so it was called the coastal city which emphasized its social economy on shipping and trade. The City Center was designed to have three main node arrangement patterns that confirm the city center. The road pattern forms the city into a form of constellation (constellation) a series of city blocks that almost have the same size in close proximity. The road pattern produced by this area in the colonial era is homogeneous, where there is only one road arrangement pattern
3	**Road and Block Patterns in 1924** 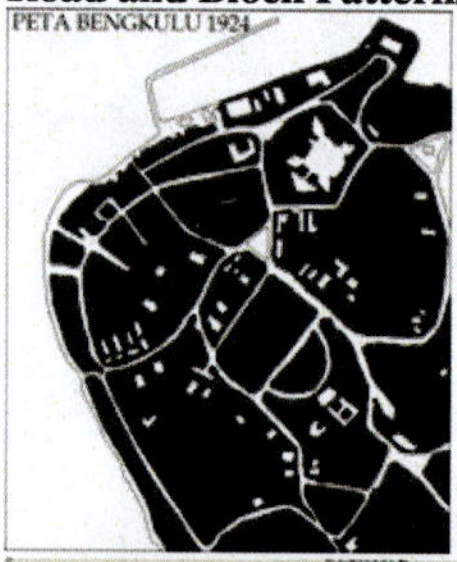 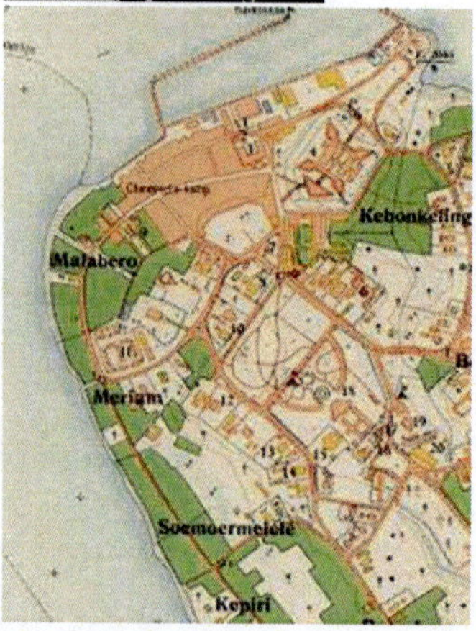	In 1924, The pattern of mass distribution has become wider and denser. The main road pattern is still maintained but the mass distribution is increasingly widespread. Radial concentric patterns have shifted not to the fortress anymore. This is because the growth of the city and the spread of the masses are expanding to the south

(continued)

Table 2 (continued)

No	Map	Identification
4	**Road Patterns and Blocks in 2018**   	At this time after the Independence of Bengkulu, there were many changes. If Bengkulu city development used to be on the beach, but now the development of the Bengkulu city moves in the direction of the north and west of the city, no longer on the beach. This is marked by the relocation of the city center in an area known as the Suprapto area which is the current trade and government area. The Urban Patterns no longer follow the pattern of British heritage where the main path is to use a north–south-oriented linear pattern. Resettlement blocks are no longer in the form of concentration. The composition of the area becomes increasingly complex. No longer homogeneous but now heterogeneous, where two or more patterns colliding because of the influence of building masses and higher building density

1. **China Town**

The shops were rebuilt around the 1980s with colorful façades
Traditional style shophouse Singapore (Figs. 5, 6 and 7).

Characteristics of buildings in Chinatown:

1. Low, consisting of two floors with an ornament that only uses plaster, the construction uses local wood material and the window is on the top floor.
2. Taller, three floors, decorated with Chinese architecture, with two windows on the upper floor façade.
3. Many ornaments, influenced by Europeans, in the form of carvings and paintings, there are three windows in the upper floor façade and maximum ventilation

Fig. 5 Model shophouse Singapore

Fig. 6 Mariaman Hindu temples Singapore

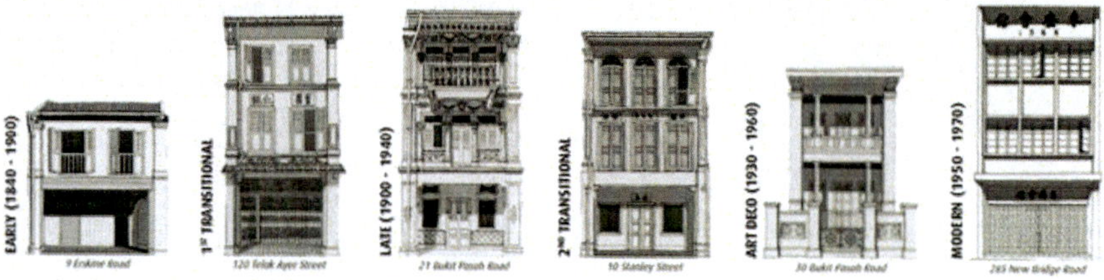

Fig. 7 Characteristics of buildings in Chinatown

4. A combination of Chinese and European architecture, simpler and slimmer
5. The design is geometric and uses precast elements, then there is the added fence at the edge of the road
6. Simple, functional and reflects the postwar situation, using steel and concrete materials, functional sun shading, and ventilation.

2. **Little India**

Little India is not included in the planned village in JacksonPlan. But this village was not intentionally developed because of the large number of workers from India who came to Singapore. Existing buildings in this village in the form of houses designed specifically for workers by adopting the local architecture of Singapore. Then it developed into a shophouse which also served as a residence for workers. This shophouse feeds shophouses in Chinatown or most shophouses in general but does not include Chinese architectural elements.

Kampung Glam

See Fig. 8.

3. **Kampong Glam**

See Fig. 9.

Fig. 8 Buildings in Little India

Fig. 9 Buildings in Kampung Glam

Kampung Glam is the Center of the Muslim Community in Singapore. The majority of buildings in this village are Ruko which is the most type of building in Singapore and is an architectural heritage that reflects the history and culture in Singapore.

Kampung Glam is older than modern Singapore, a village that existed before Raffles' arrival in Singapore, is the historical center of Malay nobility in Singapore. Then Raffles allocated Glam village to Malays, some Muslim immigrants, Malaysians, Javanese, Sumatran, Sulawesi, Indian, and Arabic. Arab traders gave important influence in Glam Village (Fig. 10).

Fig. 10 Masjid Sultan in Kampung Glam

Fig. 11 Shophouse in Kampung Glam

The buildings in this village are mostly mosques that adopt Malay and Arabic architecture in this glam village. In addition, there are also shop houses with characteristics in the form of a harmonious blend of Malay and Chinese architectural styles (Fig. 11).

Then there is also the oldest Islamic school in Singapore called Alsagoff Arab School, which is very thick and influenced by its colonial style.

Then there is also the Hajjah Fatimah Mosque and the Malabar Mosque (Fig. 12).

Fig. 12 Hajja Fatimah Mosque and The Malabar Mosque in Kampung Glam

This mosque has the strength of a tower that resembles a tower in Singapore monument and the deadpan side. And the Malabar mosque with its golden dome which is a Singaporean Indonesian mosque managed by the Malabar tribe.

Characteristics of shop houses in Singapore two or three floors. The first floor is for business and the upper floor is for residence, using a traditional style, modern style, and art deco style, influenced by traditional Malay buildings and fascia carvings and a 5-ft wide road as a pedestrian way that protects pedestrians from the heat, as a sun shading and a place to interact.

Identification of road and block patterns, typo-morphology

The pattern of roads and blocks is the second identification carried out to determine the pattern and morphological characteristics of the old city of Singapore. The analysis is done using figure and ground analysis. This analysis is used to identify road and block patterns by looking at the texture and patterns of an urban fabric.

Identification of the city morphology is done by using the figure and ground analysis technique, it can be known the pattern of existing roads and blocks by looking at the existing ones and historical information of the past. This analysis is carried out on the map of the city of Bengkulu in 1867–208. Then it will be known the changes and characteristics of the existing road and block patterns in the old city area (Table 3).

Table 3 Analysis road and block patterns in Singapore

No	Maps	Identification
1	Map of Singapura, Jackson Plan 1822 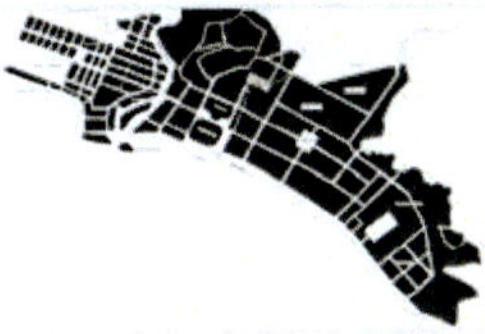	Based on the picture analysis of the ground figure, in addition, it can be seen that the road pattern in the form of a grid is the most ideal form for the city because it has the advantage of being in all directions and balanced. The formation of the city in 1822 by Jackson was with laying the foundation of city roads and zonal layout, for example, the allocation of civil institutions on the north bank of the Singapore river and locating the main commercial areas which became known as the district civic and CBD on both sides of the river, while the settlement zones are in Chinatown, Little India, and Kampong Glams are grouped by race and religion 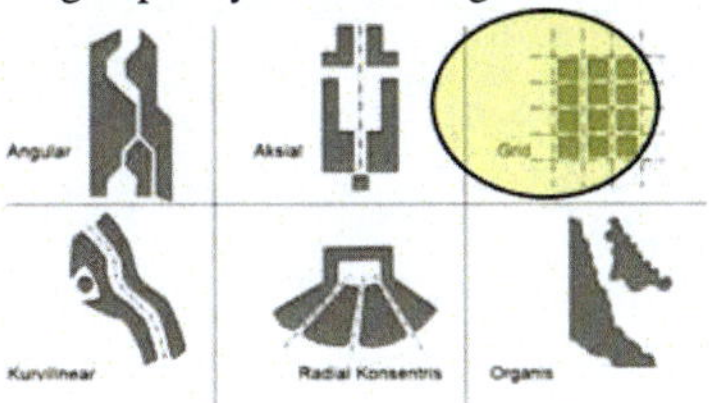

(continued)

Table 3 (continued)

No	Maps	Identification
2	Map of Singapura 2019, google earth 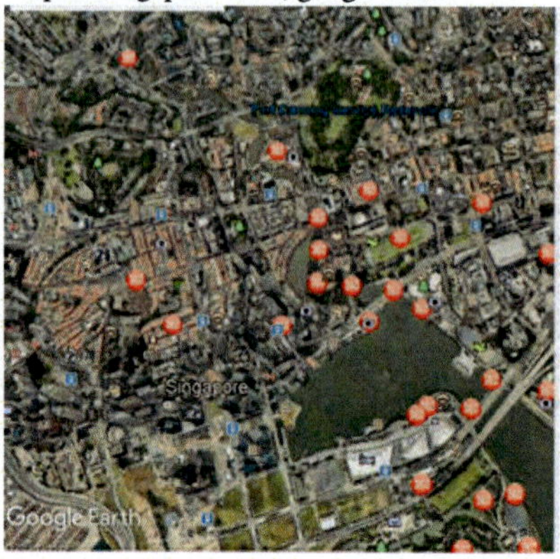 	Based on the map image, besides being seen, the old city of Singapore still uses grid patterns like the Jackson plan, but there has been the development of road and block patterns 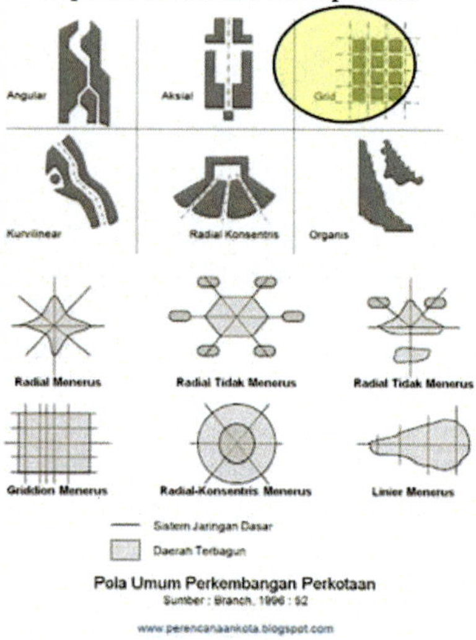 The development is in the form of road patterns and blocks from the city of Singapore forming a satellite city. The form of satellites and new centers (satellite and neighborhood plans), the main city, and small cities will be established effective and efficient functional linkages; 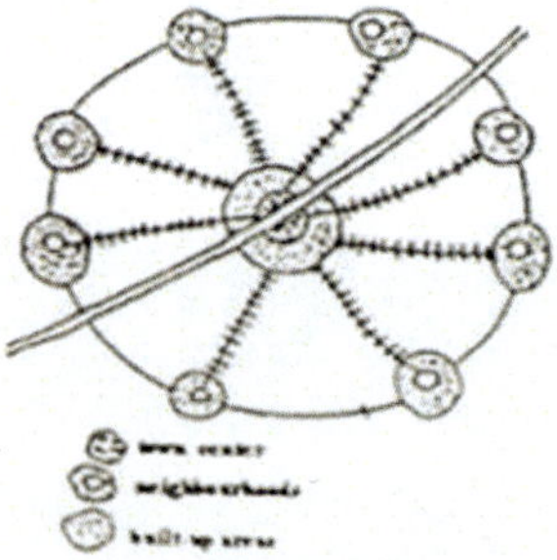 It also forms a stellar or radial. Stellar or radial plans (stellar or radial plans), each tongue is formed a second center of activity that serves to provide services in urban areas and protrudes into the planned as a green line and functions as the lungs of the city, recreation areas and sports areas body for city-dwellers;

(continued)

Table 3 (continued)

No	Maps	Identification
		The development of the city of Singapore did not leave its old city but synergized with the old city to form the CBD area which remained centered in the Marina Bay. The concept of structuring CBD areas in developed countries generally uses a simple pattern of a superblock grid. The central commercial buildings are arranged in such a way that they are squeezed together so that the line is commensurate with only a tight 0 m. Bangunan immediately met the sidewalk next to the skyscraper. Superblock pattern besides practical, rapid, and unsightly. The purpose of the Singapore city plan is to maximize land use efficiently but comfortably and to serve as many residents as possible with their respective functions, such as infrastructure, environmental preservation, space for water catchment, and land for military use all require consideration and national urban planning.

4 Conclusion

Based on the above identification, it can be concluded that the similarity of patterns and morphological characteristics of the cities of Singapore and Bengkulu is the same as the initial pattern which is laying the foundation of the city planning with zonal i.e. between commercial areas, settlements and government but the pattern of roads and blocks used by different British colonies in Bengkulu uses a pattern concentric radial and rounded cities centered in Marlborough Fort, while Singapore uses a grid pattern. The city center is equally located in the Port, namely Marina Bay and Fort Marlborough in Bengkulu. But in the development of the city of Singapore City Center is still located in the Marina Bay Area which synergizes with the old city, but the city of Bengkulu left the old city and the city center shifted toward the North and South. In terms of building the characteristics of the old city almost as many buildings in the form of shophouses that are found in the villages of China, Little India, Glam in Singapore while in Bengkulu Kampung China, Kebon Keling, and Malabero. Many and equally many adopted European Architectural Buildings.

References

Conzen MRG (2003) Morphogenesis, morphological regions and secular human agency in the historic townscape, as exemplified by Ludlow. In: Conzen MP (ed) Thinking about urban form: papers on urban morphology, 1932–1998. Oxford, Lang, pp 116–42

Smailes AE (1955) Some reflections on the geographical description and analysis of town scapes. Trans Pap of Institute of British Geograph 21:99–115

Whardhani (2019) Patterns and characteristics of urban Morphology in the old city of Bengkulu. J Architect Environ